T0327353

What people say about the book

Engineers will find this book insightful in addressing the enormous challenges of our future world dominated by data, complexity and increasing connectedness. Handling the velocity of change requires such a book for researchers, students, and practitioner's alike. – **Kerry Lunney, Engineering Director, Thales Australia and INCOSE President 2020-2021**

We are on the path of conceptualizing and building multifaceted systems in different levels of implementation that entail complex logic with many levels of reasoning in intricate arrangement, organized by web of connections and demonstrating self-driven adaptability which are designed for autonomy and exhibiting emergent behavior that can be visualized. System engineering research and practice need to rejuvenate itself to cope with these changes with a new perspective. This book provides answers to new questions that we face and directions to follow in creating new systems engineering practices for digital industrial revolution written by international authors of practicing systems engineering and academicians. It is a great book to read for understanding new systems engineering vision. – **Prof. Cihan Dagli, DoD Systems Engineering Research Center-UARC, Founder and Director Systems Engineering Graduate Program, USA**

This is a useful and timely book. The topics are well chosen and they helped me to understand better the ongoing industry transformation and the role of analytics in it. I highly recommend it. – **Dr. Emmanuel Yashchin, IBM Thomas J. Watson Research Center, USA**

In this era of increasingly complex systems, processes, and data, a systems engineering perspective can serve us well to overcome the challenges and pursue the best opportunities afforded us in this fourth industrial revolution. This book compiles useful materials for dealing with complex systems with multidisciplinary and data-based approaches. The need for a systems engineering perspective is underscored by the fact that Industry 4.0 is disruptive to every industry that uses data. Still, some things remain the same – the importance of data *in context*, the usefulness of *applied* examples, the need to better anticipate the future to take best actions versus merely reacting to events as they occur, and the ability to build on our knowledge to further understand our world. – **John Sall, Co-Founder & Executive Vice President, SAS Institute and JMP Business Division, USA**

The book clarifies many questions and concerns that I personally had when working with dozens of manufacturing industry collaborators, such as the role of human and human factor in future machine-learning-dominated industrial environments. This volume will be very valuable for readers, especially engineers and students seeking to experience the real world, unstructured problems. It will equip them with effective tools for solution. – **Prof. Ran Jin, Grado Department of Industrial and Systems Engineering, Virginia Tech, USA**

This book will certainly enrich our systems engineering (SE) literature collection, especially since it is focused upon "Industrie 4.0" the framework for "modern" SE. Modern SE will have to function in many ways different from "classical" SE and needs to question some of the "established" practices and in particular some of the associated processes. It is in this spirit that I welcome this new book, since it enters into some of the "Industrie 4.0" challenges and provides some SE answers which point in the right direction. – **Prof. Heinz Stoewer, Space Systems Engineering, TU Delft, The Netherlands.**

This book addresses a new challenge of our time, the development of Industry 4.0, which brings with it the combining of big data, complex objectives for efficiency and effectiveness of industrial systems, all in a context of continuous and rapid change. The chapters of the book lead the reader through understanding the nature of the special challenges of Industry 4.0 and providing guidance on how to do the work required to achieve, and provide assurance of achieving, the most desirable outcomes. The authors build on the existing tradition of systems engineering and lead the reader to understand how the existing methods of systems engineering can be adapted to enable achievement of the goals of Industry 4.0. – **Dr Tim Ferris, Centre for Systems Engineering, Cranfield Defence and Security, Cranfield University, UK**

Systems Engineering in the Fourth Industrial Revolution provides an important and timely perspective on how systems engineering must adapt to address the needs of evolving systems and technologies. This book explains how technological drivers such as the Internet of things (IoT) and big data fundamentally change the nature of systems, and how we must rethink how to engineer these systems to address concerns such as resilience, safety, and security. – **Sanford Friedenthal, SAF Consulting, former senior systems engineer at Lockheed Martin and Hughes Aircrafts, USA**

Reading this unique book, theory and practical examples will enable those in the advanced fields to understand and recognize the new tools required in the modern engineering world. Insights related to development, maintenance, manufacturing, reliability, and safety are needed for those who share the new world. – **Prof. (Ret. B.Gen) Jacob Bortman, ME Department, Ben Gurion University of The Negev, Israel**

The book offers a diverse set of chapters covering topics related to various aspects of "systems engineering meets big data." This provides a wealth of information that will be useful not only to those in systems engineering but also to researchers and practitioners in other big data fields. – **Prof. Galit Shmueli, Institute of Service Science, College of Technology Management, National Tsing Hua University, Taiwan**

We live in an increasingly interdependent world of ecosystems. Yet policy makers lack effective tools for twenty-first century systems thinking. This book, in 22 fascinating chapters, shows how data and technologies are impacting those who truly understand and apply systems thinking – systems engineers. – **Prof. Shlomo Maital, The Samuel Neaman Institute, Technion, Israel**

The book offers an impressive variety of twenty-first century topics for interconnected software dominant systems, written by global leaders in these fields. I especially like that systemic concepts are first explained and then shown graphically along with tools that support their exploration, such as system thinking and the Conceptagon. – **Sarah Sheard, The Software Engineering Institute (SEI), Carnegie Mellon University, USA**

The book presents a multiperspective insight unveiling the fundamental change in the contemporary nature of the term "system." Such a revolutionary approach is needed to cope with both the complex structure of components, subsystems, systems, and systems of systems; and, above all, the vast amount of data. – **Prof. Arie Maharshak, President, ORT Braude Academic College of Engineering, Karmiel, Israel**

Exciting book about Fourth Industrial Revolution era and changes around us. Kenett et al. are connecting several aspects of Industry 4.0, challenges, opportunities, benefits, and efforts. The book is pointing out to the complexity and the need of collaborative efforts of all stakeholders of global policy to implement new and innovative technologies. This is an inspiring state-of-the- art book for academia and industry about systems, data, engineering, manufacturing, technologies, and human aspects. – **Dr. Julita Panek, Siemens Digital Industrie Pharma Central and Eastern Europe**

The book represents a pioneering movement which addresses audiences who are familiar with the systems engineering area and are interested to do a "leap jump", and to be exposed to fascinating perspective which is dealing with the systems challenges of the Fourth Industrial Revolution. – **Giora Shalgi, The Samuel Neaman Institute, Technion and Former CEO of Rafael, Israel**

The book explores and clarifies many important topics for the fourth industrial revolution, including why this is driving change in the focus areas for Systems Engineering. In particular, the book highlights the importance of Applied Systems Thinking as a way of bringing together different methods to improve the probability of a successful outcome. – **Dr. Andrew Pickard, Systems Engineering, Rolls-Royce, USA**

This must read edited book highlights the essential elements of Industry 4.0 through a renewed look at the Systems Engineering discipline by leveraging the knowledge of a diversified group of leading scientists. The various contributions are well organized, insightful, present the current state of the art and transcend international boundaries completely. – **Prof. Emeritus Ehud Menipaz, Department of Industrial Engineering and Management, Ben Gurion University, Israel**

With the Fourth Industrial Revolution, "a storm is coming." It implies an enormous challenge for all engineering disciplines, especially for Software Engineering and Systems Engineering. To be able to deal with these challenges, this book provides the reader with a series of over 20 perspectives to consider. These perspectives range from Systems Thinking to Organizational Aspects, each written by a specialist or specialists in that field. Reading this book surely gives you insights for transforming your Systems Engineering efforts, so you'll get ready to deal with the fourth industrial revolution challenges. – **Paul Schreinemakers, INCOSE Technical Director, USA**

The book "Systems Engineering in the Fourth Industrial Revolution" is a fresh collection of chapters providing a window to a new era of systems engineering. In contrast to the past 70 years, where the pace of technology disruption was manageable by the adjustment of systems engineering tools, the pace in the last 10 years has called for revolutionary changes. The chapters in the book introduce a variety of issues, including methods, tools, and perspectives, that underlie this revolution. The book is an important introduction to this new era for every system engineer or those interested in the subject. – **Prof. Yoram Reich, Faculty of Engineering, Head, Systems Engineering Research Initiative, Tel Aviv University, Israel.**

Systems Engineering in the Fourth Industrial Revolution: Big Data, Novel Technologies, and Modern Systems Engineering

Systems Engineering in the Fourth Industrial Revolution: Big Data, Novel Technologies, and Modern Systems Engineering

First Edition

Edited by

Ron S. Kenett
The KPA Group,
Ra'anana, Israel
and
The Samuel Neaman Institute for National Policy Research,
Technion, Israel

Robert S. Swarz
Worcester Polytechnic Institute,
Worcester, MA, USA

Avigdor Zonnenshain
The Gordon Center for Systems Engineering and Samuel Neaman Institute for
National Policy Research,
Technion, Israel

This edition first published 2020
© 2020 John Wiley & Sons Inc

The right of Ron S. Kenett, Robert S. Swarz, and Avigdor Zonnenshain to be identified as the editors of this work has been asserted in accordance with law.

Registered Office
John Wiley & Sons, Inc., 111 River Street, Hoboken, NJ 07030, USA

Editorial Office
111 River Street, Hoboken, NJ 07030, USA

For details of our global editorial offices, customer services, and more information about Wiley products visit us at www.wiley.com.

Wiley also publishes its books in a variety of electronic formats and by print-on-demand. Some content that appears in standard print versions of this book may not be available in other formats.

Library of Congress Cataloging-in-Publication Data is applied for

9781119513896

Cover Design: Wiley
Cover Image: © Willyam Bradberry/Shutterstock

Set in 10/12pt WarnockPro by SPi Global, Chennai, India

10 9 8 7 6 5 4 3 2 1

Contents

Preface

During the last decade, companies in advanced economies have experienced significant changes in engineering and manufacturing practices, processes, and technologies. These changes have the potential to create a resurgence in their engineering and manufacturing capabilities. This phenomenon is often referred to as the Fourth Industrial Revolution (IR) or Industry 4.0, and is based on massive digitization, big data analytics, advanced robotics and adaptive automation, additive and precision manufacturing (e.g. 3D printing), modeling and simulation, artificial intelligence, and the nano-engineering of materials. This book addresses the topic from a Systems Engineering (SE) perspective. Rather than being based on scientific principles, systems engineering consists of a rich and useful set of principles, processes, practices, and lessons-learned. The contributors to this volume provide a rainbow coverage of such topics.

What is a system? Virtually all the textbooks and courses on systems engineering begin by trying to define a system. The description usually involves some combination or variations of the following themes: (i) it's an engineered response to a need, (ii) it possesses synergy (the whole is greater than the sum of its parts), (iii) it is bounded (one can "draw a box" around it), (iv) it has known internal functions, relationships, and interfaces, and (v) it has well defined and clearly understood external interfaces.

What has changed? Everything! Our notions of what a system is, and the characteristics of systems engineering, have served us well for the past 60 or 70 years, especially in the defense and aerospace industries. However, with the exponential increases in the amount of computational power, communications bandwidth, raw data, and addressable objects, things have changed! The Internet of Things (IoT) and Big Data have changed our fundamental concepts of what systems are and how we do systems engineering.

System complexity has increased so much that it is not uncommon for systems to exhibit emergent behavior, occasionally with unintended consequences, i.e. the ability to perform functions that could not be imagined by examination of its component parts. Increases in communications bandwidth and the so-called IoT obviate the boundedness of systems and of System-of-Systems (SoS). The vast amount of data presents enormous challenges to manage, store, analyze (and use!). Systems engineering is a discipline that promises to deal with complex issues in understandable and quantitative terms. The extended access to big data from sensors, together with concepts and technologies of the fourth industrial revolution, offers an opportunity for systems engineering to grow and expand. The bottom line is that systems engineering should move from being focused on documents to an approach being driven by data. This transition involves new and old disciplines such as computer simulations, statistically derived design and operational spaces, multivariate process control and monitoring, prognostic health management, complexity management, and performance evaluation methodologies. The transition from Documents to Data (from D to D) poses a range of analytic challenges, including sophisticated predictive analytics and data integration accounting for information generated at various levels of form, resolution, and speed. Questions that need to be considered regard the evaluation of high level and detailed designs, setting up of test suites in parallel with system development, and advanced support to decision making providing an indication of the impact of alternative scenarios.

This book deals with the above-mentioned challenges and opportunities for the systems engineering discipline. The chapters in the book cover several domains, with chapters written by leading experts.

These domains are (acronym in brackets):
- Systems Engineering and Systems Thinking (SEST)
- System Software and Process Engineering (SSPE)
- The Digital Factory (DF)
- Reliability and Maintainability Modeling and Analytics (RMMA)
- Organizational Aspects of Systems Engineering (OASE)

These five domains are covered by 22 state-of-the-art chapters. Some of the chapters are based on advanced and original research, others report and present advanced practices and case studies. Chapters can be read and studied in sequence or by domain area. They provide a wide-angle perspective of *Systems Engineering in the Fourth Industrial Revolution: Big Data, Novel Technologies and Modern Systems Engineering*. The titles of the chapters, a classification to the five domains (in brackets), and their brief description, are listed below.

1. Systems Engineering, Data Analytics, and Systems Thinking (SEST)
   Advanced systems engineering is described and presented as an approach to meet the challenges and opportunities of the fourth industrial revolution. The chapter presents the main elements and principles to be included in the systems engineering of the new era.

2. Applied Systems Thinking (SEST)

   This chapter provides an overview of applied systems thinking particularly as it pertains to the fourth industrial revolution. It includes a discussion of the systems thinking concept and a summary of principles and tools used in its execution. It is meant to be a guide for system engineers but can be used by experts seeking to resolve a problem or improve system performance.

3. The Importance of Context in Advanced Systems Engineering (SEST)

   In this chapter, the traditional view of "context" in systems engineering is described and challenges related to Industry 4.0 are identified. The insights from systems The resulting idea of "context of use," the interrelated technological, environmental, social, and operational conditions by which system behaviors can be fully understood, is described and positioned as an important element of advanced systems engineering.

4. Architectural Technical Debt in Embedded Systems (SEST)

   When software teams and system engineers are not synchronized, the system architecture might not be deployed because of short-term pressure to deliver. On the other hand, the system architecture might need to be updated when new requirements are presented or changed during system development. The consequence of not having this feedback loop often leads to a mismatch between the actual implementation of the system and the qualities specified in the desired system architecture. Such a phenomenon is recognized in the literature as Architectural Technical Debt. In this chapter, the causes and trends related to the accumulation of Architectural Technical Debt are presented as well as implications for system engineering.

5. Relay Race: The Shared Challenge of Systems and Software Engineering (SSPE)

   As systems become more and more software intensive, the cooperation between systems engineers and software engineers plays a critical role in system development. In this chapter, the areas of responsibility of the two disciplines are analyzed and those requiring shared effort are identified. A detailed process is proposed, mainly based upon functional analysis, through which a software-intensive system architecture is constructed, covering both structural and behavioral characteristics.

6. Data-Centric Process Systems Engineering (PSE) for Industry 4.0 (SSPE)

   PSE has over 50 years of contributions to Chemical Engineering, mostly related to the exploitation of computational power in order to make better plants and products by applying deductive models. However, the fourth industrial revolution is taking its course, and making the hidden potential for improvements and competitiveness contained in a variety of data streams readily available to be processed and analyzed in the Chemical Process Industries (CPI). This chapter describes the development of data-centric PSE for the chemical Industry 4.0.

7. Virtualization of the Human in the Digital Factory (DF)

   In the paradigm of the digital factory, human aspects must be considered as the focal point of technologies, activities and information. This chapter presents an overview of the virtual humans/digital human models (DHM) that are considered and applied along the design and manufacturing processes.

8. The Dark Side of Using Augmented Reality (AR) Training Systems in Industry (DF)

   AR systems, in which virtual information is superimposed on the real physical environment, are becoming more and more used in industry. Some of these systems serve as training systems. Surprisingly, when the purpose is training, some of the benefits for on-the-job performance support can turn to be threats. Some of these threats are identified, studied, and discussed.

9. Condition-Based Maintenance (CBM) via a Targeted Bayesian Network Meta-Model (RMMA)

   CBM methods are increasingly applied to operational systems and often lead to reduction of their life cycle costs. Nonetheless, trying to evaluate a priori the cost reduction of various CBM policies, under different scenarios and conditions, is a challenging task which is addressed in this chapter.

10. Reliability-Based Hazard Analysis and Risk Assessment: A Mining Engineering Case Study (RMMA)

    The fourth industrial revolution has created a need for enhanced reliability-based hazard analysis and risk assessment for managing the uncertainties in earth resources industries. As an example, quantification of risks due to failure of natural and engineered rock slopes is presented for supporting objective decision making in rock engineering. This quantification of risk methodologies can be applied in different domains.

11. OPCloud: An OPM Integrated Conceptual-Executable Modeling Environment for Industry 4.0 (SEST)

    The chapter integrates systems engineering with software engineering into a coherent unifying framework. To address this challenge, a Methodical Approach to Executable Integrated Modeling (MAXIM) and its implementation environment, OPCloud, are presented.

12. Recent Advances Toward the Industrialization of Metal Additive Manufacturing (SEST)

    The adoption of additive technology for metal parts opens new opportunities for flexible and decentralized manufacturing of more complex and high added value components. Today, an increasing number of companies is applying disruptive technologies to manufacture prototypes, tools, spare parts, and components in a faster and cost-effective manner.

13. Analytics as an Enabler of Advanced Manufacturing (RMMA)

    Advanced manufacturing or Industry 4.0 is based on three interconnected pillars: (i) Computerized Product Design and Smart Technology; (ii) Smart

Sensors, IoT, and Data Collectors integrated in manufacturing lines; and (iii) Analytics, Control, and Data Science. This chapter is a critical review of the third pillar of data analytics and how it is related to the other pillars. The objective is to present a context for a range of analytic challenges in Industry 4.0.

14. Hybrid Semiparametric Modeling: A Modular Process Systems Engineering Approach for the Integration of Available Knowledge Sources (SSPE)
    In this chapter, different methods are presented for inferring system descriptions. Models from data and their merits are discussed by considering data requirements and a priori knowledge. These can be described as the structure of interactions followed by a hybrid semiparametric framework that combines a priori knowledge about the system with nonparametric models where structure and parameters are inferred from the data.

15. System Thinking Begins with Human Factors: Challenges for the 4th Industrial Revolution (OASE)
    This chapter addresses possible developments in the fourth industrial revolution of the role of human factors in systems engineering. The chapter suggests that the fourth industrial revolution may involve various shifts toward Human–Machine Interaction (HMI), associated with technology, methodology, and HMI thinking and practices. These shifts may affect the people productivity, quality of life, and safety.

16. Building More Resilient Cybersecurity Solutions for Infrastructure systems (RMMA)
    The world's infrastructure is generally vulnerable to cyberattack because so much of it is dependent upon software and so many of its systems are interconnected. The chapter describes a new form of operating system that matches the exponential changes occurring around us. The objective should be to collectively transcend away from linear thinking about cybersecurity and toward a multidimensional mindset to address a multifaceted problem.

17. Closed-Loop Mission Assurance (MA) Based on Flexible Contracts: A Fourth Industrial Revolution Imperative (SEST)
    Mission Assurance (MA) is concerned with reducing uncertainty in the ability of systems to accomplish mission in the face of disruptions. This chapter presents a model-based MA approach that accounts for uncertainties arising from partial observability of the operational environment and employs reinforcement learning based on incoming data (i.e. observations) from the operational environment to progressively determine system states.

18. FlexTech: From Rigid to Flexible Human–Systems Integration (OASE)

    Human–Systems Integration (HSI) denotes an evolution of conventional Human Factors and Ergonomics (HFE) discipline that focuses on evaluation of existing systems and usages, as well as a Human–Computer Interaction (HCI) that provides methods and tools for interaction design. In the chapter, several areas related to HSI are covered, including task and activity analysis, cognitive engineering, organization design and management, function allocation, complexity analysis, modeling and human-in-the-loop simulation (HITLS). The chapter proposes a system-of-systems structure-function approach combined with a cognitive-physical distinction, and more specifically on cognitive-physical function allocation among a society of agents.

19. Transdisciplinary Engineering Systems (OASE)

    In this chapter, a system view on transdisciplinary engineering is presented. Transdisciplinary processes are most often performed in large, complex projects, aimed at creating a solution for a problem that cannot be solved by one person, nor by one discipline alone.

20. Entrepreneurship as a Multidisciplinary Project (OASE)

    Entrepreneurship is a creative process, initiated to form and manage an innovative model, product, service, business or any combination thereof. An entrepreneurship is a creative complex project, incorporating all the activities required to form and manage it. This chapter elaborates on a holistic perspective, with details on the various components and interrelations of entrepreneurship and presents a systemic model for the planning and operation of an entrepreneurship in the fourth industrial revolution.

21. Developing and Validating an Industry Competence and Maturity for Advanced Manufacturing Scale (OASE)

    In this chapter, the process of building and validating a model for assessing the maturity level of plants in advanced manufacturing is described. The assessment tool is based on four dimensions (content areas) comprising 14 subdimensions that are relevant to advanced manufacturing. The questionnaire includes statements designed for assessing the status of a manufacturing unit relative to Industry 4.0 ideals. This model is validated internally by manufacturing companies and externally by international experts.

22. Modeling the Evolution of Technologies (OASE)

    This chapter expresses some of the evolutionary dynamic life cycle characteristics of technologies based on reliability theory. The reliability mathematics is successfully implemented on a broader philosophic holistic domain rather than in the engineering domain alone; the classic product becomes the newly defined complex entity of technology or the species of related products. This model serves the fast-changing technologies in the fourth industrial revolution.

As a complementary perspective to this chapter annotation, we evaluated these descriptions with JMP 14.2 Text Explorer (www.jmp.com). A word cloud of this text is presented below.

This describes the overall content of the book. The chapters in this book are about systems, data, engineering, manufacturing, technologies, and human aspects.

The main aims of this book are to share advanced models, innovative practices, and state-of-the-art research findings on systems engineering in the fourth industrial revolution era, with the whole professional community and with all managers in relevant industries and companies in various domains. Most of the methodologies of the current systems engineering body of knowledge were developed and practiced in the defense and aerospace industries. We target here industries that are affected by the fourth industrial revolution, like manufacturing, healthcare and life science, food and agriculture, communication and entertainment, smart transportation and cities, and more. We consider this an opportunity to introduce systems engineering into such new domains.

The book can serve as textbook and learning material in academic and professional training programs of systems engineering, and especially in programs on advanced systems engineering for the fourth industrial revolution. Also, the chapters included here can serve the curious systems engineers and managers who want to implement new and innovative approaches.

The chapters in the book present material that can also stimulate research efforts and initiatives on the new methodologies and approaches for advanced systems engineering in the new and evolving ecosystem called Industry 4.0.

Final words – This book is a call for action to change the way of carrying out systems engineering in order to meet the challenges and opportunities of the new technological era. It is designed as a resource to facilitate these actions for change and reform.

Our contributing authors clearly deserve most of the credit. They are the real "heroes" of this book. They were generous in sharing their experience, knowledge, and new ideas, and investing the time to write the chapters. We wish to thank them for their meaningful collaboration and true patience through the various stages of writing this book.

Ron S. Kenett, KPA Group and The Samuel Neaman Institute for National Policy Research, Technion
Robert S. Swarz, Worcester Polytechnic Institute
Avigdor Zonnenshain, Gordon Center for Systems Engineering and The Samuel Neaman Institute for National Policy Research, Technion

List of Contributors

Eitan Adres
The Samuel Neaman Institute
Technion
Israel

Oliver Avram
University of Applied Sciences and
Arts of Southern Switzerland
Switzerland

Cristiana Rodrigues de Azevedo
Newcastle University
School of Engineering
Newcastle upon Tyne
UK

Irad Ben-Gal
Tel Aviv University
Israel

Kfir Bernstein
Technion
Israel

Jan Bosch
Chalmers University of Technology
Sweden

Guy A. Boy
CentraleSupélec and ESTIA Institute
of Technology
France

Emanuele Carpanzano
University of Applied Sciences and
Arts of Southern Switzerland
Switzerland

Victor Grisales Díaz
Newcastle University
School of Engineering
Newcastle upon Tyne
UK

Dov Dori
Technion
Israel

and

Massachusetts Institute of
Technology
USA

Sebnem Duzgun
Colorado School of Mines
USA

Robert Edson
The MITRE Corporation
USA

Nirit Gavish
ORT Braude College of Engineering
Israel

Aviv Gruber
Tel Aviv University
Israel

Avi Harel
Ergolight
Haifa
Israel

Ahmad Jbara
Technion
Israel

and

University of Connecticut
Storrs
USA

Arnon Katz
M-Bios
Haifa
Israel

Ron S. Kenett
The KPA Group, Ra'anana

and

The Samuel Neaman Institute
Technion
Israel

Hanan Kohen
Technion
Israel

Rea Lavi
Technion
Israel

Azad M. Madni
University of Southern California
Los Angeles
USA

Antonio Martini
University of Oslo
Norway

Federico Mazzucato
University of Applied Sciences and
Arts of Southern Switzerland
Switzerland

John P.T. Mo
RMIT
Melbourne
Australia

Rui Oliveira
Universidade Nova de Lisboa
Caparica
Portugal

Oscar Andrés Prado-Rubio
Universidad Nacional de
Colombia – Sede Manizales
Campus La Nubia
Colombia

Véronique Préat
Université Catholique de Louvain
Brussels
Belgium

Daniele Regazzoni
University of Bergamo
Italy

Marco S. Reis
University of Coimbra
Portugal

Caterina Rizzi
University of Bergamo
Italy

Pedro M. Saraiva
University of Coimbra and NOVA
University of Lisbon
Portugal

Yair Shai
Technion
Israel

Uri Shani
Technion
Israel

Michael Sievers
University of Southern California
Los Angeles
USA

Natali Levi Soskin
Technion
Israel

Josip Stjepandic
PROSTEP AG
Darmstadt
Germany

Moritz von Stosch
Newcastle University
School of Engineering
Newcastle upon Tyne
UK

Robert S. Swarz
Worcester Polytechnic Institute
USA

Amir Tomer
Kinneret Academic College
Israel

Anna Valente
University of Applied Sciences and
Arts of Southern Switzerland
Switzerland

Daniel Wagner
Country Risk Solutions
USA

Niva Wengrowicz
Technion
Israel

Adam D. Williams
Sandia National Laboratories
USA

Mark J. Willis
Newcastle University
School of Engineering
Newcastle upon Tyne
UK

Nel Wognum
Technical University of Delft
The Netherlands

Inbal Yahav
Tel Aviv University
Israel

Shai Yanovski
Tel Aviv University
Israel

Avigdor Zonnenshain
The Gordon Center for Systems
Engineering and the The Samuel
Neaman Institute
Technion
Israel

1

Systems Engineering, Data Analytics, and Systems Thinking

Ron S. Kenett, Robert S. Swarz, and Avigdor Zonnenshain

Synopsis

In the last decade, industries in advanced economies have been experiencing significant changes in engineering and manufacturing practices, processes, and technologies. These changes have created a resurgence in engineering and manufacturing activities. This phenomenon is often referred to as the Fourth Industrial Revolution or Industry 4.0. It is based on advanced manufacturing and engineering technologies, such as massive digitization, big data analytics, advanced robotics and adaptive automation, additive and precision manufacturing (e.g. 3D printing), modeling and simulation, artificial intelligence, and the nano-engineering of materials. This revolution presents challenges and opportunities for the systems engineering discipline. For example, virtually all systems now have porous boundaries and ill-defined requirements. Under Industry 4.0, systems have access to large types and numbers of external devices and to enormous quantities of data, which must be analyzed through big data analytics. It is, therefore, the right time for enhancing the development and application of data-driven and evidence-based systems engineering. One of the trends in data analytics is the shift from detection to prognosis and predictive monitoring in systems operations. Similar trends apply to testing and maintenance using prognostics and health monitoring (PHM). Also, an enhanced practice of evidence-based risk management as a more effective approach for managing the systems' risks is observed. System properties like reliability, safety, and security are increasingly important system attributes. Such "ilities" can now be integrated into the systems engineering processes by introducing modern methodologies based on data analytics and systems engineering theories. Modeling and simulation are also going through impressive developments. The modern systems engineer needs to exploit such advancements in system design, operation, and demonstration.

Systems Engineering in the Fourth Industrial Revolution: Big Data, Novel Technologies, and Modern Systems Engineering, First Edition. Edited by Ron S. Kenett, Robert S. Swarz, and Avigdor Zonnenshain.

Companies that are striving to upgrade to the fourth industrial revolution can apply the maturity level assessment tool introduced in this chapter and elaborated with more details in Chapter 21. This chapter provides a context and a roadmap for enhancements in system engineering that are designed to meet the challenges of Industry 4.0. It is suggested that systems engineering needs a new conceptual model, which we refer to here as "the double helix." This model emphasizes the tight connection that systems engineering must have with applied systems thinking, as well as with data analytics, security, safety, reliability, risks, resilience, etc. All these must be integrated into the systems engineering process.

1.1 Introduction

Over the past several decades, Systems Engineering has developed a rich foundation of standard models, methods, and processes. The basic goal has been to transform a set of user needs into a system. The application of these methods typically depended on the assumptions that the system had a stable, well-constructed set of requirements, a well-defined set of stakeholders with stable expectations, and well-known and controllable constraints and boundaries of the system.

Over the years, a number of things have happened to change this simple paradigm. The number and variety of stakeholders has grown, the number of requirements has grown, the boundaries of the system have become less clear, and the number of external interfaces has proliferated, making the task more complicated by orders of magnitude, but nevertheless manageable.

Today, we find ourselves in the era of the Fourth Industrial Revolution with the internet of things (IoT) and big data (BD) challenging our notions of system constraints and borders (Figure 1.1). The number of interfaces and amount of data that we have to deal with have grown exponentially. Big data has been described as the amount of data just beyond technology's capability to store, manage, and process efficiently. In this situation, organizations cannot get value out of the data because most of it is in raw form or in a semistructured or unstructured form. How and what to do in order to extract value from it may not be clear. McKinsey estimates that IoT will have a total potential economic impact of between 3.9 and 11.1 trillion dollars per year by 2025 (Manyika et al. 2015); the International Data Corporation (IDC) forecasts that machine-generated data will increase to 42% of all data by 2020, up from 11% in 2005; and Gartner estimates a potential total of 26 billion connected devices by 2020 (Middleton et al. 2013). (Morgan Stanley puts this figure at 75 billion! (Business Insider 2013)) ABI Research (www.abiresearch.com) estimates that the amount of IoT data will reach 1.6 zettabytes (1.6×10^{21} bytes) by 2020!

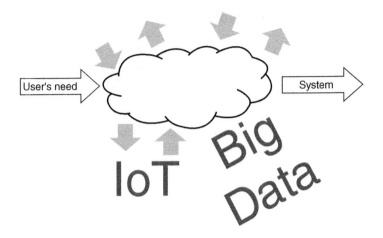

Figure 1.1 An overall perspective of systems engineering.

According to a 2016 report by Cisco (www.cisco.com), global internet traffic is expected to reach 1.3 zettabytes annually, with over 3.4 billion internet users (about 45% of the world's population) and over 19 billion network connections. That's 2.5 connections for each person on earth!

The IoT and data analytics (DA) are closely tied together. For example, data extracted from IoT devices can provide a mapping of device interconnectivity so that companies can better evaluate who and where their audience is. IoT is also increasingly being adopted as a means of gathering sensor data used in medical, manufacturing, and other contexts.

Systems engineers can help companies expand beyond traditional business intelligence and scientific applications and extend data analytics to new domains with the development of new predictive and prescriptive analytical techniques. For a 2525 vision of system engineering see INCOSE (2014).

The goal we envision for system engineering activity is to yield actionable predictive and prescriptive results that *facilitate decision-making* by going beyond data mining and statistical processing to encompass logic-based methods, qualitative analytics, and quantitative methods.

The bottom line implications are that, in the near future:

- Virtually all systems will have porous (or nonexistent) boundaries.
- Virtually all systems will have ill-defined requirements that are changed frequently.
- Virtually all systems will have access to large types and numbers of external devices.
- Virtually all systems will have access to enormous quantities of data, which have to be analyzed through data analytics to be of value.

1.2 The Fourth Industrial Revolution

The first industrial revolution was triggered by the introduction of the steam engine and the mechanization of manual work in the eighteenth century. Electricity drove mass production in the second industrial revolution in the early twentieth century. The third revolution was due to the use of electronics and computer technology for manufacturing and production automation. We are now entering the fourth industrial revolution, also called Industry 4.0, which uses advanced manufacturing and engineering (for a description of this evolution see Chapter 1 in Kenett and Zacks 2014).

We stand on the brink of a technological revolution that will fundamentally alter the way we live, work, and relate to one another. In its scale, scope, and complexity, the transformation will be unlike anything humankind has experienced before. We do not yet know just how it will unfold but one thing is clear: the response to it must be integrated and comprehensive, involving all stakeholders of the global polity, from the public and private sectors to academia and civil society. There are three reasons why today's transformations represent not merely a prolongation of the third industrial revolution but rather the arrival of a fourth, distinct one, i.e. **velocity, scope, and systems impact**. The speed of current breakthroughs has no historical precedent. When compared with previous industrial revolutions, the fourth is evolving at an exponential rather than a linear pace. Moreover, it is disruptive to almost every industry in every country. The breadth and depth of these changes herald the transformation of entire systems of production, engineering, management, and governance (paragraph adapted from Schwab 2017).

We begin by a review of the challenges and opportunities for the systems engineering discipline and community in the fourth industrial revolution ecosystem, and their consequences. Overall, the chapter presents an integrated view of systems engineering, systems thinking, and data analytics.

With the emergence of Industry 4.0 and the Big Data movement, industry is now presented with unique opportunities in terms of key enablers that move performance to new and higher levels. Performance is considered here in its widest sense, from operational, economic, and market-related aspects to process/product/system safety/reliability and environmental considerations.

Reis and Gins (2017) discuss this trend for industrial process monitoring (IPM). It also applies to systems engineering as an important process and crucial system in many companies that are developing, producing, delivering, and maintaining complex systems. Kenett et al. (2018) provide a road map for applied data sciences supporting sustainability in advanced manufacturing that builds on the information quality (InfoQ) framework introduced by Kenett and Shmueli (2016). Reis and Kenett (2018) present an application of this roadmap to chemical process engineering. Chapter 17 in Kenett and Redman (2019) is about the role of data science in the fourth industrial revolution, implying how data science and systems engineering can work synergistically.

The key enablers for this strategic shift to prognosis and improvement are big data, technology, and data analytics. Big data is possible because of the development of better, faster, inexpensive, and more informative sensing technology (sensors), which can collect information from multiple sources in order to store it in integrated databases and make it available anywhere and at any time to engineers, operators, and decision makers. New technologies also provide the computational resources (e.g. high-performance computing, cloud services, distributed, and parallel computing) that are required to process large amounts of data using the third enabler – advanced analytics capabilities – turning them into actionable information, in a timely manner.

Up to now, the detection and diagnosis of process failures and anomalies dominated processes control, but due to the advanced sensing technologies and advanced analytics capabilities, we see an important shift to *prognosis* of processes and systems. This enables us to measure and predict the health of systems far beyond preventive approaches.

Shifting into prognostic mode for systems engineering implies the following activities:

- Introducing prognostic capabilities in the development stage by using prognostics and health monitoring (PHM) in testing and experimenting components, subsystems, and systems.
- Using PHM capabilities in demonstration and life tests.
- Integrating PHM capabilities into the system development process as design for maintenance policies that enhance life cycle management and system reliability and availability.
- Introducing condition-based maintenance (CBM) for the systems under development.
- Integrating PHM capabilities into the manufacturing lines of systems and products under development.
- Supporting design for sustainability of the products and systems under development and operation using PHM capabilities.
- Enhancing systems safety and security by using PHM capabilities.

So, smart systems engineering based on PHM will have an essential positive impact on systems effectiveness through the application of advanced processes in systems engineering.

Companies that are willing to improve the way they are doing business using some of the opportunities implied by the fourth industrial revolution can assess their maturity levels in their different business areas. This serves the companies as a tool for benchmarking. A framework for maturity assessment has been developed on the basis of the Software Engineering Institute's capability maturity model integration (CMMI) approach (Kenett and Baker 2010). The model, when tailored to advanced manufacturing, will suggest several possible actions and activities of a company striving to advance its maturity level toward Industry 4.0 capabilities. For example, in the area of processes in engineering, the following activities are proposed for a capability assessment:

- There is an engineering planning system based on information technologies.
- The tools of engineering design are computerized.
- The processes of engineering design include modeling and simulations in the framework of model-based design (MBD).
- Simulations are used for statistical design of experiments (DOE) as part of the design and engineering processes.
- There is relevant use of 3D printing for fast prototyping, and for designing molds and dies.
- During the development and engineering of new products and systems, it is suggested to introduce the use of new, advanced, and innovative materials that improve the products, the systems, and the manufacturing processes.

Chapter 21 in this book is about this CMMI-like model for assessing the maturity of organizations adopting advanced manufacturing. Implementing this model supports companies interested in assessing strengths and implementation gaps to prepare improvement plans. Also, it provides organizations with a tool for assessing their actual improvements and achievements. In addition, it is an excellent benchmarking tool.

1.3 Integrating Reliability Engineering with Systems Engineering

In our experience, most systems engineers are lacking knowledge and experience in reliability engineering. This problematic situation is because:

- Reliability engineering is not always recognized by systems engineers as one of their core competencies.
- In most companies, the reliability engineers and systems engineers operate in different "silos".
- In some of the companies, reliability engineers are part of the quality organization, not systems engineering.
- In recent years, the added value of traditional reliability engineering activities for systems engineering has been in question.

To meet the challenges of the fourth industrial revolution it is necessary to integrate reliability engineering with systems engineering through two parallel efforts:

- Upgrading the effectiveness of reliability engineering for systems development by using the opportunities of the new era, like big data analytics, prognostics and health monitoring (PHM), and modeling and simulation
- Integrating reliability engineering processes into systems engineering processes

Zio (2016) proposes the big KID (Knowledge, Information, and Data) for upgrading the effectiveness of reliability engineering through several models such as:

- Degradation models (Shahraki et al. 2017).
- Integration of the physics of failure in reliability models by using multistate physics-based models.
- Accelerated degradation testing (ADT).
- Prognostics and health monitoring (PHM).

In addition to these models, reliability engineering can make use of different design for reliability tools like: DOE, robust design (RD), design for variability (DFV), and design for manufacturing and assembly (DFM/A). All these approaches use data analytics of field and testing data.

Integrating reliability engineering into systems engineering academic programs needs to be promoted through training of systems engineers in reliability engineering and through training reliability engineers in systems engineering. Systems engineering programs should also include reliability engineering programs like Reliability Case (which is a reliability program based on structured arguments, supported by evidence and data). For a model linking field data with reliability assessments at the development's stages see Halabi et al. (2017). See also Chapters 9, 10, 13, and 16 in this book.

1.4 Software Cybernetics

Modern systems such as advanced manufacturing technologies require sophisticated control strategies to operate. Cybernetics was defined by mathematician Nobert Wiener in 1948 as control and communication in the animal and the machine (Wiener 1948). The essential goal of cybernetics is to understand and define the functions and processes of systems that have goals, and that participate in circular, causal chains that move from action to sensing to comparison with desired goal, and again to action. Studies in cybernetics provide a means for examining the design and function of any system, including social systems such as business management and organizational learning, to make them more efficient and effective.

Software cybernetics explores the interplay between software/systems behavior, and control (Cai et al. 2004). The fundamental question of interest is how to adapt software behavior, software processes, or software systems to meet basic and expanded objectives in the presence of a changing environment. This challenge is typical of advanced manufacturing environments where flexibility and responsiveness are essential capabilities. Software cybernetics addresses issues and questions on: (i) formalization and quantification of feedback and self-adaptive control mechanisms in software; (ii) adaptation of control theory principles to software processes and systems, (iii) application of software engineering principles and theories to control systems, and (iv) processes and integration of the theories of software engineering and control engineering.

At the first level, software cybernetics studies software and its environment from a control theory perspective. It views software plus its environment as a system, and software mostly as a control and/or communication media of the system. Such a system could be a production plant with IoT and advanced control mechanisms. At this level, the challenge in advanced manufacturing is how to design and implement an optimal and robust system.

At the second level, the development of the above software + environment system involves many software tools, organizations, physical devices, and equipment; it forms a complex system and its product is also a complex system. Different software development methodologies employ different control and communication mechanisms and strategies. They result in software systems that are of different quality, and the processes are of different productivity and cost. Software cybernetics is focused on the control and communication in such a "system" and its behaviors. Many research efforts in research on software engineering belong to this level.

At the highest level, various software technologies and methodologies are competing. Why do some technologies and methodologies survive while others become extinguished? Is there a law of evolution as in the natural biology system? At this level, software cybernetics evaluates software/IT technology as a big, dynamic, and evolving sociotechnical system. For more on software cybernetics see Kenett and Baker (2010). For a treatment of cybersecurity see Chapter 16.

1.5 Using Modeling and Simulations

Modeling the Evolution of Technologies

An approach to predict the life cycles of technologies based on analysis of field data and reliability models is presented in Chapter 22 and also Shai (2015). This model is well in line with the third perspective of software cybernetics presented above. The *raison d'être* for this approach is that, in the modern era, systems and equipment are not replaced due to technical failures but due to obsolete technology, or due to changing of the technology by the suppliers. In this model, the technologies' life cycle characteristics and their dynamics are based on reliability theory. Instead of predicting mean time between failures (MTBF) the model predicts the end of life due to the introduction of new technologies or changes in the market.

In the fourth industrial revolution, we encounter rapid technological changes, and this approach can be a powerful business tool for predicting change cycles based on data; it can be also a support decision tool for companies and the developers of disruptive technologies. This is a systematic and holistic view of market and industry evolution, using reliability models and data analytics (see Chapter 22).

Modeling and Simulation for Systems Engineering

Today, it is possible to build simulated physical models or behavioral models for almost any system/subsystem or system process. These models are based on powerful computerized tools like MATLAB, Simulink, modeFRONTIER, LS-DYNA, and others. These models describe the system under development at any relevant stage of the life cycle. The models provide engineers with quantitative data on the performances of the systems. They are used for simulating the behavior of systems for different internal and external inputs. Through these simulations, it is possible to evaluate design alternatives to find better design solutions and to support design for robustness, reliability, and safety.

During the modeling and simulation phase, it is possible to prepare rapid prototypes based on the current model. These prototypes can be produced by traditional manufacturing or 3D printing based on the model. Testing the prototypes under different scenarios is the first step for validating the systems and its models.

In later stages of the development process, there are system tests in selected points of the design and operation of the system. The outcomes of these tests can be compared to simulation results to validate the system models and verify the system's design.

So, modeling and simulation are very crucial tools for systems engineers and should be integrated into the systems engineering process. It is worth mentioning that some of these tools (like modeFRONTIER) provide multidimensional and multidisciplinary optimization capabilities, which is one of the important roles of systems engineers.

The models provide additional advantages and benefits for the systems engineering in the new era:

- The models are the only "truth" on the system parameters and behavior – the model should be considered the only valid documentation of the system.
- The models are based on data analytics and they also create valuable data through simulations.

- The models are also the only source for manufacturing the systems. They shorten the manufacturing engineering stage, and save a lot of misunderstandings and faults along the manufacturing and integration chain.
- It is possible as well to create automatic system software code.
- The models support and ease the possible changes along the systems lifecycle.
- The models are used, too, for preparing the operation and training material for the system.

Modeling and simulation and the goal of data-driven and evidence-based systems engineering are an essential part of the fourth industrial revolution

Computer Experiments

Experimentation via computer modeling has become very common in systems engineering. In computer experiments, physical processes are simulated by running a computer code that generates output data for given input values. In physical experiments, data are generated directly from a physical process. In both physical and computer experiments, a study is designed to answer specific research questions, and appropriate statistical methods are needed to design the experiment and to analyze the resulting data. We focus here on computer experiments and specific design and analysis methods relevant to such experiments. Because of experimental error, a physical experiment may produce a different output for each run at the same input settings. On the other hand, computer experiments are capable of being deterministic, so that the same inputs will always result in the same output. Thus, none of the traditional principles of blocking, randomization, and replication can be used in the design and analysis of computer experiments data. Computer experiments consist of a number of runs of a simulation code and factor combinations that correspond to a subset of code inputs. By considering computer runs as a realization of a stochastic process, a statistical framework is available both to design the position of the experimental points and to analyze the responses. The major difference between computer numerical experiments and physical experiments is the logical difficulty in specifying a source of randomness for computer experiments.

The complexity of the mathematical models implemented in computer programs can, by themselves, build equivalent sources of random noise. In complex code, a number of parameters and model choices give the users many degrees of freedom that provide potential variability to the outputs of the simulation. Examples include different solution algorithms (i.e. implicit or explicit methods for solving differential equations) and various approaches to discretization intervals and convergence thresholds for iterative techniques. In this very sense, an experimental error can be considered in the statistical analysis of computer experiments. The nature of the experimental error in both physical and simulated experiments is our ignorance about the phenomena and the intrinsic error of the measurements. Real-world phenomena are often too complex for the experimenter to keep under control by specifying all the factors affecting the response of the experiment. Even if it were possible, the physical measuring instruments, being not ideal, introduce problems of accuracy and precision. Perfect knowledge would be achieved in physical experiments only if all experimental factors could be controlled and measured without any error. Similar phenomena occur in computer experiments. A complex code has several degrees of freedom in its implementation that are often not controllable. For more information on computer experiments, see Kenett and Zacks (2014).

1.6 Risk Management

Risk Management and Systems Engineering

Risk management is a practice used by systems engineers and project managers. However, one typically faces a dilemma in trying to answer questions such as: Is it an effective practice in developing, deploying, and operating complex systems? This is a relevant question, especially in the era of exponential technology growth, and in continuous changes of the ecosystem. We are questioning here the abilities of the project managers to effectively identify and evaluate the real risks, and to plan and deploy effective mitigations (Figure 1.2).

Soon, many of the risks will come from the system's context and not from inside the system itself. The standard matrix that exhibits likelihood vs. consequences can be misleading. A risk with an infinitesimally low likelihood could have disastrous effects. As an example, consider the Internet failure in March

Figure 1.2 Risk management heat map.

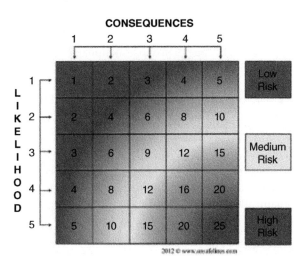

2017, when a large number of servers supporting Amazon's web hosting service went down (Newton 2017). An investigation revealed the source to be a simple typographical error made during a routine upgrade of a subset of the servers that cascaded through and brought down all of the servers. It took more than four hours to bring the system back up, even after the cause was discovered.

As another example of the risk environment in the IoT, consider the Ring doorbell commercial product. This clever device allows the owner of a home to "answer" the doorbell even when not home. A homeowner could use this feature to make it appear that a house is occupied when it is not, and it can allow one to screen visitors when one is home or away. In early versions of the device, there were several security incidents and scares with this device. The first problem was when it was discovered that the device was storing Wi-Fi passwords in plain text. This potentially allowed attackers to gain entrance to a home that was thought to be secured by a burglar alarm system. It was later found that the devices were sending packets of data to potential malefactors (Tilley 2017). It was not known how the vulnerability got there, exactly where and what type of data were being sent, how this problem got though the systems engineering process in the first place, etc. Ring's makers claimed that the bug was benign and, in fact, no malicious use ever turned up, but they did patch it. This illustrates what can and will be happening with increasing frequency in the use of IoT devices.

Wagner (2016) raises several critical aspects of man-made risks that each of them can impact the business future profit and survival, like:

- The rising threat of cyber risk and why it can no longer be ignored.
- How terrorism can impact business and what to do to mitigate its effects.
- Why climate change is on everyone's doorstep, and its future impacts.
- Why corporate social responsibility is not only the right thing to do but good for business.

It is the right time to propose an updated methodology for risk management that fits the new ecosystem.

Evidence-Based Risk Management

As noted above, risk management is traditionally practiced using subjective assessments and scenario-based impact analysis. This common approach is based on experts providing their opinions and is relatively easy to implement. In some cases, accounting firms are hired to conduct such projects at the organizational level; however, modern evidence-based management relies on data, and not only opinions, to achieve effectiveness and efficiency. In that context, risk management can exploit information from structured quantitative sources (numerical data) and semantic unstructured sources (e.g. text, voice, or video recordings) to drive risk assessment and risk mitigation strategies.

One of the major challenges of risk management are Black Swans (Taleb 2007; Kenett and Tapiero 2010). A Black Swan is a highly improbable event with three principal characteristics: (i) it is unpredictable; (ii) it carries a massive impact; and (iii) after the fact we concoct an explanation that makes it appear less random and more predictable than it was. Why do we not acknowledge the phenomenon of Black Swans until after they occur? Part of the answer, according to Taleb, is that humans are hardwired to learn specifics when they should be focused on generalities. We concentrate on things we already know, and time and time again fail to take into consideration what we don't know. We are, therefore, unable to truly estimate opportunities, too vulnerable to the impulse to simplify, narrate, and categorize, and not open enough to rewarding those who can imagine the "impossible." Taleb has studied how we fool ourselves into thinking we know more than we actually do. We restrict our thinking to the irrelevant and inconsequential, while large events continue to surprise us and shape our world. We suggest that proper exploitation of organizational data can help prevent some of the Black Swans and/or blind spots.

Modern systems collect huge amounts of data. We showed how such data can be incorporated into risk management and governance of organizations. By "data" we refer to information of all type. It can consist of text manually entered in a call center where operators help internal of external customers overcome problems and difficulties. It can also use information recorded in web logs where various topics are discussed or data warehouses where all transactions and tasks leave a trail. To exploit unstructured semantic data, one needs to combine with the data collection text annotation tools for data extraction and to devise ontologies for organization of the data. These areas have been developed by experts in artificial intelligence and natural language processing. Combining these data with quantitative data provides a powerful infrastructure for effective and efficient risk management. The combination requires advanced statistical models such as Bayesian Networks and nonlinear analysis. For examples and methodologies achieving this see Kenett and Raanan (2010).

1.7 An Integrated Approach to Safety and Security Based on Systems Theory

Despite an enormous amount of efforts and resources applied to security assurance and enhancement in the recent years, significant progress seems to be lacking. At the same time, changes in systems and in systems engineering are making traditional safety analysis and assurance techniques increasingly less effective. Most of these techniques and methodologies were created more than 30 years ago, when systems were composed of mechanical, electromechanical, and electronic components and were orders of magnitude less complex than today's software-intensive systems. Also, the fourth industrial revolution creates innovative autonomous systems like driverless vehicles, drones, and multipurpose robots.

Leveson (2012) and her team at MIT have developed the systems approach to safety engineering (STAMP). The STAMP model supports the development, systems engineering, and operation of software intensive complex systems with human interactions. The principles of the STAMP model include:

- Treat accidents as a **control problem** not a failure problem.
- Note that accidents are more than just a chain of events – they involve complex dynamic processes.
- Prevent accidents by enforcing constraints on component, software behaviors, and **interactions.**
- This model captures many causes of accidents, such as:
    - Component failure accidents.
    - Unsafe interactions among components.
    - Complex human and software behaviors.
    - Design errors.
    - Flawed and incomplete requirements.

This methodology has been used successfully for several years for different systems and projects. It is recommended to integrate this methodology into the systems engineering process to encounter the systems safety threats and challenges.

Traditionally, practitioners treat safety and security as different system properties. The safety people and security people generally work in isolation from each other, using their respective language and methodologies. Safety experts see their mission as preventing losses due to unintentional actions by benevolent actors. Security experts see their role as preventing losses due to intentional actions by malevolent ones. Professor Leveson's team demonstrated that the STAMP approach is suitable in solving, analyzing, and preventing security problems by focusing on loss prevention strategies, regardless of actor intent. As recommended above, systems engineering should integrate security engineering into its processes by using STAMP. Systems engineering can be a pioneer in "breaking down the walls" between the safety and security communities.

The systems developed, deployed, and operated in the fourth industrial revolution are complex and software intensive, use the cloud very extensively, apply Big Data, and feature autonomous, robotic, and IoT capabilities, so they are highly vulnerable to safety and security threats. The integrative approach of safety and security engineering based on the STAMP model through systems engineering is essential for the success of systems in the new era.

1.8 Applied Systems Thinking

Systems thinking is a critical topic. Expressed simply, systems thinking focuses on how a system interacts with the other constituents of the larger system of which it is a part and with all aspects of that larger context. Consider, for example, a hydrogen-powered car. (Such vehicles are currently available in select markets from Toyota, Hyundai, and Honda). On the surface, this would be a great thing because it would enable an efficient and clean zero-emissions vehicle.

Using a systems thinking approach, however, one would ask questions like: (i) Where does the hydrogen come from? (ii) What resources are involved in its production delivery? (iii) What are the byproducts? (iv) How will it be stored and delivered? (v) What are the safety issues? (vi) What are the security issues? (vii) How reliable will the system be? and (viii) How available does it need to be?

Today, 95% of the hydrogen is produced from natural gas, a nonrenewable resource (Wikipedia n.d.). Extracting hydrogen from sea water instead, using electrolysis, is an expensive and inefficient process. Additional energy would have to be expended in compression, transportation, and the delivery infrastructure.

Systems thinking extends beyond strictly engineering considerations. For example, the Trans-Alaska Pipeline, which was built to transport crude oil from northern Alaska (where it is produced) to southern Alaska (where it can be loaded onto ships), a distance of over 800 miles (1300 km), involved many legal, environmental, and cultural issues (Swarz 2016). Another example, from integrated healthcare systems is provided in Kenett and Lavi (2016).

The importance of Systems Thinking in the new systems environment cannot be understated. The reason, of course, is that virtually all systems will soon operate in a broad context that transcends the usual boundaries, extending to the involvement of science and engineering, computation and communication, and myriad social, legal, cultural, environmental, and business factors.

It is a great opportunity for applying practical systems thinking as an integral component of the systems engineering. Edson (2008) presents a useful primer for Applied Systems Thinking. This primer includes practical viewpoints, methodologies, and tools for practical and applied systems thinking. For details see Chapter 2.

Systems Engineering as a Data-Driven Discipline

Through this chapter and through other chapters in this book we observe that one of the important features of the fourth industrial revolution is being Data Driven. Kenett and Redman devote their whole book for applying Data Science and how to turn data into information and better decisions. Systems engineering should be data driven, as it is based on engineering requirements, engineering calculations, testing, and modeling and simulations – all of them should be based on data or should create data, but all too often the systems engineers are making decisions based on intuition and/or qualitative assessments. We claim that most of the systems engineering methodologies are not data driven. The fourth industrial revolution, with the proliferation of sensors of various types and big data analytics, creates a great opportunity to be data driven. In the above-mentioned book by Kenett and Redman, there are several traits for organizations to be "Data Driven." These traits can be applied for systems engineering as a discipline and/or the systems engineers as decisions makers. For example:

- Bring as much diverse data and as many diverse viewpoints to any situation as is possible.
- Use data to develop deeper understanding of the business context and the problem at hand.
- Develop an appreciation for variation, both in data and in the overall business.
- Deal reasonably well with uncertainty, which means recognizing that mistakes may be made.
- Recognize the importance of high-quality data and invest in trusted sources and in making improvements.

Relevant academic research in systems engineering is rare. Partially, that is due to the fact that systems engineering is not driven by data and is not always based on quantitative evidence. So, introducing quantitative data and evidence into the systems engineering discipline will promote and support the research in this area. This is an important issue for developing systems engineering as a applied scientific discipline.

Systems Engineering – A Discipline in the Making

Zonnenshain and Stauber (2015) is based on research conducted at the Gordon Center for Systems Engineering at the Technion, Israel entitled "The many faces of systems engineering." This research was conducted by interviewing experts and practitioners from all over the world on their views and their experience in practicing and studying systems engineering. This included case studies of two complex projects: The IAI Lavi project and The Iron Dome project. The findings of this research led to various insights, among them:

- There many faces of systems engineering, such as: engineering aspects, managerial aspects, technological aspects, human aspects, and leadership aspects. When dealing with developing and educating systems engineers, it is necessary to consider all these aspects. Also, when dealing with developing and practicing systems engineering in projects and organizations, it is important to consider aspects for successful implementation.
- Systems engineering, as a discipline, is still in the making and the development stage as a professional discipline.

Kenett and Shmueli (2016) provide a comprehensive framework for designing and evaluating analytic work aiming at generating information of quality. The eight InfoQ dimensions are applied in the context of the fourth industrial revolution in Kenett et al. (2018) and in chemical process engineering in Reis and Kenett (2018).

These insights can serve us while driving changes in systems engineering in response to the challenges and opportunities of the fourth industrial revolution. Such changes should consider and deal with all faces of systems engineering. For more on changes requires in systems engineering see Zonnenshain and Stauber (2015). For the role of analytics and data science in meeting the fourth industrial revolution challenges see Chapter 17 in Kenett and Redman (2019).

1.9 Summary

The Fourth Industrial Revolution presents large challenges to the "traditional" systems engineer, to be sure. These include: unanticipated behavior, lack of centralized control, cyber security, scalability, sustainability, resilience, dependability, growth, and the definition and management of interfaces and communication channels. Most of these problems, however, can be viewed as *opportunities*, too. On the basis of the above challenges, it is suggested that there is a need for a new model[1] based on the double helix (Figure 1.3). This model emphasizes the tight connection that Systems Engineering must have with Systems Thinking, as well as the new properties and processes, such as data analytics, security, safety, reliability, risks, and resilience, that must be integrated into the process.

Systems engineers need to be the *leaders* and the *internal entrepreneurs* in their companies to initiate and facilitate the needed and recommended changes in the systems engineering processes in view of the challenges and the opportunities of the fourth industrial revolution as presented in this chapter.

Figure 1.3 The systems engineering double helix.

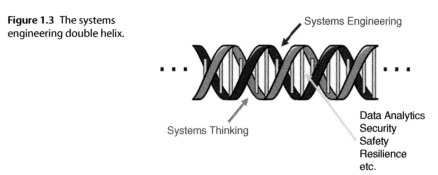

Systems Engineering

Systems Thinking

Data Analytics
Security
Safety
Resilience
etc.

---

1 At least new to systems engineering!

Throughout this chapter we present changes and shifts needed in the way of practicing systems engineering. Table 1.1 summarizes these shifts of OUT–IN patterns:

Table 1.1 Advanced systems engineering in and out patterns.

Out	In
Waterfall model	Iterative and agile models
V model, T&E scheme	Modeling and simulations
The boundaries of the system are well defined	The boundaries of the system are vague, porous and changing
Strict requirements management	Requirements are changing as an opportunity
Risk management based on qualitative assessments	Risk management based on evidences, data and models
Conservative approaches and avoiding changes	Innovation and entrepreneurship as a culture
Rare integration of human factors engineering	Human systems integration as a developed, advanced and meaningful discipline
Limited integration with reliability engineering and Reliability, Availability, Maintainability, and Safety (RAMS)	An integrated approach of RAMS in systems engineering (SE)
A process approach for safety	A system approach for safety – STAMP
Procedural protection and filters for external threats	System protection for cyber threats (by STAMP)
Linear thinking	Applied and advanced systems thinking
Preventive maintenance policy	Predictive maintenance policy – CBM, PHM
Passive design for environmental issues	Proactive design for sustainability
Limited integration of ethical aspects	Enhanced integration of ethical aspects, like: privacy rights, transparency
SE is based on documents	MBSE-model based systems engineering
Internal development and internal demonstration	Using demonstrated elements in the internet and in the virtual market like GitHub
Working with established companies	Working with start-ups
The business aspects are not part of SE	The business model is essential part of SE
SE is based on documents	Data driven and evidence-based SE

References

Business Insider (2013). Morgan Stanley: 75 Billion Devices Will Be Connected To The Internet Of Things By 2020. http://www.businessinsider.com/75-billion-devices-will-be-connected-to-the-internet-by-2020-2013-10 (Last accessed 7 July 2019).

Cai, K., Cangussu, J.W., DeCarlo, R.A., and Mathur, A.P. (2004). An overview of software cybernetics. In: *Eleventh International Workshop on Software Technology and Engineering Practice*, 77–86. Amsterdam: IEEE Computer Society.

Edson, R. (2008). *Systems Thinking-Applied*. ASysT Institute.

Halabi, A., Kenett, R.S., and Sacerdotte, L. (2017). Using dynamic Bayesian networks to model technical risk management efficiency. *Quality and Reliability Engineering International* 33: 1179–1196. https://doi.org/10.1002/qre.2186.

INCOSE (2014), A World in Motion, Systems Engineering Vision 2025. https://www.incose.org/docs/default-source/aboutse/se-vision-2025.pdf (Last accessed 7 July 2019).

Kenett, R.S. and Baker, E. (2010). *Process Improvement and CMMI for Systems and Software: Planning, Implementation, and Management*. Taylor and Francis, Auerbach CRC Publications.

Kenett, R.S. and Lavi, Y. (2016). Integrated models in healthcare systems. In: *Systems Thinking: Foundation Uses and Challenges* (eds. M. Frank, S. Kordova and H. Shaked). Nova.

Kenett, R.S. and Raanan, Y. (2010). *Operational Risk Management: A Practical Approach to Intelligent Data Analysis*. Chichester, UK: Wiley.

Kenett, R.S. and Redman, T. (2019). *The Real Work of Data Science: Turning Data into Information, Better Decisions, and Stronger Organizations*. Wiley.

Kenett, R.S. and Shmueli, G. (2016). *Information Quality: The Potential of Data and Analytics to Generate Knowledge*. Wiley.

Kenett, R.S. and Tapiero, C. (2010). Quality, risk and the Taleb quadrants. *Risk and Decision Analysis* 4 (1): 231–246.

Kenett, R.S. and Zacks, S. (2014). *Modern Industrial Statistics: With Applications in R, MINITAB and JMP*, 2e. Chichester, UK: Wiley.

Kenett, R.S., Zonnensahin, A., and Fortuna, G. (2018). A road map for applied data sciences supporting sustainability in advanced manufacturing: the information quality dimensions. *Procedia Manufacturing* 21: 141–148.

Leveson, N. (2012). *Engineering a Safer World, Systems Thinking Applied to Safety*. MIT Press.

Manyika, J., Chui, M., Bisson, P., et al. (2015). Unlocking the Potential of the Internet of Things. McKinsey Global Institute. http://www.mckinsey.com/insights/business_technology/The_Internet_of_Things_The_value_of_digitizing_the_physical_world (Last accessed 7 July 2019).

Middleton, P., Tully, J., and Kjeldsen, P. (2013). Forecast: The Internet of Things, Worldwide, 2013. Gartner. https://www.gartner.com/doc/2625419/forecast-internet-things-worldwide- (Last accessed 7 July 2019).

Newton, C. (2017). How a Typo Took Down S3, the Backbone of the Internet. The Verge. https://www.theverge.com/2017/3/2/14792442/amazon-s3-outage-cause-typo-internet-server (Last accessed 7 July 2019).

Reis, M.S. and Gins, G. (2017). Industrial process monitoring in the big data/industry 4.0 era: from detection, to diagnosis, to prognosis. *Processes* 5 (3): 35.

Reis, M.S. and Kenett, R.S. (2018). Assessing the value of information of data-centric activities in the chemical processing industry 4.0. *AIChE, Process Systems Engineering* 64 (11): 3868–3881.

Schwab, K. (2017). *The Fourth Industrial Revolution*. Portfolio Penguin.

Shahraki, A.M., Yadav, O.P., and Liao Habitao, L. (2017). *A review on degradation modelling and its engineering applications. International Journal of Performability Engineering* 13 (3): 299–314.

Shai, Y. (2015) Reliability of Technologies, PhD Thesis, Technion, Haifa, Israel

Swarz, R.S. (2016). The Trans-Alaska pipeline system: a systems engineering case study. In: *Proceedings of the Seventh International Conference on Complex Systems Design and Management* (eds. Fanmuy et al.). ISBN: 978-3319491028.

Taleb, N.N. (2007). *The Black Swan: The Impact of the Highly Improbable*. Random House.

Tilley, A. (2017). This Smart Doorbell Was Accidentally Sending Data To China, Until People Started Freaking Out. Forbes. https://www.forbes.com/sites/aarontilley/2017/03/22/this-smart-doorbell-was-accidentally-sending-data-to-china-until-people-started-freaking-out/#b84194859840 (Last accessed 7 July 2017).

Wagner, D. (2016). *Global Risk Agility and Decision Making*. Palgrave, Macmillan.

Wiener, N. (1948). *Cybernetics: Or Control and Communication in the Animal and the Machine*. New York: Wiley.

Wikipedia (n.d.) Hydrogen Vehicle. https://en.wikipedia.org/wiki/Hydrogen_vehicle (Last accessed 7 July 2019).

Zio, E. (2016). Some challenges and opportunities in reliability engineering. *IEEE Transactions on Reliability* 65 (4): 1769–1782.

Zonnenshain, A. and Stauber, S. (2015). *Managing and Engineering Complex Technological Systems*. Wiley.

2

Applied Systems Thinking

Robert Edson

Synopsis

With advancement of technology and social process, the ability to generate new knowledge and global relationships has increased. These changes, however, have also created more complex systems and wicked problems. To help address these problems and challenges, the world is moving into the Fourth Industrial Revolution.

The first industrial revolution (circa 1760–1840) was a time of large machines powered by steam and wind. The systems moved tasks carried out by hand to machines and were designed to replicate manual labor. The second industrial revolution (the late 1800s to mid 1900s) ushered in large scale production and assembly lines. These systems integrated several activities and were powered through electricity. The third industrial revolution involved computer technology, electronics, and the use of information in new ways. System automation became widespread and smart systems arose based on human-generated rule sets. We are now at the time of the fourth industrial revolution, expanding industry and systems through further advances of the digital revolution. Artificial intelligence and autonomous systems are growing and the lines between biology, cyber, and physical systems are blurring.

New challenges and new complexity. From replicating hand tasks and integrating multiple activities, to incorporating computers and artificial intelligence. Systems engineers are now faced with new systems. They must also think through problems in new ways. New systems and new thinking; how is this achieved?

★The author's affiliation with The MITRE Corporation is for identification purposes and is not intended to convey or imply MITRE's concurrence with, or support for, the positions, opinions or viewpoints expressed by the author. Approved for public release. Distribution unlimited. Case Number: 18–1498-002.

Systems Engineering in the Fourth Industrial Revolution: Big Data, Novel Technologies, and Modern Systems Engineering,
First Edition. Edited by Ron S. Kenett, Robert S. Swarz, and Avigdor Zonnenshain.
© 2020 John Wiley & Sons, Inc. Published 2020 by John Wiley & Sons, Inc.

This chapter provides an overview of applied systems thinking, particularly as it pertains to the fourth industrial revolution. It presents a discussion of the systems thinking concept and a summary of principles and tools used in its execution. It is meant to be a guide for system engineers but can be used by a variety of thinkers seeking to resolve a problem or improve system performance.

2.1 Systems Thinking: An Overview

Systems thinking provides a way of looking at problem situations and an approach to problem solution. First, approach problem investigation and assessment in a systemic way. Understand the problem from the stakeholders' points of view. Second, evaluate the related system of interest as a whole, as part of a larger system. Problem assessment and solution function as a learning system; the system becomes the lens through which the problem is viewed.

Despite the size and complexity of the problems the world currently faces, many people persist in the machine-age thinking of the first and second industrial revolution, taking apart the system to understand how it works. This approach no longer works in a system-age, where the individual parts do not equate to the system, where the system's external environment is just as important as its component parts. We must approach problem situations with a new way of thinking if meaningful solutions are to be developed. Systems thinking is one of these new ways.

Systems thinking is both a world view and a process; it can be used for both the development and understanding of a system and for the approach used to solve a problem. Systems thinking is the view that systems and problem situations cannot be addressed through reducing the systems to their component parts. The uniqueness and behavior of the system are only present when the system is together – it is not a sum of the individual components. System behavior comes about as a result of the interactions and relationships amongst the parts. In addition, systems thinking acknowledges the strong interactions between the system components, and the emergent behaviors and unintended consequences that may result from these interactions.

Figure 2.1 provides a simplified diagram to help in the understanding of systems thinking. The figure comprises three elements or activities: synthesis, analysis, and inquiry. These are not "steps" or "phases" – the elements are strongly related and support each other. Synthesis and analysis reflect the first part of the systems thinking definition: problem situations must be assessed as systems that exist in a larger environment. As discussed by Ackoff (1999), synthesis and analysis are both essential, complementary aspects of systems thinking. **Synthesis** is putting together, assessing the system as a whole, and understanding it in its environment. In synthesis, the "containing whole" is the point of focus. In **analysis**, the system is taken apart and understood from

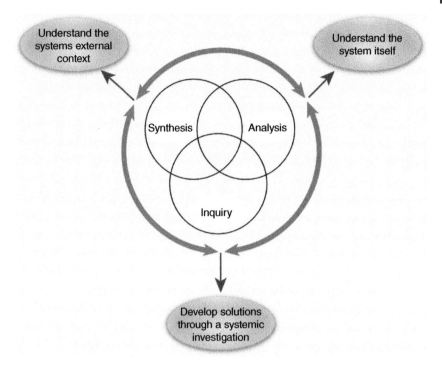

Figure 2.1 Systems thinking as three synergistic activities operating in an iterative manner.

its component parts, behavior and activities, and, importantly, relationships. Analysis is the traditional, machine-age way of thinking but remains of value in the systems age *if combined with synthesis.*

The third element in the diagram, *inquiry*, reflects the second part of the systems thinking definition: obtaining solutions through systemic investigation. The arrows surrounding the figure denote the interrelationships of the elements and iterative nature of the construct. Further detail on executing these three elements is contained in a later section that develops an applied systems thinking framework. In the midst of the complexity and confusion of the problem situation, systems thinking advances an understanding that solutions can be obtained through a learning system (Checkland 1993; Senge 2006).

In order for systems thinking to be effectively utilized, both perspectives – the full, contextual view of the problem and the systems approach to solutions – must be applied. These perspectives are synergistic and without both one reverts to ineffective, machine-age thinking.

Systems thinking is also a methodology (Figure 2.2): an ordered, methodical approach to understanding problem situations and identifying solutions to these problems. The initial focus is on the problem, understanding the

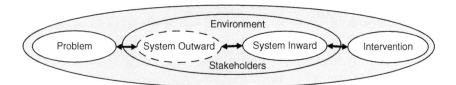

Figure 2.2 The systems thinking approach.

behavior that needs change. Identifying the specific problem behavior is necessary to focus efforts and to monitor improvement. The behavior over time also provides insight into the underlying system structure. The systems thinking approach then moves to assessing the system within its environment, the "system outward," taking the external context and stakeholders into consideration (Senge 2006). You then move to looking "system inward," assessing the system transformation, parts and relationships, structure and processes, and control mechanisms. These two steps, system outward and system inward, support the concept that systems thinkers see and assess both the "forest and the trees" (Richmond 1994).

The final stage in the systems thinking approach is the identification of intervention points to improve or fundamentally change system performance. It is in this step that the deep understanding of the larger containing whole and internal relationships pays off, expanding the possible solution space.

2.2 The System in Systems Thinking

The system in systems thinking can be technology or human activity based. It can be cyber or biological. It can be natural or anthropogenic. In fact, the difficult challenges faced today take the form of hybrid systems incorporating elements from many of these systems. This hybridization is the challenge faced in the Fourth Industrial Revolution.

The "system" can also take the form of a problem. A problem has an external context and internal structure. It should be addressed through a process of systemic inquiry. A problem is essentially a system with a critical value proposition. The problem is the system behavior which should be changed.

Finally, the "system" in systems thinking refers to the actual process of assessment. It is the path used to gain understanding of the problem; it is a structure for the thinking itself. All of these perspectives are legitimate and appropriate views of "system." The "system" in systems thinking is a flexible construct to help in understanding and solution.

The analysis and understanding of each of these aspects of the systems, and how these aspects are related, are critical to systems thinking. Without a clear perspective on the system of interest itself, any systems thinking application will be likely to fall short of its potential.

2.3   Applied Systems Thinking

In the analysis of systems there have emerged two broad schools or approaches, loosely labeled for discussion here as "engineering" and "management." The engineer seeks to understand and design through quantitative, well defined, or "hard" methods, while the management analyst leverages more qualitative, collaborative, or "soft" methods. The "applied" denotes an emphasis on addressing current problems of national significance utilizing tools and techniques that bridge the hard and soft system approaches.

Often there is confusion on the value or importance of hard versus soft systems methodologies in systems thinking. This is also a point of frustration for the engineer who feels that systems thinking does not provide meaningful, quantitative actions for problem solving. Checkland (1999) provides an interesting perspective with his assertion that hard methods serve to design systems, while soft methods provide a "systemic process of inquiry." Checkland's definitions are included in Table 2.1.

This is the traditional view of hard versus soft systems. In a departure from this separation, applied systems thinking leverages a combination of these two approaches, with the soft systems providing a better insight into the problem and thus allowing for a more effective hard system design and development.

Another perspective on applied systems thinking is shown in Figure 2.3. This figure provides the emphasis and approaches of each of the hard and soft systems communities. Both approaches fall within the larger systems thinking construct, but applied systems thinking specifically applies to the intersection or overlap region between the two approaches.

Many practitioners would agree that the soft methods provide significant insight into the problem definition and the stakeholder opinions, while the hard methods provide quantitative evaluations, solutions, and metrics for success. It is not necessary to establish the precise relationship between hard and soft systems methods, but it is important that the applied systems thinker *leverages both schools* in seeking to understand and solve the problem situation. This combination provides the power in applied systems thinking.

Table 2.1 Definitions of hard and soft systems methodologies.

Hard systems methodology	Systems-based methodology, also known as "system engineering," for tackling *real-world problems* in which an objective or end-to-be-achieved can be taken as given. A system is then engineered to achieve the stated objectives.
Soft systems methodology	Systems-based methodology for tackling *real-world problems* in which known-to-be-desirable ends cannot be taken as given. Soft systems methodology is based upon a phenomenological stance.

Source: adapted from Checkland 1999.

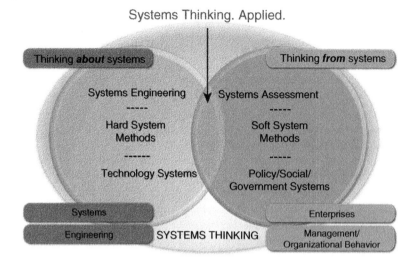

Figure 2.3 The hard and soft systems community and the overlap with applied systems thinking.

2.4 Applied Systems Thinking Approach

Systems thinking provides a framework (Figure 2.2) for utilizing tools from multiple areas both hard and soft including:

- Operations Research
- Systems Engineering
- Systems Dynamics
- Decision Analysis
- Management Sciences
- Business Process Reengineering.

These methods and tools (Figure 2.4) span the spectrum of hard and soft systems tools. Example hard system tools comprise: the Pugh Matrix; Quality, Function, Deployment Diagrams – House of Quality; functional decomposition; Integrated Definition (IDEF) Method; Functional Flow Block Diagram; N2 diagrams; Systems Modeling Language (SysML); Universal Modeling Language (UML); and a broad range of modeling and statistical tools.

Soft systems methods include Checkland's Soft System Methodology (1993) and traditional management analysis tools such as the responsibility, accountability, consulting, informed (RACI) matrix and swim lane charts. Rosenhead and Mingers (2001) and Pidd (2003, 2004) provide excellent summaries of several systems thinking analytical methods and include examples of their application.

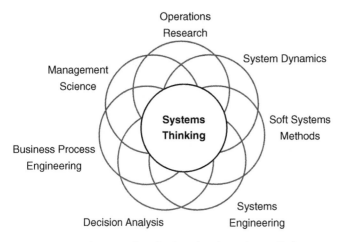

Figure 2.4 A wide range of methods and tools can be applied to systems thinking.

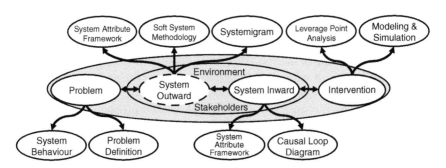

Figure 2.5 The Systems Thinking Approach with representative methods.

While all of these tools are valuable, several methods and tools are used extensively in the systems thinking process and are discussed in more detail below: the system attribute framework, the soft systems method, systemigrams, causal loop diagrams, and leverage point analysis. Figure 2.5 provides *representative* methods aligned with the various steps of the systems thinking approach. Other methods and tools are also possible.

2.5 Problem Definition: Entry Point to Applied Systems Thinking

The applied systems thinking approach begins with problem definition. Figure 2.6, derived from Ackoff (1974) and Pidd (2003, 2004), illustrates

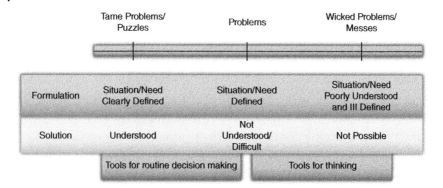

Figure 2.6 The degree of problem formulation and the understanding of the solution. Source: adapted from Ackoff (1974) and Pidd (2003, 2004).

the continuum of problems addressed with the associated formulation and solution space. The spectrum starts with tame problems or puzzles. These problems are well understood and well defined. The method for solving these problems is well known and the solution obvious.

In the middle ground of the spectrum are the common, but significant, problems one is challenged with on a regular basis. These problem situations are defined but the solution is not obvious, either in process or endpoint.

The final area on the spectrum is filled with wicked problems and messes. Wicked problems were first described by Rittel and Webber (1973) in their discussion of public planning. Wicked problems cannot be definitively described; responses cannot be meaningfully correct or false; and there is no solution in the sense of definitive and objective answers.

This spectrum and these distinctions are important because they help direct the systems practitioner toward a better understanding of the problem situation and the solution approach. The scale also helps to direct the systems practitioner to tools appropriate for the problem situation. In some cases, tools for decision making are appropriate, where in other cases tools for thinking should be applied.

In problems and systems where the behavior is well understood, where there is strong regularity and reproducibility, decision support tools and models can be developed. In this category are many of the rigorous mathematical system models and simulations. At the other end of the scale, the tools serve to provide information about the problem or system itself, and do not necessarily provide a decision directly. Thinking tools help to better define the problem and to communicate amongst diverse stakeholders. Such tools include the systemigram and soft systems method discussed later in this chapter. The tools may also include mathematical models when they are used for better problem understanding and consensus building.

In the case of the tame problem there may be recipes and techniques readily available to support a complete understanding of the problem and a rapid solution. For the middle ground, "regular" problems, the practitioner may want to look for patterns and relationships from other problems that will help to guide the solution. In the final case, wicked problems, solutions are not obvious and, in fact, may not even be possible. This distinction is important to know. The system practitioner must be willing to embark on solutions to the wicked problems with the understanding that these solutions will not be complete and will have to be continually adjusted as more becomes known about the problem and acceptable compromises.

Using multiple viewpoints and a variety of tools, the applied systems thinker begins the problem assessment and solution by first defining the problem. This is the context for the analysis that follows and the first step in the understanding necessary for the problem solution. However, problem definition is not done once and then never considered again. While it is the entry point to the applied systems thinking approach and the first step in the more complete assessment discussed in the framework of the next section, problem definition may be refined as the assessment progresses. The systems thinker must understand that problem definition may change as a more complete understanding of the system is obtained through application of different viewpoints and tools, and through the completion of the various actions of the system attribute framework noted in the following section.

2.6 The System Attribute Framework: The Conceptagon

Systems thinking derives its power from the discipline and order of its approach. Returning to the two-part definition of systems thinking, this approach includes both an assessment of the problem situation as a system (synthesis and analysis) *and* the problem solving method as a systemic process of inquiry. This approach is facilitated through the use of a system attribute framework called the Conceptagon (Boardman and Sauser 2008; Boardman et al. 2009) to assess the system or problem situation, its context, and solution development.

The Conceptagon is an analytical framework for ordered thinking about the system and problem situation. It provides seven triplets of system characteristics that are useful for assessment and understanding of the system or problem and its solution. This seven-sided framework (Figure 2.7) assures a complete assessment of the system with aspects that address synthesis, analysis, and inquiry.

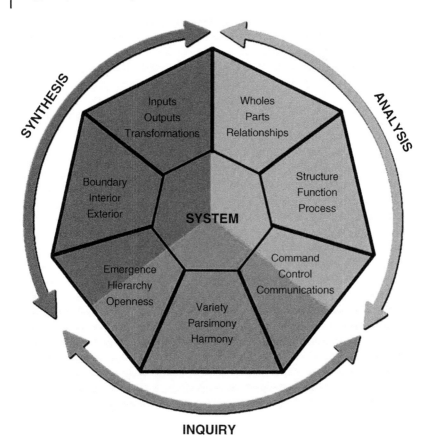

Figure 2.7 The Conceptagon analytical framework for system assessment. Source: adapted from Boardman and Sauser 2008.

The Conceptagon provides a detailed lens into the system, but it also provides an overview of the system at a conceptual or design level (Figure 2.8). By leveraging the Conceptagon for our thinking, we address the system context, design principles, and operations.

Each of the triplets will be addressed in turn. They are presented here in an order that generally follows a progression of gaining knowledge about the system from the outside in. However, the triplets are usually assessed in a concurrent, iterative matter. They are also not mutually exclusive to one area (i.e. synthesis, analysis, or inquiry) but are related to and build upon each other.

Boundary, Interior, Exterior

Setting the boundary of the problem or system of interest is one of the most important aspects of the systems thinking assessment. This triplet is a pivotal

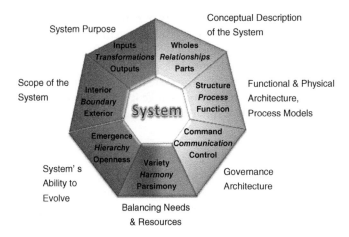

Figure 2.8 Scope and coverage of the Conceptagon.

part of synthesis since it denotes the interface and separation from the outside environment.

The boundary provides further definition of the system by identifying the interior and exterior of the problem situation, as well as the system of interest. The interior contains those parts of the problem that the solution system can and must address. It also contains the components and resources under the control of the system. The exterior has the systems and resources outside the system's control, but which are important for the system of interest. Just as importantly, the exterior contains those problems or issues that will be excluded from the current problem solution. Inside the boundary you can control and change. Outside the system are things not controlled by the system. Importantly, however, things outside can be influenced and the system of interest should be designed to facilitate this influence.

The boundary is often permeable, allowing resource inputs and system outputs. The boundary can be specified temporally, as well as across other dimensions, such as organizational, issues, or definitions. It is important not to limit one's consideration to just the systems analog of the "fence around a yard."

The boundary is specific to the problem situation or system and its location can dramatically change the system under consideration. When setting the boundary, the systems thinker must consider two seemingly contradictory constraints:

1. The boundary should be set narrowly enough as to make the problem being assessed tractable.
2. The boundary should be set broadly enough to capture all the salient components and relationships.

Sometimes the compromise between these two constraints is simple, sometimes it is complex. The novice systems thinker is tempted to set the boundary broadly to ensure that nothing is missed, but this often makes the problem so large and ill-defined that it is impossible to assess and solve. Care should be taken in setting the boundary and the systems practitioner must be willing to readdress the boundary, location and change it if necessary as more information is obtained on the problem situation.

Inputs, Outputs, Transformations

This triplet also supports synthesis; it speaks to what the system is taking in from and returning to the outside and is also the outward manifestation or purpose of the system. Both the engineer and the scientist understand input, output, and transformation. Energy or resources transfer into the system, a transformation occurs, and products or byproducts move out. Both the inputs and outputs exist outside the system boundary, again illustrating the importance of where the boundary is set.

What if the system under consideration does not produce things, or if it is a social system? The concept of transformation is now more difficult to grasp. Take for example the US Congress, which is a legislative system. One interpretation is that societal problems and needs are inputs to the Congress system, a transformation occurs through problem study and analysis and legislative activity, and the outputs are laws and regulations to guide societal actions and resource allocation. Thus, the transformation can be considered analogous to the system purpose. The input becomes the current situation and the transformation results in the new situation, or output.

Understanding the inputs and outputs is essential, as these are the resources and goals of the system. Understanding the transformation is critical – it relates to system actions and purpose. Understanding of the complete transformation helps to direct the development of system requirements and helps provide metrics for evaluation of system success.

Wholes, Parts, Relationships

The wholes, parts, relationships triplet addresses analysis and, to a lesser extent, synthesis as the whole manifests outward. The systems thinker is looking into the system seeking to understand its component parts and how they relate. The systems thinker also seeks to understand the relationships between the parts and the whole. There are several important observations associated with the wholes, parts, relationships triplet. First, when a system is taken apart it loses its essential properties and the behavior of interest. Second, though the whole is greater than the sum of the parts, the parts do exist and it is the relationships between the parts that make the whole so much greater. The systems thinker must understand the parts, their relationship to the whole, and their relationship with each other to make the system whole successful. Finally, the whole is not always "greater" in the traditional sense. Often high performing parts come together to produce something dysfunctional. Consider the example of a group of employees who individually provide superior work, but when they come together, the team fails. The team members cannot agree on leadership and path forward, making the whole so much less.

An additional observation on the whole system: maximum efficiency of the independent parts does not equate to maximum system effectiveness. "If each part of a system, considered separately, is made to operate as efficiently as possible, the system as a whole will not operate as effectively as possible" (Ackoff 1999). A corollary to this principle is specific realization that "the performance of a system depends more on how its parts interact than on how they act independently of each other" (Ackoff 1999). Thus, the systems thinker must also understand the compromises necessary at the part level to make the system perform well.

Structure, Function, Process

Structure, function, and process are inherently an analytical construct – how the system is put together and how it works. In many ways, this triplet equates most strongly to classic systems engineering and the related hard systems methodologies. The systems thinker must develop physical, functional, and process architectures, which together provide a complete understanding of the system operations.

Significant thinking is required with this triplet. Many structures can produce the same function and one structure can have many functions. What is the optimal configuration for the system purpose? What is the most elegant solution to the problem? Wasted effort is neither appreciated nor credited. This relates strongly to a consideration of the whole and parts and the compromise that has to be made there to get an effective system. It is easy to see the dependencies between structure, function, and process, and the tension in the design and solution that results.

This triplet also brings to the analysis a sense of the various levels, or scales, within the system. The structure, function, and the related processes may differ at different levels in the system, particularly as various components are aggregated and integrated. Attention must be paid to these factors: "scale, moving across scale, and discovering new behaviors as we go higher in scales" (Boardman and Sauser 2008).

There are a variety of ways to illustrate and assess structure, function, and process. Most architecting and business process (re)engineering tools work well in these areas. In many cases, the traditional process flow chart is the best mechanism for system assessment and design.

Command, Control, Communications

This triplet actually addresses two aspects of the systems thinking construct: analysis of the system and how it operates, and systemic inquiry into the system and the reasons for the problem situation. Command, control, and communication are perhaps the easiest triplet for the traditional systems engineer to understand; on the surface, this triplet fits the machine age. First, there is the traditional process consistent with the military perspective. There is a command structure within the system that exerts control on the system through communication. This command structure can be a central computer, distributed intelligence of a hardware system, or the hierarchy or collaborative structure of social systems. The governance structure for the system needs to be clearly understood and consideration given to its appropriate form. Is hierarchical control structure appropriate or is a decentralized, distributed structure a better option? The decentralized organization, management, and operation, with its ability to survive strong perturbations, may have advantages over the strong control and alignment of the centrally managed system.

Second, there is the concept of feedback. This concept is particularly important to systems thinking. Processes within the system are not linear; there are relationships within the processes that lead to control and both positive and negative change. These feedback loops must be identified and understood so that meaningful change can be affected.

Therefore, this triplet speaks to two related but distinct issues. First, the governance and command structure of the system must be well designed and appropriately architected; and, second, the control loops in the system must be understood and accounted for. An understanding of the traditional command structure in the system as well as the control loops is imperative to understanding the system and improving the problem situation.

Variety, Parsimony, Harmony

Variety, parsimony, and harmony provide guidance and considerations for the inquiry process. This triplet speaks to the elegance of the possible problem solutions and system design. There should be sufficient variety to cover all cases and possible needs, yet not so many options and paths that resources are wasted and confusion arises. Variety provides resilience but is ultimately limited by complexity and budget. Parsimony, or simplicity, is equally important. Simplicity facilitates effective design and operations, and may help keep in budget, but can also lead to a fragile, inflexible system. All options and paths must exist in accord or harmony, working together toward problem resolution. Ashby's

Law of Requisite Variety (Ashby 1957) should always guide design activities: the variety in the control system must be equal to or larger than the variety of the perturbations in order to achieve control.

Openness, Hierarchy, Emergence

This triplet is perhaps one of the most difficult to grasp but is the most important to consider. It combines aspects of inquiry and synthesis, presenting both a way to think about systems and an understanding of the system's interactions with the external environment. Openness speaks to the solidity of the system boundaries and the level of interaction the system has with the external environment. How open is the system to taking in components from the outside?

As already mentioned, all systems exist to perform a transformation, whether it is the solution of a problem or the creation of a product. This necessitates that the system be open to the environment. The porosity of the boundary and the freedom of movement into and out of the system is a measure of system openness. Openness may be more than this, however. Does the boundary move or flex as the system operates? Do only consumables or problem situations enter the system, or do new system components also come and go, effectively allowing the system to restructure itself as needed? What is the strength of the relationship and communication with the outside? One observation from biology: a closed system is a dead one.

Hierarchy encompasses the structure and organization of the system components and the system itself to the outside world. In some respects, this is also strongly related to command, control, and communication as discussed previously. The relationships can be highly structured, top-down or bottom-up. They can also be flat and equalitarian in nature. The organizational structure, whether dictated or self-motivated, also relates to the flexibility of the system and its ability to adapt and change. Regardless of the type of structure, *as new components enter into the system, the structure will change: a new system will form.*

As the hierarchy and openness is established in the system, behaviors start to emerge that were not previously in evidence. This behavioral emergence is expected. New behaviors are desired – otherwise why change the system by bringing in new components? New behaviors and properties, not readily identifiable through component analysis, emerge and redefine the system and problem solution. These emergent characteristics become even more complex when dealing with human-activity systems, and the power and the unanticipated actions of groups (Surowiecki 2004). The system thinker must identify, account for, and adapt to such behaviors, and must be aware of emergence and the important role it plays in complex systems. The system itself must be designed to facilitate the growth of good behaviors and to damp out the bad behaviors (e.g. through command, control, and communication).

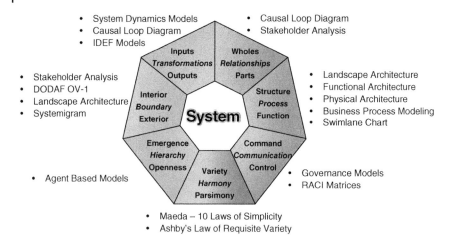

Figure 2.9 The Conceptagon with example artifacts and methodologies for each triplet.

Framework Summary

The Conceptagon provides a framework for assessing the complete set of attributes for the system. Often the framework guides the assessment implicitly but does not result in a specific set of artifacts. The Conceptagon exists in the systems thinker's mind as a constant reminder of the system aspects that should be considered. There are times, however, when additional rigor and fidelity is necessary, and there are specific artifacts associated with each of the triplets. Figure 2.9 provides examples of possible tools and methods associated with each triplet.

The seven triplets of the Conceptagon guide the systems thinker through the three aspects of systems thinking: synthesis, analysis, and inquiry. Through the Conceptagon framework, a complete understanding of the system, problem situation, and solution space is possible. In an actual analysis, the viewpoints and tools would be coupled with the problem definition and this framework to gain a complete picture of the problem. An "as-is" systems depiction and solution options would be developed in the context of the external environment. The solution options would be socialized and refined through soft systems methods and the selected solution advanced, implemented, and evaluated.

2.7 Soft Systems Methodology

Checkland's Soft Systems Methodology (1993) provides a logical flow for assessment and improvement of human activity systems in particular, but the process can be applied to technology and hybrid systems as well. The approach has three phases:

1. Develop an understanding of the operational picture of the system to gain insight into the problem. This would include understanding system constraints and goals. It further necessitates an understanding that everything exists within a social and political context.
2. Utilize systems tools to determine potential solutions to the problem. Multiple potential solutions should be developed, so that they can be compared to determine the best possible solution.
3. Bring solutions into the as-is space to determine their feasibility within the system's operational constraints and desired end state.

In the context of systems engineering, the soft systems approach provides several particularly valuable concepts. For example, root definitions of relevant systems is a framework that can be employed in any system activity. The root definitions provide a framework for assessing the system from each stakeholder's perspective. The framework includes six system characteristics: customers, actors, transformation, worldview, owners, and environment (CATWOE). Using this framework provides a consistent way to understand the stakeholders and their differences. It also provides a mechanism for identifying possible interventions for system improvement.

Checkland's soft system method is invaluable in gaining an understanding of the interaction between the problem or situation being assessed and the larger external context. It is important to ensure that the external context – political, social, and otherwise – is continually considered in the problem assessment and solution development. Otherwise, an excellent solution that solves the problem may not be accepted due to political and social exigencies.

2.8 Systemigram

Developed by Boardman (1994) and Boardman and Cole (1996), systemigrams provide a powerful tool for the analysis of systems first described in written form. The systems thinker takes lengthy documentation and distills it down to "concentrated" prose covering all the salient points of the system. This reduced narrative represents a description of the problem situation, the system of interest, or any aspect of the system assessment. The concentrated prose is itself a system, with the nouns representing the system components and the verbs defining the relationships, together making the whole. Based on this system, a visual systemigram is then constructed, decomposing the prose into individual, but related, threads, showing the flow of information, resources, and actions. Systemigrams are powerful constructs to facilitate complete understanding of a system and to provide a common foundation for group discussions. However, it should be noted that the power of the systemigram is in the storytelling. It is difficult to utilize this tool in a static context. Such use is possible, however, through breaking the larger diagram into its component threads.

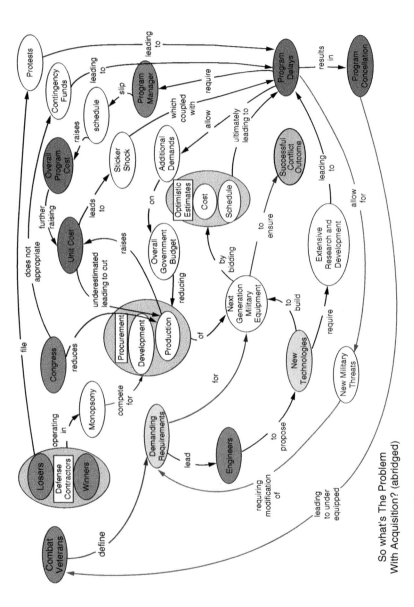

Figure 2.10 A systemigram of the challenges facing the US Department of Defense acquisition system.

Figure 2.10 provides an example systemigram of the process and challenges associated with the US Department of Defense acquisition system. When used during an actual stakeholder engagement, this systemigram would be presented thread by thread, in a storytelling mode, ensuring that the stakeholders have a shared understanding of the system.

A more thorough discussion of the evolution of systemigrams and the methodology can be found in Blair et al. (2007) and Boardman and Sauser (2008).

2.9 Causal Loop Diagrams

Causal loop diagrams are an integral part of systems thinking and the specific discipline of system dynamics. System dynamics provides both the qualitative method of causal loops as well as the numerical modeling technique based on stocks and flows. System dynamics modeling has been used for assessment of systems as diverse as the manufacturing production system, urban renewal, and climate change (Forrester 1961, 1969a,b, 1971). Causal loops assist in seeing beyond the linear to understanding feedback loops and control structure. Senge (2006) provides an illustration of the underlying construct of feedback loops in the simple act of filling a glass of water. The linear perspective is simply "I am filling the glass." The more complete system includes five variables organized in a loop, or feedback process. The initial faucet position results in a specific rate of water flow. This flow starts to fill the glass and our eye measures the current water level. We evaluate the current level against the desired water level and calculate a perceived gap. This gap leads to an adjustment in the faucet position, a change in the flow, and so on.

Figure 2.11 provides an example causal loop diagram of a more complex system, the US Department of Defense acquisition system. This example is an extension of the same system shown in Figure 2.10. The causal loop diagram provides insight into operation and feedback in the systems and highlights the problems associated with the unit cost and schedule slip. The arrows in the diagram show the direction of the causal relationship. The positive ("+") sign at the head of the arrow denotes the variables move in the same direction, while the negative ("−") sign shows that the variables move in the opposite direction.

Additional information on causal loop use is found in Kim (2000) and Sterman (2000). Without understanding the feedback and closed loop structure of the system, leverage points are difficult to identify and implement. Real change and problem solution are unlikely.

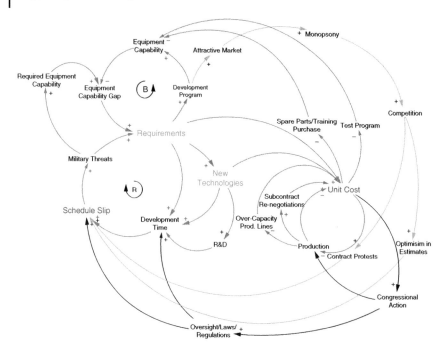

Figure 2.11 A causal loop diagram illustrating the components and relationships in the US Department of Defense acquisition system.

2.10 Intervention Points

The last stage in the systems thinking approach is identifying intervention points. In simple systems, beneficial change is easy to identify. In more complex systems, leverage point identification can be challenging. Meadows and Wright (2008) provide a list of 12 possible intervention points (Figure 2.12). Often multiple intervention points are possible and necessary. Selecting the best set of interventions may require significant qualitative analysis and modeling.

When working to identify intervention points, it is important to differentiate between high and low leverage change. High leverage, long-term interventions are preferred, but low leverage short-term changes may be necessary to address immediate need and requirements. The systems thinker must understand that a combination of both long and short term, high and low leverage, interventions may be necessary. Do not assume that low leverage will meaningfully change the system.

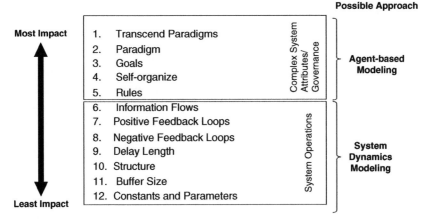

Figure 2.12 Intervention point types with relative impact and possible approaches to identification.

2.11 Approach, Tools, and Methods – Final Thoughts

All the tools of the hard and soft systems communities have value in applied systems thinking. Four specific methods are presented here but any of the qualitative tools, such as brainstorming or consensus building, or more quantitative tools, such as mathematical modeling or statistics, can be used within the various stages of applied systems thinking. The applied systems thinker will leverage a variety of these tools throughout problem assessment and solution development. These tools and methods, along with the viewpoints previously presented, are instruments for completing the problem definition and the systems thinking approach.

2.12 Summary

> Any fool can make things bigger, more complex, and more violent. It takes a touch of genius—and a lot of courage—to move in the opposite direction.
>
> (Albert Einstein)

The world is an increasingly complicated and overwhelming place. Systems are getting larger, more interconnected, and more complex. In the age of the fourth industrial revolution, systems thinking provides the tools to make sense of the chaos. Systems thinking improves understanding and facilitates solution of tough problems through the combination of synthesis, analysis, and inquiry. This discipline assesses the problem as a whole system in a larger world and approaches problem investigation and assessment in a systemic way.

This chapter has presented a set of systems thinking concepts and principles, including the concept of problem definition and the importance of multiple viewpoints. The broad range of applicable tools has also been discussed, with several specific tools and methods presented in detail. The Conceptagon has been presented as an analytical framework for system assessment. These concepts, methods, and framework provide an approach to system and problem assessment and solution.

Systems thinking can be used in all areas, including development and innovation, industry, national security, homeland security, energy, environment, education, healthcare, and business. It is particularly important that policy analysis and support be addressed though systems thinking.

In today's world of interconnectivity, interdependence, and globalization (Friedman 2005), traditional, reductionist approaches to problem solving are inadequate, driving solutions in wrong directions, costing millions, and potentially causing critical unforeseen damages. To adequately address problems spanning from border protection to environmental protection, food safety to appropriate healthcare delivery, counterterrorism to education, industrial revolution to human resources challenges, our problem-solving approach *must* change. Systems thinking provides one such change.

References

Ackoff, R.L. (1974). *Redesigning the Future: A Systems Approach to Societal Problems*. New York: Wiley.

Ackoff, R.L. (1999). *Ackoff's Best: His Classic Writings on Management*. New York: Wiley.

Ashby, W.R. (1957). *An Introduction to Cybernetics*. London: Chapman & Hall Ltd.

Blair, C.D., Boardman, J.T., and Sauser, B.J. (2007). Communicating strategic intent with systemigrams: application to the network-enabled challenge. *Systems Engineering* 10 (4): 309–321.

Boardman, J.T. (1994). A process model for unifying systems engineering and project management. *Engineering Management Journal* 4 (1): 25–35.

Boardman, J.T. and Cole, A.J. (1996). Integrated process improvement in design manufacturing using a systems approach. *IEE Proceedings-Control Theory Applications* 143: 171–185.

Boardman, J.T. and Sauser, B.J. (2008). *Systems Thinking: Coping with 21st Century Problems*. Boca Raton: Taylor & Francis.

Boardman, J.T., Sauser, B.J., John, L., and Edson, R. (2009). The Conceptagon: a framework for systems thinking and systems practice. In: *Proceedings of the IEEE Conference on Systems, Man and Cybernetics*, San Antonio, TX (11–14 October). IEEE.

Checkland, P. (1993). *Systems Thinking, Systems Practice*. Chichester, UK/New York: Wiley.

Checkland, P. (1999). *Soft Systems Methodology: A 30-Year Retrospective*. Chichester, UK/New York: Wiley.

Forrester, J.W. (1961). *Industrial Dynamics*. Cambridge, MA: MIT Press.

Forrester, J.W. (1969a). *Principles of Systems*. Cambridge, MA: Wright-Allen Press.

Forrester, J.W. (1969b). *Urban Dynamics*. Cambridge, MA: MIT Press.

Forrester, J.W. (1971). *World Dynamics*. Cambridge: Wright-Allen Press.

Friedman, T.L. (2005). *The World is Flat: A Brief History of the Twenty-First Century*. New York: Farrar, Straus and Giroux.

Kim, D.H. (2000). *Systems Thinking Tools*. Waltham, MA: Pegasus Communications.

Meadows, D.H. and Wright, D. (2008). *Thinking in Systems: A Primer*. White River Junction, VT: Chelsea Green Publishing.

Pidd, M. (2003). *Tools for Thinking: Modelling in Management Science*. Chichester, England/Hoboken, NJ: Wiley.

Pidd, M. (2004). *Systems Modelling: Theory and Practice*. Chichester, England/Hoboken, NJ: Wiley.

Richmond, B. (1994). Systems thinking/system dynamics: let's just get on with it. *System Dynamics Review* 10 (2–3): 135–157.

Rittel, H.W.J. and Webber, M.M. (1973). Dilemmas in a general theory of planning. *Policy Sciences* 4: 155–169.

Rosenhead, J. and Mingers, J. (2001). *Rational Analysis for a Problematic World Revisited: Problem Structuring Methods for Complexity, Uncertainty and Conflict*. Chichester, UK: Wiley.

Senge, P.M. (2006). *The Fifth Discipline: The Art and Practice of the Learning Organization*. New York: Doubleday/Currency.

Sterman, J. (2000). *Business Dynamics: Systems Thinking and Modeling for a Complex World*. Boston, MA: Irwin/McGraw-Hill.

Surowiecki, J. (2004). *The Wisdom of Crowds: Why the Many Are Smarter Than the Few and How Collective Wisdom Shapes Business, Economies, Societies, and Nations*. New York: Doubleday.

3

The Importance of Context in Advanced Systems Engineering

Adam D. Williams

Synopsis

Traditionally, systems engineering is predicated on a well-defined, mutually agreed upon operational environment. This perspective assumes a clear boundary between the interacting set of components of interest, precise definitions of external interfaces, known data limits, and anticipated patterns of use. Yet, the wave of change associated with the Fourth Industrial Revolution directly challenges these assumptions on how engineered systems will operate. In this chapter, we describe the traditional view of "context" in systems engineering and identify challenges to this view related to "the Fourth Industrial Revolution". We then offer insights from systems theory and organization science to address the blurring of the lines between "system" and "environment." The resulting idea of "context of use," the interrelated technological, environmental, social, and operational conditions by which system behaviors can be fully understood, is described and situated as an important element of advanced systems engineering. We conclude with a representative context of use example, a discussion of implications, and conclusions.

3.1 Introduction to Context for Advanced Systems Engineering

Systems engineering is predicated on understanding the interactions between components working toward a common goal. Traditionally, this perspective assumes a clear boundary between interacting sets of components,

★Sandia National Laboratories is a multimission laboratory managed and operated by National Technology and Engineering Solutions of Sandia, LLC, a wholly owned subsidiary of Honeywell International, Inc., for the US Department of Energy's National Nuclear Security Administration under contract DE-NA-0003525. SAND2019-4268B.

Systems Engineering in the Fourth Industrial Revolution: Big Data, Novel Technologies, and Modern Systems Engineering, First Edition. Edited by Ron S. Kenett, Robert S. Swarz, and Avigdor Zonnenshain.

precise definitions of external interfaces, and known data limits to establish a well-defined, mutually agreed upon description of the conditions, settings, and circumstances in which systems are expected to operate. Oftentimes, these clear boundary definitions lead to a perception of the "environment" and the system's stakeholders as unidirectional, exogenous forces on the interactions within the system. This understanding assumes that system engineers' expectations of human behavior perfectly align with how systems are *actually used*. Yet, this perspective fails to account for the role of human, social, and organizational influences on system behavior. Moreover, this perspective cannot account for legacy effects (e.g. the difficulty of using digital replacement parts in analog nuclear power plant safety systems [National Research Council, Committee on Application of Digital Instrumentation and Control Systems to Nuclear Power Plant Operations and Safety 1997]), historical memories (e.g. recollections of Challenger and Columbia NASA projects (Hall 2003)), or long-standing assumptions about how the world works (e.g. Moore's Law of computational power evolution (Waldrop 2016)) on interactions within systems. In short, "although traditional systems engineering does not completely ignore context influences on systems problem formulation, analysis, and resolution, it places context in the background" (Keating et al. 2003, p. 38).

A traditional conception of environment in systems engineering fails to explain how "identical technologies [or systems] can occasion similar dynamics and yet lead to different structural outcomes" (Barley 1986, p. 105) or "why different groups enact different…interaction(s) with a particular set of technological properties, in similar and different contexts" (Orlikowski 2000, p. 420). The inability to explain how the *same* system employed in *similar* environments results in *different* performance outcomes suggests a limitation in traditional systems engineering approaches. This limitation, combined with the wave of technological changes associated with the Fourth Industrial Revolution suggests that traditional perspectives are insufficient for advanced systems engineering. Big data, emerging technology, and increasingly complex systems will likely result in the blurring of lines between use environments and system boundaries, changing roles of (and interactions with) stakeholders, and increasing variability in human behaviors in system use. For example, in comparing the different outcomes of implementing the same information management system at two extremely similar companies, Robey and Rodriquez-Diaz (1989) argued that

> The social context of [system] implementation includes the specific organizational setting which is the target of the implementation and the wider cultural and national setting within which the organization operates.
>
> (p. 230)

This outcome demonstrates the need to better understand – and account for – the importance of *context* in systems engineering.

In approaching "advanced systems engineering," there is a need to relax traditional assumptions of boundary clarity, exogenous influences, unidirectional requirements, and environment-independent definitions of risk or success. This is especially salient as operating environments and stakeholders move from simply being exogenous forces *on* a system to being constituent parts *within* a system-of-systems (SOSs). Here, perhaps the "novelty" of novel technologies is how they challenge traditional assumptions of the context in which systems operate. Taking insights from systems theory, organization science, and engineering systems results in an emphasis on the "context of use" for advanced systems. Context of use, consistent with the "worldview" perspective offered by the INCOSE fellows (Silitto et al. 2018), is defined as the conditions, settings, and circumstances in which systems operate. Including context of use in advanced systems engineering can help to navigate the challenges posed by the Fourth Industrial Revolution. After reviewing perspectives of context in traditional systems engineering, this chapter explores gaps in traditional views, introduces nontraditional approaches to context for systems, and provides more detail on the "context of use" concept for advanced systems engineering. The chapter ends with a representative context of use example using physical security systems for high consequence facilities, a discussion of implications, and conclusions.

3.2 Traditional View(s) of Context in Systems Engineering

Context is defined by Meriam-Webster's Dictionary as "the interrelated conditions in which something exists or occurs," while the Oxford Dictionary defines it as "the circumstances that form the setting for an event, statement, or idea, and in terms of which it can be fully understood." For systems engineering, conditions, settings, and circumstances are often defined by establishing boundaries that separate the set(s) of interrelated components of concern from their environment. While such boundaries consist of physical (or digital) limitations (e.g. property borders of a manufacturing facility or restrictions on computing power), they also include perceptions of performance requirements, stakeholder involvement, and functional outcomes. This concept is evident in how the official INCOSE definition for systems engineering (International Council on Systems Engineering (a), n.d.) focuses

> ...on defining customer needs and required functionality...[that] considers both the business and the technical needs of all customers with the goal of providing a quality product that meets the user needs.

Establishing system boundaries to describe the conditions, settings, and circumstances for system operations has provided a common and useful framework for improving system design and performance. Organizing systems analytical and implementation thinking around component behavior requirements and performance outcomes has allowed systems engineering to advance the state-of-the-art solutions provided to customers across a range of professional domains. This approach, which is common across engineering disciplines, leverages a "value-free" engineering paradigm where the "focus is almost wholly on the abstract concepts, principles, and methods of the domain" (Bucciarelli 2010, p. 10). In this paradigm, stakeholders are considered exogenous forces on an open system or assumed out of a closed system.

Systems engineering also includes assumptions of the ability to *control* system boundaries. For example, in one clear example of traditional view(s) of context, Kenett et al. (2018) identified that current systems engineering typically depends

> on that fact that the assumption that the system had a stable, well-constructed set of requirements, a well-defined and manageable set of stakeholders and their expectations, and well-known and controllable constraints on and boundaries of the system.
>
> (p. 1609)

From this perspective, defining conditions, settings, and circumstances for systems engineering is an exercise in clearly articulating what a system has control over (e.g. "stable," "well constructed," "well defined," "manageable," or "well known") and what it does not (e.g. everything else). In many ways, focusing on identifying and developing such crisp, clear, and controllable aspects of a system boundary results in a deeper understanding of the range of potential behaviors that can emerge from component interactions. When variability in a system boundary arises, efforts are focused on reducing (or removing) the associated uncertainty by changing performance requirements or adjusting the limits on component behaviors. While there are numerous advantages to using "open system" or "closed system" principles, solely relying on these approaches is insufficient to adequately account for the technological changes associated with the Fourth Industrial Revolution.

3.3 Challenges to Traditional View(s) of Context in the Fourth Industrial Revolution

As discussed in earlier chapters of this book, several aspects of the Fourth Industrial Revolution, such as big data, emerging technologies, and increasingly complex systems, are driving a need to evolve toward advanced systems engineering. Yet, these drivers also directly challenge assumptions in traditional systems engineering on the ability to control the conditions, settings, and circumstances in which increasingly complex systems must operate. Consider the impact of "big data" on vehicle manufacturing systems. Current manufacturing systems define their operating context in terms of individual purchasing histories, geographical purchasing patterns, and vehicle-specific sale trend data streams. Here, purchasing dynamics act as unidirectional signals to trigger changes in manufacturing system performance outputs – for example, production of electrical cars may increase in response to new, more restrictive emissions regulations. Big data, however, can couple these traditional data streams with new, nontraditional data to help forecast modifications in manufacturing necessary to meet, and perhaps even *change*, customer needs. Incorporating nontraditional data, such as trends in public transportation usage and changes in average distances between homes and places of employment, can improve the efficiency and effectiveness of vehicle manufacturing systems meeting customer needs. In addition, vehicle manufacturing systems – by producing *more* electrical vehicles – can, in turn, *influence* the operating environment by increasing societal support of more restrictive emissions regulations. By affording opportunities to influence a system's operating environment, "big data" challenges the efficacy of traditional systems engineering views of context.

Or, consider autonomous vehicles as an example of emerging technology that challenges conventional systems engineering approaches to context. The systems engineering assumptions on controlling data presented to automobile drivers in the 1990s are insufficient to provide the same system-level performance in today's artificial intelligence-based vehicles. Vehicles developed in the 1990s provided a limited amount (and type) of data to the human driver about vehicle performance. Data taken in by the vehicle was similarly limited. Systems engineers could define operating conditions, settings, and circumstances based on the human driver as the components of the vehicle interacted to provide transportation and responded to the exogenous influences of the driver on component interactions (e.g. braking). These same assumptions that humans act as the boundary between the vehicle and the environment are inadequate for developing autonomous vehicles. These vehicles require an immense amount of data types to flow *between* the vehicle and the driving environment. For example, emergency braking or lane correction technologies send signals out to identify other vehicles or the road itself and receive data on which to make driving decisions. As autonomous vehicles direct components to change their individual and interdependent behaviors to move the vehicle, the movement changes the environment in which the vehicle operates. The ability for emerging technologies to affect dynamic change in their operating environments challenges traditional unidirectional perceptions of context.

Increasingly complex systems pose perhaps the clearest challenge to traditional systems engineering perspectives on context. One approach to addressing increasing complexity is a SOSs approach where isolated systems are interpreted as interacting components of metasystems to achieve desired goals. Keating et al. (2003) argue that this perspective "dictate[s] increasing appreciation for contextual influence on all aspects" of SOSs and further assert that "assumptions based on stability in…contextual domains must be considered suspect in SE" (p. 41). A similar approach to addressing system complexity argues that "actualizing the engineered [system] solutions…within the real system…[involves] working with decision makers (change the world), other researchers (change the field) and people affected by the change (understand the impact)…through a [system] solution's lifecycle" (Glass et al. 2012, p. 12). Developed at Sandia National Laboratories, this *complex, adaptive systems of systems* (CASoSs) engineering approach adopts a wider view of systems engineering by shifting away from "[technological system] solutions that assume isolation (at smaller scales)" because complexity stems from socio-economic-ecologic-technical "feedbacks from outside the (narrowly) idealized system" (Glass et al. 2012, p. 14). For these two approaches, increased complexity is addressed by relaxing rigid system boundary definitions and including more external interactions with other systems.

Rather than calculating complexity by looking at the interplay between systems, the *complex, large-scale, interconnected, open system* (CLIOS) perspective defines systems as "consisting of a physical domain embedded (conceptually) in an institutional sphere" (Sussman 2014, p. 13). With CLIOS, complexity is partially addressed by "explicitly includ[ing] the institutional world as part of the system" (Sussman 2014, p. 13) and acknowledging that changes in these institutional worlds can drive both enhanced system performance and unwanted system behaviors. Different approaches toward capturing increasing complexity in systems represent common characteristics that gave rise to a new academic discipline at the Massachusetts Institute of Technology called *engineering systems*. This perspective addresses increasing system complexity by the linking of "technological artifacts, enabling networks, the natural environment, and human agents" (de Weck et al. 2011, p. 15). De Weck et al. (2011, p. 37) also characterized engineering systems by how the "configuration of the entire system – its properties, elements, and interrelationships – are always fluid, always changing with time," allowing for a bidirectional relationship between technological systems and human use(s) of those systems.

Each of these approaches to increased system complexity – which is only expected to increase *further* in the *Fourth Industrial Revolution* – establish that "both the technical aspects and the social context within which the systems are operating play a central role" (Sussman 2014, p. 7). This demonstrates the need to address changes in how systems engineering traditionally treats the relationship between the system and its operating environment. More pointedly, in the development of their argument for a "system-of-systems engineering" framework, Keating et al. (2003, p. 38) argue that as systems increase in complexity "it is naïve to think that problem definitions and requirements will be isolated from shifts and pressures stemming from highly dynamic and turbulent development and operational environments."

If it is naïve to think that complex systems are isolated from their operating environments, then how *should* system context be perceived in the Fourth Industrial Revolution? For tomorrow's more complex systems, isolation should be traded for a more nuanced and comprehensive understanding of the operational environment surrounding system performance. In his support of the engineering systems approach described above, Long (2018) argued that "if we are to advance [systems engineering] in the right direction, we must advance, informed and guided by our *greater context*" (emphasis added). In response to technological developments, systems engineering should move beyond its long-held assumptions of clear boundaries and well-defined operating environments to derive system performance. Advanced systems engineering should move toward embracing variation, dynamism, and uncertainty in a system's operational environment. This logic suggests that the "novelty" of the Fourth Industrial Revolution-based challenges to traditional systems engineering – increased autonomy, big data, and increasing complexity – may be related to how they change the relationship between systems, boundaries, and environments. For example, de Weck et al. (2011) identify temporality as a key feature of complex systems. In so doing, system engineers are encouraged to examine what *may* change over the operational lifetime of the system – thereby explicitly addressing legacy issues, historical memories, and changes in assumptions about how the world works around systems.

Simply equating context with defining and meeting the business/technical needs of customers or as "well-known and controllable constraints" cannot account for variability in how humans will operate systems nor the potential impacts of changes in operational conditions, settings, or circumstances. For advanced systems engineering, this suggests that context is more than a set of variables to be held as constant or unidirectional, exogenous forces on interacting component behavior. If, according to INCOSE Fellows (International Council on Systems Engineering (b), Undated), the added value of systems stems from the interconnectedness of their *component parts*, perhaps the value of advanced systems will stem from the interconnectedness of their *component parts* and *their operating conditions, settings, and circumstances*. Context for systems engineering would then evolve from an assumed static set of variables in closed systems and exogenous forces on component interactions in open systems to design and implementation variables interacting with components in advanced systems. These arguments echo claims that "systems engineering problems are evolving in ways suggesting contextual aspects of a complex system problem must be moved to the foreground" (Keating et al. 2003, p. 38), to include better addressing the conditions, settings, or circumstances in which advanced systems will be operated.

3.4 Nontraditional Approaches to Context in Advanced Systems Engineering

In response to technological evolution(s), advanced systems engineering should seek to more clearly and comprehensively describe operating environments – to include accounting for contextual descriptions consisting of the interrelated human behavior, social, and organizational factors that impact system performance and success. Recent research in safety (Leveson 2012; Stringfellow 2010) and security (Anderson 2008; McFarlane and Hills 2013; Ross et al. 2016; Williams 2018) for complex systems illustrates the importance of including contextual influences in system design, operation, and evaluation. Yet, there is still a need to improve the ability of systems engineering to account for the role of context. Three academic literatures – systems theory, organization science, and engineering systems – offer insights to better understand and incorporate context into advanced systems engineering.

Systems-Theoretic Concepts and Insights

Returning to core systems-theoretic concepts – and exploring their modern applications – lays a foundation which more comprehensively incorporates context into advanced systems engineering. Expanding on the Aristotelian argument that "the whole is more than the sum of its parts," general systems theory describes how observed behavior(s) are not always explained by the behavior(s) of the related component parts. In an attempt to address *organized complexity* – defined by Weaver (1948) as "problems which involve dealing simultaneously with a sizable number of factors which are interrelated into an organic whole" – systems theory provides a nonstatistical, nonrandom logic to describe the behaviors of "many, but not infinitely many" (Von Bertalanffy 1972, p. 415) components. From this perspective, a system constitutes a hierarchical order of processes in dynamic equilibrium driven by initial conditions, boundary constraints, and external disturbances (Von Bertalanffy 1950). These concepts have been observed in both the laws of physics (Von Bertalanffy 1972) and sociology (Rasmussen 1979). Additional observations across these disciplines indicate that systems naturally migrate toward states of greater disorder unless there are counteracting forces to maintain desired system behaviors. Observed advanced systems performance, then, can be described as system states resulting from initial conditions, boundary constraints, and external disturbances interacting with natural tendencies toward states of higher risk. This suggests two postulates for advanced systems performance. First, the same end state – or desired system performance – can be achieved with different internal dynamics (e.g. technological systems design) and initiating conditions (e.g. conditions, settings, and circumstances surrounding system use). Second (and similar to the results of the comparative study of CT scanners in hospitals by Barley 1986), *different* end states can result from the *same* initiating conditions and internal dynamics.

Hierarchy and Emergence

A new take on two concepts from general systems theory – hierarchy and emergence – is instructive on how to account better for context in advanced systems engineering. Hierarchy refers to understanding the fundamental differences (and relationships) between levels of complexity within a system, including identifying what generates, separates, and links each level (Von Bertalanffy 1950). Therefore, identifying what generates, separates, and links hierarchical levels is a vital aspect of system engineering to ensure that desired performance is achieved. A better understanding of the conditions, settings, and circumstances in which systems operate can more comprehensively describe the dynamics between hierarchical levels. Consider how new emissions regulations could change what separates and links how electric vehicles interact within the transportation system, for example.

Emergence refers to the phenomenon by which behaviors at a given level of complexity are irreducible to (and thus cannot be explained by) the behavior or design of its subordinate component parts. Part of this irreducibility of observed system-level behaviors is driven by the system's interaction with the conditions, settings, and circumstances in which it operates. Consider how the impact of the Internet as an information transfer system on current social practices cannot be explained by the protocols for transmitting digital packets of data, for example.

Hierarchy and emergence offer well known systems-theoretic concepts on which to build a better understanding of context in advanced system engineering. For both the electric car/transportation system and data packet/information transfer system examples, the context in which the designed systems are used serves an important role in achieving desired outcomes. Invoking these concepts relates advanced system performance to understanding how both component reliability and component interactions are influenced by the conditions, settings, and circumstances in which they operate.

The importance of contextual influences on advanced systems behavior resonates with growing research related to system-theoretic causality models for emergent system properties. For example, recent work has leveraged the advances of Systems-Theoretic Accident Model and Process (STAMP) which argues that safety[1] of complex systems can be redefined as control over system behavior that eliminates losses resulting from systems migrating into hazardous states and experiencing worst case environmental events (Leveson 2012). Related studies include system safety for the aviation (Fleming and Leveson 2014), automotive (Placke et al. 2015), medical (Pawlicki et al. 2016), and nuclear power (Electric Power Research Institute 2013) domains, as well as system security for cyber (Bakirtzis et al. 2017), transportation (Williams 2015), and nuclear (Williams 2013) applications. These studies argue that safety and security are not protecting against random failure problems but preventing the loss of control over desired system performance resulting from component interactions, dynamics, and (potentially) nontechnical causes. Additional studies have also indicated that the system-theoretic causality

---

1 And, by extension, other similar *emergent* system properties such as security.

model can capture human, social, and organizational influences – each representing elements of the conditions, settings, and circumstances of use – on system behaviors. Marais (2005) argues that these influences can be described as challenges to the effective implementation of controls over desired system behavior(s). This study describes organizational risk factors as safety control flaws that violate design decisions intended to constrain unsafe system behavior. Building on this approach, Stringfellow (2010) demonstrated how incorporating human and organizational behavior-generated guide-words helped identify additional violations of STAMP-derived safety control actions – illustrating the role of context in complex systems performance.

The 2018 handbook for systems-theoretic process analysis (STPA) – the analytical technique based on the STAMP causality model – expands this incorporation of contextual influences (Leveson and Thomas 2018). According to the handbook, identifying particular loss scenarios that "describe the causal factors that can lead to the unsafe control actions and to hazards" (Leveson and Thomas 2018, p. 42) is the last step for evaluating complex systems for safety. More specifically, this step can include changes to a controlled process itself, to the path for communicating control, or to how control is administered. This perspective of systems safety – which has shown improvements in safety of complex systems in the aforementioned domains – clearly articulates the importance of including the conditions, settings, and circumstances of use into advanced systems engineering. Applying these systems-theoretic concepts expands advanced systems engineering beyond technology-centric approaches and supports the role of contextual factors on system behaviors – particularly social, organizational, political, and cultural influences on human use of complex systems.

Organization Science Concepts and Insights

If systems theory provides hierarchy and emergence of *technical* components to explain systems behaviors, then perhaps describing *nontechnical* components in similar terms (or with similar concepts) can capture the importance of context in advanced systems engineering. Organization science, according to Robbins (1990), is an academic discipline[2] that studies the structure, behavior, design, and internal dynamics of organizations – which can be loosely thought of as "human systems." Advanced systems engineering should explore how individuals construct institutions, processes, and practices to achieve a common goal while implementing (or operating within) systems. For example, Argyis (1976) and Cyert and March (1963) argue that differences between designed and as-built organizations can lead to unexpected outcomes – which

---

2 "Organization science," "organization theory," and "organizational studies" are often used interchangeably to describe this academic domain. For example, the INFORMS journal *Organization Science* – a leading social science journal – explores organizational behavior and theory from strategic management, psychology, sociology, economic, political science, information systems, technology management, communications, cognitive sciences, and systems theory perspectives. For more, see: https://pubsonline.informs.org/journal/orsc.

suggests that understanding the relationship between daily work practices (as-built) and performance assumptions (design) can better explain the context in which advanced systems must operate.

A more recent approach for reconciling as-built versus designed organizations is Carroll's (2006) argument that it is useful to investigate organizations from three distinct perspectives: the strategic design lens, the cultural lens, and the political lens. Each of these lenses represents shared ideas about human nature, the meaning of organizing, interpretation of collective goals, and the information required to make sense of an organization. Therefore, these three lenses also share descriptions of the context in which advanced systems operate. The *strategic design lens* argues that with the right plan, information flow, and resource distribution, the organization can be rationally optimized to achieve its goal. This lens aligns well with a traditional systems engineering understanding of context. The *cultural lens* describes organizational behavior in terms of the tacit knowledge of "this is how we do things around here" and the processes used to share this knowledge with newcomers. This lens emphasizes that systems are only understood in how they are actually used. Lastly, the political lens interprets organizations as diverse coalitions of stakeholders with different (and sometimes conflicting) interests whose performance is influenced by ever-changing power dynamics that influence decisions. This lens more explicitly identifies multidimensional dynamics between components (or stakeholders) within a system and those in the environment. The *three lens* approach suggests that system performance assumptions underlying complex systems designs–including the conditions, settings, and circumstances in which they operate– are influenced by both the independent focal areas of and the interactions between each lens.

Two primary philosophical perspectives undergird the organization science analytical domain can help frame conceptual options for better understanding the context of use. The first is the *functionalist* perspective, which is an overall approach that seeks to rationally explain organizational phenomena based on observed relationships that can be identified, studied, and measured in a similar manner to natural science (Burrell and Morgan 1979). As applied to the complex systems expected in the Fourth Industrial Revolution, this perspective reduces the introduction of technology into an organization as an exogenous force that changes the behavior of the organization itself or its attendant employees. In other words, the system and operating environment have a distinct boundary and there is a unidirectional causality from the system to its performance in its operating environment – much like traditional systems engineering approaches to context. Other characteristics of the functionalist perspective include an expectation of design dominance in performance, a deterministic view of system operations, and designs based on the assumption that change in organizational behavior is driven by technology (or technological systems). Yet, only assuming that technological changes results in unexpected organizational behaviors dismisses the idea that these observed organizational behaviors might actually result in changes to the technological system itself.

The second primary perspective within organization science is the *constructivist* approach, which seeks to explain organizational phenomena in terms of individual perspectives and in terms of processes that emerge from the resultant interpretive flexibility between these individual perspectives (Burrell and Morgan 1979). From this perspective, humans are seen as creating technological systems under the influence of a set of socially constructed norms, assumptions, and beliefs. Once implemented, the effect of these technological systems on organizations is understood as a set of (not necessarily the same) socially constructed norms, assumptions, and beliefs. Here, systems and operating environments have similarly distinct boundaries, but the unidirectional influence is from the external operating environment to the system – a perspective diametrically opposed to a functionalist approach. Other characteristics of this perspective include an expectation of system design stabilization, interpretive flexibility of system operations over time, and assumptions that organizational change is driven by socially (re)structured relationships between human users and technological systems. Where the constructivist perspective does consider the (often-assumed as unnecessary or extraneous) social context surrounding technological systems as vital, there does not seem to be much room for these emergent social constructs to be influenced by the technological system itself–as anticipated in the Fourth Industrial Revolution.

Between these two primary philosophical perspectives of organization science lies a theory that argues organizational behaviors are not *either* driven by technological systems *or* interpreted by users of technological systems, but rather are a result of both. Structuration theory asserts that organizational behavior emerges from recurrent human action that is both (and simultaneously) shaped by technological artifacts and constructed by their interpretation (Giddens 1984). Consistent with expectations for advanced systems engineering in the Fourth Industrial Revolution, this perspective incorporates more ambiguity in and interactions between systems and their operating environments. The origins of Giddens' (1984) structuration theory (or the enactment of structure) expanded into a spectrum of descriptions of the recursive interactions between humans, technology, and organizations. Where functionalist approaches argue that change is driven by technological systems and constructivist approaches argues that change is driven by interpretation of technological systems in use, structurational approaches assert that change is driven by recursive and recurrent use of technological systems in operational environments. Other characteristics of this perspective include an expectation of recurrent use of designs, dynamic understanding of a system's operating environment, and an assumption that change results from regular patterns of system use and observed performance. Structuration theory offers a perspective that accounts for both the effects of the technological system on its operating context and the effects of its operating context on the technological system itself.

Structuration theory further asserts that organizational structure is a dynamic process seeking equilibrium via recurrent human actions. This assertion is typified by Giddens' (1984) argument that, despite the circumstances or context, an individual always has an option to change their response to the current organizational structure. This perspective also argues that structure always both enables and constrains performance and outcomes. Organizational structure, then, either reinforces or transforms desired performance, suggesting that both continuity and change require a balance between supporting current designs and adapting to changes in the operating environment. By replacing "organization" with "system" and "individual" with "component," structuration theory provides a useful logic incorporating context into advanced systems engineering.

The interactions between technology and organizations in structuration theory align well with the general systems theory phenomenon, where interdependence among components is influenced by, and influences, the operating environment. For example, in a study on implementing self-served Internet technologies in the health insurance domain, Schultze and Orlikowski (2004, p. 88) show "how macrophenomena are constituted by micro-interactions, and how these micro-interactions, in turn, are shaped by macro-influences and effects." Structuration and systems theories share additional conceptual commonalities. Consider the similarities between the *enactment* (e.g. resulting from recurrent human use) of technological systems in the former with *emergence* of systems behavior (e.g. from regularly interacting components) in the latter. They also share a lexicon, including dynamism, complexity, equilibrium, and interdependence. Both theories also assert that a focus on transitioning between *steady* state A and *steady* state B – as opposed to *stable* state A and *stable* state B – better addresses the complexity present when social and technological components interact.

Orlikowski's (1992) structurational model of technology (SMOT) serves as an instructive example of the importance of context in explaining emergent behaviors. SMOT describes recursivity in how *human agent* actions situate the use(s) of *technology*, which then shapes the enacted organizational structure that produces *institutional properties* that enable or constrain those *human agent* actions (Figure 3.1). By replacing *technology* with *system*, the traditional systems engineering perspective gives way to a perception of system performance as resulting from the interaction(s) of technological components as used by human agents under the influence of institutional properties. For example, from the SMOT perspective, performance of vehicle manufacturing systems is not a singular product of production rates but rather a descriptor of the capability of human agents to use the system (under the influence of institutional properties) to produce the appropriate number of cars to meet societal needs and regulatory requirements. The technologies that compose a complex system also shape the enacted structure of the organization and the resulting institutional properties influencing its performance. For example, nuclear power plants that rely on advanced digital controllers for safety critical subsystems have different enacted structures than those that rely on more traditional analog controllers. According to the SMOT, institutional properties can either reinforce or oppose desired human use of technology – further supporting the important role that use context plays in ensuring complex systems achieve their designed objectives.

The logic of the SMOT offers two useful insights for describing the relationship between systems and the conditions, settings, and circumstances in which they operate. First is the direct and recursive relationship between human agents and technology (or technological systems). More explicitly understanding these dynamics is necessary as systems continue to advance and grow in complexity to meet increasingly difficult – and constantly changing – social needs. For example, the Internet as a mass communication and information sharing system looks markedly different today than its designers intended "because if we had known what it would turn into, we would have designed it differently."[3] The second insight is the role of institutional properties in directly influencing humans and being directly influenced by the technology (or technological system) itself. Thus, advanced systems engineering needs to address sources of institutional properties – which can range from cultural norms to societal needs to regulatory requirements to organizational policies. Additionally, advanced systems engineering has the potential to transform these institutional properties. For this second insight, the Internet as a complex system is again instructive when considering current debates on net neutrality, cryptocurrencies, and exploitation of the "deep web" for malicious purposes. As such, structuration theory helps crystallize some key insights from organization science that support the case for – and provide mechanisms to incorporate – the importance of context in advanced systems engineering.

Engineering Systems Concepts and Insights

Nevertheless, there is still a need to more formally integrate these organizational science insights into advanced systems engineering. Engineering systems[4] is a growing academic discipline that seeks to develop theory and practice

---

3 This quote is from Dr. David Clark, chief protocol architect in the development of the Internet, during his lecture to the Fall 2012 offering of *Engineering Systems Doctoral Seminar* (ESD.83) at the Massachusetts Institute of Technology. For more about the designing of the internet, see Clark, 2018.
4 This academic discipline was pioneered by the *Engineering Systems Division* (ESD) at the Massachusetts Institute of Technology, which aimed to focus on the science and engineering of large and complex socio-technical systems by leveraging insights across academic disciplines. In 2015, ESD was subsumed by the *Institute on Data, Systems,* and *Society* – for a thoughtful obituary of this program (de Weck 2016).

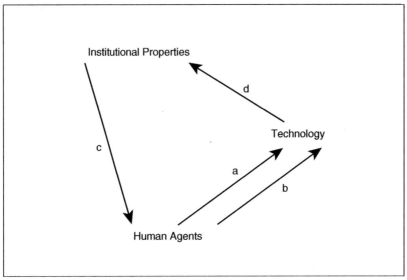

ARROW	TYPE OF INFLUENCE	NATURE OF INFLUENCE
a	Technology as a Product of Human Action	Technology is an outcome of such human action as design, development, appropriation, and modification
b	Technology as a Medium of Human Action	Technology facilitates and constrains human action through the provision of interpretive schemes, facilities, and norms
c	Institutional Conditions of Interaction with Technology	Institutional properties influence humans in their interaction with technology, for example, intentions, professional norms, state of the art in materials and knowledge, design standards, and available resources (time, money, skills)
d	Institutional Consequences of Interaction with Technology	Interaction with technology influences the institutional properties of an organization, through reinforcing or transforming structures of signification, domination, and legitimation

Figure 3.1 Orlikowski's structurational model of technology (SMOT). Source: Orlikowski (1992).

related to characterizing and analyzing the complex causality describing large, technical system behaviors in their social and political contexts. With this perspective and several key characteristics established in de Weck et al. (2011), engineering systems offer a way to reconcile these insights to explain advanced system performance as "characterized by a high degree of technical complexity, social intricacy, and elaborate processes, aimed at fulfilling important functions in society" (de Weck et al. 2011, p. 31). More specifically, Table 3.1 summarizes

Table 3.1 Definitions of key characteristics of engineering systems[a].

Function	"action for which a thing is specifically fitted or used, or the reason for which a thing exists" [p. 51]
Scale	"geography, demography, numbers of components, number of people, and any other aspect that can be used to assess the size of the system *quantitatively*" (emphasis in original) [p. 50]
Scope	"the number of aspects that need to be considered when defining the system" [p. 50]
Structure	"the way in which elements of the system are interconnected...also includes assignments of sub-functions to the elements of the system"[52]
Temporality	"dynamic; [the systems] change with time...[the] time scale[s] that agents, mass, energy, and information flow through complex engineering systems" [p. 55]

a) Definition from de Weck et al. (2011).

the definitions of the primary characteristics that distinguish an engineering systems perspective from traditional system engineering approaches.

In an essay supporting this perspective of systems engineering evolution, Long (2018) noted the importance of correctly understanding the operational context when he underscored "a team of *systems engineers working in concert with the right subject matter experts* to understand the problem and characterize the solution can be a thing of beauty delivering elegant solutions to complex problems" [emphasis added]. This discipline's incorporation of technology, management and social science highlights the importance of context, or the confluence of dynamic, exogenous factors that influence how system behaviors propagate, shape how stakeholders interact, and bound how systems operate. For example, an engineering systems approach explicitly includes the impact of changes over time and different timescales of these changes on system design, implementation, and evaluation. Consider US nuclear power plants built 60+ years ago that are undergoing operating life extension license renewals, are replacing analog components with digital controllers, and are simultaneously faced with competing against historically low natural gas prices. This characteristic of engineering systems highlights the interactive role of context on advanced system performance. Reframing the characteristics of engineering systems provides organizing principles by which to reconcile the different aspects of systems theory and organization science to develop a better understanding of the *context of use* for advanced systems.

3.5 *Context of Use* in Advanced Systems Engineering

If human (inter)actions can influence the structure, boundaries, and dynamics of how systems operate, then systems engineering needs new mechanisms for addressing this new understanding of context. Leveraging insights from the systems theory, organization science, and engineering systems disciplines can help develop how to better account for context in advanced systems engineering. Context of use is important because, as argued by Bucciarelli (2003, p. 13), with advanced systems "...different participants, with different technical responsibilities and interests, see the object of design differently." Further, from his philosophy of science perspective, Bucciarelli (2010) also introduces two different fundamental descriptors of a designed object – either "value free" or "context of use." Though Bucciarelli was writing about technological widgets in general, replacing "designed object" with "advanced system" is helpful. From a "value-free" engineering design standpoint, focusing on the "formal structure" removes such sticking points as social, political, and cultural issues and aids in more expedient system implementation. This perspective seems to support the logic that a well-designed operating system can operate independently of contextual influences and, thus, can be an element of common ground among multiple stakeholders. Or, as Bucciarelli (2010, p. 21) puts it, "There is the mutual trust in abstraction itself".

Conversely, Bucciarelli (2010) asserts that "context of use"-based design can translate into better solutions. This perspective seeks to more explicitly include social, political, and cultural influences on system design and implementation, allowing systems-theoretic trade-offs to be better manipulated to address the problem at hand. This perspective can also allow an already existing technology (or system) to be more effective. Consider how well Apple products sell, even though oftentimes the technology on which they are based has been around for years (e.g. the touch screen). Ultimately, the use of "formal structures" can help shape and improve design discussions within systems engineering but "context, the circumstances and conditions within which a complex systems problem is embedded, can constrain and overshadow technical analysis determining system solution success" (Keating et al. 2003, p. 38). As such, the *context of use* for advanced systems engineering can be described as the conditions, settings, and circumstances in which a system operates, including the dynamic, interdependent factors of a system's embedded environment that bound how it operates, influence how its behaviors propagate, and shape how its stakeholders interact.

To better incorporate the context of use into advanced systems engineering requires more than shifting from a closed to an open systems perspective. In the former, a system's environment is simply a universal constraint with an immutable boundary. In the latter, context is not completely discounted, as a system's environment is considered assets of bounding conditions or guidelines to satisfy exogenous forces on desired system behaviors. Yet, neither of these perspectives capture how "contextual, human, organization, policy, and political system dimensions that ... ultimately shape the decision space and feasible solutions for the technical system problems" (Keating et al. 2003, p. 39). Discounting these influences reduces design and development to predetermined "correct" uses of advanced systems – drastically limiting their success to an artificially narrow sliver of the problem space. Therefore, the context of use for advanced systems engineering leverages insights from structuration theory to move toward capturing how "everyday actions are consequential in producing the structural contours of everyday life" (Feldman and Orlikowski 2011, p. 1241). As such, the conditions, settings, and circumstances in which systems operate are more than constraints affecting advanced systems, they are now important variables impacting system performance.

The context of use perspective addresses how the impact of systems engineering (including the associated problem and potential solution spaces) is always larger than the associated technological basis of the system itself. For instance, this context of use perspective can help explain the engineering systems paradigm from MIT where advanced systems are "socially transformative" (de Weck et al. 2011). Similarly, the engineering systems emphasis on better understanding of emergent system properties – the so-called "ilities"[5] – seemed to (at least partially) emerge directly from this larger context of use concept in systems engineering. The role of context of use is also seen in how a system's network structure can be designed to propagate diffusion of desired behaviors, how contextualized understanding of sociotechnical systems influences stakeholder interactions, and how context of use is a driving force for technological changes over time. Context of use is also captured in assertions made by Keating et al. (2003, p. 41) that "metasystems are themselves comprised of multiple autonomous embedded complex systems that can be diverse in technology, context, operation, geography, and conceptual frame". Therefore, for SOSs to achieve their objectives, they argue, metasystems act as the conditions, settings, and circumstances for the embedded systems – not a static or unidirectional operating environment but an interactive and interdependent context of use.

Better understanding a system's context of use can also help alleviate the increase in system performance uncertainty stemming from increases in system complexity. Where emergence is a vital characteristic, more systems complexity expands possibilities for unanticipated outcomes that can be missed by traditional systems engineering approaches. Rather than allowing uncertainty to limit system success, advanced systems engineering should track changes in the context of use to anticipate any "unexpected" shifts in system performance. For big data, this idea argues that it is not the quantity of information available that helps mitigate increasing system complexity but the capacity to use that information in meaningful ways to achieve desired system performance within the conditions, settings, and circumstances in which it operates.

---

5 Per de Weck, *et al.* (2011, p. 66), the *ilities* are "are desired properties of systems…[that] are not primary functional requirements of a system's performance, but typically concern wider systems impacts with respect to time and stakeholders than are embedded in those functional requirements". Examples include flexibility, reliability, and adaptability.

3.6 An Example of the Context of Use: High Consequence Facility Security

To further make the case for including the context of use in advanced systems engineering, this section explores improving systems engineering approaches for security at high consequence facilities. Here, the objective of security is to protect the high consequence (or high value) facility from malicious or intentional acts that could result in such unacceptable losses as human fatality or injury, theft or loss of assets, damage to property/infrastructure, or environmental destruction or contamination (e.g. nuclear or chemical facilities). Security at high consequence facilities – be they chemical plants, oil refineries, nuclear power plants, or other pieces of critical infrastructure – is typically provided by a combination of sensors, cameras, barriers, and other technologies called a *physical protection system* (PPS), with the objective to detect, delay, and respond to malicious acts. Adequately selecting, arranging, installing, operating, and maintaining these interconnected PPS components are major challenges to ensuring security of high consequence facilities. To demonstrate the value of the context of use in advanced systems engineering, two approaches for resolving these challenges to security will be compared. The first approach is based on traditional systems engineering approaches, whereas the second approach incorporates the context of use to expand the solution space for security at high consequence facilities.

A Traditional Systems Engineering Approach: Design Evaluation Process Outline (DEPO)

Building on generic systems engineering concepts such as feedback processes and hierarchical design, high consequence security becomes the process of appropriately determining, designing, and evaluating PPS. One of the most popular – and most prolific – high consequence security methodologies is the design evaluation process outline (DEPO) developed at Sandia National Laboratories. More specifically, DEPO is the current standard for security analysis at nuclear facilities around the world. DEPO methodology is popular for its clarity and stochastic modeling paradigm that concludes when balance is achieved between the level of risk and upgrade impact to facility budget and operations concerns (Garcia 2008; Biringer and Danneels 2000). The traditional systems engineering basis of DEPO also borrows a philosophical foundation from applying probabilistic risk assessment (PRA) (Sandoval 2014) to nuclear safety. The accident timeline is replaced by the competing timelines of required adversary action to achieve a malicious act and the response force actions necessary to protect the high consequence facility. The DEPO methodology describes security as probabilistic influences on this timeline – the probability that a particular sensor alarms when an adversary

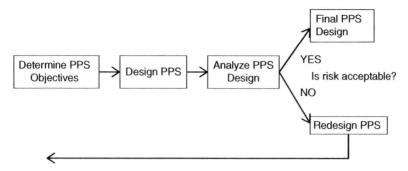

Figure 3.2 Illustration of the DEPO methodology. Source: Recreated from Figure 1 in Garcia (2008).

enters a prohibited area or the probability that the response force is able to intercept the adversary, for example.

As illustrated in Figure 3.2, the DEPO consists of three primary functions: determining the PPS objectives, designing the PPS, and evaluating the PPS (Garcia 2008). Determining PPS objectives consists of four aspects: *characterizing the facility* (e.g. What is its mission? Where is it located? Who works at it?); *identifying undesired events* and *critical assets* (e.g. What are the potential targets?); *determining consequences of undesired events* (e.g. What is the facility liability or regulatory responsibilities?); and, *defining threats to the facility* (e.g. What are the capabilities of the adversary according to national threat assessments and local factors?). Once the objectives are defined, DEPO methodology is used to design the PPS itself. Here, the PPS design is analyzed for system effectiveness using different heuristics to identify, down-select, arrange, and optimize the characteristics of security-related technologies to achieve delay, detection, and response goals. Lastly, DEPO requires an evaluation of the PPS, which includes *estimating the risk* (as related to system effectiveness against malicious acts) and *comparing this risk to the acceptable risk levels*. Consistent with traditional systems engineering concepts, if the risk is sufficient, the (re)design is complete; if it is insufficient, modifications (or upgrades) to the PPS should be suggested and reevaluated. By this process DEPO uses the "detect, delay, and respond" paradigm to fully describe – and engineer PPS to meet – the necessary functions of strong security for high consequence facilities (Garcia 2008).

More specifically, DEPO calculates the ability of an arranged collection of technologies to defeat a specific adversary along a specific attack path. This methodology relies on two probabilistic descriptors of security performance. The first is the *probability of interruption (P_I)*, or the conditional probability that detection and delay system components will *assess* an adversary in time for response forces to arrive onsite to engage. The second is the *probability of* *neutralization (P_N)*, or the conditional probability that, upon arriving, response force capabilities can kill, capture or cause the adversary to flee. P_I and P_N are nested, conditional probabilities related to security system component performance derived from performance testing (at best) or (at worst) expert opinion. DEPO defines the ability to adequately protect high consequence facilities as the *system effectiveness (P_E)* of the PPS – itself the product of P_I and P_N. More specifically, DEPO employs the risk formula $R = P_A*(1-P_E)*C$. Here, P_A is the assumed probability of attack; C is a quantitative approximation of qualitative consequence descriptions; and P_E is the system effectiveness. Given the difficulty in of accurately quantifying P_A, DEPO best practice is to use a "conditional risk" (where $P_A = 1$) that simplifies the equation to $R_C = (1-P_E)*C$.

The traditional focus on technological system solutions in DEPO is based on the implicit assumption that any PPS can achieve desired levels of performance regardless of the operational context. This is an oversimplified and untenable assumption at the core of DEPO-based approaches that negates the impacts of contextual influences on security performance at high consequence facilities. Consider, for example, a visit by Dr. Matthew Bunn[6] to a Russian nuclear institute in the mid-2000s where he recounted that:

> …inside the hallway leading to the vault where a substantial amount of weapons-grade nuclear material was stored, there were two portal monitors that personnel had to pass through, one after the other, an American machine and a Russian machine. When asked why, the site official conducting the tour said that the building next door made medical isotopes, and on Thursdays, when the chemical separations were done to get the desired isotopes from the remainder, so much radiation went up the stack that it set off the American-made portal monitor. So on Thursdays, they turned off the American-made monitor and relied on the less sensitive Russian one. Of course, every insider was aware of this practice, and would know to plan an attempted theft for a Thursday, making the existence of the American portal monitor largely pointless.
>
> (Bunn and Sagan 2014, p. 11)

Relying on traditional systems engineering-based approaches – like DEPO – would indicate the same estimate for PPS effectiveness even *after* this observation of its use. In other words, there is still a need to better understand the relationship between the technical system (e.g. American portal monitors of the PPS to detect unexpected radioactivity), the technical system's context of use (e.g. turning off the American portal monitors to avoid

---

6 Former nuclear security advisor to the Office of Science and Technology Policy; current Professor of Practice at Harvard University's Kennedy School of Government and co-Principal Investigator for the Project on Managing the Atom.

high levels of false alarms from medical isotope production on Thursdays), and security performance (e.g. reduced detection capability and increased opportunity for a malicious act by an adversary). More specifically, a National Academies of Science (2015) report described "pressures, both positive and negative, at all times" that impact nuclear security performance.

An Advanced Systems Engineering Approach: Systems-Theoretic Framework for Security (STFS)

Based on the previous anecdote, part of a security system's context incudes the capability of the security organization to oversee the correct use of the PPS in pursuit of strong security performance. In response, the systems-theoretic framework for security (STFS) argues the need to include *some* measure of *context* in security for high consequence facilities (Williams 2018). Using a nontraditional perspective to build on traditional systems engineering concepts, STFS provides an approach to better explain security performance. First, consider that actual security performance is impacted by individual actions during daily security work practices, PPS operations, and the security organization decisions – and their interactions. Second, the current level of performance influences both the PPS itself and the security organization (e.g. feedback). Here, if daily work practices help explain these feedback and interaction dynamics, then security performance emerges from how individual security actions are influenced by the technology within the PPS itself and the organizational factors that affect its use context. The technological availability relates to how well the PPS meets its detection, delay, and response objectives *within* a given operational context. The organizational factors relate to aspects of the conditions, settings, and circumstances in which the PPS operates that affect daily work practices. Security performance, then, results directly from how well the PPS design operates (e.g. how it meets the P_E objectives) and indirectly from how the security organization affects individual security actions.

As illustrated in Figure 3.3, STFS describes security performance in terms of five *elements* (e.g. denoted with bold text), eleven *links* (denoted with numbered arrows), and four *feedback loops* (denoted with Roman numerals). These elements represent both tangible and intangible attributes that interact and result in security performance. More specifically, these elements are *security performance* (completion and quality of detection, delay, and response functions), *individual security actions* (work practices that support achieving PPS security functions), *security task completion expectations* (contextual conditions for work practices to support PPS operations), *security organization* (entity taking actions and making decisions to operationalize the PPS); and the *PPS design* itself. The STFS elements are connected by *links* that describe causality and functional roles within the framework. The causality of each link is expressed in terms of how the sending STFS element interacts with the receiving STFS

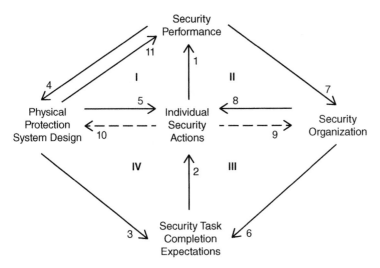

Figure 3.3 Overview of the systems-theoretic framework for security. Solid arrows ⟶ indicate currently recognized influences and dashed arrows ⇢ indicate influences not recognized by current state-of-the-art approaches.

element. For example, the monitoring and evaluation role of Link 4 should be read as "Security performance *provides* assessment of technical performance as observed by inspection results, corrective actions or response to a security event *to* the security organization." These links, summarized in Table 3.2, partially describe the conditions, settings, and circumstances in which the PPS operates – explicitly incorporating the *context of use* into an advanced systems engineering approach to security at high consequence facilities.

The links describing causality also represent the dynamic, interactive relationships between elements that identify interdependency between the PPS (as a technical system) and its context of use. Connected sets of links form *feedback loops* – relationships wherein output of a system (or process) returns as new input into the same system (or process). These feedback loops help describe the complex, nonlinear behaviors observed in high consequence facility security performance. In addition, interacting feedback loops provide an opportunity to describe more complex security performance behaviors, including the effect of the contextual influences on security performance described in the Bunn account above. These four feedback loops (summarized in Table 3.3) capture the dynamics often assumed with security design for high consequence facilities (Loop I), offer deeper explanations of contextual influences (Loops II and III), and illustrate the interactions between systems and their operational environments (Loop IV).

The STFS uses *elements* (key attributes), *links* (descriptions of causality), and *feedback loops* (dynamic relationships) to build on traditional systems

Table 3.2 Summary of link descriptions for the systems-theoretic framework for security.

#	Name	Description
1	Security task completion	Which (e.g. actual P_E) and how well (e.g. quality indicators) security functions are achieved
2	Influence of expectations	CONOPs, procedures and behavioral performance requirements defining necessary security tasks
3	PPS-based requirements	Necessary actions expected by designers to achieve estimated system effectiveness ($\sim P_E$) for a particular PPS design
4	Feedback on technical performance	Assessment of technical performance as observed by inspection results, corrective actions or response to a security event
5	Technical availability for security	Security-related technologies implemented to achieve expected security functions[a] (e.g. $\sim P_E$)
6	Organizational expectations	Decisions on the necessary and sufficient actions (e.g. resources provided) to support the expected PPS performance
7	Feedback on organizational performance	Assessment of organizational performance as observed by inspection results, corrective actions or response to a security event
8	Behavioral quality for security	Operational context in place to achieve expected security functions (e.g. data-derived organizational influences)
9	Human behavior effects (Quality)	Recurrent security-related actions of individuals that describe how the PPS is used (e.g. patterns of security practice)
10	Human behavior effects (Structure)	Recurrent security-related actions of individuals that describe what in the PPS is used (e.g. patterns of security practice)
11	Actual PPS Capacity	Capability and reliability with which technical components achieve expected security functions (e.g. traditional DEPO)

a) The term *security functions* relates to traditional security performance measures: probability of detection (P_D), delay time (t_D) and response force time (RFT).

engineering approaches (e.g. DEPO) to include the role of the context of use in security for high consequence facilities. In addition, the STFS provides the ability to ask why actions were taken or decisions made in the operational environment and trace their ramifications to PPS performance, which can illuminate additional causal factors affecting security performance. By doing so, the STFS provides a mechanism for describing how the interactions between technical systems and use contexts can better describe emergent properties in advanced systems. By including the context of use, the STFS offers several new insights for understanding nuclear security performance. First, this framework articulates that security performance is a system-level property based on both the technical ability of the PPS and how it is operated (e.g. behavioral quality).

Table 3.3 Summary descriptions of feedback loops for the systems-theoretic framework for security.

Feedback loop	Summary description
Link 5 Link 1 I Link 4	Technical feedback based on formal monitoring and oversight (capturing traditional DEPO-based approaches and dynamics)
Link 8 Link 1 II Link 7	Organizational feedback based on formal monitoring and oversight (capturing traditional security culture model-based dynamics)
Link 6 Link 2 III Link 9	Observational feedback based on patterns of security behavior (capturing the dynamics between organizational decisions, security task completion, and individual security actions)
Link 3 Link 2 IV Link 10	Observational feedback based on patterns of PPS use (capturing dynamics between the PPS, security task completion, and individual security actions)

The STFS captures the dynamics of traditional DEPO-based approaches (that emphasize PPS ability) in Feedback Loop I, but also helps explain how different interactions between technical ability and behavioral quality result in a range of security performance outcomes.

This STFS context of use is not intended to describe operational reality *perfectly* but rather to serve as a signal directing attention toward possible security vulnerabilities and PPS performance inadequacies not included in traditional DEPO system effectiveness calculations. For some PPS components, the potential for contextual factors to challenge acceptable performance and change observed behaviors will be limited – the ability of a 0.5 m concrete thick wall of an underground storage vault to provide delay, for example. For others, though, contextual factors present greater opportunities to challenge acceptable performance and change observed behaviors – turning off the portal monitors in the Bunn anecdote degraded the ability to meet detection objectives, for example. Similarly, where DEPO-based approaches are based on the analytic assumption that security personnel will use the PPS as expected by designers, the STFS captures how the context of PPS use may deviate from expectation.

In maintaining the role of technical PPS in security performance, an advanced systems engineering approach (e.g. the STFS) extends traditional systems engineering approaches (e.g. DEPO) for security at high consequence facilities. Such approaches also better account for the dynamic interactions with their context of use and the resultant effects on traditional performance metrics (e.g. P_D, t_D, and RFT). Though not attempting to explain all aspects of security at high consequence facilities, the STFS provides an advanced systems engineering framework that captures the use context of security in terms of simple feedback loops and observable patterns of system behavior. The implication that security performance is not a static attribute of such facilities indicates a need to shift from optimizing PPS designs toward equilibrating between technical systems and operational contexts toward desired levels of performance–a paradigm shift that could significantly benefit advanced systems engineering for the Fourth Industrial Revolution.

3.7 Summary

Summarizing the context of use, the effects of system engineering solutions are wider ranging than their intended technological problem space. Systems can be better designed, better implemented, and more effective when accounting for the conditions, settings, and circumstances in which they operate. As the challenges and changes of the Fourth Industrial Revolution manifest, the relationship between "system" and "operating environment" will be increasingly important – especially as social, political, and cultural influences continue to evolve. The difference between the STFS and DEPO-based approaches to security at high consequence facilities demonstrates how the ability of a PPS to adequately achieve desired levels of performance can be affected – both positively and negatively – by dynamic interactions with its context of use. This is illustrative of how advanced systems engineering can incorporate the context of use to improve system performance. Advanced systems engineering should explore the commonalities between key concepts in systems theory, organization science, and engineering systems to develop additional analytical capabilities to address the role of context and improve the increasingly complex systems of the future.

In this regard, Keating et al. (2003) offer four questions around which to craft new analytical techniques for including the context of use into advanced systems engineering:

(1) What are the relevant contextual aspects of the SOSs engineering effort?
(2) Has the engineering process accounted for the contextual nature of the effort?

(3) Has an appropriate balance been struck between the technical and contextual aspects of the problem?
(4) Can this balance shift over the SOSs engineering effort?

Providing the basis for advanced systems engineering capabilities to address these questions is the foundation of the context of use concept – and the benefits of doing so were demonstrated in the comparison of traditional and advanced systems engineering approaches to security at high consequence facilities. Describing security as a result of patterns of practice and feedback loop dynamics aids in identifying potential interventions for improving performance. For example, advanced system engineering approaches could seek to redirect oppositional contextual influences – and undesirable sociotechnical interactions – to support desired system-level behaviors. By invoking systems theory, organization science, and an engineering systems approach, the context of use helps to redefine desired complex system properties in terms of aligning interacting technical systems and the conditions, settings, and circumstances in which they operate.

Moving forward, advanced systems engineering will likely need to more explicitly incorporate patterns of use into system design. As described, the context of use offers to the ability to better ensure consistency in performance of the *same* complex system in *similar* operating environments with *different* contexts – overcoming the challenges experienced in Barley's (1986) analysis of why implementing the same set of CT scanners in similar hospitals had drastically different results or Williams' (2018) evaluation of different security performance outcomes from the same PPS being implemented a similar high consequence facilities. Similar to Leveson's (2012) argument that evaluating emergent system properties can help explain unexpected systems behaviors (especially in the absence of component failure), including the context of use offers an additional perspective by which to explain such unexpected system behaviors. To the extent that the context of use paradigm is beneficial, it also serves as a mechanism by which to incorporate new ideas and analytical frames from other disciplines. Here, two examples include advances in applying structuration theory to technology (DeSanctis and Poole 1994) and the use of assumption-based planning concepts in developing leading indicators for complex system behaviors (Leveson 2015). Further, this context of use perspective offers a wider range of potential explanations for – and levers to correct – unexpected system behaviors and reducing opportunities for "unintended consequences." In short, the need to understand and incorporate the context of use in design and analysis of complex, sociotechnical systems is one of the most unique characteristics facing advanced systems engineering–to better address the challenges of the Fourth Industrial Revolution.

References

Anderson, R. (2008). *Security Engineering: A Guide to Building Dependable Distributed Systems*, 2e. Wiley.

Argyis, C. (1976). Single-loop and double-loop models in research on decision making. *Administrative Science Quarterly* 21 (3): 363–375.

Bakirtzis, G., Carter, B. T., Fleming, C. H., and Elks, C. R. (2017). MISSION AWARE: Evidence-Based, Mission-Centric Cybersecurity Analysis. *Computing Research Repository (CoRR),* abs/1712.01448.

Barley, S.R. (1986). Technology as an occasion for structuring: Evidence from observations of CT scanners and the social order of radiology departments. *Administrative Science Quarterly* 31: 78–108.

Biringer, B. and Danneels, J. (2000). Risk assessment methodology for protecting our critical physical infrastructures. In: *Risk-Based Decision Making in Water Resources IX* (eds. Y.Y. Haimes, D.A. Moser, E.Z. Stakhiv, et al.), 33–43. Reston, VA: ASCE.

Bucciarelli, L.L. (2003). Designing, like language, is a social process. In: *Engineering Philosophy*, 9–22. Delft, The Netherlands: Delft University Press.

Bucciarelli, L. L. (2010). From function to structure in engineering design. Cambridge, MA: Massachusetts Institute of Technology. https://dspace.mit .edu/handle/1721.1/51789 (Last accessed 28 May 2019).

Bunn, M. and Sagan, S. (2014). *A Worst Practices Guide to Insider Threats: Lessons from Past Mistakes*. Cambridge, MA: American Academy of Arts and Sciences.

Burrell, G. and Morgan, G. (1979). *Sociological Paradigms and Organizational Analysis.* London: Heinemann.

Carroll, J.S. (2006). *Introduction to Organizational Analysis: The Three Lenses.* Cambridge, MA: Unpublished Manuscript.

Clark, D. (2018). *Designing an Internet.* Cambridge, MA: MIT Press.

Cyert, R. and March, J. (1963). *A Behavioral Theory of the Firm.* Englewood Cliffs, NJ: Prentice-Hall.

de Weck, O. (2016). MIT Engineering Systems Division R.I.P.: Eulogy for a successful experiment 1998–2015. *MIT Faculty Newsletter* XXVIII (4): 12–16, Massachusetts Institute of Technology. Retrieved from http://web.mit.edu/fnl/ volume/284/deweck.html (Last accessed 28 May 2019.

de Weck, O., Roos, D., and Magee, C. (2011). *Engineering Systems: Meeting Needs in a Complex Technical World.* Cambridge, MA: The MIT Press.

DeSanctis, G. and Poole, M.S. (1994). Capturing the complexity in advanced technollogy use: Advanced structuration theory. *Organization Science* 5 (2): 121–147.

Electric Power Research Institute. (2013). *Hazard Analysis Methods for Digital Instrumentation* and *Control Systems.* Technical Report 3002000509. Electric Power Research Institute.

Feldman, W.S. and Orlikowski, W.J. (2011). Theorizing Practice and Practicing Theory. *Organization Science* 22 (5): 1240–1253.

Fleming, C.H. and Leveson, N.G. (2014). Improving hazard analysis and certification of integrated modular avionics. *Journal of Aerospace Information Systems* 11 (6).

Garcia, M.L. (2008). *The Design and Evaluation of Physical Protection Systems*, 2e. Butterworth-Heineman.

Giddens, A. (1984). *The Constitution of Society: Outline of the Theory of Structuration*. University of California Press.

Glass, R. J., Beyeler, W. E., Ames, A. L., et al. (2012). Complex Adaptive Systems of Systems (CASoS) Engineering and Foundations for Global Design (SAND2012–0675). Albuquerque, NM: Sandia National Laboratories.

Hall, J.L. (2003). Columbia and challenge: organizational failure at NASA. *Space Policy* 19: 239–247.

International Council on Systems Engineering (a). (n.d.). *What is Systems Engineering?* INCOSE.org: https://www.incose.org/about-systems-engineering (Last accessed 28 May 2019).

International Council on Systems Engineering (b). (n.d.). *A Consensus of INCOSE Fellows: Definition of a System*. INCOSE.org: https://www.incose.org/about-systems-engineering. (Last accessed 28 May 2019).

Keating, C., Rogers, R., Unal, R. et al. (2003). System of systems engineering. *Engineering Management Journal* 15 (3): 36–45.

Kenett, R.S., Zonnenshain, A., and Swarz, R.S. (2018). Systems engineering, data analytics, and systems thinking: moving ahead to new and more complex challenges. *INCOSE International Symposium* 28 (1): 1608–1625.

Leveson, N. (2015). A systems approach to risk management through leading safety indicators. *Reliability Engineering and System Safety* 134: 17–34.

Leveson, N.G. (2012). *Engineering a Safer World: Systems Thinking Applied to Safety*. Cambridge, MA: MIT Press.

Leveson, N., and Thomas, J. P. (2018). STPA Handbook. https://psas.scripts.mit.edu/home/get_file.php?name=STPA_handbook.pdf (Last accessed 28 May 2019).

Long, D. (2018, June 8). MIT was Right – Focusing on EoS Rather than SE. Retrieved from ViTechCorp.com: http://community.vitechcorp.com/index.php/mit-was-right-focusing-on-eos-rather-than-se.aspx (Last accessed 28 May 2019).

Marais, K. (2005). A new approach to risk analysis with a focus on organizational risk factors. PhD Dissertation. Massachusetts Institute of Technology.

McFarlane, P. and Hills, M. (2013). Developing immunity to flight security risk: prospective benefit from considering aviation security as a socio-technical eco-system. *Journal of Transportation Security* 6: 221–234.

National Academy of Sciences (2015). *Brazil–U.S. Workshop on Strengthening the Culture of Nuclear Safety and Security: Summary of a Workshop*. Washington, DC: The National Academies Press.

National Research Council (1997). Committee on application of digital instrumentation and control systems to nuclear power plant operations and safety. In: *Digital Instrumentation and Control Systems in Nuclear Power Plants: Safety and Reliability Issues*. Washington, D.C.: National Academies Press.

Orlikowski, W.J. (1992). The duality of technology: rethinking the concept of technology in organizations. *Organization Science* 3 (3): 398–427.

Orlikowski, W.J. (2000). Using technology and constituting structures: a practice lens for studying technology in organizations. *Organization Science* 11 (4): 404–428.

Pawlicki, T., Samost, A., Brown, D. et al. (2016). Application of systems and control theory-based hazard analysis to radiation oncology. *Journal of Medical Physics* 43 (3): 1514–1530.

Placke, S., Thomas, J., and Suo, D. (2015). *Integration of Multiple Active Safety Systems Using STPA*. SAE Technical Paper 2015-01-0277. SAE.

Rasmussen, J. (1979). *On the Structure of Knowledge - A Morphology of Mental Models in a Man-Machine System Context (RISO-M-2191)*. Roskilde, Denmark: Riso National Laboratory.

Robbins, S.P. (1990). *Organization Theory: Structures, Designs, and Applications*, 3e. Prentice Hall.

Robey, D. and Rodriquez-Diaz, A. (1989). The organizational and cultural context of systems implementation: case experience from Latin America. *Information and Management* 17: 229–239.

Ross, R., McEvilley, M., and Oren, J.C. (2016). *Systems Security Engineering: Considerations for a Multidisciplinary Appraoch in the Engineering of Trustworthy Security Systems*. NIST Special Publication 800–160. Gaithersburg, MD: National Institute of Standards and Technology.

Sandoval, J. (2014). The Use of a Modified PRA Risk Equation in DOE Security: A Chronological History. (SAND2014-1003C). Risk Informed Security Workshop, Stone Mountain, GA (11–12 February). Institute of Nuclear Materials Management, Mount Laurel, NJ.

Schultze, U. and Orlikowski, W.J. (2004). A practice perspective on technology-mediated network relations: the use of internet-based self-serve technologies. *Information Systems Research* 15 (1): 87–106.

Silitto, H., Griego, R., Arnold, E. et al. (2018). What do we mean by "system"? – System beliefs and worldviews in the INCOSE community. *INCOSE International Symposium* 28 (1): 1190–1206.

Stringfellow, M. V. (2010). Accident Analysis and Hazard Analysis for Human and Organizational Factors. Cambridge, MA: PhD Dissertation. Massachusetts Institute of Technology.

Sussman, J.M. (2014). *The CLIOS Process: Special Edition for the East Japan Railway Company.* Cambridge, MA: Massachusetts Institute of Technology.

Von Bertalanffy, L. (1950). The theory of open systems in physics and biology. *Science* 111 (2872): 23–29.

Von Bertalanffy, L. (1972). The history and status of general systems theory. *The Academy of Management Journal* 15 (4): 407–426.

Waldrop, M.M. (2016). More than Moore. *Nature* 530 (7589): 144–147.

Weaver, W. (1948). Science and complexity. *American Scientist* 36 (4): 536–544.

Williams, A.D. (2013). System security: rethinking security for facilities with nuclear materials. *Transactions of the American Nuclear Society* 109 (1): 1946–1947.

Williams, A.D. (2015). Beyond a series of security nets: applying STAMP and STPA to port security. *Journal of Transportation Security* 8 (3–4): 139–157.

Williams, A. D. (2018). Beyond Gates, Guards, and Guns: The Systems-Theoretic Framework for Security of Nuclear Facilities. PhD Dissertation. Massachusetts Institute of Technology.

4

Architectural Technical Debt in Embedded Systems

Antonio Martini and Jan Bosch

Synopsis

With the advent of the fourth industrial revolution, large industries developing cyber-physical systems strive to reach continuous deployment. They try to make their development processes fast and more responsive, minimizing the time between the identification of a customer need and the delivery of a solution. Given the growth in size and value related to software in cyber-physical systems, the trend in the last decade has been the employment of agile software development (ASD) (Dingsøyr et al. 2012). However, for companies developing long-lasting systems, the responsiveness in the short term needs to be balanced with the ability to maintain the system for several years and with the ability of being able to deliver new features in the long run. Such a goal relies on the system architecture qualities.

ASD has been employed effectively in the development of embedded systems, contributing to speeding up the delivery of value to the customers (Eklund and Bosch 2012). However, some obstacles have been found, limiting the application of ASD to such a domain (Eklund et al. 2014). In this chapter we focus on one such challenges, the degradation of the system architecture over time, which is called, in literature, Architectural Technical Debt (ATD).

To better introduce the problem we show Figure 4.1: ASD is mostly implemented on a team level, which is restricted to the design and test of a single module rather than of the whole system extracted from (Eklund et al. 2014); the current implementation of agile processes is limited to the software team only (which is usually responsible for the implementation within a single module of the system). The exclusion of the system and architecture design from the agile loop is due to the lack of agile process for such purposes. On one hand, there are no agile practices to be applied by system engineers, and, on the other hand, the current practices related to ASD do not include practices for synchronizing with system engineers.

Systems Engineering in the Fourth Industrial Revolution: Big Data, Novel Technologies, and Modern Systems Engineering,
First Edition. Edited by Ron S. Kenett, Robert S. Swarz, and Avigdor Zonnenshain.

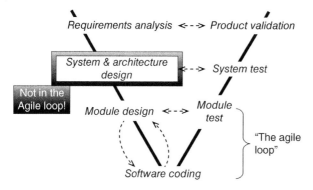

Figure 4.1 Agile software development is mostly implemented on a team level, which is restricted to the design and test of a single module rather than of the whole system.

When software teams and system engineers are not synchronized, the system architecture might not be followed because of the short-term pressure to deliver. On the other hand, the system architecture might need to be updated when new requirements are discovered or changed during the development. The consequence of not having this feedback loop often leads to the mismatch between the actual implementation of the system and the qualities specified in the desired system architecture. Such phenomenon is recognized in literature as Architectural Technical Debt. In this chapter, we report an empirical investigation conducted over seven years of research in collaboration with large companies developing embedded systems. Through several qualitative and quantitative inquiries and document analysis, we found what are the causes and trends related to the accumulation of Architectural Technical Debt and what the implications for system engineering are.

4.1 Technical Debt and Architectural Technical Debt

To illustrate the phenomenon, a financial metaphor has been coined; it relates taking suboptimal solutions in order to meet short-term goals to taking a financial debt, which has to be repaid with interests in the long term. Such a concept is referred to as technical debt (TD), and recently it has been recognized as a useful basis for the development of theoretical and practical frameworks (Kruchten et al. 2012). Tom et al. (2013) have explored the TD metaphor and outlined a first framework in 2013.

A more specific kind of technical debt is architectural technical debt (ATD, categorized together with design debt in Tom et al. 2013). A further classification can be found in Kruchten et al. (2012), where ATD is regarded as the most challenging TD to be uncovered, as there is a lack of research and tool support in practice. Finally, ATD has been further recognized in a recent systematic mapping on TD (Li et al. 2015).

Analogously to TD, architectural technical debt (ATD) is regarded (Tom et al. 2013) as suboptimal solutions with respect to an optimal architecture for supporting the business goals of the organization. Specifically, we refer to the architecture identified by the software and system architects as the optimal trade-off when considering the concerns collected from the different stakeholders. In the rest of the paper, we call the suboptimal solutions inconsistencies between the implementation and the architecture, or violations, when the optimal architecture is precisely expressed by rules (for example for dependencies among specific components). However, it is important to notice that (in the cases studied in this paper) such optimal trade-off might change over time due to business evolution and information collected from implementation details. Therefore, it's not correct to assume that the suboptimal solutions can be identified and managed from the beginning.

Architectural Technical Debt Theoretical Framework

Studies on TD are quite recent. Some models, empirical (Nugroho et al. 2011) or theoretical (Schmid 2013), have been proposed in order to map the metaphor to concrete entities in system and software development. Furthermore, a recent systematic literature study by, among others, the authors of this chapter has summarized the main scientific contribution on the architecture level (Besker et al. 2017, 2018). We present here a summary of the conceptual model comprehending the components of the TD framework, useful to understand the metaphor and its practical application.

Debt

The debt is regarded as the technical issue. Related to the ATD in particular, we consider the ATD item as a specific instance of the implementation that is suboptimal with respect to the intended architecture to fulfill the business goals. For example, a possible ATD item is a dependency between components that is not allowed by the architectural description or by the principles defined by the architects. Such dependency might be considered suboptimal with respect to the modularity quality attribute (ISO – International Organization for Standardization 2010), which, in turn, might be important for the business when a component needs to be replaced in order to allow the development of new features.

Principal

This is considered to be the cost for refactoring the specific TD item. In the example explained before, in which an architectural dependency violation is present in the implementation, the principal is the cost for reworking the source code in order to remove the dependency and make the components no longer being directly dependent on each other. The principal might include other costs related to the refactoring, for example the delay of delivering new features.

Interest

A suboptimal architectural solution (ATD) might cause several effects, which have an impact on the system, on the development process or even on the customer. For example, having a large number of dependencies between a large number of components might lead to a big testing effort (which might represent only a part of the whole interest in this case) due to the spread of changes. Such effect might be paid when the features delivered are delayed because of the extra time involved during continuous integration. In this paper, we treat accumulation and refactoring of ATD as including both the principal and the interest. The interest payment can vary but, in our well-documented studies (Besker et al. 2018), companies spend on average from 28 to 37% of development time in waste caused by Technical Debt interest.

The Time Perspective

The concept of TD is strongly related to time. Contrary to having absolute quality models, the TD theoretical framework instantiates a relationship between the cost and the impact of a single suboptimal solution. In particular, the metaphor stresses the short-term gain given by a suboptimal solution against the long-term one considered optimal. Time wise, the TD metaphor is considered useful for estimating if a technical solution is actually suboptimal or might be optimal from the business point of view. Such risk management practice is also very important in the everyday work of software architects, as mentioned in Kruchten (2008) and Martini et al. (2014). Although research has been done on how to take a decision on architecture development (such as Architecture Tradeoff and Analysis Method (ATAM), Architecture Level Modifiability Analysis (ALMA), etc. (Babar and Gorton 2004)), there is no empirical research about how suboptimal architectural solutions (ATD) are accumulated over time and how they can be continuously managed.

4.2 Methodology

We conducted a series of case studies. The studied phenomena are strongly rooted in a very complex industrial context. Case study research, defined as "investigating contemporary phenomena in their context" (Yin 2009) fits this purpose. The main design choices related to case study aim at "flexibility while maintaining rigor and relevance," as explained in the following and recommended in (Runeson and Höst 2008).

Case study research is a widely used approach in social science research (Yin 2009) but in the last years it has gained appreciation as a suitable tool for investigating system and software engineering topics (Runeson and Höst 2008), especially related to research concerning the study of complex systems including humans. In fact, with system and software engineering a highly cognitive-intense discipline, many research questions involve phenomena that are influenced by a wide number of sociotechnical factors. This is even more evident when considering phenomena that present themselves in large organizations involving a large number of employees with different backgrounds, expectations, and perceptions.

4.3 Case Study Companies

Our research results have been obtained in collaboration with several large companies developing embedded systems. Below we describe some of the companies that we have worked with (renamed A–F). In some of the studies, other companies were also involved, with similar background, which we don't report here. Additionally, we investigated different topics with different units within the companies below.

Company A carried out part of the development out by suppliers, some by in-house teams following Scrum. The surrounding organization follows a stage-gate release model for product development. Business is driven by products for mass customization. The specific unit studied provides a software platform for different products. The internal releases were short but needed to be aligned, for integration purposes, with the stage gate release model (several months).

In **company B**, teams work in parallel in projects; some of the projects are more hardware oriented while others are related to the implementation of features developed on top of a specific Linux distribution. The software involves in house development with the integration of a substantial amount of open source components. Despite the agile set up of the organization, the iterations are quite long (several months), but the company is in transition toward reducing the release time.

Customers of **Company C** receive a platform and pay to unlock new features. The organization is split in different units and then in cross-functional teams, most of which with feature development roles and some with focus on the platform by different products. Most of the teams use their preferred variant of ASD (often SCRUM). Features were developed on top of a reference architecture, and the main process consisted of a pre-study followed by few (c. 3) sprint iterations. The embedded cases studied slightly differed; C3 involved globally distributed teams, while the other units (C1 and C2) teams were mostly colocated.

Company D is a manufacturer of a product line of embedded devices. The organization is divided in teams working in parallel and using SCRUM. The organization has also adopted principles of software product line engineering, such as the employment of a reference architecture. Also, in this case, the hardware cycle has an influence.

Company E developed field equipment responsible for several large subsystems, developed in independent projects. Implementation of agile processes had been initiated, while team composition still followed formal and standard project processes due to legal responsibilities. Projects are characterized by a long-lasting product maintained for several years, by a strict formalization of the requirements and by the need of substantial testing effort.

Company F is a company developing software for calculating optimized solutions. The software is not deployed in embedded systems. The company has employed SCRUM with teams working in parallel. The product is structured in a platform entirely developed by F and a layer of customizable assets for the customers to configure. Company F supports also a set of APIs for allowing development on top of their software.

All the companies have adopted a component-based software architecture, where some components or even entire platforms are reused in different products. The language that is mainly used is C and C++, with some parts of the system developed in Java and Python. Some companies use a Domain Specific Language (DSL) to generate part of their source code.

All the companies have employed SCRUM, and have a (internal) release cycle based on the one recommended in SCRUM. However, the embedded companies (A–D) depend on the hardware release cycles, which influence the time for the final integration before the releases. Therefore, some of the teams have internal, short releases and external releases according to the overall product development.

4.4 Findings: Causes of ATD

Business Evolution Creates ATD

The amount of customizations and new features offered by the products brings new requirements to be satisfied. Whenever a decision is taken to develop a new feature or to create an offer for a new customer, instantaneously the desired architecture changes and ATD is automatically created. The number of configurations that the studied companies need to offer simultaneously seems to be growing steadily. If for each augmentation of the product some ATD is automatically accumulated when the decision is taken, the same trend of having more configurations over time implies that the corresponding ATD is also automatically accumulated faster.

Uncertainty of Use Cases in Early Stages

The previous point also suggests the difficulty in defining a design and architecture that has to take in consideration a lot of unknown upcoming variability.

Consequently, the accumulation of inconsistencies toward a "fuzzy" desired design/architecture is more likely to take place in the beginning of the development (for example, during the first sprints).

Time Pressure: Deadlines with Penalties

Constraints in the contracts with customers, such as heavy penalties for delayed deliveries, make the attention to manage ATD less of a priority. The approaching of a deadline with a high penalty causes both the accumulation of inconsistencies due to shortcuts and the low-prioritization of the necessary refactoring for keeping ATD low. The urgency given by the deadline increases with its approaching, which also increases the amount of inconsistencies accumulated.

Priority of Features over Product Architecture

The prioritization that takes place before the start of the feature development tends to be mainly feature oriented. Small refactorings necessary for the feature are carried out within the feature development by the team, but long-term refactorings, which are needed to develop "architectural features" for future development, are not considered necessary for the release. Moreover, broad refactorings are not likely to be completed in the time a feature is developed (e.g. few weeks). Consequently, the part of ATD that is not directly related to the development of the feature at hand is more likely to be postponed.

Split of Budget in Project Budget and Maintenance Budget Boosts the Accumulation of Debt

According to the informants, the responsibility associated only with the project budget during the development creates a psychological effect; the teams tend to accumulate ATD and to push it to those responsible for the maintenance after release, which rely on a different budget.

Design and Architecture Documentation: Lack of Specification/Emphasis on Critical Architectural Requirements

Some of the architectural requirements are not explicitly mentioned in the documentation. This causes the misinterpretation by the developers implementing code that is implicitly supposed to match such requirement. According to the informants, this also threatens the refactoring activity and its estimation; the refactoring of a portion of code for which requirements were not written (but the code was "just working," implicitly satisfying them) might cause the lack of such requirements satisfaction.

As an example, three cases have mentioned temporal-related properties of shared resources. A concrete instance of such a problem is a database, and the design constraint of making only synchronous calls to it from different modules. If such a requirement is not specified, it may happen that the developers will ignore such a constraint. In one example made by the informants, the constraint was violated in order to meet a performance requirement important for the customer. This is also connected with the previous point.

Reuse of Legacy/Third Party/Open Source

Software that was not included when the initial desired architecture was developed contains ATD that needs to be fixed and/or dealt with. Examples included open source systems, third party software, and software previously developed and reused. In the former two cases, the inconsistencies between the in-house developed architecture and the external one(s) might pop up after the evolution of the external software.

Parallel Development

Development teams working in parallel automatically accumulate some differences in their design and architecture. The agile-related empowerment of the teams in terms of design seems to amplify this phenomenon. An example of such phenomenon mentioned as causing efforts by the informants is the naming policy. A name policy is not always explicitly and formally expressed, which allows the teams to diverge or interpret the constraint. Another example is the presence of different patterns for the same solution, e.g. for the communication between two different components. When a team needs to work on something developed by another team, this nonuniformity causes extra time.

Uncertainty of Impact

ATD is not necessarily something limited to a well-defined area of the software. Changing part of the software in order to improve some design or architecture issues might cause ripple effects on other parts of the software depending on the changed code. Isolating ATD items to be refactored is difficult, and especially calculating all the possible effects is a challenge. Part of the problem is the lack of awareness about the dependencies that connect some ATD to other parts of the software. Consequently, there exists some ATD that remains unknown.

Noncompleted Refactoring

When refactoring is decided, it's aimed at eliminating ATD. However, if the refactoring goal is not completed, this not only will leave part of the ATD but it will actually create new ATD. The concept might be counterintuitive, so we will explain with an example. A possible refactoring objective might be to have a new API for a component. However, what might happen is that the new API is added but the previous one cannot be removed, for example because of unforeseen backward compatibility with another version of the product. This factor is related to other two; time pressure might be the actual cause for this phenomenon, when the planned refactoring needs to be rushed due to deadlines with penalties and the effects uncertainty, which causes a planned refactoring to take more time than estimated because of effects that have been overlooked when the refactoring was prioritized.

Technology Evolution

The technology employed for the software system might become obsolete over time, both for pure software (e.g. new versions of the programming language) and for hardware that needs to be replaced together with the specific software that needs to run on it. The (re)use of legacy components, third party software and open source systems might require the employment of a new or old technology that is not optimal with the rest of the system.

Lack of Knowledge

Software engineering is also an individual activity and the causes for ATD accumulation can be related to suboptimal decision taken by individual employees. New employees as well are more subjected to accumulating ATD due to the natural noncomplete understanding of the architecture and patterns.

4.5 Problem Definition: Entry Point to Applied Systems Thinking

In the following the costly consequences (interest) of specific classes of ATD are described. The results are summarized in Figure 4.2, which is an adaptation from Martini and Bosch (2017a).

ATD1. Suboptimal Reuse

Reuse is a good strategy to reduce costs and to increase speed. There are two main problems related to reuse, which can be considered ATD:

- P1. The lack of reuse (when such strategy is possible) represents ATD. The presence of very similar code (if not identical) in different parts of the system, which is managed separately and is not grouped into a reused component,

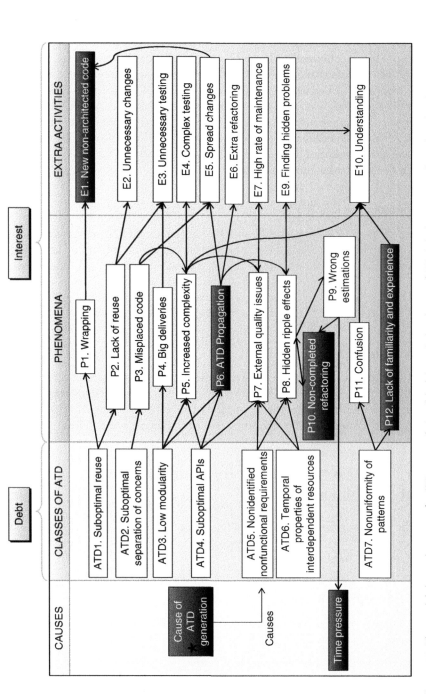

Figure 4.2 The model shows the causes for ATD accumulation (black boxes), the classes of ATD (which represent the Debt), the phenomena caused by the items, and the final activities (which together represent the interest to be paid).

leads to paying interest in duplicated activities. Such duplicated activities lead to unnecessary changes (E2) and unnecessary testing (E3).

- P2. Another phenomenon related to reuse is called "wrapping"; this phenomenon is also called "glue code." Such code is needed to adapt the reused component to the new context with new requirements. Such code is usually unstructured and suboptimal, because it is not part of the architectural design but rather developed as a workaround to exploit reuse (E1). The consequences are not direct, but the risk is that, with the continuous evolution of the system around the reused component, such glue code would grow, becoming a substantial part of the (sub)system containing nonarchitected code and therefore ATD (of different kinds). This is why the box is black, representing the fact that the consequence of wrapping can be considered the cause for new forms of ATD.

It is important to notice that some of the developers from different companies mentioned the fact that the lack of reuse (citing interviewees, "copy-paste") is not always considered a bad solution but, in some cases, it can be more convenient to copy-paste and evolve the new code. In such cases, though, it seems better, in order to avoid the wrapping phenomenon, to re-architect the component together with the new code in order to avoid propagation of ATD.

ATD2. Suboptimal Separation of Concerns

In this case, the code is not well allocated in the components composing the system. A typical example is a layered architecture where a generic business logic layer has nonallowed dependencies (e.g. functions calling) to the user interface layer. Such code is then misplaced (P3), which causes the changes to be spread in the system (E5), requiring a larger and riskier change applied by the developers. Also, such ATD causes an increment of the complexity (P5). Increased complexity leads to several extra-activities; testing the modules and their behavior in a complex architecture is also complex (E4), and a more complex system requires developers to spend more time in understanding it. Also, we found that complexity increases the error-proneness of the system, which leads to external quality issues (P7).

ATD3. Low Modularity

This category includes the items that are related to the presence of many architectural dependencies among the components. The components can be said to be not loosely coupled or cohesive enough. In one of the cases, the interviewees mentioned a concrete example of the phenomenon: the large amount of dependencies among the components in one of the subsystems caused, each time a new release involved a small change, the test of the whole subsystem

(Big deliveries or P4 in Figure 4.2). Such an event hinders agile practices such as continuous integration, in which the high modularity of the system allows the fast test of small portions of the code (for example a single or a small set of components). In the picture, we called this E3 – unnecessary testing. Also, having many dependencies among the components increases complexity (P5) and increases the chances of existing TD spreading (P6).

ATD4. Suboptimal APIs

The APIs (interfaces) of a component are the point of access among different parts of the system or different systems. The suboptimal design of APIs or the misuse of them is considered ATD. As an example of the first class we can consider methods that contain too many parameters or that are difficult to use because they return too generic or unstructured values, which need to be interpreted and might create confusion. As an instance of the second kind, components might call private APIs (that are not supposed to be called outside the component) instead of public and well-designed ones; this, in practice, makes public private and not well-designed APIs.

The effects of these kinds of problems might lead to increased complexity (P5) and, since they are the access point among components, are the possible gateway for the spread of ATD (P6). In fact, the more components access an API, the more likely it is that the suboptimality would be propagated to the code using suboptimal code, for example workarounds. Also, once the API is used by many components, it becomes very costly to refactor the components in case the API changes (which means the cost of refactoring would grow, E6). Finally, suboptimal APIs are considered more bug-prone (P7, External quality issues) by the developers, leading to extra maintenance work (E7).

ATD5. Nonidentified Nonfunctional Requirements

Some nonfunctional requirements (NFRs), such as performance, scalability, and signal reliability, need to be recognized before or early during the development and need to be tested. The ATD items represent the lack of an implementation that would assure the satisfaction of such requirements and also the lack of mechanisms for them to be tested. Some cases were mentioned by the informants; for example, one case mentioned the lack of a fault handling mechanism, which turned out to be very expensive when added afterwards. Another case reported frequent struggles with nonfunctional requirements regarding memory consumption and communication bus allocation. Yet another company mentioned the difficulties in anticipating the future use cases for a given feature.

The introduction of such debt causes a number of quality issues (P7), which are difficult to test. The informants argue that it was difficult to repair this kind of ATD afterwards, especially for requirements orthogonal to the architectural structure, when the changes would affect a large part of the system. In such case, it was difficult for the developers to know all the affected parts in advance (P8, Hidden ripple effects). Therefore, quantifying the change and estimating the cost of refactoring has always been reported as a challenge (P9, Wrong estimation of effort). Consequently, teams experience additional time pressure when the estimations are wrong. Also, the hidden ripple effects might lead to developers not recognizing parts that cannot be refactored and, therefore, to the dangerous presence of noncompletely refactored code (P10). Noncompleted refactoring, in turn, leads to ATD that is not known.

ATD6. Temporal Properties of Interdependent Resources

Some resources might need to be accessed by different parts of the system. In these cases, the concurrent and nondeterministic interaction with the resource by different components might create hidden and unforeseen issues. This aspect is especially related to the temporal dimension; as a concrete example from one of the companies, we mention the convention of having only synchronous calls to a certain component. However, one of the teams used (forbidden) asynchronous calls (which represents the ATD).

Having such a violation brings several effects; it creates a number of quality issues (P7) directly experienced by the customer, which triggers a high number of bugs to be fixed (and, therefore, time subtracted to the development of the product for new business value, E7). However, the worse danger of this problem is related to its temporal nature. Being a behavioral issue, it is difficult, in practice, to test for it with static analysis tools. Also, once introduced and creating a number of issues, it might be very difficult for the developers to find it or understanding that it's the source of the problems. A developer mentioned a case in which an ATD item of this kind remained completely hidden until the use case changed slightly during the development of a new feature. The team interacting with such an ATD item spent quite some time figuring out issues rising with such suboptimal solution. The hidden nature of these ATD and their ripple effects (P8) create a number of other connected effects.

ATD7. Nonuniformity of Patterns

This category comprehends patterns and policies that are not kept consistent throughout the system. For example, different name conventions applied in different parts of the system. Another example is the presence of different design or architectural patterns used to implement the same functionality. As a concrete example, different components (ECUs in the automotive domain) communicated through different patterns.

The effects caused by the presence of nonuniform policies and patterns are of two kinds: the time spent by the developers in understanding parts of the system that they are not familiar with (P11) and in understanding (E10) which pattern to use in similar situations (it causes confusion, P11). For example, the developers experienced difficulties in choosing a pattern when implementing new communication links among the components, since they had different examples in the code.

Contagious Debt and Vicious Circles

We have previously described the connection of classes of ATD items with phenomena that might be considered dangerous for their costs when they occur during the development. Such cost represents the interest of the debt and the previously explained categories of debt have been associated with a high interest to be paid. Such association can be considered as fixed, which means that each item brings a constant cost. However, our analysis of the relationships among the different phenomena has brought to light patterns of events that are particularly dangerous, because they create loops of causes–effects that lead to linear and potentially nonlinear accumulation of interest; they are vicious circles.

Such vicious circles are well visible in our model in Figure 4.2. the column on the left includes the possible causes of ATD accumulation (explained earlier), which we have represented with black boxes (Cause of ATD generation). Among the phenomena triggered by some classes of ATD, we can find also causes of ATD (also represented by black boxes). This means that when a path starts with a cause (all ATD items have a cause), pass through some ATD items and ends in a phenomenon that is also a black box, such a black box also causes the creation of additional ATD. This loop represents a vicious circle.

The implications of vicious circles are important, since their presence implies the constant increment of the ATD items and, therefore, of their effects over time. Which means that for each moment that passes with the ATD items involved in the vicious circles remaining in the system, the interest increases. Such phenomenon causes the ATD to become very expensive to be removed afterwards, together with the effects that hinder the development speed; as shown for each cycle of the vicious circle, the cost might remain constant (for example the principal of fixing the ATD item), linear but with low increment over time (low interest), linear but with a high steepness (linear interest) or it might even reach nonlinearity (nonlinear interest). In the end, such accumulation might lead to a crisis, as explained later on. It's important to notice that the worst case might not be consist of nonlinearity; if the linear accumulation is steep enough, the crisis might happen earlier than in the nonlinear case.

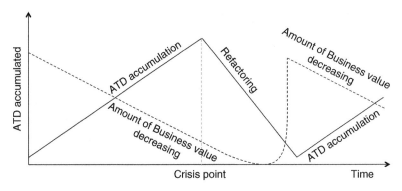

Figure 4.3 Crisis point model.

4.6 Findings: Long-Term Implications of ATD Accumulation

As presented in the previous sections, the presence of costly growing interest (impact) of ATD is due to several causes. As time passes, more suboptimalities are present in the system as it grows, and vicious circles boost this accumulation. Consequently, we found that the aggregation of these several factors leads the system being quite difficult to manage, both for maintenance and also for releasing new features. In other words, when the ATD and its interest are too much in the system, it becomes very costly to perform maintenance tasks and to change the system in order to add new features. When this happens, we say that there is a crisis, as described in Figure 4.3. In such a case, the only solution is a big refactoring or a redesign of the system. In both cases, it is a quite expensive operation, which might stop the companies delivering value for significant periods of time and, therefore, limiting their revenues.

4.7 Solutions for ATD Management

In order to avoid the accumulation of ATD, it is necessary to apply several strategies and to cover different aspects, from the technical domain to the organizational structure. List here are three approaches that are needed in practice and that have preliminarily been employed at the studied companies. The evaluation of such approaches is ongoing and we are collecting very positive results on the benefits of such approaches. However, a complete description of the approaches and the evaluation of data obtained for space constraints are not reported here.

Architectural Retrospectives to Manage ATD

The dangerous classes, interest activities, vicious circles, and the factors for the increment of the interest that we propose here can be useful to support iterative architectural retrospective. In such sessions, the models here serve as a guideline for practitioners who might recognize the presence of dangerous classes of ATD items and who have to verify the phenomena. The cases show how some ATD items belonging to some classes (for example the contagious one) usually creep in the system because the interest is not evaluated continuously. When, finally, the interest becomes visible as a software development crisis, it's usually too late to refactor the ATD item.

Organizational Structure to Connect System Architects with Product Owners and Development Teams

We developed an organizational framework called CAFFEA, including roles, teams, and practices. CAFFEA has been statically validated by 40 employees from seven different sites and has also been introduced in four large companies. Such a framework is better described elsewhere (Martini et al. 2015). It represents a set of practices, architecture roles, and virtual teams (including the involved stakeholders) that are needed in order to manage ATD. The main focus of the framework is to combine ASD with system architecture practices. See Dingsoyr et al. 2012; Eklund and Bosch 2012 and Eklund et al. 2014.

Tool to Track and Estimate ATD

Figure 4.4 shows how the stakeholders involved in the ATD management can synchronize. The information about quality and the consequent risk in terms of threats for the achieving of long-term business goals is assessed in the technical solutions developed by the development teams. As suggested in previous research (Guo and Seaman 2011), an effective way of visualizing and utilizing the TD information is to create a Portfolio (or Backlog) with TD items. We use the same idea: including the information about ATD in an Architectural Technical Debt Map. Such a map is then shared between product owners, architects and development teams. The information thus contributes to alignment among the stakeholders. They need to take into consideration the risk of unbalancing the development in favor to short-term delivery by neglecting qualities necessary for achieving long-term business goals (Figure 4.4).

4.8 Solution: A Systematic Technical Debt Map

Building a map of technical debt is not trivial. It requires (i) automated tools and (ii) manual tracking and estimations in a portfolio fashion. In fact, not all

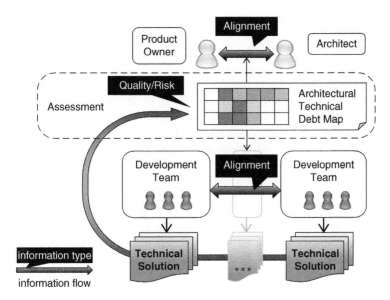

Figure 4.4 Need for an Architectural Technical Debt map in order to assess the quality and risk related to the current implementation of the system (technical solution) developed by the teams.

the architectural problems can be automatically quantified. In addition, future costs and impacts are even more difficult to calculate automatically. In the next section, we present a quantitative approach to measure and visualize architectural debt, while here we discuss a method and tool, whose usage is being evaluated at some companies, that have recently been developed by Antonio Martini (one author of this chapter).

A portfolio approach to tracking TD is a well-known practice, defining how items are ranked in a backlog. We have carried out a study with 15 organizations to understand what tools have been used; backlogs, together with static code analyzers (described in a section later), were found to be the most effective ways of track TD (Martini et al. 2018a). Therefore, we have actively researched and developed solutions to build the technical debt map. First, we have empirically developed a method, called AnaConDebt (Analysis of Contagious Debt) (Martini and Bosch 2016), together with several industrial partners, to analyze the interest of large ATD items. The method was found helpful by the practitioners to estimate the impact of ATD.

The Evolution of AnaConDebt into a Tool

Having developed the methodology, we developed a tool of the same name, AnaConDebt, as it builds on top of such work. However, it presents many differences from the original method, which has been considerably evolved and refined over time according to new experiences gathered from practice. This was presented at the Technical Debt Conference in 2018 (Martini 2018).

In a first attempt to create a prototype tool, the original method was modified and simplified to allow the efficient assessment of several kinds of TD items. The prototype was then evaluated and the positive results were published in a study at the Ninth Workshop on Technical Debt (Martini and Bosch 2017b). This tool allowed the practitioners to assess the negative impact of TD by systematically assessing several aspects, including the future growth of interest. This prototype was then used multiple times during consultancy assignments with large software companies, a step that allowed the author to evaluate and refine the tool further, understanding better the requirements and the useful features.

After the second evaluation, the prototype was found useful by the users; however, the tool needed to be re-implemented to be available as a usable tool. For example, requirements such as portability, security, and usability needed to be taken into account. The author then received funds from the Swedish innovation agency Vinnova, in collaboration with Chalmers Innovation Office, to refactor the tool to make it practically usable and ready for commercialization.

AnaConDebt

AnaConDebt is a management tool that consists of a TD-enhanced backlog. The backlog allows the creation of TD items and TD-specific operations that are currently not available in other existing tools to be performed on the items.

AnaConDebt supports the users in the following activities:

- Tracking technical debt items in a dedicated repository. The TD items can be further characterized with attributes and properties such as name, description, etc.
- Assessing technical debt principal (cost of refactoring) and interest (current and future extra costs) systematically, using an approach that was previously evaluated, both scientifically and in practice. The values are input in a simple and intuitive way, using advanced widgets. Additionally, the values and how they are used are explained. This is important, as stakeholders find many tools difficult to use.
- Estimating growth of principal and interest with respect to future scenarios; for example, in the short, medium and long term. The time related to the scenarios is customizable.
- Assessing and aggregating the amount of total technical debt, principal, and interest for the whole backlog of technical debt items. This is useful for practitioners to have an overview of their TD.
- Reviewing and assessing previous TD estimations. In fact, it is important to show if the expected principal and interest of technical debt were correctly estimated at the initial time of estimation.

- Comparing and ranking technical debt items based on the convenience of refactoring, calculated by weighting different parameters related to principal and interest. Such parameters are customizable according to the company's context. For example, for some companies the same cost might be more or less important with respect to other factors.
- Visualizing and comparing technical debt items in a cost/benefits graph.
- Locating the technical debt items in the system. For example, linking a technical debt item with a specific file or component.
- Visualizing the areas of the system with more technical debt items and with more interest to pay (more convenient to refactor). This can be useful to group tasks related to the same part of the system.
- Relating technical debt items among each other by defining dependencies; for example, it might be important to specify that one TD item should be refactored before another. Eventually, this feature would help grouping and planning the refactoring of technical debt items.
- Thanks to the previous point, calculating an aggregated principal and interest based on the dependencies among the technical debt items.
- Presenting an easy-to-understand report on the convenience of refactoring technical debt items to stakeholders that are not technical (e.g. project managers or product owners), to allow an assessment with respect to other project or product aspects.
- Integrating the repository and the TD-specific operation to other tools via a REST API on top of a portable web service. In particular, the architecture of the tool separates the GUI from a web service. This way, the data related to TD can be used in other tools and visualizations; for example, the aggregated data related to the TD items can be added in a dashboard containing other project- and product data that needs to be compared with the current information on TD.

Main Factors for Calculating the Interest

The tool is based on the assessment of the following main factors:

- **Reduced development speed.** If the speed is reduced, the interest of the ATD hinders the ability of the system to evolve and be maintained.
- **Bugs related to the TD item.** If many bugs are generated from the ATD, this greatly affects the maintainability of the system and the time wasted to fix bugs instead of developing new features.
- **Other qualities compromised.** There are several qualities that can be affected by ATD. Here we used the ones suggested in the ISO standard (ISO – International Organization for Standardization 2010) but excluding maintainability and evolvability (as they are already covered by the previous factors).

- **User affected** ATD might not involve a large part of the system, but can still affect a large number of developers. In this case, the extra cost of the interest would be multiplied by all the "victims."
- **Frequency of the issue.** The more frequent the negative impact occurs, the worse the interest.
- **Future growth of interest.** To understand the overall negative impact of TD, it is important to assess the current negative impact, but we need to understand its future growth as well.
- **Other extra costs**. There might be other context-dependent extra costs to be considered when assessing the interest.
- **Spread in the system.** The larger the portion of the system affected by the TD, the more ripple effects the interest might have on the organization and on the newly added code.

4.9 Solution: Using Automated Architectural Smells Tools for the Architectural Technical Debt Map

Quantitative evaluation of software architecture can be difficult. However, recently, automated tools are gaining interest and companies are trying to use them in their development processes. One example is Arcan, a tool that has been developed at the university of Milan Bicocca (Fontana et al. 2017). The tool helps in automatically identifying antipatterns in the dependency graph of different classes and modules by computing and visualizing different metrics. Three examples of antipatterns that have been studied are:

- **Unstable Dependency (UD)**: describes a subsystem (component) that depends on other subsystem that is less stable than itself, with a possible ripple effect of changes in the project.
- **Hub-Like Dependency (HL)**: occurs when an abstraction has (outgoing and ingoing) dependencies with a large number of other abstractions.
- **Cyclic Dependency (CD)**: refers to a subsystem (component) that is involved in a chain of relations that breaks the desirable acyclic nature of a subsystems dependency structure. The subsystems involved in a dependency cycle can be hardly released, maintained or reused in isolation.

In a recent study we evaluated such a tool in practice, to understand how developers can use an automatic approach to identify ATD. We looked into four large projects in a large company developing embedded systems for the automotive domain. We ran the automatic tool on the codebases and discussed the outcome of such antipatterns with the developers of such projects. We analyzed the code related to the suggested smell together with the developers and they estimated a cost and impact to fix such smell.

The following results can be summarized from the experience:

| Name | Description | Metrics | | Architectural Smells | | | |
| | | | | | Class | Package | |
		NoP	NoC	UD	HL	CD	CD	AS
A	Product Data Management (PDM)	269	10171	476	1	199	31	707
B	After market	240	7261	98	0	7	6	111
C	Audit project	220	3250	34	1	31	8	74
D	Warehouse management project	166	3067	53	0	49	7	109

Legenda: NoP: Number of Packages, NoC: Number of Classes, Unstable Dependency (UD), Hub-Like Dependency (HL), Cyclic Dependency (CD), Total Architectural smells (AS)

Figure 4.5 Projects analyzed, number of packages, number of classes, and number of different architectural smells.

- Approximately **50% of the smells were found to be problematic**.
- Approximately 50% of these (**25%** of the total evaluated) **were not known** by the developers, which increased their awareness of the problems.
- Unstable dependencies were **not recognized as very impactful**.
- Cyclic dependencies seem to be the **most dangerous** in terms of bugs but some of them are false positives, depending on the kind of software analyzed. For example, such antipatterns in GUI software are inevitable and shouldn't be regarded as problems.
- Hub-like dependencies are rare but they should be refactored, and their **impact tends to grow** together with the growth of software.
- Dependencies at the package level were very difficult to understand for developers. It would require quite a long investigation to understand what to refactor.
- **Developers liked to use such a tool** to be informed about possible architectural problems and they would use it during development if such tool was to be usable

In summary, a quantitative approach is feasible, but the available tools need some polishing and need some effort to be integrated into the development process (Figure 4.5).

4.10 Solution: Can We Calculate if it is Convenient to Refactor Architectural Technical Debt?

Decisions on design and refactoring of the system architecture with respect to short-term and long-term goals might be hard to take. However, a quantitative approach can aid when difficult decisions have to be made. A quantitative approach, although prone to estimation bias, can give a good rough idea if an approach can work or not. In a recent study (Martini et al. 2018b), we studied the difference in maintenance and development effort in a component and its

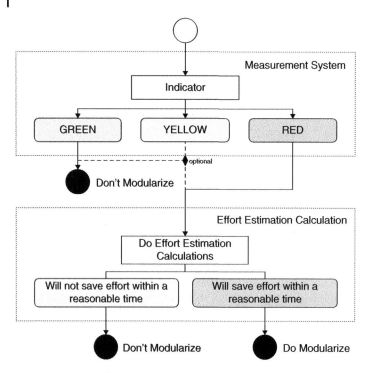

Figure 4.6 Process flow model to decide on refactoring initiatives.

refactored counterpart. In particular, the second component was refactored to factor-out a specific functionality to achieve better modularization.

The overall semi-automated method to calculate TD and interest followed the architecture outlined in (Figure 4.6):

The approach combines an automatic measurement system and the reported effort estimation calculated using a formula including all the important variables. This allows the stakeholders to estimate the need for modularizing a component and provides a quantification of how much effort can be saved per month. In particular, the measurement system was useful for identifying the ATD that needs to be refactored, while the effort estimation points out the business value of doing such a refactoring. Their combination aids the refactoring decision.

Measurement System

The full measurement system is described elsewhere (Martini et al. 2018b). It would be too extensive to report the whole system here, so just a few characteristics of what was calculated are mentioned here.

Measuring the Debt

ATD was considered the presence of complex files dependent both from the functionality and the component, which were often changed. It is important to note that the ATD was not considered the code complexity itself, but the lack of a modular structure that would encapsulate it. To choose a complexity measure for the selected files, we interviewed developers involved in the refactoring of the component. The mentioned goal regarding complexity was that the stakeholders wanted to reduce the number of "if" and "for" statements in the code that was modularized in the refactoring. A complexity measure that calculates the complexity with regards to this is the McCabe Cyclomatic Complexity (MCC). We complemented the MCC with Halstead's delivered bugs metric.

Measuring the Interest

The extra effort paid as interest of the ATD is related to the complexity; as mentioned by the stakeholders (and supported by various references), it takes more effort to change more complex code. However, if the complex code is not changed, it is less likely that repaying the debt will give a return in the form of avoided effort, making the refactoring not worth its cost. This is in line with the TD concept of refactoring the debt that generates more interest. For this reason, those complex files that were also more commonly changed were weighted higher, since they had debt but they also generated interest. To quantify the changes, we measured the committed Lines of Code (LOC), a simple measure that is widely used in the literature.

Evaluation of Convenience of Refactoring

The key effort factors for modeling the calculation were selected to calculate the convenience of refactoring. The first two are related to the effort spent in the development and maintenance before the refactoring: the old development effort (E_1) and the old maintenance effort (E_2). Similarly, it was important to estimate the development and maintenance after the refactoring: the new development effort (E_4) and the new maintenance effort (E_5). Finally, we need to take into consideration the refactoring effort (E_3), which includes the effort spent on refactoring.

These factors were inserted into Eq. (4.1). All the factors, excluding E_3, are related to a specific time t (for example, expressed in months). It was necessary to specify such a variable as we needed to calculate the effort (the interest of the ATD) over a specific time span. The time span had to be the same in order to be able to compare the quantities (see the following sections). The difference between the old development effort E_1 and the new development effort E_4 defines the development effort gain from the refactoring. The difference between the old maintenance effort E_2 and new maintenance effort E_5

defines the maintenance effort gain from a refactoring. These gains needed to be balanced considering the refactoring effort E_3.

In addition, the data used in the two components needed to be normalized; for example, the older component was more complex than the new one, as more applications were using the old component. We used a simple measure of complexity with respect to the number of objects handled by the two components, which was suggested by the architects.

Overall formula to calculate the effort gain:

$$E = (E_1 - E_4)^*t + (E_2 - E_5)^*t - E_3 \tag{4.1}$$

Results from the Application of the Approach and Evaluation

With the evaluation of the approach we learned that:

- The McCabe complexity measured in the two solutions differed greatly, with the refactored component encapsulating complex files. This gave a quantification that the technical solution was sound.
- We surveyed the developers, who mentioned that the refactoring was successful in terms of complexity and effort saved.
- According to the final calculations, the refactoring would see a gain in development effort after 23 months. The total effort on the refactoring, including the double effort of maintaining two solutions for a period of time, was of 53.34 man-month. After 23 months the refactoring, the developers of would be able to save 2.279 man-month per month. The 2.279 man-month was also the interest that was payed each month due to the ATD.

In conclusion, quantifying ATD was possible and useful for decision making. However, it took some effort to setup the measurement system and to collect the relevant data, so such cost should be calculated when an exercise such as this is made by companies. However, when the stakes are high, for example when a re-architecting decision is to be made, the additional information can be worth the time.

4.11 Summary

Industry 4.0 takes computers and automation to the next level of cyber-physical systems. Different from earlier generations, where system design was dictated by the mechanical parts, an Industry 4.0 system design is dictated by digital technologies, including software, data, and artificial intelligence. One key factor in this context is that systems will be subject to continuous evolution throughout their economic lives, meaning that software gets deployed continuously (at least once per agile sprint) and also that the electronics as well as some of the mechanical parts of systems are likely to be upgraded multiple times during the lifetime of the system.

As we are entering an era of continuous evolution, the concept of technical debt, both at the system level and at the software level, becomes increasingly important, as it will accumulate for each iteration unless it is explicitly managed. Our experience shows that, especially in the context of Industry 4.0, of all the types of technical debt, architectural technical debt is the most important to manage in a structured, continuous and systematic manner.

Consequently, decisions concerning the design and refactoring of the system architecture with respect to short-term and long-term goals need to be balanced and need to rely on the knowledge of the underlying phenomenon of ATD. The current management of ATD is an ongoing topic and empirical evidence shows how large companies are struggling with this phenomenon. In this chapter, we have summarized the causes, trends, consequences, and possible solutions related to the management of ATD. The results are all based on a strong collaboration with several software-intensive systems industries located in northern Europe and, consequently, relevant to all Industry 4.0 companies.

References

Babar, M.A. and Gorton, I. (2004). Comparison of scenario-based software architecture evaluation methods. In: *Proceedings of the 11th Asia-Pacific Software Engineering Conference*, , Busan, South Korea (30 November– 3 December), 600–607. IEEE https://doi.org/10.1109/APSEC.2004.38.

Besker, T., Martini, A., Bosch, J. (2017) The Pricey Bill of Technical Debt – When and by whom will it be paid?. 2017 IEEE International Conference on Software Maintenance and Evolution (ICSME) , Shanghai, China.

Besker, T., Martini, A., and Bosch, J. (2018). Technical debt cripples software developer productivity: a longitudinal study on developers' daily software development work. In: *The 2018 IEEE/ACM International Conference on Technical Debt (TechDebt)*, 105–114.

Dingsøyr, T., Nerur, S., Balijepally, V., and Moe, N.B. (2012). A decade of agile methodologies: towards explaining agile software development. *Journal of Systems and Software* 85: 1213–1221. https://doi.org/10.1016/j.jss.2012.02.033.

Eklund, U. and Bosch, J. (2012). Applying agile development in mass-produced embedded systems. In: *Agile Processes in Software Engineering and Extreme Programming*, 31–46. Springer.

Eklund, U., Olsson, H.H., and Strøm, N.J. (2014). Industrial challenges of scaling agile in mass-produced embedded systems. In: *Agile Methods. Large-Scale Development, Refactoring, Testing, and Estimation*, Lecture Notes in Business Information Processing (eds. T. Dingsøyr, N.B. Moe, R. Tonelli, et al.), 30–42. Springer International Publishing.

Fontana, F.A., Pigazzini, I., Roveda, R. et al. (2017). Arcan: a tool for architectural smells detection. In: *The 2017 IEEE International Conference on Software Architecture Workshops (ICSAW)*, 282–285. https://doi.org/10.1109/ICSAW .2017.16.

Guo, Y. and Seaman, C. (2011). A portfolio approach to technical debt management. In: *Proceedings of the 2nd Workshop on Managing Technical Debt (MTD'11)*, 31–34. New York, NY: ACM https://doi.org/10.1145/1985362 .1985370.

ISO (International Organization for Standardization), 2010 System and software quality models http://www.iso.org/iso/catalogue_detail.htm?csnumber=52075 (accessed 3.8.15).

Kruchten, P. (2008). What do software architects really do? *Journal of Systems and Software* 81: 2413–2416. https://doi.org/10.1016/j.jss.2008.08.025.

Kruchten, P., Nord, R.L., and Ozkaya, I. (2012). Technical debt: from metaphor to theory and practice. *IEEE Software* 29: 18–21. https://doi.org/10.1109/MS.2012 .167.

Li, Z., Avgeriou, P., and Liang, P. (2015). A systematic mapping study on technical debt and its management. *Journal of Systems and Software* 101: 193–220. https://doi.org/10.1016/j.jss.2014.12.027.

Martini, A. (2018). AnaConDebt: a tool to assess and track technical debt. In: *Proceedings of the 2018 International Conference on Technical Debt (TechDebt'18)*, 55–56. New York, NY: ACM https://doi.org/10.1145/3194164 .3194185.

Martini, A. and Bosch, J. (2016). An Empirically Developed Method to Aid Decisions on Architectural Technical Debt Refactoring: AnaConDebt. In: *ICSE'16 Proceedings of the 38th International Conference on Software Engineering*, 31–40. ACM Press https://doi.org/10.1145/2889160.2889224.

Martini, A. and Bosch, J. (2017a). On the interest of architectural technical debt: uncovering the contagious debt phenomenon. *Journal of Software Evolution and Process* 29 https://doi.org/10.1002/smr.1877.

Martini, A. and Bosch, J. (2017b). The magnificent seven: towards a systematic estimation of technical debt interest. In: *Proceedings of the XP2017 Scientific Workshops (XP'17)*, 7:1–7:5. New York, NY: ACM https://doi.org/10.1145/ 3120459.3120467.

Martini, A., Pareto, L., and Bosch, J. (2014). Role of architects in agile organizations. In: *Continuous Software Engineering* (ed. J. Bosch), 39–50. Springer International Publishing.

Martini, A., Pareto, L., and Bosch, J. (2015). Towards introducing agile architecting in large companies: the CAFFEA framework. In: *Agile Processes, in Software Engineering, and Extreme Programming*, Lecture Notes in Business Information Processing (eds. C. Lassenius, T. Dingsøyr and M. Paasivaara), 218–223. Springer International Publishing.

Martini, A., Besker, T., and Bosch, J. (2018a). Technical debt tracking: current state of practice: a survey and multiple case study in 15 large organizations. *Science of Computer Programming* 163: 42–61. https://doi.org/10.1016/j.scico .2018.03.007.

Martini, A., Sikander, E., and Madlani, N. (2018b). A semi-automated framework for the identification and estimation of architectural technical debt: a comparative case-study on the modularization of a software component. *Information and Software Technology* 93: 264–279. https://doi.org/10.1016/j .infsof.2017.08.005.

Nugroho, A., Visser, J., and Kuipers, T. (2011). An empirical model of technical debt and interest. In: *Proceedings of the 2nd Workshop on Managing Technical Debt (MTD'11)*, 1–8. New York, NY: ACM https://doi.org/10.1145/1985362 .1985364.

Runeson, P. and Höst, M. (2008). Guidelines for conducting and reporting case study research in software engineering. *Empirical Software Engineering* 14: 131–164. https://doi.org/10.1007/s10664-008-9102-8.

Schmid, K. (2013). A formal approach to technical debt decision making. In: *Proceedings of the 9th International ACM Sigsoft Conference on Quality of Software Architectures*, 153–162. ACM.

Tom, E., Aurum, A., and Vidgen, R. (2013). An exploration of technical debt. *Journal of Systems and Software* 86: 1498–1516. https://doi.org/10.1016/j.jss .2012.12.052.

Yin, R.K. (2009). *Case Study Research: Design and Methods*. SAGE.

5

Relay Race: The Shared Challenge of Systems and Software Engineering

Amir Tomer

Synopsis

As systems become more and more software intensive, the collaboration between systems engineers and software engineers plays a critical role in system development. While many standards (e.g. CMMI, ISO 15288) address this inevitable pairing, they still make distinctions between the two disciplines, with less focus on how to make them work together continuously throughout the system's life cycle. The sensitive points are the interfaces. There are specific responsibilities that are allocated to each party, but the area of shared responsibilities becomes wider. Interdisciplinary interfaces cannot anymore be implemented as "throwing over the fence" but should rather be perceived as a relay race, where the baton is handed over from one actor to the other through perfect coordination, while both are running very fast. Needless to say that winning the race is a shared challenge, and therefore a baton drop is a shred failure. In this chapter we analyze the areas of responsibility of the two disciplines and identify those requiring shared effort. Then we propose a detailed process, mainly based upon Functional Analysis, through which a software-intensive system architecture is constructed, covering both structural and behavioral characteristics.

5.1   Introduction

As systems become more and more software intensive, the cooperation between systems engineers and software engineers plays a critical role in system development. The interrelationships and joint effort of the two disciplines have been the concern of many researchers, practitioners, and standards, such as capability maturity model integration (CMMI) (Chrissis et al. 2004), the International Organization for Standardization (ISO)/International

Systems Engineering in the Fourth Industrial Revolution: Big Data, Novel Technologies, and Modern Systems Engineering,
First Edition. Edited by Ron S. Kenett, Robert S. Swarz, and Avigdor Zonnenshain.

Electrotechnical Commission(IEC)/Institute of Electrical and Electronics Engineers (IEEE) standard 15288 (ISO 2015) the systems engineering body of knowledge (SEBOK) (INCOSE 2013), and many more. However, the sensitive points are, as always, the interfaces. There are specific responsibilities that are allocated to each party, but the area of shared responsibilities becomes wider. In that area, interdisciplinary interfaces cannot be implemented anymore by "throwing the ball over the fence" but should rather be perceived as a Relay Race, where the baton is handed over from one discipline to the other in perfect coordination, while both parties are running very fast. Needless to say, winning the race is a shared challenge, and therefore a baton drop is a shared failure. This feature is of special importance in the development of Industry 4.0 systems that combine a range of application domains, such as mechanical engineering, optics, robotics, and sensor technologies.

In this chapter, we highlight the common challenge of systems and software engineering in two major aspects – the life cycle aspect and the system decomposition aspect – and show that the main "friction zone" is building the system/software architecture. This is where the baton must be transferred smoothly in order to ensure the common goal of delivering a successful system. Our starting point is very fundamental – we rely on basic software engineering standards, mainly the ISO/IEC/IEEE 15288 standard, but elaborate their elementary definitions into a joint-effort process, adapted to nowadays systems and the Relay Race principles.

The chapter is structured as follows:

Sections 5.2–5.4 describe the place of the software in a system and discuss the question of whether the software is just embedded within the system, or it is a system by itself. It also identifies the principles of a relay race, as applied to building a software-intensive system;

Section 5.5 looks at the entire life cycle of system development and maintenance in various contexts, such as continuous delivery, single system evolution, product line evolution, etc., and identifies the "baton hand-off" area in view of the life cycle;

Section 5.6 proposes a unified look at a software-intensive system and its decomposition into levels-of-interest, pointing out the roles of systems engineers and software engineers and their mutual responsibility allocation following the relay race principles;

Section 5.7 concentrates on the practice of *functional analysis*, which plays a major role in the shared responsibility for architectural design, being a practical implementation of handing off the architectural baton;

A summary section concludes the chapter by proposing a few means that may help in enabling the joint endeavor;

Appendix 5.A contains a detailed architecture example, in Universal Modeling Language (UML) notation, of an elevator system.

5.2   Software-Intensive Systems

The Role of Software in a System

Henry Ford's first car had no single line of software, and actually all the cars by the 1950s, were merely mechanical systems. The use of electricity in production lines made it the second industrial revolution. With the development of electronics, programmable electronics, and powerful processors and sophisticated sensors, supported by wide communication channels and extensive and fast storage devices, car development and manufacturing became a multidisciplinary endeavor, involving mechanical, electronics, electro-optics, and software engineers, as well as other disciplined engineers, such as safety, materials, and ergonomics. With the growing role of computing power and the internet of things (IoT), the industry moved to the third and fourth industrial revolution. Although the basic functionality of a car – conveying a group of people from one location to another – has not changed, the way this functionality is achieved (i.e. the nonfunctionality or the quality) has improved in many ways, aiming for the ultimate goal of a car to be fully autonomous. The amount of software in a car, estimated to be over 100 million lines of code (Charette 2009), leads sometimes to defining a car as "software on wheels." Although much of this software comes from reusing open-source software, use commercial off-the-shelf (COTS) software and standard infrastructure (e.g. operating system), the effort of developing and integrating software for a specific car is still a significant effort, which might, sometimes, exceed the effort of the other engineering disciplines. This categorizes a car as a software-intensive system, according to the definition: "A system for which software is a major technical challenge and is perhaps the major factor that affects system schedule, cost, and risk" (Chrissis et al. 2004). Cars are only a single example of software-intensive systems, although they are considered at the high end of the lines-of-code scale (Desjardins 2017), exceeding other sophisticated systems (e.g. the F-35 fighter) by an order of magnitude.

Systems are often classified (Pyster et al. 2015) into three classes: physical systems, computational systems, and cyber-physical systems. The first of these classes refers to "traditional" mechanical or electromechanical systems that require little or no software; therefore, they are out of the scope of our concern in this chapter. The other two classes are software-intensive systems, as defined above. The major distinction between these two classes is that the main functionality of computational systems is obtained by software, with the hardware serving merely as a computational platform, whereas cyber-physical systems' functionality, as applied in Industry 4.0, is obtained by a more complex integration and interoperability of software with sophisticated hardware. Examples of such systems are "smart automobiles, power grids, robotic manufacturing systems, defense and international security systems, supply-chain systems, the so-called internet of things, etc." (Pyster et al. 2015).

Although this is all valid at the principle level, software is still implemented by programmers, and for that sake, it is important to take a closer look at software and at the roles it plays in software-intensive systems.

Software exists in a system in two distinct contexts:

- **Component software** is the specific set of programs installed or burnt onto a specific component, e.g. the engine controller, the breaking system controller, the audiovisual system, etc. Each of such components is usually based upon a dedicated processor. A modern car, for example, usually contains dozens of such processors (Charette 2009).
- **Network software** is the "nerve system" of a complex system; it usually includes the "brain" (a main computer) and a communication infrastructure that enables command, control, and coordination among the set of components. The main task of this software is to make the components incorporate in order to achieve the main goals of the system.

From these two contexts emerge two relationships between "system" and "software": (i) the **software in the system**, comprising the set of programs installed in the entire system and (ii) the **software system**, which implements the interconnection and coordination of the entire system. This second relationship highlights the main concern: the entire functionality of a system can be achieved only by incorporated effort of both software engineering and systems engineering.

In addition, as more sophisticated data processing techniques emerge and progress, such as big data, machine learning, etc., the role of the individual components is extended from simple hardware command-and-control to more sophisticated data processing; but, even more than that: each becomes a data consumer and a data producer. The data are collected from the components by the software system and are distributed back to relevant components, as well as to the driver and, probably, to other external systems and stakeholders (e.g. other cars, a fault diagnosis system, etc.). This view blurs even more the distinction between cyber-physical systems and computational systems, giving rise to adapting more and more computational techniques and architectures (e.g. microservices) to cyber-physical systems (e.g. remote intelligence systems).

Privacy and security techniques, previously of concern mainly by finance and commerce systems, are now also extensively applied to cyber-security systems, as they become more software intensive and communication dependent.

5.3   Engineering of Software-Intensive Systems

The incorporation of systems engineering and software engineering has long been the concern of many researchers and practitioners. Moreover, standards and handbooks have been harmonized and integrated in order to enable best this inevitable joint endeavor. These include CMMI (Chrissis et al. 2004), the SEBOK (INCOSE 2013), ISO/IEC/IEEE 15288 (2015), and more. However, the entire scope of the joint endeavor – from stakeholders needs down to the tiniest piece of code – is huge and cannot be allocated as a whole to any discipline individually. In this chapter we suggest two aspects two refine this scope – the life cycle aspect and the system decomposition aspect – and try to locate more closely the most significant overlapping area, where the cooperation and coordination between the two disciplines is the most critical.

At the early stages of developing a complex system, the main concern is of principal systems engineers, who work close to the customers, users, regulators, market analysts, and other stakeholders in eliciting the business goals and requirements, analyzing the business processes, and proposing a system architecture to support these processes. They usually work in a top-down fashion, proposing further decomposition into subsystems. However, it is very rarely that systems engineers will be involved in the development process down to the programming of software units, which is the "kingdom" of the software engineers. The question, however, is how far down they should go.

On the other side, programmers and software engineers are mainly concerned with code and its functional performance. They perceive their role in providing reliable software that performs the functionality allocated to it. The software they produce is integrated with the hardware platform, providing the clients, users, and other stakeholders the entire system functionality. Programmers, however, do not always see the business and the business goals concerned by stakeholders. The software development processes is a bottom-up process by nature, and therefore the higher levels are in many cases beyond their scope.

According to the description above both engineering disciplines apply the human nature of "far from the eyes, far from the heart," and the inevitable risk, therefore, is that the middle part will stay out of both scopes. In practice, the scope of each of the disciplines relies on the knowledge and experience of the individual engineers: A systems engineer with strong software background would dare to dive deeper into detailed software design considerations, and software engineers with strong system background will get more involved in higher levels of system and business consideration. Organizations try, sometimes, to set a clear borderline between these disciplines, although encouraging both to work closely together, but the responsibility assigned for a certain set of work-products, handed over from one discipline to the other, is usually solely defined. The inevitable result is the phenomenon of "throwing over the fence," which causes undesirable rework rounds, in the best base, or defects in the final product, resulting from misunderstandings, in the worst case.

5.4 Role Allocation and the Relay Race Principles

As mentioned above, a better way to enhance the coordination between systems and software engineers is to identify a zone of joint work, rather than a borderline, in which the system level work-products are transferred to the software level engineers smoothly and seamlessly. This view resembles the baton hand-off action in the sport of Relay Race.[1]

The first principle in a relay race is that winning the race is a team challenge. Although each member is expected to perform at their best, the prize is awarded to the entire team, regardless of their individual contribution. The winning driver is, therefore, not the individual capability of each team member but rather their coordination capability, and the willingness of the most capable team member to compensate for the least capable one. In this analogy, the entire performance of the mix of systems and software engineers depends sometimes on either the excellent software capabilities of systems engineers or the excellent system capabilities of software engineers.

A second principle is the shared responsibility for avoiding the baton being dropped. Both parties must be alert to recognize such situations and make every effort to care for the guaranteed delivery of the baton. This means that the current holder of a work-product will never unleash it until being absolutely sure that the receiver holds it securely and can, therefore, use it and maintain it properly.

Another principle is that the two participants run in parallel in the hand-off zone. The receiver starts to run before the baton holder arrives and the participant handing over continues to run after delivering the baton. The meaning is that even if the hand-off action is short, the involvement of both parties goes beyond this time period – before it starts and after it is completed.

In Section 5.7 the functional analysis process that yields the creation of a software-intensive system architecture following the relay race principles is detailed.

5.5 The Life Cycle of Software-Intensive Systems

The Fundamental Nature of the Software-Intensive System Development Process

The development process of a system takes various forms and natures and may be described in a variety of process models, as described, for example, under

---

1 https://www.britannica.com/sports/relay-race

the topic Applying Life Cycle Processes of the SEBOK (INCOSE 2013). As systems are in use for the long term, development, and maintenance[2] cycles are iterated, yielding modified versions of the system or derived systems. Nevertheless, whichever development model is chosen, any development cycle has two clear phases, which together form a "V" shaped process:

- The top-down refinement phase, which starts from stakeholders' new needs, from which system requirements are derived, yielding design specification and detailed requirements down to the "atomic" units from which the system is assembled.
- The bottom-up implementation phase, which starts by creating and testing the units, and continues by gradually integrating parts of the system into greater parts, throughout verification and validation, until the entire system is complete.

Not all parts of the product need to be developed from scratch and may be reused from previous products or obtained from other sources, e.g. open-source code.

Sometimes, maintenance cycles are not intended to provide solutions for new stakeholders' needs but, rather, are done for the purpose of fixing defects in the system in operation. In such cases, the cycle does not necessarily start from the very top; however, even the smallest amendment needs to be integrated into the entire product without causing any regression in its existing capabilities.

Software-intensive systems' life cycle models are usually placed on a scale between plan-driven development (with the waterfall model at its extreme) and change-driven development (with agile and DevOps at its other extreme). In fact, during the entire life cycle one may find plans of agile cycles (sprint/release planning) besides change-driven endeavors of existing systems that were never planned in advance.

A Unified Approach to the System Life Cycle

We introduce here a unified and holistic view that covers all types of development and maintenance cycle over the fundamental "V" model. Moreover, the scope and the frequency of each life cycle iteration is determined by its end product. The approach, which is visualized in Figure 5.1, is detailed in the following subsections. It is emphasized that most of this discussion refers to software-intensive systems, i.e. the iterations refer usually to changes in the software.

---

2 We exclude here hardware defect repairs and refer only to maintenance that incurs redelivery or reinstallation of the system. Note, that this is always the case for software.

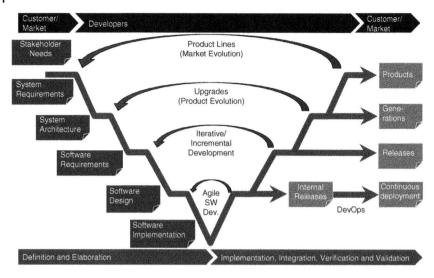

Figure 5.1 An holistic view over development/maintenance iterations throughout the entire life cycle of a software-intensive system.

New Product Cycles

Most companies do not develop just one kind of a system, but rather span the development effort over a family of systems, in a product-line fashion. New products evolve through market evolution in which emerging stakeholders express needs for systems that differ from those already available. Technology advancements boost such endeavors, but, in most cases, new product development is driven by new needs. The new products, alongside existing products, may be further enhanced and developed side by side, in individual courses of evolution.

In the product-line practice, new products are developed on the basis of existing assets accumulated in the company, such as architecture, generic components, design blueprints, etc.

New Generation Cycles

A new generation of a system usually refers to a major progress in the system, aimed mainly at the same market share, but also provides an opportunity to attract new customers. The development of a new generation is usually driven by technological advancements that yield better solutions for stakeholder needs. What mainly changes are the system requirements, which may incur changes in the architecture. A product is not kept still during the time interval between generations and is usually going through cycles of maintenance. Regardless of whether the maintenance is corrective (fixing defects), perfective (introducing new features or other improvements) or adaptive (fitting the product for a changed environment), upgrades to the product are likely to be released.

Release Cycles

A release is an amended version of a product that is usually already in use by the customer. In most cases, the releases are delivered to the customers as upgrade files that just need to be installed on the existing hardware platform. Iterative life cycle models, e.g. the Unified Software Development Process (Jacobson et al. 1998), replaced the "once through" development style of the waterfall model, showing correlation between development iterations and risk reduction. In such models the product is delivered to the customer incrementally, i.e. each release contained an improved version of the previous release, with additional functionality and other properties. Yet, the iterative/incremental life cycle model is usually plan driven, since the number and content of each release is planned in advance. Moreover, the system architecture is usually designed in advance, as a preliminary iteration, whereas the detailed specification and detailed design are deferred to the cycle of each iteration.

Version/Sprint Cycles

Software is the only part of a system that can be modified or duplicated without additional materials. Therefore, amendments and modifications in the software are easier to implement and integrate. The Agile Manifesto (www.agilemanifesto.org) set out the principles under which software should be developed and delivered in short code-integrate-test-deliver cycles. Sometimes minor changes to the design occur, referred to as *refactoring*. In SCRUM (www.scrum.org), a common implementation of an agile process, such cycles are called "sprints," through which a set of features (the "sprint backlog") is selected from the entire set of required features (the "product backlog") as the next candidates for implementation. The end product of a sprint is a version of the software, with new functionality, and maybe some new nonfunctional properties, such as improved performance, easier maintainability, etc.

Although the resultant version is delivered by the software team, from the entire system point of view "delivery" may have different meanings:

In multidisciplinary systems development, the software is usually delivered to the system group, where it would be integrated with other parts of the system. The integrated release is not necessarily delivered to the customer, but rather may be used for field tests, stress tests, hybrid simulations, and other internal uses. Nowadays systems, mostly internet-based, have adopted the DevOps approach (https://aws.amazon.com/devops/what-is-devops), where every new version of the software is directly delivered and immediately deployed at the customers.

Responsibility Allocation in the Life Cycle

A product line may develop multiple products simultaneously. The engineers of each of the products usually develop the latest generation, but at the same time also maintain multiple past generations. Every instance of the product is delivered through multiple releases, and each release comprises multiple versions, some of which may be produced in parallel by multiple software teams. The implication of this view is that systems engineers usually care for the major cycles, such as new products, product generations, and major product releases, while software engineers usually care for the details of software implementation and integration of versions to be released. Therefore, the iterative/incremental development cycle is the one where the common responsibility is most significant. Software engineers can work on the detailed design of each iteration, and implement the incremental builds of the product, but in order to do this properly a system/software architecture should be in place, as the stable skeleton of the product, over which the other parts will be built. The architecture constrains and influences other engineering decisions, and therefore it is the baton which should be smoothly handed off from systems to software engineers, as is detailed in Section 5.7.

5.6   Software-Intensive System Decomposition

Recursive Decomposition of a System

For our discussion we use here the fundamental definition of a system from the ISO/IEC/IEEE 15288 standard (ISO 2015): "Combination of interacting elements organized to achieve one or more stated purposes." This definition implies that a system is decomposed into smaller elements. Moreover, a closer look at the definition yields that a system has four fundamental properties:

- It has one or more *purposes*.
- It comprises a *set of elements.*
- The elements are organized (i.e. interconnected or placed in a certain *structure*).
- Elements interact with other elements (i.e. follow certain interactive *processes*).

In addition, since each of the elements of a system may be a (sub)system by itself, these four properties are induced recursively throughout the system decomposition tree. At the bottom layer of such a tree are "atomic" elements (or components), which are not broken further down to smaller elements. In this case, they possess only the first property, i.e. every element has at least one purpose. This concept may be described best by the composite design pattern (Gamma et al. 1994, pp. 163–173) in UML Class Diagram notation, as depicted

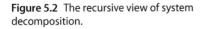

Figure 5.2 The recursive view of system decomposition.

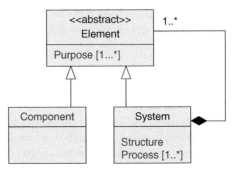

in Figure 5.2. The diagram presents an Element as an abstract entity (i.e. has no real instances), whose fundamental property – a set of one or more Purposes – is inherited, as denoted by the hollow-head arrows, by two subclasses: a System and a Component. It is also denoted (by the black diamond arrowhead) that a System (unlike a Component) is composed of one or more other Elements. Moreover, a System has two additional properties: a Structure over the set of elements and a set of one or more Processes, which are implemented by the interaction among the elements, as well as by interaction with the outside environment. The set of processes is often referred to as the *behavior* of the element.

This unified presentation of a system, regardless of the depth of the decomposition tree, gives rise to the principle of unification of development processes over levels and disciplines in a multidisciplinary complex system development.

The decision whether an element is a system by itself or an elementary component is not arbitrary but depends on the scope of the stakeholder concerned with the decomposition. In terms of the IEEE/IEC/IEEE 15288 standard (ISO 2015) this scope is defined as the system-of-interest, i.e. the system whose life cycle is under consideration.

Principle Levels of Interest

Based on the description above, a system's decomposition tree can be of any depth and can comprise any number of levels. However, we would like to show next that, regarding software-intensive systems, it is possible to define five principle levels of interest with distinct characteristics. This level of categorization is later used for the understanding of separate and shared roles between systems engineering and software engineering.

The five levels of interest are:

- The **business** level: the environment in which the system is installed.
- The **system** level: software-intensive systems installed in the business.
- The **item** level: software and hardware configuration items comprising a system.

- The **component** level: software components comprising a software item.
- The **unit** level: software modules from which software components are built.

These levels are detailed in the following subsections, which also show how the four properties are interpreted in each of the levels.

The Business Level

A system does not exist for itself – its purposes are directly derived from the goals of the business in which it is installed and which it serves. The business goals may be achieved in various ways, but installing a certain system enables those goals to be achieved by operating that particular system as part of the business procedures. For example, a medical business' goal is to provide some medical treatment to patients. The treatment is conducted following a medical procedure that may include patient preparation, tool sterilization, anesthetics, performing the treatment procedure, recovery, etc. This medical procedure can be performed with or without a specific system, as well as with alternative systems.

A business contains people, (noncomputerized) equipment, and (computerized) systems – these comprise the set of elements of the business, including its business model. The business structure is the set of interfaces through which these elements are connected, whereas the business processes are achieved by interaction among these elements and with the external world (e.g. the patient). These processes are performed in order to achieve the business goal, e.g. curing an ill patient.

Any of the particular systems to be developed for the business is the concern of the next level of interest – the system level.

The System Level

A business may be broken down into subbusiness, e.g. a hospital may include medical departments, operating theaters, administration, etc. However, as long as these subelements comprise people, equipment, and systems they are still categorized in the business level of interest. The next level – the system level – is the one in which only a computerized physical system, built out of hardware and software, is concerned.

The purposes of the system are directly derived from its business goals, subject to the business model, but they are described in terms of system operation. For example, when a medical treatment is concerned, the purpose of the system is to enable the medical team to assist in accomplishing the medical procedure in a safe and precise way, by operating the provided medical system appropriately.

The structural elements of the system are hardware items (computers, processors, etc.) and the software items (applications) installed in them. The separation into hardware and software is twofold: on the one hand, since software cannot work without a hardware platform, and a processor is practically useless without software programs, a piece of combined hardware and software operates as a united functional block in the system. On the other hand, software and hardware are usually developed separately, thus the separation makes sense. Moreover, the physical development of hardware has very little effect on the functional behavior of the system, which is almost ultimately determined by the software, either loaded onto or burnt into the hardware. This fact sometimes misleads systems engineers to think that, from here on, software development, as well as hardware development, is the sole concern of the respective discipline engineers. This problem and its impacts are discussed in the following chapters.

The physical structure of the system is described in the system architecture, specifying the hardware nodes with their installed software applications and the physical interfaces among them. Physical links are specified by the physical communication carrier (e.g. cables, connectors, LAN, Internet, etc.) together with their low-level communication protocols (e.g. RS232, TCP/IP, etc.).

The operational processes are often described by use cases, or other behavioral models, describing the interaction among these elements and with the environment, which is performed during the operational processes. For example, a diagnostic process may be specified as an interactive process between the human operator, the computed tomography (CT) scanner, the image database, the diagnosis workstation, and the human diagnostician, based on the scanner software, the database engine, and the diagnosis-support software installed in each of these nodes, respectively.

A hardware-software system may by further decomposed into smaller subsystems, but as long as these include hardware and software together they are still categorized in the system level of interest. However, since the behavior of the system, as mentioned above, is mostly determined by its software, the next level of interest, the item level, refers to software items only.

The (Software) Item Level
Under this level, we categorize pieces of software that may be executed individually. Software cannot operate without hardware, but in the software development process we want to first disconnect their mutual influence, in order to implement and test the software in a "sterile" environment, i.e. a development environment that is assumed to be totally free of hardware problems. For the same reason, the interaction with the environment is often implemented using appropriate simulators. One of the common definitions of a software item is CSCI (Computer Software Configuration Item): an "aggregation of software that is designated for configuration management and treated as a single entity in the configuration management process" (ISO 2017b).

The term "aggregation of software," mentioned in the definition above, refers to the set of software components comprising the software item. The CT scanner software, for example, is built from software components such as the operator's user interface, the moving bed controller, the camera controller, the radiation source controller, the image repository, the preliminary image stitching algorithm, etc. This set of components are connected to each other by means of logical interfaces, defined by the information passed between one another. This information may include command and control signals and data associated with them. The sequences of this information exchange, in cooperation with all the relevant elements, are the implementation of the system processes by means of component interaction, whose shared purpose is to achieve the operational goals of the system. For example, a three-dimentional image of the defect organ with the diagnostician's marking will be achieved by an appropriate sequence of information exchange among the relevant participating components, namely, the operator's user interface, the moving bed controller, the camera controller, the radiation source controller, the image repository, the preliminary image stitching algorithm, and the diagnostician's user interface (UI).

The set of components and the logical interfaces interconnecting them describe the logical structure of the software, which may be referred to as the software architecture. Here again, systems engineers may mistakenly assume that building a software architecture is solely the role of software architects. As will be shown later, the involvement of systems engineers in the software architectural design is critical.

The Component Level
Components at this level are functional components implemented in software. The purpose of a component is to provide the functionality allocated to it during the functional analysis procedure, which is described in Section 5.7. Here again we may rely on the definition of CSC (Computer Software Component): a "functionally or logically distinct part of a computer software configuration item, typically an aggregate of two or more software units" (ISO 2017b).

Thus, the set of elements comprising a component is a collection of software units, in the form of pieces of source code or classes, which are incorporated into the whole components by means of a "build" process (e.g. compilation and link). These units are structured and interact according to a procedure-call or function-call mechanism, which is defined in the source code by means of function definition, including the function name, the number, order, and type of the parameters and the type of the returned value, if applicable. The set of all functions that are exposed to other components (aka "public methods") are considered as the component's API (Application Program Interface). The functionality of one component is utilized by another component at run-time; however, their binding may occur either at the implementation process (when, for example, a source module calls a function from a library that is compiled/linked together with the code) or at run-time (when, for example, a component calls a function available in a DLL – Dynamic Linked Library, which is available at the execution environment).

Software items and software components seem to be of a very close nature and yet we categorize them into two separate levels of interest. The main reason for the different perspective lies in their creation; while software items are mission oriented and are designed and implemented in direct relation with the operational and business goals, software components are usually designed and implemented as generic components, which may be reused as building blocks for constructing various items (applications), regardless of their operational and business goals. Moreover, software components do not necessarily need to be developed: a huge variety is available today as open-source, and may be reused, with or without specific modifications.

The Unit Level

Units are the "leaves" of the decomposition tree of a software-intensive system and are defined as an "atomic-level software component of the software architecture that can be subjected to standalone testing" (ISO 2017a). The purpose of a unit, similar to that of a component, is to provide the functionality allocated to it at a more detailed functional analysis. However, the main reason to separate units from components into a different level of interest is that the writing and testing of software units is usually the task of a single programmer; integration and testing of higher-level elements is usually done by a team.

The five levels of interest and their properties are summarized in Table 5.1.

A question that may be raised in view of the above description is where is a System of System (SoS) concerned? The SEBOK (INCOSE 2013) provides a

Table 5.1 The five levels of interest in a software-intensive system decomposition.

Level of interest	Purpose(s)	Elements	Structure	Processes/ behavior
Business	Business goals	Systems, people, equipment	Business structure	Business processes
System	Operational goals	Hardware items, software items	System architecture	Operational processes
Software (SW) Item	Functional goals	Components	Software architecture	Functional processes
SW Component	Functional goals	Units	Software design	Functional processes/logic
SW Unit	Functional goals	—	—	—

thorough discussion on the meaning of SoS and the differences between systems engineering and SoS engineering. These are based on the definition of ISO/IEC/IEEE 15288 (Appendix G) (ISO 2015): "A system of systems (SoS) brings together a set of systems for a task that none of the systems can accomplish on its own. Each constituent system keeps its own management, goals, and resources while coordinating within the SoS and adapting to meet SoS goals." Therefore, a SoS is an instance of a business, at our decomposition's top level of interest, as described above. Furthermore, if the individual systems are developed just for the sake of a particular SoS, then, according to the recursive nature of system decomposition, the SoS is simply a higher level of system composition, which blurs the difference between a SoS and a "big system." A common perception of a SoS, however, is the one in which each of the systems has been developed independently with its own business purposes, but by interconnecting them over a standard communication they can interoperate and achieve new goals that could not be achieved by each system separately. A typical example is the interoperability between a car and a traffic-light system. Under this perception, each of the individual systems has been previously developed for a separate business, which was the top level of interest when the system was initially developed. In this sense, the SoS endeavor does not usually involve further development of the individual systems but merely interconnection and interoperability within their existing capabilities, in order to achieve new business models, new services, etc.

As mentioned in Section 5.3, the practitioners at the top level of interest are usually principal systems engineers, who work close to the customers, users, regulators, market analysts, and other stakeholders in clearly eliciting the business goals and requirements, analyzing the business processes, and proposing a system architecture to support these processes. Programmers and software engineers, on the other hand, are mainly concerned with code and its functional performance. Thus, they have almost ultimate responsibility over the two lowest levels of interest – the unit level and the component level. The structure and behavior of the set of components is the software architecture, which is integrated onto the system architecture – both are constituents of the entire software-intensive architecture. The creation of that shared work product, in a relay race fashion, is discussed in the next Section.

5.7 Functional Analysis: Building a Shared Software-Intensive Architecture

The Hand-off Zone

As already discussed, the systems engineers' concern is to make sure that the system developed fits the business needs, and therefore they usually work with the customer, users, and other external stakeholders at the upper level of interest, allocating business tasks to people and systems. In the next level of interest, the hardware platform of the system in focus should be chosen. Information management systems, including web services, use (COTS) platforms, comprising servers, workstations, storage, and communication equipment, utilizing standard network and Internet communication. In these cases, the design of system architecture does not leave much freedom of choice. Moreover, in certain application domains there exist standard communication network architectures, such as the CAN (Controller Area Network) bus (BOSCH 1991) utilized in the vehicle industry. The penetrating technology of the internet of things (IoT) is expected to standardize system's internal and external communication even further (Schlegel 2017). The challenge is, therefore, to apply the required functionality, derived from the business processes, to the architecture. Where multidisciplinary systems development is concerned, the case is sometimes easier, since the functionality of each hardware part is well defined, and the debate on which software item should be installed on which hardware item is pretty much predetermined. For example, since the engine and the air-condition nodes in a car have totally different functions, no one would suggest, for any reason, installing part of the engine control onto the air-conditioning, and vice versa. Although the latter description refers to cyber-physical systems, we mentioned above that in modern computerized systems hardware nodes often have more general functionality as data producers or data consumers. The functionality of the system's software, therefore, constitutes two types: operational functionality, implemented as hardware command and control, and data processing functionality, implemented by information management techniques, such as database management, big data, automated reasoning, etc. Therefore, allocation of system functionality to system elements in software-intensive systems is not a trivial task and requires significant considerations both of system issues and of software issues.

At the system level of interest (Table 5.1) the functional allocation is to hardware nodes and the software items installed in them. However, when these nodes become more sophisticated with greater software capabilities, the allocation considerations get one step down, to the item level of interest, influencing the software architecture as well. The resulted product – a software-intensive architecture, comprising both physical and logical architectures – is the baton to be handed off from systems engineering to software engineering, and therefore the architectural design process is their shared concern.

Nevertheless, functional allocation seems to be the easier part of system architectural design, whereas achieving the other quality attributes of a system (sometimes referred to as "the -ilities") is much more complicated. The reason is that functions are usually assigned to specific functional parts, whereas quality attributes are achieved depending on the implementation of those parts and their interrelationships. For example, the acceleration function of a car will be naturally assigned to the engine and the deceleration will be assigned to the breaks. However, the performance attributes (e.g. how long would it take the car to accelerate from 0 to 100 km h^{-1}), will be achieved not only by the engine but rather by a combination of engine (power supply), tiers (road friction), body (aerodynamics shape), materials (weight), and others. In software-intensive systems not only functional but many quality attributes also depend on software characteristics, such as algorithm complexity, message-transferring protocols, data processing and distribution, etc. This is another good reason for the joint effort of the two disciplines.

But the functional and nonfunctional allocation is not a single-direction process. Take, for example, two functionally-identical systems, A and B, with just one differentiating requirement: System B should provide nearly 100% availability (aka 24/7 service), while system A may tolerate periods of inoperability. If system A has a client-server architecture with a single server, the straightforward solution for system B will be to add a secondary ("redundant") server, which will continue the service in case the first server fails, until it recovers. It seems that the two systems remain functionally identical, whereas they differ from each other by their architecture. This might be a sufficient solution from the viewpoint of systems engineers. The software engineers' point of view might be different, since the redundancy solution imposes further functionality, such as backup, indication of failure, synchronization and data persistency, switching between servers, etc. Considering these new functionalities may impose new constrains on the entire system architecture, such as process capacity, broadband, and more. This "ping pong" process must be done by and within the two disciplines together.

Software-Intensive Architectural Design

In Chapter 17 of the book *Software Architecture in Practice* (Bass et al. 2012), the authors describe a methodological approach to architecture construction, named Attribute-Driven Design. The method is an iterative process, applied to the system decomposition, in which at each iteration a set of (refined) functional and nonfunctional requirements/attributes, which have been allocated to the whole part, are assigned to subparts. The decomposition of a part into its subparts is considered "an hypothesis," which should be validated by testing. Once validated, the process can continue to the next subpart in the decomposition. If not validated – a new hypothesis is generated and the process iterates. At the starting point of this process there is a set of Architecture Significant Requirements (ASRs), which are generated by a method described in Chapter 16 of that book, taking into account both business and architectural considerations. Although the system requirements specification is usually generated by systems engineers at the business level of interest, the two types of considerations, used to identify ASRs, imply that this process will be performed jointly by systems and software engineers. The resulting architecture may then be evaluated using an intensive Architecture Tradeoff and Analysis Method (ATAM), described in Chapter 21 of the book mentioned above.

Although the attribute-allocation action is just a part of the described architectural design, we perceive it as the most critical one for both engineering disciplines, and therefore we elaborate it next.

Functional Analysis

Functional Analysis is defined in ISO/IEC/IEEE 24765:2017 (ISO 2017b) as: "examination of a defined function to identify all the subfunctions necessary to accomplish that function, to identify functional relationships and interfaces (internal and external) and capture these in a functional architecture, to flow down upper-level performance requirements and to assign these requirements to lower-level subfunctions."

This definition incorporates a whole process comprising the following steps:

- Decomposing a function into subfunctions.
- Identifying functional relationships and interfaces.
- Capturing these in a functional architecture.
- Flowing down upper-level performance requirements and assigning these requirements to lower-level subfunctions.

It is clearly concluded from the above that functional allocation comes before nonfunctional allocation (of performance and other quality attributes). System specifications, as described by stakeholders, are functional in nature; quality attributes are, in fact, constrains upon the way the functionality will be implemented. Therefore, considering functionality before nonfunctionality seems straightforward. Moreover, we described every system element as possessing both structure and processes. Since the processes reflect the functionality, whereas the structure is driven by nonfunctional attributes, it makes sense to do the functional allocation first, followed by nonfunctional allocation/distribution. Under this view, we will next elaborate the steps of the functional analysis process:

Decomposing a Function into Subfunctions

In the system level of interest, system functions are expressed in terms of functional requirements. However, a set of functions, by itself, does not provide the full functionality of the system; functions are applied during the operation of the system, while the system executes its processes. A guided missile's "interception" function, for example, is implemented as a process of applying a set of functions, sequentially or concurrently, such as *discovery* (of a potential target), *identification* (e.g. friend or foe), *acquisition, maneuvering, tracking,* and *destruction*. Although all these functions may be allocated to the parts of a single physical missile (such as homing head, maneuvering subsystem, warhead, etc.) this functional decomposition is logical by nature. This means that various functions, e.g. *discovery*, *identification*, and *tracking*, may be performed by separate physical subsystems, such as a radar system for discovery, an intelligence center for identification, and a guiding drone. Obviously, while the physical architecture is totally different, the logical architecture is very similar. A specific system might benefit when functional decomposition is done with a given physical architecture in mind. However, in the context of a product line or product evolution, the meaning might be that the architectural design would have to be repeated for every product/generation, and reuse of subfunctions in a different architectural context will be very hard, if at all possible.

This view, with the system's structural and behavioral properties in mind, may be generalized by asserting that although a system architecture must have both properties, the structure is intended to carry the processes, and not vice versa. This is clearly reflected by the steps of the functional analysis process: The first step analyzes the functional behavior and the third step allocates the composed functions to structural elements. However, between those comes the second step, which is described in the next subsection.

Identifying Functional Relationships and Interfaces

When a functional element cannot perform its allocated function all by itself, it needs to get assistance from another element, by requesting it to perform another function. An ATM, for example, is incapable of deciding whether its user is eligible to withdraw the requested cash amount. Therefore, the ATM calls a central bank server to check the user's balance and to approve or reject the transaction. In order to do so, the element providing the "approve transaction" function must expose an interface (aka "provided interface"), through which the element who requires this service can issue the request. The function provider does not need to know who is requesting the function. In fact, every element who has an appropriate interface (aka "required interface"), i.e. "knows" how to request the function, may apply for it.[3] These provided-required interface couplings are the enablers of the functional relationships among the elements comprising a system. In other words, the structure of a system, i.e. the organization of its elements, is derived from the behavior of the system, i.e. the interaction among those elements.

Functional interfaces, in contrast to physical interfaces, are usually the concern of software engineering. Moreover, physical interfaces usually provide only the communication platform over which functional interfaces may be realized, and in many cases a single hardware interface may carry several functional interfaces. For example, a user, equipped with an online desk

---

3 The terms provided and required interfaces come from UML's sequence diagrams, but they can be used out of the scope of modeling, as written here.

computer may perform several tasks, e.g. receive and send e-mails, surf the web, and upload backups to the cloud. While the Internet connection provides the communication link for all these purposes, the various tasks are performed in different couplings between the desktop on one end and a server on the other end. Namely, an e-mail client application is coupled with an e-mail server application, an internet browser is coupled with a web site, and a backup application is coupled with a backup storage manager.

Besides the provided-required interface coupling, a major decision is whether the communication over this interface should be synchronic or a-synchronic. Synchronic communication means that the requesting side remains inactivated and waits until the result of the request arrives, whereas a-synchronic communication means that the requester does not wait and continues its activity. It should be noted that if the result of the request is delivered by the service performer, another interface-coupling (in the reverse direction) is needed.

At this point in the functional analysis process, when upper-level functions are decomposed into subfunctions, the interfaces between these subfunctions may be kept at their simplest form of "function calls." This will enable the functional elements to virtually connect and test the process sequences they are expected to perform by means of a model or a simulation. Although this does not represent the entire architecture, it may be an important step to verify the functional decomposition in a software-only environment.

Capturing the Functional Relationships in a Functional Architecture
Functional architecture is defined in ISO/IEC/IEEE 24765 (ISO 2017b) as the "hierarchical arrangement of functions, their internal and external (external to the aggregation itself) functional interfaces and external physical interfaces, their respective functional and performance requirements, and their design constraints." Thus, the architecture is supposed to support all the functional relations revealed in the previous step, subject to performance and other constrains.

Capturing the functional relationships in an architecture means that the architecture already exist. In the architectural design process the architecture exists as an hypothesis, which needs to be validated. However, in contexts such as a new product in a product line, a new generation of an existing product or a new product increment during an iterative development, there already exists an architecture. The task is, therefore, to either allocate the revealed subfunction to existing elements of the architecture, while verifying that the existing interfaces can support the new functional relationship, or propose new system elements, with appropriate interfaces, to host the allocated functions. These are critical decisions that may have significant effects on the entire performance and other qualities of the system and involve both system and software considerations.

In the previous step, we could build a virtual functional architecture, where functional elements could interact by "function calls" and then verify the system's functional behavior. In the current step, when the decision of where to physically locate these functional elements is taken, three different cases need to be considered:

- **Case 1: The two functional elements reside in the same code.** This is the case, for example, of various source-code modules, or libraries, which are compiled and linked together, yielding a single run-time component (e.g. an executable file). In this case the "function call" interfaces remain at their original form, i.e. as written in the source code.
- **Case 2: The two functional elements are separate run-time components, but coexist in the same execution environment (e.g. a computer).** This might be the case, for example, of an application using a DLL. The development and installation of the two elements have been performed independently, and they are bound to each other only at run-time. A simple "function call" cannot hold here and, therefore, their functional relationships should be implemented over interaction mechanisms imposed by the execution environment (e.g. the operating system).
- **Case 3: The two functional elements reside in separate execution environments.** This might be the case, for example, of a distributed service over a client-server architecture. In this case the functional relation can only be realized over a physical link, which means that in order to enable the requesting element to apply for a service from its function provider, the logical function call must be captured in a message transmitted over a communication link, with the help of a communication mechanism, e.g. TCP/IP protocol.

The result of this step is a (modified) system architecture, which should include the following two ingredients:

- The physical structure, containing physical nodes and physical communication links (physical interfaces);
- The functional (logical) structure, reflecting the functional elements connected by functional interfaces, regardless of their physical carriers;
- A set of operational sequences showing how the system processes are implemented by interaction of functional elements over their functional interfaces.
- A combined structure, reflecting the installation/allocation of functional elements over physical elements and the allocation of functional interfaces to physical interfaces, in cases of cross-platform communication;

A detailed example of an elevator system, modeled in UML diagrams, may be found in Appendix 5.A.

Flowing Down Upper-Level Performance Requirements

Performance is indeed not the only attribute that needs to be flown down in a system decomposition. This is a common practice for systems engineers regarding power consumption, weight, heat omission, etc. Functional performance, however, seems to be at the capacity of software engineers. They, on the other hand, are usually educated to deliver a working bug-free software, with much less concern about performance. Moreover, a software engineer might request a more powerful processor or a bigger storage, in order to meet performance requirements, without being aware of the changes in power supply, weight, and heat omission that such a request may impose on the entire system. Performance allocation, therefore, needs to be done with a wider system perspective. From their point of view, systems engineers are not always aware of issues such as algorithm complexity and computational time/space trade-offs. This is a good reason to include this practice in the hand-off zone to be dealt with by engineers of both disciplines.

Once performance requirements are flown down and allocated to subfunctions, the architecture (or the architectural hypothesis) can be tested and validated.

It was already mentioned that performance in not the only attribute that requires attention of both parties. Functional safety (in contrast to physical safety) is another one; so are (cyber) security, usability, reliability, availability, and others. These are run-time quality attributes, which affect the way the system behaves during its operation. In view of the various contexts of system development life cycle, non-run-time (or development time) quality attributes, such as modifiability, portability, maintainability, testability, etc., may affect significantly the difficulty of generations modified products, as well as the cycle-time of iterations.

We have previously demonstrated how an availability issue may be resolved by changing the architecture from a single server to two redundant servers. We mentioned there that an apparently simple structural solution may impose further functional and nonfunctional requirements. This example shows that a proposed solution to performance, or other quality attribute, may cause the architectural construction to reiterate. During this period, which is still in the hand-off zone, systems engineers and software engineers must take joint responsibility for the architecture, in order not to let this precious baton fall off.

5.8 Summary

In this chapter we discussed the development of software-intensive systems and showed that the critical work-product of the system – the architecture – should be constructed through a joint endeavor of both systems and software engineers, following the principles of a relay race. This is not an easy challenge and cannot be achieved intuitively. The two disciplines need to prepare well for this complex task. As a conclusion of the discussion we propose, in the following subsections, a number of enablers that may be helpful for such preparation.

Awareness of Global Changes

The rapid technological changes in the world have much wider impact on other aspects of the entire world, such as cultural, social, political, economic, and others. The overwhelming use of smartphones, for example, changed entirely the concept of "user interface (UI)": since almost most users carry the client side of a system in their pocket, the system designer should divert their UI concept from "bringing the user to the system" to "bringing the system to the user." Another example is the cloud, which changed the concept of "proprietary systems" to "public systems," where a system cannot anymore hold all the information it needs within its own boundaries but must depend upon cloud systems, while sharing information with many others. This new perception raises many problems of privacy and security that need to be captured in proposed system solutions. Therefore, while the architecture of the system is better technically understood by software engineers, the other aspects and implications of the architecture on system users is mostly in the capacity of systems engineers. Both parties need to be aware of these frequent changes and make sure that they are all captured in their commonly-designed architecture.

Education

Software engineers are usually educated through a bachelor's degree, which is mostly based on computer science. Moreover, for software engineering positions the industry hires graduates of various curricula, such as computer science, software engineering, computer engineering, information systems engineering, and more. Besides, many nonacademic programmers are hires for the same job positions. Therefore, very few software engineers are exposed during their academic education to engineering knowledge in general, and systems engineering knowledge in particular. In addition, software engineers who were promoted to the position of "software architects" achieved this promotion after long experience as programmers; formal software architecture education rarely exists. This phenomenon may be compensated for by educating software engineering in systems context, as described, for example, in (Tomer 2012).

Systems engineering is usually a master's degree, provided to engineers holding bachelor's degree in any engineering discipline. Since the weight of software in software-intensive systems is most significant, we would expect software-intensive systems engineers to come from the software engineering field. Engineers from other engineering disciplines have studied very little about software (usually a fundamental programming course) during their original undergraduate studies, and very little, if any at all, during their systems engineering graduation. The case is that many systems engineers do not have enough background to contribute significantly to the joint effort of creating a software-intensive architecture. The resolution to this phenomenon may be by recognizing *software-intensive systems engineering* as a unique profession by itself and adapt their course of education accordingly.

Model-Based Engineering – Language and Methodology

The educational gap between systems and software engineers, as described in the previous subsection, implies a deeper cultural gap that also suffers from the lack of a common language. In order to overcome this gap, the two parties should adopt a language that can be understood and applied by both. UML (www.uml.org) and SysML (www.sysml.org) are good examples of visual modeling languages that may cater for the joint challenge. Unfortunately, these two languages have not been exploited to their full potential power. One reason is that UML is perceived as an object-oriented software tool, whose greatest promise is automatic code generation. Programmers, therefore, reject the use of UML, since they can generate and maintain better code without bothering about its visual representation. Systems engineers, on the other hand, also perceive UML as a software tool, and don't bother to use it. The truth is, that UML contains a set of tools that are utilized best for software-intensive architectures, as is demonstrated in Appendix 5.A.

But the knowledge of a common language is not sufficient: using the language may yield meaningless models that are not properly related to each other, and therefore become useless in short time. What is needed on top of the language is a methodology, which sets the rules on how to generate meaningful models, how to derive models from other models, and how to relate models with each other, comprising together a cohesive architecture, which is maintainable and kept updated throughout the system's life cycle. This should be an important part of modern systems engineering; the system model should include all parts of the system: hardware, software, interfaces, communication, data. Such an holistic model can support the ideas and processes that are described in this chapter and may support the shared responsibility of software and systems engineers.

Architecture Patterns

The vast experience gained in building software-intensive architectures is captured in a large set of architecture patterns, available in many sources (e.g. volume 1 of the Pattern-Oriented Software Architecture [POSA] book

[Buschmann et al. 1996], and its sequels). Many architectural patterns, such as the client-server, broker, peer-to-peer, publish-subscribe, and model-view-controller – just to name a few – emerged from web systems, but as modern cyber-physical systems become more distributed and rely on big amounts of data, these patterns become relevant to them too. Architecture patterns serve, sometimes, not only the run-time quality attributes of a system but also its life cycle. The microservices architecture pattern (Richardson 2018), for example, emerged from the need to deploy new features as soon as possible, following the DevOps life-cycle model. Naturally, this pattern is penetrating also into classical computer embedded and real-time systems, such as defense systems (OMG 2015).

Architecture patterns should not be mixed with design patterns, which are object-oriented design techniques applied usually to programming. Design patterns are included, in many cases, in the software engineering curriculum, whereas architecture patterns are usually part of an advanced distributed systems' course. Systems engineers, coming from engineering disciplines other than software, rarely have vast knowledge in software architecture patterns and, therefore, rely on software architects, without being aware of the full impact of the software architecture on the entire system. A software-architecture complementary course, including architecture patterns, could be very beneficial to systems engineers.

The relationship between architecture patterns and quality attributes is well studied (Bass et al. 2012, Chapter 13) and therefore they are an important source of considerations in the architecture design process, avoiding the reinvention of the wheel.

Essence – A Common Engineering Ground

In the last decade, Ivar Jacobson and a group of other people decided to redefine software engineering from essential principles (Jacobson et al. 2012). One of the results was an OMG standard (OMG 2015), which sets up a kernel of fundamental concepts (named *alphas*) and their interrelationships, which are essentially applicable to any software development and maintenance endeavor. The kernel, with the alphas partitioned into customer's, solutions, and endeavors concerns, is presented in Figure 5.3.

The Essence standard, aka SEMAT (Software Engineering Method and Theory), also defines a common visual and textual language in which all these alphas may be elaborated and detailed in terms of *alpha states*, *activities* associated with every alpha, resulted *work products*, and *competencies* required from the participants of those activities. The software engineering essence can be utilized as-is for the joint development effort of software intensive systems, as suggested in (Simonette and Spina 2013). However, there is an ongoing activity, e.g. by Jacobson et al. (2015), to extend/generalize/specify a similar Essence for systems engineering.

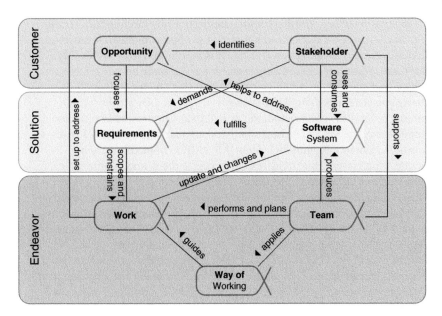

Figure 5.3 The software engineering essence kernel.

References

Bass, L., Clements, P., and Kazman, R. (2012). *Software Architecture in Practice*, 3e. Westford, MA: Addison-Wesley.

BOSCH GmbH (1991), *CAN Specification*, Version 2.0, http://esd.cs.ucr.edu/webres/can20.pdf (Last accessed 4 June 2019).

Buschmann, F. et al. (1996). *Pattern Oriented Software Architecture*, vol. 1. Wiley.

Charette, R.N. (2009). This car runs on code. *IEEE Spectrum*, February, https://spectrum.ieee.org/transportation/systems/this-car-runs-on-code (Last accessed 4 June 2019).

Chrissis, M.B., Konrad, M., and Shrum, S. (2004). *CMMI – Guidelines for Process Integration and Product Improvement*. Addison-Wesley.

Desjardins, J. (2017). How Many Millions of Lines of Code Does It Take? *Visual Capitalist*, February, http://www.visualcapitalist.com/millions-lines-of-code (Last accessed 4 June 2019).

Gamma, E., Helm, R., Johnson, R., and Vlissides, J. (1994). *Design Patterns – Elements of Reusable Object-Oriented Software*. Addison-Wesley.

INCOSE (International Council on Systems Engineering) (2013), *Guide to the Systems Engineering Body of Knowledge (SEBoK)*, v.1.1.2.

ISO (International Organization for Standardization) (2015). *ISO/IEC/IEEE 15288:2015 Systems and software engineering – System life cycle processes*.

ISO (International Organization for Standardization) (2017a). *ISO/IEC/IEEE 12207:2017 Systems and software engineering – Software life cycle processes.*

ISO (International Organization for Standardization) (2017b). *ISO/IEC/IEEE 24765:2017 Systems and software engineering-Vocabulary.*

Jacobson, I., Booch, G., and Runbaugh, J. (1998). *The Unified Software Developments Process.* Addison-Wesley.

Jacobson, I., Ng, P.-W., McMahon, P.E. et al. (2012). The essence of software engineering: the SEMAT kernel. *Communications of the ACM* 55 (12): 42–49.

Jacobson, I., Lawson, H., and McMahon, P.E. (2015). Towards a systems engineering essence. In: *Software Engineering in the Systems Context* (eds. I. Jacobson and H. Lawson), 359–395. College Publications.

OMG (Object Management Group) (2015) *Essence – Kernel and Language for Software Engineering Methods,* http://semat.org/documents/20181/57862/ formal-15-12-02.pdf/e7ba1188-c477-4585-b18a-06937f0e62f3 (Last accessed 4 June 2019).

Pyster, A., Adcock, R., Ardis, M. et al. (2015). Exploring the relationship between systems engineering and software engineering. *Procedia Computer Science* 44: 708–717.

Richardson, C. (2018) *Pattern: Microservices Architecture,* https://microservices .io/patterns/microservices.html (Last accessed 4 June 2019).

Schlegel, C. (2017) *The role of CAN in the age of Ethernet and IOT,* CAN in Automation, iCC 2017, https://www.can-cia.org/fileadmin/resources/ documents/proceedings/2017_schlegel.pdf (Last accessed 4 June 2019).

Simonette, J. and Spina, E. (2013). Software & systems engineering interplay and the SEMAT kernel. In: *Recent Advances in Information Science, Proceedings of the 4th European Conference of Computer Science (ECCS '13)* (eds. M. Margenstern, K. Psarris and D. Mandic), 267–271. WSEAS Press.

Tomer, A. (2012). Applying system thinking to model-based software Engineering. In: *Proceedings of the 2012 IEEE International Conference on Collaborative Learning & New Pedagogic Approaches in Engineering Education (EDUCON 2012),* 1–10. Marrakesh, Morocco: Institute of Electrical and Electronics Engineers.

5.A Appendix

This appendix contains an example architectural model of an elevator system, in UML notation. The model contains all four ingredients of a software-intensive architecture.

Figure 5.A.1 reflects the physical structure, in which 3D boxes represent physical hardware nodes. The physical links between nodes are, by default, cables. The notation 1.* reads as "one or more." ≪device≫ refers to hardware nodes that may have embedded component software, but not as part of the entire system's software, which is captured in the two software items. These items (denoted as ≪CSCI≫) are to be installed in the two computers (Central

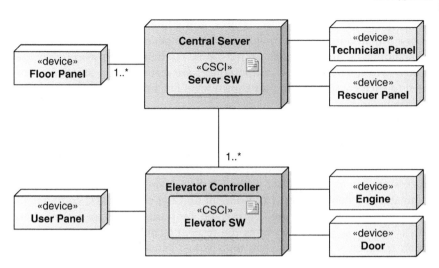

Figure 5.A.1 The physical structure model as a UML deployment diagram.

Server and Elevator Controller) and are the concern of the item level of interest of the system decomposition.

Figure 5.A.2 reflects the functional structure. Each rounded block is a functional component, which is associated with a set of functional interfaces: a short line ending with a circle represents a provided interface, through which

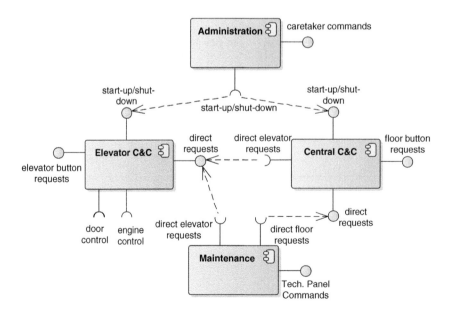

Figure 5.A.2 The functional structure model as a UML component diagram.

other components may request services, and a line with a half-circle represents a required interface, through which a component can request a service from another component. Internal functional relations are represented by the dashed arrow, whereas unrelated interfaces are external. The functions allocated to each component are not shown here, but may be reasoned from the interface names.

Figure 5.A.3 shows the interactivity between the entities implementing the system process of *calling an elevator*. The process is initiated by a user (represented by a floor panel) and terminates when an elevator arrives to the requested floor and the floor button is switched off. The participants of this process are the (external) user, using the floor panel, and both the elevator command and control (C&C) component and the central C&C component. The functional interfaces are represented by horizontal arrows ("function calls"), where each arrow activates the receiving component, until this activation terminates (activations are represented as narrow vertical triangles). Solid arrowheads represent synchronic communication, which means that the calling component waits until it gets an answer from the called component, whereas open arrowheads denote asynchronous communication.

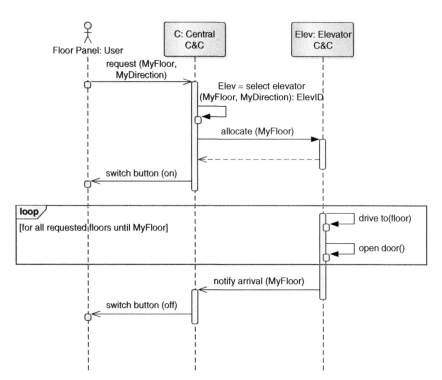

Figure 5.A.3 The interaction model between functional components, performing the process of calling an elevator, as a UML sequence diagram.

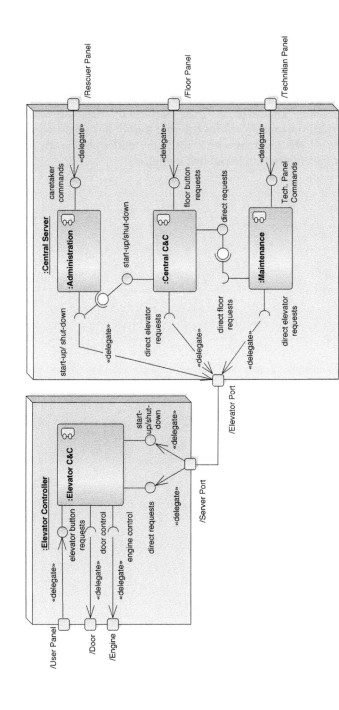

Figure 5.A.4 A combined physical and functional structure model as a UML composite diagram.

Figure 5.A.4 shows a combined functional-physical architecture. The components of Figure 5.A.2 are installed within the physical nodes of Figure 5.A.1, representing the item (CSCI) decomposition performed in the item level of interest. The physical links are represented here by *ports* (small squares on the node's circumference carrying the names of the nodes they are physically connected to). All the external interfaces of Figure 5.A.2 are associated (by *delegate* relation) with physical ports. Cross-platform interfaces are associated (*delegated* to/from) the cross-platform communication link (elevator-server), whereas functional relationships within the same nodes are shown here as *assembly connectors* (paired circle and half-circle).

6

Data-Centric Process Systems Engineering for the Chemical Industry 4.0

Marco S. Reis and Pedro M. Saraiva

Synopsis

Process Systems Engineering (PSE) has now made over 50 years of contributions to chemical engineering, mostly related to the exploitation of computational power to make better plants and products (e.g. safer, more efficient, with higher quality or smaller environmental fingerprints). Although never neglecting the power of data, one must, however, recognize that PSE has emerged and evolved closer to the paradigm of chemical engineering science, and the corresponding view that first-principles models were the way to push chemical engineering forward. Therefore, PSE has been, to no small extent, essentially a deductive discipline that makes use of well-established knowledge from industrial systems and mechanistic models to deduce the best designs for equipment, processes, and products, as well as the optimal operation policies and conditions to be adopted.

However, the fourth industrial revolution is taking its course and making the hidden potential for improvements and competitiveness contained in the immense variety of data streams readily available to be processed and analyzed very clear, namely in the Chemical Process Industries (CPI). This context creates the necessary conditions to further develop and adopt powerful inductive approaches that complement the traditional deductive practices. Therefore, this may also be the right time to rethink and update the paradigms, scope, and tools of PSE, and to devise its positioning in chemical engineering for the next 50 years.

Systems Engineering in the Fourth Industrial Revolution: Big Data, Novel Technologies, and Modern Systems Engineering, First Edition. Edited by Ron S. Kenett, Robert S. Swarz, and Avigdor Zonnenshain.
© 2020 John Wiley & Sons, Inc. Published 2020 by John Wiley & Sons, Inc.

6.1   The Past 50 Years of Process Systems Engineering

"This chapter draws from *inductive and analogical learning* ideas in an effort to develop systematic methodologies for the extraction of structured new knowledge from operational data of manufacturing systems. These methodologies do not require any *a priori* decisions and assumptions either on the character of the operating data (e.g. probability density distributions) or the behavior of the manufacturing operations (e.g. linear or nonlinear structured quantitative models), and make use of *instance-based learning* and *inductive symbolic learning* techniques, developed by artificial intelligence."

At first sight, one might think that this quote is taken from a recent publication or is even an integral part of this chapter. However, the above sentences were taken from a book published almost 25 years ago (Saraiva 1995), entitled "*Intelligent Systems in Process Engineering.*" The work provides an overview of research conducted in the 1980s and early 1990s at the LISPE – Laboratory for Intelligent Systems in Process Engineering, at the MIT Chemical Engineering Department – where one of the authors obtained his PhD at the time, with a thesis named "*Data-Driven Learning Frameworks for Continuous Process Analysis and Improvement,*" (Saraiva 1993), well before the popularity now associated with Industry 4.0, Machine Learning and Artificial Intelligence, emerged.

Chemical engineering is a well-established domain of knowledge, with over 100 years of accumulated knowledge dealing mostly with the design, operation, improvement, and optimization of chemical products and plants. It has relied on data collection, either at laboratory, pilot plant or industrial scales, since its early beginnings. Indeed, chemical plants have always been run and had their operations managed and optimized taking into account frequent measurements of variables such as temperatures, pressures, concentrations, or stream flows. It is therefore not surprising to realize that many developments in applied statistics have been made in close connection with these types of industrial units.

Looking back at the twentieth century, it is thus understandable why unarguably one of the most profound and influential contributors to applied statistics, George E. P. Box, had a strong background in the Chemical Process Industries CPI. Born in 1919, he performed undergraduate work in chemistry at the London University, before earning a BSc and then a PhD degree in mathematical statistics and later on ended up working for Imperial Chemical Industries. It was based on this quite sound knowledge, very close to chemical engineering, that he moved to the USA, where he created an outstanding career mostly at the University of Wisconsin, with countless contributions to the field (Tiao et al. 2000). Some of the members of the George Box academic tree have also bridged the gap between chemical engineering and applied statistics, such as is the case of John MacGregor, from McMaster University, in Canada, with additional significant contributions, namely regarding the application of multivariate statistical approaches to CPI.

However, despite such historical connections between data analysis and chemical engineering, for the second half of the twentieth century many academic contributions to the field of chemical engineering became strongly aligned with the paradigm of chemical engineering science, under the belief that first principles knowledge was the path to follow and would allow us to model, understand, and improve the world of chemical engineering and CPI to solve most of its problems.

This apparent permanent tension of first principles versus empirical learning modes, or deductive versus inductive learning, has been and will probably continue to be a never-ending discussion, although two prominent researchers, who share their first name among many other qualities (George Box and George Stephanopoulos), have illustrated the complementary role played by each of these two facets of discovery and quality improvement (Stephanopoulos 2017; Tiao et al. 2000).

With the appearance of computers during the third industrial revolution and the corresponding digital transformation that faced CPI, both of the above schools of thought went through strong transformations. Among these metamorphoses are the use of first-principles-based process simulators to design, understand, and improve the performance of chemical plants or the exploration of a wider variety of digitally collected data that became available to be treated appropriately by progressively advanced statistical software. With the emergence of digital systems and computers, data collection indeed became much broader and intensive. The "old" pneumatic control systems, where operators would register the values being collected from several process variables by hand, were replaced by sophisticated and complex control rooms, where operating conditions are converted to digital streams, and this clearly widened the scope of industrial data available, as well as methods to handle such data.

When it became possible to foresee the potential benefits of using computers in chemical engineering, this opportunity resulted in the creation of Process Systems Engineering (PSE) as a new field within the area of chemical engineering, as well as the appearance of journals dedicated to it. Perhaps the most well-known and recognized as such is *Computers and Chemical Engineering*, first published in 1977, described in its first issue as being "intended primarily as a journal of record for new developments in the application of computers to chemical engineering problems (…) including: (i) process synthesis, analysis, and design; (ii) dynamic analysis and control of chemical processes; (iii) design methods for chemical engineering equipment, including chemical reactors, distillation columns, extractors, etc.; (iv) applications of computing and numerical analysis in chemical engineering science."

Table 6.1 Percentage of "data-driven" papers published in the first issue of the journal *Computers and Chemical Engineering* for selected years.

Year	Data-driven papers (%)
1977	10
1987	13
1997	14
2007	20
2017	16

This description reveals that there was somewhat of a bias in PSE and the journal editorial philosophy, in the sense that the deductive use of computers appears to receive much more attention than the corresponding counterpart of data-driven or inductive approaches. The very same pattern seems to have been kept along the years, as illustrated by perusing issues of the same journal across the decades. We closely examined the titles of the articles published in the first issue of every 10 years and concluded that the percentage of articles found that can be mostly related to data analysis and analytics is quite small and remained practically stable between 1977 and 2017 (Table 6.1).

Although many interesting data-driven approaches related to CPI have been published elsewhere, including journals related with applied statistics, quality, or chemometrics, to date in the core PSE conferences and journals there remains a quite evident unbalanced situation, with more than 80% of contents dedicated to deductive approaches and less than 20% being mostly data driven. This unbalanced situation can also be explained by the historical roots and origins of PSE, as discussed in the next paragraphs.

As opposed to what often happens in other fields of science and engineering, where a number of pioneers are usually pointed out as being their "fathers" or "mothers," in the case of PSE there is undisputed recognition that Roger W. H. Sargent played that role. Born in 1926, he obtained both a bachelor's and a PhD degree in chemical engineering from Imperial College. Sargent gained industrial experience (something he has in common with George Box, showing how a practical exposure to chemical plants can support the creation of extraordinary academic careers and contributions) at the early stage of use of computers in chemical plants at Air Liquide, in France. In later years he returned to his *alma mater* to pursue an astonishing career, where he founded the Center for PSE. His relentless efforts to apply computers to chemical engineering date to as early as the 1960s and the school of thought he created at Imperial College was centered mostly around first principles modeling and optimization. This philosophy ended up being so influential that, to a large extent, this helps to explain why in the forthcoming decades data-centric PSE came to play a somewhat limited role, since the same predominance was also inherited by other parts of Sargent's academic tree, such as the one established by Arthur Westerberg at Carnegie Mellon University (CMU), in the USA.

PSE now has over 50 years of vibrant activity, mostly related to using computational tools to develop chemical engineering and chemical plants further. Although never neglecting the power of data, one must, however, recognize that PSE has emerged within the scope of what is predominantly known as being the era of the third paradigm of chemical engineering, typically associated with the view that first principles models and chemical engineering science were the ways to go. This situation is changing today, mostly in response to the evolution of industry and digital transformation, leading to data intensive environments that need to create added value, and this also calls for an updated view of PSE priorities, scopes, and approaches.

6.2 Data-Centric Process Systems Engineering

Industry is currently undergoing unprecedented conditions and challenges related to data availability (structured and unstructured), technology (communications, cloud storage, and high-performance computing), and analytical solutions (new methods, artificial intelligence, free, and commercial software) (Figure 6.1). Either alone or synergistically combined, these drivers are changing the industry landscape. New opportunities arise everywhere in existing industrial processes, but new businesses are being created as well, around companies' ecosystems, providing them with specialized services for data handling, storage, high-performance computing, advanced analytics, artificial intelligence, online visualization, etc.

This new data-intensive context is usually characterized by the well-known 5 Vs of Big Data (Qin 2014; Reis et al. 2016):

Volume. The existence of large amounts of data, either stored or in transit.
Velocity. The increasing pace at which data are created and transferred.
Variety. The plethora of data formats for conveying information, either structured (low or high-dimensional arrays of numbers) or unstructured (e.g. text, sound, images).
Veracity. The assessment of data quality and accuracy for the intended purpose, often not easily established.
Value. The associated benefits and competitive advantage for companies and stakeholders.

Even though these issues were only recently recognized, they have been around in industry for some time, creating many challenges to process engineers and analysts. In the next paragraphs, examples are provided of

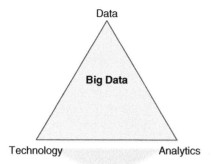

Figure 6.1 The basic ingredients behind the emergence of big data initiatives.

concrete cases where the big data 5 Vs have been addressed by the authors, under the context of data-centric process system engineering. Hopefully, this short list of concrete examples may contribute to consolidating the idea that the new paradigm shift, now widely recognized and coined as Industry 4.0, corresponds to an accelerating trend whose beginning started a long time ago, when analytical solutions began to be developed for addressing the challenges that CPI face when moving to data-intensive environments.

Volume

The number of process sensors collecting data from industrial units in real time has increased dramatically. Products, whether in-process or final, also have their quality characterized by multiple dimensions that can be assessed using advanced analytical devices that rigorously acquire the relevant data for each quality dimension. Process engineers have now to deal with high-dimensional data to perform standard tasks such as statistical process monitoring. For instance, in the case study reported by Reis and Delgado (2012), a given product was characterized by more than 15 000 measurements, based on which process stability needed to be quickly assessed. More specifically, this example corresponds to the initial stages of the assembly process of microelectronic devices, where typically thousands of Solder Paste Deposits (SPDs) are rapidly and accurately placed over Printed Circuit Boards (PCBs), at different and predetermined positions, to support and functionalize electronic components for positioning at a later stage.

The cost associated with each device in the process of assembly is significant and increases steeply as it progresses along the production line. Therefore, any malfunction should be detected as early as possible to avoid further processing on faulty devices and to identify the fault origin and rectify the root cause as fast as possible. For that purpose, it is necessary to inspect all the SPDs placed during the first assembly stage and assess their shape and placement. Inspection is conducted using Phase Moiré Interferometry, based on which a complete three-dimensional profile of each deposit is obtained and then summarized in five variables: height (h), area (a), and volume (v), as well as the position offsets in the X and Y directions with respect to the target positions in the reference coordinate system. Overall, the number of variables generated for each device under analysis amounts to a total of $5 \times 3084 = 15\,420$. This factor represents a "volume" issue, given the high number of measurements collected, but a second challenge in this case study relates to the reduced number of samples available for implementing a standard Statistical Process Control (SPC) approach.

SPC corresponds to a family of methods focused on maintaining the process under a stable state of normal operational conditions (NOC), close to the desired target, or returning to this state as fast as possible if an abnormal event takes place. A process operating in such stable conditions is said to be under "statistical control," i.e. its variability is only due to "common" causes of variation (Shewhart 1931). This process status is usually assessed through control charts, which are designed to signal situations where some sort of abnormality occurs (called "special" or "assignable" causes of variation), that pushes the process out of its normal state toward an "out of control state." Classical control chart procedures address situations with small numbers of monitoring variables (Hotelling 1947; Page 1954; Roberts 1959; Shewhart 1931) but procedures have also been devised for high-dimensional situations (Jackson 1959; Kourti and MacGregor 1996; Wise and Gallagher 1996). The challenge, in our application, was also to extract the essence of high-dimensional NOC variability with very low sample sizes – the ones available at the beginning of each new product run. Finally, a third challenge faced was the pressing need to make swift decisions, at the same level of speed as the pace of the production process.

These three challenges (high number of variables to monitor, reduced number of NOC samples, and short decision times) together represent the new sort of statistical quality monitoring problems likely to be found in modern Industry 4.0 scenarios. The methodology developed consisted of a two-stage latent variable framework that is able to efficiently capture the sources of NOC variability underlying SPD deposition (which are usually connected to rigid body movements and a small number of other variability sources). This dynamic resulted in a low dimensional latent variable space, comprising most of the variation exhibited by the 15 000 highly correlated measurements (usually less than five latent variables are retained). The framework not only monitored the variability projected into the Principal Component Analysis (PCA) subspaces but also around them and used these residuals to identify the SPDs with specific problems, whenever an out of control situation was signaled. The detection stage is made with resort to only a pair of monitoring statistics and is, therefore, simple to implement and quick to run (the same applies to the diagnostic stage based on residuals); more information about this application can be found elsewhere (Reis and Delgado 2012).

Velocity

Data collection as well as their distribution across the plants is happening increasingly fast. An example of the relevance of being able to cope with the speed at which data are collected comes from the paper production industry. This example concerns the monitoring of a critical paper quality dimension, regarding its surface: "paper formation" (Reis and Bauer 2009). Paper formation is the designation used to qualify the level of homogeneity in the distribution of fibers in the surface of paper. A high-quality paper presents a uniform distribution of fibers, without major visible flocks or fiber-dense regions. This property not only impacts the esthetic perception of quality by the customer but also affects paper strength (papers with poor formation are weaker, since there are regions where cracks can originate and start propagating when submitted to stress, and they are, therefore, more prone to break under tension during printing and conversion operations) and printing (quality is inferior for papers with poor formation), as well as leading to process problems (e.g. runability and higher consumption of coating agents).

For these reasons, paper formation is routinely assessed in the quality control laboratories of modern paper making units, usually located near the process from where samples are taken. The assessment is made by visual inspection upon comparison with a series of templates typifying different quality grades of formation or with recourse to instrumentation that scans the paper sheet, measuring the variation of light transmitted through it. The periodicity of this monitoring scheme is a few times per day, depending on the specific routine testing plans of each paper mill. This step introduces a very significant monitoring delay, considering that modern paper machines produce paper at linear speeds over $100\,\text{km}\,\text{h}^{-1}$ (therefore, in an eight-hour shift more than $800\,\text{km}$ of paper can be produced). This offline process monitoring scheme thus presents strong limitations, being ineffective in promptly detecting problems in paper formation and likely to lead to significant losses of quality in cases where a process upset arises. A solution that increases the velocity of data acquisition from the process was devised to overcome this limitation, avoiding the need to only rely on collected samples that are analyzed offline. This solution consists of installing a camera in the process, equipped with the required mechatronics (a rotating housing) to avoid drops sticking onto the camera and a backlight elimination system. This apparatus is able to collect gray-level images at very high speeds, which then need to be processed to assess paper formation.

The methodology developed was based on texture analysis, namely Wavelet Texture Analysis (WTA; Figure 6.2). WTA is an image analysis technique consisting of the application of a wavelet transform to collected images and subsequent processing of the resulting wavelet coefficients to classify the paper formation quality grade. The methodology scales very well with the

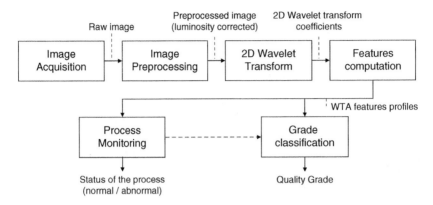

Figure 6.2 Scheme of the WTA methodology used for assessing paper formation.

number of images, as the wavelet transform has complexity of $O(N)$ and not all coefficients turn out to be necessary for the analysis; in fact, a PCA analysis reveals that the finer scale coefficients, which dominate in number, do not carry any relevant discrimination power to the analysis, and can be discarded without any detrimental effect in the outcome of the grading task. The implication being that lower resolution images can be collected instead, thus increasing the speed of data transfer and analysis, and reducing the memory requirements of the system. Therefore, WTA meets the speed challenges for this application. The study performed clearly demonstrates that the proposed image-based monitoring system is able to rapidly discriminate between good formation images and those with some fiber distribution problems.

Variety

With the development of sensing technology, the amount and diversity of measurements available are proliferating, this being one of the most visible outcomes of Industry 4.0. It is becoming common practice to have access to scalar quantities (i.e. zeroth order tensors, such as the well-known industrial sensors and univariate quality parameters), first order tensors (i.e. 1D profiles, such as spectra, chromatograms, particle size distributions, vibration records), second-order tensors (e.g. gray-level and thermographic images), third-order tensors (e.g. hyperspectral images, hyphenated Process Analytical Technology – PAT – data), etc. Mainstream industrial data analysis methods tacitly assume that all variables have the same structure. Integrating all these different sources of data flexibly for process monitoring, prediction, and control, is still beyond the current state-of-the-art of process analytics. However, work is being carried out regarding the fusion and integration of different measurement sources. For instance, Campos et al. (2017) report advanced data analysis methods for exploring the synergies and complementarity between different analytical measurements, namely gas chromatography coupled with mass spectrometry (GC-MS), high-performance liquid chromatography combined with photodiode array detection (HPLC-DAD) and UV-Vis absorbance spectra, the objective here being to predict a target wine property (aging time).

The structure and dimensionality of the sources are distinct, and the integrity of each data block should be preserved as much as possible during model building and analysis. This operation is not possible with classical data analysis methodologies but can be achieved through the application of multiblock approaches. Several methods were tested in this case study, such as the Concatenated PLS method, Multiblock PLS (MBPLS), Hierarchical PLS (HPLS), Network-Induced Supervised Learning (NI-SL), and Sequential Orthogonalized Partial Least Squares (SO-PLS). The interpretational capabilities were also considered during the comparative evaluation of all the multiblock methods. In general, these methods proved to be able to deal with large volumes of data and to explicitly include the variety of structures in the analysis, leading to richer insights and granting access to valuable information. The best results were obtained using SO-PLS and the concatenated PLS method (with a new preprocessing method proposed by the authors), both showing equivalent relative performances, followed by the NI-SL method. A comparison between these methods and the best linear single block approaches (partial least squares (PLS) and principal component regression (PCR)) also confirmed their superior prediction accuracy.

This research line assumes special relevance in a time when the scope of the systems under analysis is evolving from the spatial scale of the unit/plant, to the entire supply chain, and the time scales of interest evolve from those typical of process operations (seconds to days) to the entire lifecycle of the products (seconds to dozens/hundreds of years). As the systems frontiers widen, data heterogeneity (variety) increases and methodologies should be available to fuse all the relevant sources of information in order to tackle critical problems. For instance, sustainability issues, a long concern of the PSE community (Bakshi and Fiksel 2003; Cabezas and Mercado 2016; Daher et al. 2018), will greatly benefit from the progress to be made in this field.

It is, therefore, essential to further pursue this line of research and increase the spectrum of solutions available to the users in order to capitalize from the synergistic combination of multiple measurement sources, and even decide which ones are required and add predictive value and which ones should be discarded, because they are redundant or do not carry valuable information over the response. Thus, saving resources of specialized personnel, time, and money, which are critical aspects in the management of the always busy process laboratories, as also demonstrated by Campos et al. (2018).

Veracity

One of the critical issues when collecting massive amounts of data is their inherent quality and fit to the purpose. In the realm of industrial data, hitherto strongly dominated by structured data streams, this aspect is well addressed under the scope of measurement uncertainty. According to the Guide to the Expression of Uncertainty in Measurement (usually known as GUM), measurement uncertainty is a "parameter, associated with the result of a measurement, that characterizes the dispersion of the values that could reasonably be attributed to the measurand" (JCGM 2008). This reference provides clear guidelines on how to assess uncertainty using a series of repeated measurements (Type A evaluation, aligned with a frequentist statistical estimation process) or some other appropriate means (Type B, inherently Bayesian in nature, as explained by Kacker and Jones (2003)). Reis et al. (2015) discussed two levels of challenges regarding data Veracity in CPI: (i) specification of measurement uncertainty; (ii) integration of data uncertainty in the analysis as an additional source of information. The previous challenge refers to the assessment of uncertainty in those situations that are not covered by standards such as the GUM. Some of these are rather common in industry and have been addressed, namely:

- *Estimating measurement uncertainty near physical limits.* This happens for instance when measuring very low concentrations of some analytes in process streams, a common situation in the chemical, pharmaceutical, semiconductor, and food industries, where trace concentrations are frequently found. As the physical limit is approached, measurement distributions become asymmetrical and the usual procedures for specifying uncertainty intervals lead to regions containing unfeasible values for the measurand.
- *Estimating measurement uncertainty in heteroscedastic scenarios.* The use of replicates (Type A evaluation in the GUM) or Gauge R&R studies (ASQ/AIAG 2002) provides protocols for estimating the uncertainty of measurements, under certain well-established conditions. However, when uncertainty depends on the level of the measurand, these procedures are only able to specify it at a certain level. This situation is common in analytical equipment, such as spectrophotometers, chromatography columns, weighing balances, etc.

The second challenge is also important and much less explored. It regards the integration of uncertainty information in data analysis, together with the actual measurements. Most classical methods make the tacit assumption of homoscedasticity and are unable, by design, to accommodate this extra source of information. However, being available, methods should be adopted that are able to incorporate it in the analysis. Some works have already addressed the potential added value of including uncertainty information in different data analysis tasks, such as predictive modeling (Reis and Saraiva 2004, 2005; Wentzell et al. 1997a; Wentzell et al. 1997b), process monitoring (Reis and Saraiva 2006b), and process design and optimization (Reis and Saraiva 2005; Rooney and Biegler 2001).

In CPI, the variables most affected by uncertainty are typically those related to the assessment of the final quality of the products (e.g. mechanical and chemical properties) or regarding the analysis of compositions in process streams (main product and subproducts, reactants). These are usually the "response" when developing predictive models, and therefore a common situation corresponds to the need to incorporate heteroscedasticity in the response when developing predictive models and soft sensors. This issue was analyzed by Reis et al. (2015) and several modeling strategies were tested. It was thus possible to confirm that uncertainty-based methods, such as Principal Component Regression using Weighted Least Squares, lead to improved prediction performances under fairly general conditions (Bro et al. 2002; Martínez et al. 2002; Reis and Saraiva 2004, 2005; Wentzell et al. 1997b).

Value

Ultimately, organizations aim to extract value from data-driven initiatives and turn them into effective sources of competitive advantage and to increase the assertiveness of decision-making processes, at all organizational levels. This objective requires managing empirical studies much like any other company's processes, for which a critical requirement is the availability of a suitable Key Performance Indicator (KPI). How to assess "effectiveness" or "assertiveness" in a rigorous and consistent way is by itself a major challenge: "If you cannot measure it, you cannot improve it" (as stated by Lord Kelvin many centuries ago), and clearly there was a gap in CPI regarding this fundamental issue. Therefore, the InfoQ framework (Kenett and Shmueli 2014, 2016), was adapted and developed for assessing, analyzing, and improving the quality of information generated in data-driven activities found in CPI (Reis and Kenett 2018). InfoQ is defined as *"the potential of a dataset to achieve a specific (scientific or practical) goal by using a given empirical analysis method."* It depends upon a set of structuring aspects present in any data-driven project, called the InfoQ-components, namely: specific analysis goal, g; available dataset, X; empirical analysis method, f; utility measure, U. These elements are related to each other through the following analytical expression:

$$InfoQ(f, X, g) = U\{f(X \mid g)\} \tag{6.1}$$

where *InfoQ* is the level of Utility, U, achieved by applying the analytical method f to the dataset X, given the activity goal, g.

The InfoQ components depend upon several assessment dimensions, as schematically depicted in Figure 6.3.

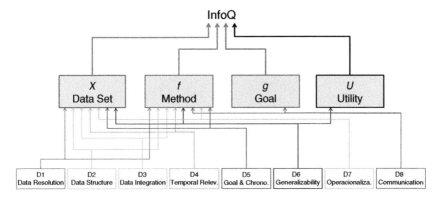

Figure 6.3 The decomposition of InfoQ into its components *(X, f, g, U)* and then on the eight dimensions that determine their quality. Also shown are the connections between dimensions and the components in which their assessment takes part.

The proposed approach provides a questionnaire template to guide the user in the assessment and formulas to establish the scores of each InfoQ component as well as the overall InfoQ scores. Such a systematic framework can be applied irrespective of the context and the specific goals to be achieved. It can either be used to provide a preliminary assessment of the project risk, by analyzing the adequacy of the dataset and analysis methods to achieve the intended goal, as well as to perform a SWOT analysis on an ongoing project, to improve it and increase the quality of information generated, i.e. increasing its InfoQ (Reis and Kenett 2018).

6.3 Challenges in Data-Centric Process Systems Engineering

Data collected from modern industrial facilities present a variety of challenges, most of them not commonly found in other application scenarios, that require the development of new solutions by the PSE community, in order to provide consistent and rigorous solutions to the specific CPI analysis goals. In this context, we consider the existence of three major levels of challenges to be faced to meet the present and future demands imposed by advanced manufacturing to PSE (Figure 6.4): *core, tangible,* and *intangible* challenges.

Core Level Challenges

This level is related to the inner mechanistic behavior of industrial processes and is essentially independent of the way data are collected. It concerns the complexity of phenomena that structure the variability of data at a very fundamental level. Data analytics must be able to conform to the structure of the

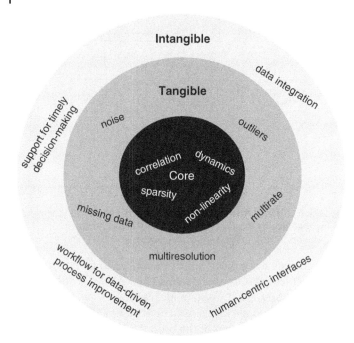

Figure 6.4 The three levels of challenges of data analytics for advanced manufacturing.

inner systems, in order to extract contextualized and useful insights from them. Common features characterizing this level are:

- High-dimensionality – many variables are required for characterizing the behavior of the process (Reis and Delgado 2012; Reis and Saraiva 2008).
- Correlation – including the inner complex network of causal relationships (Rato and Reis 2014, 2015b, 2015c, 2017a; Reis 2013b).
- Sparsity – not all variables that are collected are relevant for the analysis goal.
- Modularity – variables are organized in functionally related blocks (Reis 2013a, 2013b).
- Nonlinearity – complex nonlinear behavior may exist between variables (Choi and Lee 2004; Fourie and de Vaal 2000; Liu et al. 2000; Shao et al. 1999; Sjöberg et al. 1995; Zhang and Qin 2007; Zhao et al. 2009).
- Autocorrelation (or dynamics) – either stationary (Rato and Reis 2013a, 2013b, 2013c) or nonstationary (Louwerse and Smilde 2000; Nomikos and MacGregor 1994, 1995; Rato et al. 2015, 2017; Rendall et al. 2017; Van den Kerkhof et al. 2012; Vanlaer et al. 2013) and occurring at a single scale or at multiple scales (Rato and Reis 2015a; Reis 2009; Reis and Saraiva 2006d).

All these features can be present with different levels of intensity or some of them may be absent (for instance, in well-controlled continuous processes nonlinearity is usually not visible or perceived, given the tight operational windows used to run the plant). The most pervasive feature of industrial data is undoubtedly the existence of many variables with mutual correlations. Correlation is just a natural consequence of the well known fact that the number of sources of process variation is (much) smaller than the number of sensors collecting data. Correlation in industrial observational data is a consequence of process redundancy, which is further promoted by the existence of underlying fundamental laws (e.g. mass, energy, and momentum conservation) that increase the connectivity between variables, and the existence of redundant sensors, motivated by considerations of accuracy and reliability, or the use of control loops. A challenge for data analytics involves the development of dynamic models for these high-dimensional correlated contexts, able to handle stationary and nonstationary dynamic systems. Some solutions tested in the past include dynamic latent variable models and multiscale empirical models. Nonlinear frameworks that are scalable with data dimensionality are needed here, such as the class of tree-based ensemble approaches (Breiman et al. 1984; Dieterich 2000; Friedman et al. 2001; Strobl et al. 2009) and kernel-based methods (Choi and Lee 2004; Lee et al. 2004; Nikolaou and Vuthandam 1998; Rosipal and Trejo 2001; Tian et al. 2009; Zhang and Qin 2007; Zhao et al. 2009).

Tangible Level Challenges

This level is closely connected with the difficulties created by the data acquisition process – the tangible or transactional mechanism that affects how the values are collected and how they will be made available for analysis. This second level can be a consequential source of complexity in data analysis, originating features such as noise, outliers, missing data, multirate data (data with different acquisition rates – for instance, quality variables are usually made available at much lower rates than process variables, as discussed by Kadlec et al. (2009); Rato and Reis (2017b); Reis et al. (2015)), multiresolution data (data with different levels of granularity – e.g. pointwise process data together with time averages computed with different aggregation periods or with samples that are composed along batches, shifts, etc. as shown by Reis and Saraiva (2006a, 2006d)), and heterogeneous data (structured and unstructured, with sensor data, spectra, images, text, etc. (Reis 2015; Reis and Bauer 2009; Reis and Saraiva 2006c, 2012)). Hitherto, no method has been able to cope with all the features appearing in the core and tangible levels, but approaches have been developed to handle subsets of them, with increasing compatibility and resilience with respect to even the wildest scenarios of data collection, such as noisy (Reis et al. 2015; Reis and Saraiva 2004, 2005), multiresolution (Bassevile et al. 1992; Rato and Reis 2017b; Reis and Saraiva 2006a, 2006d; Willsky 2002), and missing data scenarios (Arteaga and Ferrer 2002; Little and Rubin 2002; Nelson et al. 1996; Reis and Saraiva 2006b; Walczak and Massart 2001).

Still, a long journey is yet to be made before these methodologies turn into effective industrial solutions ready to be used on the shop floors of Industry 4.0. For that to happen, the next challenge must also be properly addressed.

Intangible Level Challenges

This level regards the interface between humans and data or advanced analytics. It mostly concerns the virtual tools (software, interfaces) that facilitate and make it possible for users to retrieve, integrate, analyze, communicate, and act upon the data and the outcomes from data analytics. Unless the methods and tools developed to address Level 1 and Level 2 challenges are designed in such a way that can be easily used and understood by those who will actually need them the most – industrial practitioners, not academic researchers – their benefits will not be fully exploited. Therefore, we believe that, with so much data around and so many methods to be used, appropriate human-centric platforms must be developed that consider the amount of information humans are able to process and act upon, in real time and under industrial settings (Pacaux-Lemoine et al. 2017). In this regard, engineering should be combined with psychology insights, with past experiences from aviation and other fields being used to help us solve the challenges at this intangible level. That requires a multidisciplinary effort, which so far has not been fully addressed or even acknowledged. Furthermore, new workflows for data-driven process improvement should be developed, where humans are allowed to use their intelligence and creativity while fully harnessing the technology and data resources available and generalizing their findings to the benefit of other units and processes in the companies.

6.4 Summary

Process system engineering is a well-established field devoted to the development of systematic and holistic solutions to complex industrial systems, through computational, mathematical, and algorithmic approaches. This field of knowledge started in the 1970s with the fast development of computing and electronics, which made it possible to express detailed engineering knowledge in computer code and thus develop the model-based approaches for process simulation, control, and optimization. In the dawn of the twenty-first century, data abound more than ever before in industrial processes and constitute a major source of information that can improve the aforementioned tasks and address other activities, such as process monitoring, fault detection, diagnosis and troubleshooting, process visualization, improvement, etc. These activities are mostly inductive and have the power to complement the deductive approaches, strongly dependent upon preexisting knowledge about the processes. This scenario therefore sets the stage and the scope for data-centric PSE, the inductive branch of PSE, which may lead to a solid, systematic body of knowledge able to cope with the present and future challenges of Industry 4.0 and its future x.0 versions.

Depending on the specific goals, knowledge that is already available and data to be generated or that it is possible to collect, one should be able to come up with the most appropriate combination of paradigms and tools in data-centric PSE, namely through the integration of statistical, machine learning, and optimization techniques together with chemical engineering science (Stephanopoulos 2017). A wonderful occasion, where many of such different perspectives were discussed, was the "2040 Visions of Process Systems Engineering" symposium, on the occasion of George Stephanopoulos's 70th birthday and retirement, which took place in 2017 at MIT. With many prominent participants from all over the world, two days of deep thinking, strong interaction, and inspiration comprised a variety of views about the future of PSE. Different new application scopes for PSE were discussed, as well as perspectives that were presented and relate to the evolution of optimization and models ("The Future of Process Optimization," "New Optimization Paradigms for Formulation, Solution, Data and Uncertainty Integration, and Results Interpretation," "Process Modeling: where to next?"). Nevertheless, some of the inspiring presentations made also pointed out how more data-centric PSE is expected to become ("PSE in the Information Age," "The Promise of AI in PSE: is it there, finally?", "Asset Optimization for the Process Industries: from data to insights to actions," "Process Operations: a roadmap from the present to a hyper-connected future").

A wide spectrum of possibilities was discussed, all the way from a priori mechanistic models to the idea that with enough data one may take decisions directly from them without needing to build explicit models ("No Equations, No Variables, No Parameters, No Space, and No Time: data and the modeling of complex systems"). Addressing such a topic within the scope of a PSE symposium with the participation of its leading thinkers would have been unimaginable 10 years ago, and it is irrefutable that in PSE there will be still room for some equations, variables, or parameters. However, they will have to be fused with more and more data, which can indeed prevail for several kinds of applications, as well as create room for hybrid solutions, where both schools of thought can be integrated. This was also illustrated by the titles chosen by some other speakers at the event ("Bayesian Process Engineering," "The Internet of Things and Process Control: using real and statistical intelligence," "Real-Time Decision-Making Under Uncertainty: an adventure in big data analytics or an opportunity for mathematical programming?"). By looking closely at the list of speakers, their backgrounds and academic roots, it does seem to be the case that data-centric PSE is indeed happening, supported by a powerful combination and balance of paradigms, including theory and practice, combining synthesis and analysis, data and science, as wonderfully suggested by George Stephanopoulos in his farewell lecture, entitled "Synthesis and Computing in PSE: 45 year travelogue of an unindoctrinated academic."

As shown in this chapter, and also in the event just mentioned, data-centric PSE is here to stay. It takes into account the past decades of successful PSE but significantly enlarges its scope and provides a more diversified and better-balanced set of tools that can be used to address the current new challenges that the CPI are facing.

References

Arteaga, F. and Ferrer, A. (2002). Dealing with missing data in MSPC: several methods, different interpretations, some examples. *Journal of Chemometrics* 16: 408–418.

ASQ/AIAG. (2002). Measurement Systems Analysis 3e. DaimlerChrysler Corporation, Ford Motor Company, General Motors Corporation.

Bakshi, B.R. and Fiksel, J. (2003). The quest for sustainability: challenges for process systems engineering. *AIChE Journal* 49 (6): 1350–1358.

Bassevile, M., Benveniste, A., Chou, K.C. et al. (1992). Modeling and estimation of multiresolution stochastic processes. *IEEE Transactions on Information Theory* 38 (2): 766–784.

Breiman, L., Friedman, J., Stone, C.J., and Olshen, R.A. (1984). *Classification and Regression Trees*. CRC press.

Bro, R., Sidiropoulos, N.D., and Smilde, A.K. (2002). Maximum likelihood fitting using ordinary least squares algorithms. *Journal of Chemometrics* 16: 387–400.

Cabezas, H. and Mercado, G.R. (2016). *Sustainability in the Design, Synthesis and Analysis of Chemical Engineering Processes*. Butterworth-Heinemann.

Campos, M.P., Sousa, R., Pereira, A.C., and Reis, M.S. (2017). Advanced predictive methods for wine age prediction: part II – a comparison study of multiblock regression approaches. *Talanta* 171: 121–142.

Campos, M.P., Sousa, R., and Reis, M.S. (2018). Establishing the optimal blocks' order in SO-PLS: stepwise SO-PLS and alternative formulations. *Journal of Chemometrics* 32 (8): e3032.

Choi, S.W. and Lee, I.-B. (2004). Nonlinear dynamic process monitoring based on dynamic kernel PCA. *Chemical Engineering Science* 59 (24): 5897–5908.

Daher, B., Mohtar, R.H., Pistikopoulos, E.N. et al. (2018). Developing socio-techno-economic-political (STEP) solutions for addressing resource nexus hotspots. *Sustainability* 10 (2): 512. 1–14.

Dietterich, T.G. (2000). Ensemble Methods in Machine Learning. In: *Multiple Classifier Systems*, 1–15. Springer.

Fourie, S.H. and de Vaal, P. (2000). Advanced process monitoring using an on-line non-linear multiscale principal component analysis methodology. *Computers and Chemical Engineering* 24: 755–760.

Friedman, J., Hastie, T., and Tibshirani, R. (2001). *The Elements of Statistical Learning*, Springer Series in Statistics Springer, vol. 1. Berlin: Springer.

Hotelling, H. (1947). Multivariate quality control, illustrated by the air testing of sample bombsights. In: *Selected Techniques of Statistical Analysis* (eds. C. Eisenhart, M.W. Hastay and W.A. Wallis). New-York: McGraw-Hill.

Jackson, J.E. (1959). Quality control methods for several related variables. *Technometrics* 1 (4): 359–377.

JCGM (2008). *Evaluation of Measurement Data – Guide to the Expression of Uncertainty in Measurement* (JCGM 100:2008, GUM 1995 with minor corrections). Paris: JCGM (Joint Committee for Guides in Metrology).

Kacker, R. and Jones, A. (2003). On use of Bayesian statistics to make the guide to the expression of uncertainty in measurement consistent. *Metrologia* 40: 235–248.

Kadlec, P., Gabrys, B., and Strandt, S. (2009). Data-driven soft sensors in the process industry. *Computers and Chemical Engineering* 33 (4): 795–814.

Kenett, R.S. and Shmueli, G. (2014). On information quality. *Journal of the Royal Statistical Society A* 177 (1): 3–38.

Kenett, R.S. and Shmueli, G. (2016). *Information Quality: The Potential of Data and Analytics to Generate Knowledge*. Wiley.

Kourti, T. and MacGregor, J.F. (1996). Multivariate SPC methods for process and product monitoring. *Journal of Quality Technology* 28 (4): 409–428.

Lee, J.-M., Yoo, C., and Lee, I.-B. (2004). Fault detection of batch processes using multiway kernel principal component analysis. *Computers and Chemical Engineering* 28 (9): 1837–1847.

Little, R.J.A. and Rubin, D.B. (2002). *Statistical Analysis with Missing Data*, 2e. Hoboken, NJ: Wiley.

Liu, G.P., Billings, S.A., and Kadirkamanathan, V. (2000). Nonlinear system identification using wavelet networks. *International Journal of Systems Science* 31 (12): 1531–1541.

Louwerse, D.J. and Smilde, A.K. (2000). Multivariate statistical process control of batch processes based on three-way models. *Chemical Engineering Science* 55 (7): 1225–1235.

Martínez, À., Riu, J., and Rius, F.X. (2002). Application of the multivariate least squares regression method to PCR and maximum likelihood PCR techniques. *Journal of Chemometrics* 16: 189–197.

Nelson, P.R.C., Taylor, P.A., and MacGregor, J.F. (1996). Missing data methods in PCA and PLS: score calculations with incomplete observations. *Chemometrics and Intelligent Laboratory Systems* 35: 45–65.

Nikolaou, M. and Vuthandam, P. (1998). FIR model identification: parsimony through kernel compression with wavelets. *AIChE Journal* 44 (1): 141–150.

Nomikos, P. and MacGregor, J.F. (1994). Monitoring batch processes using multiway principal component analysis. *AIChE Journal* 40 (8): 1361–1375.

Nomikos, P. and MacGregor, J.F. (1995). Multi-way partial least squares in monitoring batch processes. *Chemometrics and Intelligent Laboratory Systems* 30: 97–108.

Pacaux-Lemoine, M.-P., Trentesaux, D., Rey, G.Z., and Millot, P. (2017). Designing intelligent manufacturing systems through human-machine cooperation principles: a human-centered approach. *Computers and Industrial Engineering* 111: 581–595.

Page, E.S. (1954). Continuous inspection schemes. *Biometrics* 41 (1–2): 100–115.

Qin, S.J. (2014). Process data analytics in the era of big data. *AIChE Journal* 60 (9): 3092–3100.

Rato, T.J. and Reis, M.S. (2013a). Advantage of using decorrelated residuals in dynamic principal component analysis for monitoring large-scale systems. *Industrial and Engineering Chemistry Research* 52 (38): 13685–13698.

Rato, T.J. and Reis, M.S. (2013b). Defining the structure of DPCA models and its impact on process monitoring and prediction activities. *Chemometrics and Intelligent Laboratory Systems* 125: 74–86.

Rato, T.J. and Reis, M.S. (2013c). Fault detection in the Tennessee Eastman process using dynamic principal components analysis with decorrelated residuals (DPCA-DR). *Chemometrics and Intelligent Laboratory Systems* 125: 101–108.

Rato, T.J. and Reis, M.S. (2014). Sensitivity enhancing transformations for monitoring the process correlation structure. *Journal of Process Control* 24: 905–915.

Rato, T.J. and Reis, M.S. (2015a). Multiscale and megavariate monitoring of the process networked structure: M2NET. *Journal of Chemometrics* 29 (5): 309–322.

Rato, T.J. and Reis, M.S. (2015b). On-line process monitoring using local measures of association. Part I: detection performance. *Chemometrics and Intelligent Laboratory Systems* 142: 255–264.

Rato, T.J. and Reis, M.S. (2015c). On-line process monitoring using local measures of association. Part II: design issues and fault diagnosis. *Chemometrics and Intelligent Laboratory Systems* 142: 265–275.

Rato, T.J. and Reis, M.S. (2017a). Markovian and non-Markovian sensitivity enhancing transformations for process monitoring. *Chemical Engineering Science* 163: 223–233.

Rato, T.J. and Reis, M.S. (2017b). Multiresolution soft sensors (MR-SS): a new class of model structures for handling multiresolution data. *Industrial and Engineering Chemistry Research* 56 (13): 3640–3654.

Rato, T.J., Schmitt, E., de Ketelare, B. et al. (2015). A systematic comparison of PCA-based statistical process monitoring methods for High-dimensional, time-dependent processes. *AIChE Journal* 2016 (62): 5.

Rato, T.J., Blue, J., Pinaton, J., and Reis, M.S. (2017). Translation invariant multiscale energy-based PCA (TIME-PCA) for monitoring batch processes in semiconductor manufacturing. *IEEE Transactions on Automation Science and Engineering* 14 (2): 894–904.

Reis, M.S. (2009). A multiscale empirical modeling framework for system identification. *Journal of Process Control* 19 (9): 1546–1557.

Reis, M.S. (2013a). Applications of a new empirical modelling framework for balancing model interpretation and prediction accuracy through the incorporation of clusters of functionally related variables. *Chemometrics and Intelligent Laboratory Systems* 127: 7–16.

Reis, M.S. (2013b). Network-induced supervised learning: network-induced classification (NI-C) and network-induced regression (NI-R). *AIChE Journal* 59 (5): 1570–1587.

Reis, M.S. (2015). An integrated multiscale and multivariate image analysis framework for process monitoring of colour random textures: MSMIA. *Chemometrics and Intelligent Laboratory Systems* 142: 36–48.

Reis, M.S. and Bauer, A. (2009). Wavelet texture analysis of on-line acquired images for paper formation assessment and monitoring. *Chemometrics and Intelligent Laboratory Systems* 95 (2): 129–137.

Reis, M.S. and Delgado, P. (2012). A large-scale statistical process control approach for the monitoring of electronic devices assemblage. *Computers and Chemical Engineering* 39: 163–169.

Reis, M.S. and Kenett, R.S. (2018). Assessing the value of information of data-centric activities in the chemical processing industry 4.0. *AIChE Journal* 64 (11): 3868–3881.

Reis, M.S. and Saraiva, P.M. (2004). A comparative study of linear regression methods in noisy environments. *Journal of Chemometrics* 18 (12): 526–536.

Reis, M.S. and Saraiva, P.M. (2005). Integration of data uncertainty in linear regression and process optimization. *AIChE Journal* 51 (11): 3007–3019.

Reis, M.S. and Saraiva, P.M. (2006a). Generalized multiresolution decomposition frameworks for the analysis of industrial data with uncertainty and missing values. *Industrial and Engineering Chemistry Research* 45: 6330–6338.

Reis, M.S. and Saraiva, P.M. (2006b). Heteroscedastic latent variable modeling with applications to multivariate statistical process control. *Chemometrics and Intelligent Laboratory Systems* 80: 57–66.

Reis, M.S. and Saraiva, P.M. (2006c). Multiscale statistical process control of paper surface profiles. *Quality Technology and Quantitative Management* 3 (3): 263–282.

Reis, M.S. and Saraiva, P.M. (2006d). Multiscale statistical process control with multiresolution data. *AIChE Journal* 52 (6): 2107–2119.

Reis, M.S. and Saraiva, P.M. (2008). Multivariate and multiscale data analysis. In: *Statistical Practice in Business and Industry* (eds. S. Coleman, T. Greenfield, D. Stewardson and D.C. Montgomery), 337–370. Chichester, UK: Wiley.

Reis, M.S. and Saraiva, P.M. (2012). Prediction of profiles in the process industries. *Industrial and Engineering Chemistry Research* 51: 4524–4266.

Reis, M.S., Rendall, R., Chin, S.-T., and Chiang, L.H. (2015). Challenges in the specification and integration of measurement uncertainty in the development of data-driven models for the chemical processing industry. *Industrial and Engineering Chemistry Research* 54: 9159–9177.

Reis, M.S., Braatz, R.D., and Chiang, L.H. (2016). Big data – challenges and future research directions. *Chemical Engineering Progress*, Special Issue on Big Data (March): 46–50.

Rendall, R., Lu, B., Castillo, I. et al. (2017). A unifying and integrated framework for feature oriented analysis of batch processes. *Industrial and Engineering Chemistry Research* 56 (30): 8590–8605.

Roberts, S.W. (1959). Control charts tests based on geometric moving averages. *Technometrics* 1 (3): 239–250.

Rooney, W.C. and Biegler, L.T. (2001). Design for model parameter uncertainty using nonlinear confidence regions. *AIChE Journal* 47 (8): 1794–1804.

Rosipal, R. and Trejo, L.J. (2001). Kernel partial least squares regression in reproducing Kernel Hilbert space. *Journal of Machine Learning Research* 2: 97–123.

Saraiva P. (1993). *Data-Driven Learning Frameworks for Continuous Process Analysis and Improvement*. PhD thesis, Massachusetts Institute of Technology.

Saraiva, P. (1995). Inductive and analogical learning: data-driven improvement of process operations. In: *Intelligent Systems in Process Engineering, Advances in Chemical Engineering*, vol. 22 (eds. G. Stephanopoulos and C. Han), 377–435. Academic Press.

Shao, R., Jia, F., Martin, E.B., and Morris, A.J. (1999). Wavelets and non-linear principal components analysis for process monitoring. *Control Engineering Practice* 7: 865–879.

Shewhart, W.A. (1931). *Economic Control of Quality of Manufactured Product* (Vol. Republished in 1980 as a 50th Anniversary Commemorative Reissue by ASQC Quality Press). New York: D. Van Nostrand Company, Inc.

Sjöberg, J., Zhang, Q., Ljung, L. et al. (1995). Nonlinear black-box modeling in system identification: a unified overview. *Automatica* 31 (12): 1691–1724.

Stephanopoulos, G. (2017). Synthesis and Computing in PSE: 45 Year Travelogue of an Unindoctrinated Academic – Farewell Lecture. 2040 Visions of Process Systems Engineering Symposium, MIT, Cambridge, MA (1–2 June).

Strobl, C., Malley, J., and Tutz, G. (2009). An introduction to recursive partitioning: rationale, application, and characteristics of classification and regression trees, bagging, and random forests. *Psychological Methods* 14 (4): 323.

Tian, X.M., Zhang, X.L., Deng, X.G., and Chen, S. (2009). Multiway kernel independent component analysis based on feature samples for batch process monitoring. *Neurocomputing* 72 (7–9): 1584–1596.

Tiao, G., Bisgaard, S., Hill, W. et al. (eds.) (2000). *Box on Quality and Discovery with Design, Control and Robustness*. Wiley.

Van den Kerkhof, P., Vanlaer, J., Gins, G., and Van Impe, J.F.M. (2012). Dynamic model-based fault diagnosis for (bio)chemical batch processes. *Computers and Chemical Engineering* 40: 12–21.

Vanlaer, J., Gins, G., and Van Impe, J.F.M. (2013). Quality assessment of a variance estimator for partial least squares prediction of batch-end quality. *Computers and Chemical Engineering* 52: 230–239.

Walczak, B. and Massart, D.L. (2001). Dealing with missing data. *Chemometrics and Intelligent Laboratory Systems* 58 Part I: 15–27, Part II: 29–42.

Wentzell, P.D., Andrews, D.T., Hamilton, D.C. et al. (1997a). Maximum likelihood principal component analysis. *Journal of Chemometrics* 11: 339–366.

Wentzell, P.D., Andrews, D.T., and Kowalski, B.R. (1997b). Maximum likelihood multivariate calibration. *Analytical Chemistry* 69: 2299–2311.

Willsky, A.S. (2002). Multiresolution Markov models for signal and image processing. *Proceedings of the IEEE* 90 (8): 1396–1458.

Wise, B.W. and Gallagher, N.B. (1996). The process chemometrics approach to process monitoring and fault detection. *Journal of Process Control* 6 (6): 329–348.

Zhang, Y.W. and Qin, S.J. (2007). Fault detection of non-linear processes using multiway Kernel independent analysis. *Industrial and Engineering Chemistry Research* 46: 7780–7787.

Zhao, C.H., Gao, F.R., and Wang, F.L. (2009). Nonlinear batch process monitoring using phase-based kernel independent component analysis-principal component analysis. *Industrial and Engineering Chemistry Research* 48: 9163–9174.

7

Virtualization of the Human in the Digital Factory

Daniele Regazzoni and Caterina Rizzi

Synopsis

In the paradigm of the digital factory, human beings have to be considered as the focal point of technologies, activities, and information. This requires work environments that consider the needs and coexistence of users with different levels of experience, age, habits, multicultural environments, and sociocultural conditions that vary from company to company. The proper integration of workers with a high variety of skills, competencies, educational levels, and cultures is considered a great challenge for the factory of the future. In such a context, the development of a population of human avatars that interact with the environment and the products by assuming postures, supporting musculoskeletal loads, and repeating activities over time, permits the impact of new technologies and industrial processes to be effectively analyzed. This chapter presents an overview of virtual humans/digital human models (DHMs) that can be considered along the design and manufacturing processes. First, a brief historical description is given together with the main problems and the potential applications of such virtual human models. Then, a taxonomy that subdivides DHMs into five main categories is presented. The integration of DHMs with virtual/augmented reality technology and motion capture systems is also considered, in order to improve the level of interaction and realism within the virtual environment and facilitate the evaluation of comfort levels and prediction of injuries that could arise when executing a task. Applicative examples are described for both design and manufacturing domains. Finally, conclusions are drawn as well as a description of future development trends.

Systems Engineering in the Fourth Industrial Revolution: Big Data, Novel Technologies, and Modern Systems Engineering, First Edition. Edited by Ron S. Kenett, Robert S. Swarz, and Avigdor Zonnenshain.

7.1   Introduction

In the era of the digital factory, products, technologies, and processes are conceived, developed, and implemented to be at the service of human beings, whether they are workers, managers, or consumers. The goal is to improve the quality of life and work conditions and operate in an attractive and safe environment during interactions with complex systems, products, and data. Technologies are no longer only a support for product development processes but are integrated into components, machines, management systems, and environments that become *cyber-physical production systems*, in which the operator is not just an element of the system or a user of services and information but the center around which digital technologies converge with data and activities of the industrial processes. Therefore, in the new paradigm of the digital factory, the human beings have to be considered as the focal point of technologies, activities, and information (Figure 7.1).

This drives the creation of work environments that consider the needs and coexistence of users with different levels of experience, age, habits, multicultural environments (e.g. ethnicity and religion), and sociocultural conditions that can vary from company to company always more geographically distributed and, therefore, located in very different socioeconomic contexts. The integration of workers with a high variety of skills, competencies, educational levels, and cultures is considered a great challenge for the factory of the future (Romero et al. 2016). Human skills, competencies, and needs should be considered from the early stages of the development process and continuously updated throughout the product and process life cycle. Their integration, within system design, can be formalized through a Human-Systems Integration (HSI) program and, with such a perspective, the designers should be able

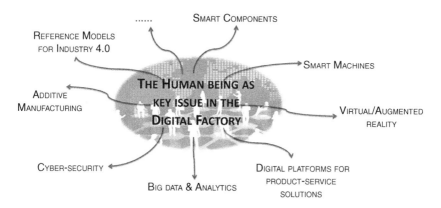

Figure 7.1 The human being at the center of the digital factory.

to quantify human performance in order to include humans as an element in a "system."

It is common practice to create virtual prototypes of products, machines, and systems, but the creation of human virtual models is not so widespread. The development of a population of avatars that interacts with the environment and the product by assuming postures, supporting musculoskeletal loads, and repeating activities over time permits the impact of new technologies (e.g. collaborative robots) and industrial processes to beanalyzed in the most effective way.

This chapter presents an overview of the virtual humans/digital human models (DHMs) that can be considered along the design and manufacturing processes. First, a brief historical description will be introduced, as well as the main problems to be faced and the potential of such models. Then, a taxonomy that subdivides DHMs into five main categories will be presented highlighting the different levels of human model details and their applications. The integration of DHMs with virtual/augmented reality technology and motion capture systems will be also considered respectively to improve the level of interaction and realism within the virtual environment and to drive the virtual human and facilitate the evaluation of comfort and prediction of injuries that could rise when executing a task. Applicative examples of each type of DHM will be also described for both design and manufacturing. Finally, conclusions are drawn and future development trends described.

7.2 The Problem

Before introducing the taxonomy of DHM systems, the main problems related to either design or manufacturing processes are contextualized.

Design

Designers have to consider that the engineering systems, which are becoming more and more complex and intelligent, can be used by human beings, acting either as workers along the production process or as final users, interacting with them in different ways according to their habits, goals, and needs. This requires both the development of products centered on human beings and suitable for the widest range of populations characterized by different size, gender, age, culture, preferences, and abilities (Ortengren and Sundin 2006), and the introduction of virtual ergonomics, i.e. an organized set of strategies and tools to simulate and evaluate ergonomic aspects from the conceptual design stage.

Adopting a human-centered design approach, which includes the human perspective in all steps of the design process, is necessary to move toward a custom-based design considering:

- Usability and comfort.
- Human behavior (physical and physiological).
- Population and aging.
- Different categories of workers/users.
- Cultural and social contexts.

In this context, DHM systems can enable designers to solve ergonomics and human factors along the whole product life cycle and face several ergonomic issues, such as:

- Simulate and evaluate product/system-human interaction considering different morphologies and populations according to the applicative context;
- Assess ergonomics issues using analysis systems such as RULA (Rapid Upper Limb Analysis), NIOSH (US National Institute for Occupational Safety and Health), and OWAS (Owako Working Posture Analysis System).
- Evaluate the body response to the interaction with the product, for example with auxiliary and rehabilitation devices specifically designed around the user's body shape.
- Size product according to the target population considering the anthropometric measures, as required in many industrial contexts (e.g. automotive, clothing, and transportation).

Manufacturing

Analogously, adopting a human-centered manufacturing approach requires the design of manufacturing process, plants, and workspaces considering the operators and their tasks; it is expected that human beings have control of the work process and technology.

The digital factory paradigm is transforming design and manufacturing together with the industrial workforce and the work environment (BCG 2015). Romero et al. (2016) presented the Operator 4.0 typology and how new technologies (e.g. virtual and augmented reality, collaborative robot, and wearable devices) can help operators to become "smarter workers" in future factory workplaces. For example, an operator with an exoskeleton becomes a super-strength worker, plus augmented reality augmented worker and plus collaborative robot (cobot) collaborative worker.

Despite the increasing automation of the production processes, there is still a need for highly-skilled operators of complex manual tasks. The design of manual operations is often lacking or not optimized with respect to the person, and the human intervention is not evaluated over the entire product/system life cycle but is limited to the company. External operators and end users are often excluded. Moreover, poor design of manual operations can cause inefficiencies due to an excess of precautions and chronic musculoskeletal diseases (MSD).

In such a scenario, DHMs can be used to:

- Simulate and validate the workplace (production plant) considering different populations, cultures and customs, age and disabilities.
- Simulate and evaluate postures according to ergonomics dictates and occupational medicine.
- Plan operations.
- Evaluate human-machine interaction to ensure that workers avoid injuries and have the best and safe working conditions. For example, in scenarios where the human and cobot perform the same task in a shared workspace, simulation enables detecting possible collisions.

7.3 Enabling Technologies

Introducing DHMs to design and simulate complex systems implies considering three main aspects (i.e. modeling, interactions and real motion) and enabling technologies that are not only related to the specific goal of the application but also to the specific context.

The primary and most important issue is modeling, which means creating a digital model of a human body or part of it, characterized by different level of details, starting from a simplified skeleton model up to a detailed representation including internal and external parts, such as ligaments, tendons, liver, lungs, and so on. We can distinguish three main approaches (Figure 7.2):

1) DHM tools (Polášek et al. 2015; Case et al. 2016). These permit the generation of standard human models (usually called virtual manikins) based on national and international databases and surveys of anthropometric measures (e.g. NHANES survey, North America automotive working population or Chinese database). These standards models can be modified to generate a specific target population.
2) Derived from a 3D scanner (Vitali and Rizzi 2018). This approach permits the capture of the external 3D shape of the human being or part of it (e.g. hands, head, legs) and the 3D model and related 3D anthropometric measures to be obtained. Although mainly developed for reverse engineering

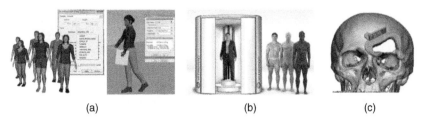

(a)   (b)   (c)

Figure 7.2 (a) A population of virtual manikins; (b) human model from body scanner; (c) skull 3D model from medical imaging.

applications, it can be used to design custom-fit products (e.g. boots, helmets, orthoses) around the digital model of the operators or final users (Colombo et al. 2018). A typical application is clothing design where we can find dedicated full body scanners but, due to the noninvasive character and ease of use, it is becoming appealing also for clinical applications and large-scale epidemiological surveys.

3) Derived from medical imaging. This approach is typical of the medical field and is used to acquire both external and internal parts of the human body using different equipment (Bertetti et al. 2006) from Computed Tomography (CT) to Magnetic Resonance Imaging (MRI) to Single Photon Emission Computerized Tomography (SPET) and Positron Emission Computerized Tomography (PET).

Differently from the DHM tools, the last two approaches require specific modeling tools to generate the 3D geometric model of the acquired body or of the anatomical districts. In addition, the human model provided by the DHM tools is substantially a cinematic chain, while the other two are mainly static.

We can also create hybrid representations merging models from different sources. For example, we can enhance the virtual manikin typical of DHM tools integrating a more detailed model of some parts of the body (e.g. head or leg) derived from a 3D scanner or medical imaging (Colombo et al. 2013).

The second aspect concerns the possibility to integrate DHM within virtual or augmented reality environments to enhance the level of interaction and realism. During the last two decades, various devices that simulate the human senses (e.g. vision, touch, and smell) have been developed with different performances, also at low cost (Figure 7.3). We may have head mounted displays (e.g.

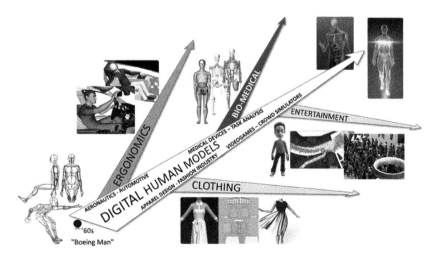

Figure 7.3 Evolution of DHMs into different fields of applications.

Oculus Rift) (Kot and Novák 2014; Oculus Rift 2019), hand tracking devices (e.g. Leap Motion and Intel Duo3D) (Leap Motion 2019, Bassily et al. 2014; Duo3D 2019), haptic devices (Colombo et al. 2010), and olfactory systems (Carulli et al. 2016). An example could be to provide a tactile feedback when simulating grasping tasks or during the simulation of maintenance operations to train operators. In such a way, as mentioned in Section 7.2.2, we can create the "virtual or augmented workers" as introduced by Romero et al. (2016).

The last issue concerns the use of motion capture systems to guide the human model using the motion of real people, such as workers or final users. The diffusion of low cost motion capture (Mocap) systems, pulled by the quick evolution of entertainment systems, boosted the use of tracking techniques together with DHMs. In fact, the evolution of Mocap systems as well as of DHMs allows a tight integration between the two and, nowadays, in several applications virtual humans are driven by data grabbed from real scenes with people playing a specific role. It is possible to acquire the postures assumed by the real workers in order to analyze task execution, identify diseases (such as musculoskeletal ones), and avoid injuries.

According to literature and tools available on the market, we can identify four main categories based on the working principles: mechanical, inertial, magnetic, and optical. Optical systems can be considered high-end technology and are the most used and discussed in the last few years. Optical systems rely on cameras or other optical sensors to track light sources or reflections or to identify profiles from video frames. There are two main techniques: with or without markers (Sharifi et al. 2017; Ong et al. 2017; Kraft et al. 2017). In the first case, the actor must wear markers in particular positions of the body (e.g. elbow, knee, etc.) to facilitate detection. The marker-based technology is widespread but has some disadvantages, such as high costs and usually a large installation space. Today, the most established solutions are infrared cameras and marker-based systems, such as Vicon (2019) and Qualysis (2019).

Marker-less mocap systems do not use markers but exploit algorithms for identifying human shape and create virtual skeletons following the acquired motions. The costs of these systems are more accessible than the previous ones. Among marker-less systems, RGB cameras were the first to be used but, in the last few years, depth cameras, i.e. RGB-D, based on infrared technology are becoming more and more interesting in the research community, as well as in industry, thanks to their low cost, simplicity, and portability. In fact, low cost cameras can be simply connected to a generic PC to collect data that can be post-processed with free or low cost software. The most diffused are Microsoft Kinect v1 and v2, even if they are now out of production and new solutions are emerging like the active camera that can be used outdoor (Bernal et al. 2017). Literature reports applications in several contexts, for example for virtual ergonomics (Krzeszowski et al. 2013) and preventing of MSDs (Dutta

2012; Colombo et al. 2013a) or for optimization in the automotive industry (Gomez-Levi and Kozak 2013).

7.4 Digital Human Models

Historical Overview

Digital humans started being a significant subject in the 1960s, even if in that period it was related only to elementary posture assessment with simple models in aerospace and automotive industries.

The first digital model, named "First Man" (also known as Boeman), was introduced by Boeing in 1959 to analyze pilot accommodation in the cockpit of a Boeing 747. Academia and private research organizations first, and then industry and medical organizations in later decades, developed more and more accurate models and ways to apply digital models of a human body as a means to fulfill their goals. Since the 1970s, automotive industry demand for DHMs encouraged the development of various virtual mannequins aimed at supporting the car design process and manufacture. Afterward, a variety of DHM solutions for a broad range of applications appeared, showing new functions and better performance, due also to the growth of computing power.

Academia and software providers have been mainly responsible for the improvement, which is far from an end, of the available solutions concerning level of detail, functionality, and degree of integration with other design solutions. Since the beginning of the 1980s, several research groups and companies have started to produce movies using virtual humans (Magnenat-Thalmann and Thalmann 2004). Among others, we can quote "Virtual Marilyn" by N. Magnenat-Thalmann and D. Thalmann in 1985.

DHM tools have been specialized on the basis of the features and needs of any application field, either in those in which models were first developed, such as automotive and aeronautics, or in all the other fields in which they started being used later on, such as constructions and building design, bioengineering and healthcare, entertainment and education. For instance, in the automotive field, applications vary from ergonomic design of car interiors to the design and assessment of manufacturing processes. D. Chaffin at the University of Michigan, N. Badler at the University of Pennsylvania, N. Magnenat-Thalmann and D. Thalmann contributed considerably to the foundation of today's human modeling (Ortengren and Sundin 2006). As an example, in the mid-1980s N. Badler and his research team developed Jack, a computer manikin software, now part of Siemens PLM Solutions suite. In addition, military authorities continuously funded researches for the development of more sophisticated digital humans. An example is Santos, a virtual soldier, developed by the Virtual Soldier Research Program at Iowa University since 2003 (Virtual Soldier 2019).

It acts as an independent agent and, apart from common features, it includes realistic skin deformation and contracting muscles.

The path toward the modeling of a complete human body does not have an end in the near future. Existing tools are sophisticated and mature but they cover a limited area of the human simulation, related only to the behavior of the musculoskeletal apparatus. Each other apparatus, beyond shape and movement, demands a deep chemical and biological modeling capability. Moreover, due to the complexity of each part of the body and, even more, to their mutual interactions we are still far from being able to describe all functions in a single multiscale model. Nevertheless, from an engineering perspective, we can easily find many applications to the models available that can bring real benefits. Actually, if the human body is considered as an hierarchic assembly of anatomical districts, each of which to some extent can be analyzed separately, we obtain models replicating or predicting reality with marginal mistakes. This is typically done, for instance, with musculoskeletal apparatus that allows simulating the kinematics and dynamics of any activity, forecasting or validating human postures, movements, and muscular stress.

Figure 7.3 shows the main branches of development of DHM tools according to their domain of applications.

A Taxonomy

By analyzing the state of the art (scientific literature and commercial solutions) and the types of applications, the DHMs have been grouped into five main categories as shown in Figure 7.4.

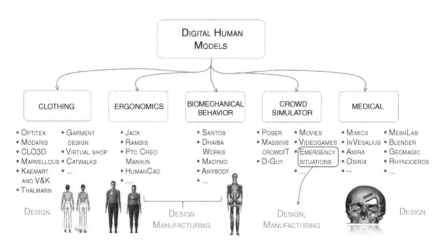

Figure 7.4 Virtual human classification with field of applications and some examples of available software tools.

The first category refers to human mannequins for the clothing industry. This particular kind of human model is not required to resemble the real human shape in any detail but only in those parts that are relevant from the tailor's point of view. Actually, it is more correct to talk about virtual mannequins in the case of a static model simulating the physical mannequin. Complete human models are required to satisfy the challenging demand for a virtual garment assessment. In order to tell if a virtual cloth is well fitting and good looking we need a digital human model able to stand still and to move respecting body volume changes. This allows creating clothing design tools and simulating cloth behavior and appearance on a virtual catwalk. There are also notable examples of commercial applications of DHMs for clothing and footwear, which are, for instance, virtual try-on systems (Jiyeon and Forsythe 2008; Yang et al. 2015) and magic mirrors (Kim and Cheeyong 2015).

DHMs for ergonomics are the most widespread and used. They are used both in design and manufacturing. For design, human-machine interaction scenarios can be created and assessed in a digital environment where the DHM deals with a digital prototype of the product. For instance, if we consider the notable example of a car driver, the interaction includes:

- Visibility of all the indicators of the cockpit, road visibility including signs, traffic lights, and pedestrians crossing the road.
- Reachability of the steering wheel and of the switches in the cockpit with one or two hands, reachability of the pedals with feet.
- Posture and comfort of the seat, including the proper positioning of the safety belt.

The results of the interaction of the digital human with the virtual prototype of the car prevents design mistakes from being discovered only when physical prototypes have been built, reducing iteration time and cost. Moreover, the simulation can be done with a set of digital humans of different size and body mass index to represent as closely as possible the target population of final users.

For the manufacturing, DHMs can be exploited to evaluate man-machine interaction, where the human is no longer the final user but a worker taking part in the manufacturing process. In this kind of scenario, in addition to the parameters listed for the design, it is possible to simulate also the physical effort and muscular stress of workers. Actually, repeated actions, handling of heavy loads or wrong postures can be the origin of injuries or chronic MSD. The creation of a scenario in which working procedures are reproduced allows the detailed effort of workers to be determined and eventual dangerous conditions to be predicted. Commercial DHM tools are generally provided with specific modules to evaluate the exposure of individual workers to ergonomic risk factors and calculate related indexes (e.g. NIOSH, RULA, or OWAS) used in standards for workload assessment.

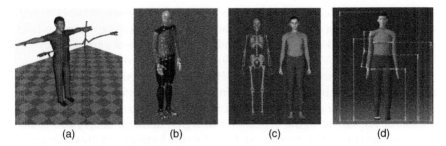

(a)   (b)   (c)   (d)

Figure 7.5 Examples of commercial mannequins for ergonomics analysis, respectively: (a) Jack (Siemens); (b) Ramsis (Human Solutions); (c) HumanCAD (NexGen Ergonomics); (d) Enovia (Dassault Systemes).

The most diffused DHM systems in industry belong to this category. Jack (Siemens 2019), Ramsis (Human Solutions 2019), HumanCAD (NexGen Ergonomics 2019), and Enovia (Dassault Systemes n.d.) are among the most known and used solutions (Figure 7.5), including ergonomics analysis for both design and manufacture of devices. They are all based on a hierarchical model of joints representing the articulations connected by rigid segments. Limbs are easily modeled according to anatomical morphology while the spine is simplified to reduce the computational effort required by a detailed modeling of every single vertebra. Each joint has some degrees of freedom and the rotations are limited within a physiological range.

In certain domains simplification of the models used for ergonomics analysis is not acceptable. This is the reason why detailed models were developed to study the biomechanical behavior of the human body. The digital humans belonging to this category are accurate and complex and they comprehend a plausible anatomic skeleton and muscular system, including ligaments and tendons.

This kind of model is used in all cases in which the human performs a complex movement or gesture that must be optimized and refined to be performed in a precise manner. Thus, the main applications are for professional sports activities, for motor rehabilitation, and for the simulation of soldier tasks. Sports gestures can be optimized in order to be as efficient as possible in terms of energy expenditure (e.g. for a marathon runner) or in terms of instant performance (e.g. for a high jumper). Rehabilitation after a trauma or a surgery requires a model of the body that describes with a high level of detail an anatomical district so that physiotherapists can rely on the DHM for assessment of the patient or simulation of the healing procedure. For instance, the shoulder is a typical example of a complex biomechanical system that benefits from a detailed modeling and for which ergonomic models are not suitable. Several national army institutions, as far as we can be aware of, are exploiting detailed biomechanical models of the human body to simulate extreme conditions soldiers could be facing during fighting. The most advanced literature in this field refers to the Virtual Soldier Project and to the Santos tool developed by the University of Iowa (Virtual Soldier 2019).

A completely different category of DHM refers to the domain of entertainment, where digital humans started being used for videogames and for movies instead of real characters. The large funds available in this field allowed a notable advancement in the appearance of digital humans, thanks also to the computer graphic performances reached since the beginning of this century. A noteworthy consequence of the development of simulation of a large number of characters, i.e. for emergency situations, is the creation of tools to simulate crowd behavior. These systems allow predicting the way a large number of individuals behave in dangerous conditions, such as the evacuation of a plant for a fire or an earthquake. Simulating the behavior of a mass by defining the parameters of single humans is extremely valuable when designing emergency exits or signs in order to avoid dangerous situations. Moreover, animation and simulation of crowd behavior can be used for training or education purposes. A research group contributing to this field of research since the 1990s has been the Miralab directed by Nadia Magnenat-Thalmann (www.miralab.ch/).

The last category refers to the medical field in which DHMs have shown their potential in several ways in recent years. The digital avatar can be partially or completely derived from a patient's data, so that it can be used for specific studies and simulations. 3D models of the body, or of a district, can be created starting from digital images captured with CT or MRI. Thus, the digital model represents the digital twin of the patient and can be used to simulate several conditions, e.g. a surgery or the implant of a prosthesis. Future enhancement will provide tools to simulate the evolution of a disease or the response of the body to pharmacotherapy.

The evolution of DHM is differentiating the models depending on the domain of application, of the functions they perform, and on the output they must generate. From a research point of view the most relevant focus is on the behavior of multiscale models of different apparatus, mainly for medical purposes. From an industrial perspective, diffusion in design and manufacturing departments requires a lot of work to reach a good level of DHMs. This could be achieved by fostering the knowledge about the benefits of using a DHM, by improving the integration with existing design tools and processes, and by decreasing the costs of software licenses.

Finally, DHMs can also play a significant role in the context of HSI (Demirel et al. 2016). Actually, HSI implies a multidisciplinary approach, which refers to several user-related areas, such as human factors engineering, training, manpower and personnel, safety and occupational health, habitability, personnel survivability. As an example, Demirel et al. (2016) describe an integration of existing tools, including DHMs, to consider a holistic approach on analysis of the human-system integration. Two novel DHM toolkits, respectively related to air quality and energy expenditure, have been introduced to evaluate the work environment and assess the effects of air quality.

7.5 Exemplary Applications

Products and machinery are becoming more and more complex and multifunctional. Therefore, it is important to be able to consider how the human being can interact with them ensuring proper use and a safe working environment and considering the needs of different typologies of user. The use of DHMs is not limited to the early stage of the product/system life cycle but can be profitably used also in the following phases up to service and maintenance applications (Ortengren and Sundin 2006). A DHM permits several ergonomic issues to be faced, such as comfort and posture prediction for envisaged groups of users, task evaluation and safety, visibility of products (machinery, equipment, vehicles, etc.), reach and grasp of devices (buttons, shelves, goods, etc.), multiperson interaction to analyze if and how multiple users can interact among them and with the product/system. They can be used for product design, crash testing, product virtual testing, workplace design, and maintenance, allowing a fast redesign and reducing the need and realization of costly physical prototypes, especially for those applications dealing with hazardous or inaccessible environments.

In the following sections some exemplary applications related to both design and manufacturing are presented.

Design

We can find several applications of DHMs to support design activity since its early stage especially in automotive (Colombo and Cugini 2005) and aeronautics (Green and Hudson 2011) fields. We briefly introduce two applications in different contexts, namely apparel and refrigeration.

The first example concerns clothing design. During last two decades several efforts have been made to virtualize as much as possible the clothing design process, starting from body acquisition through dedicated body scanners (Wang 2005; Apeagyei 2010; Hsiao and Tsao 2016) up to 3D modeling and simulation of garment wearability. In addition, attention is moving toward made-to-measure garments and this means that the customer's avatar should be considered the key element to design the virtual cloth using 3D garment modelers, such as Clo3D and Optitex.

Two types of DHM can be considered (Figure 7.6): (i) standard manikins based on anthropometric database embedded within the 3D garment modeler or (ii) customized virtual humans derived from body scanners.

(a) (b)

Figure 7.6 (a) Standard manikins (Source: Image courtesy Optitex); (b) a body scanner (Source: Image courtesy Vitronic) and an example of customer's avatar.

At present, various research work is being carried out to develop low cost solutions to acquire the human body and generate the customer's avatar (Wang et al. 2003, 2012; Wuhrer et al. 2014; Yang et al. 2016). For example, Wang et al. (2012) use a single RGB camera to acquire the customer's point cloud. The person has to turn him/herself in order to permit the acquisition of the whole body. Others aim at reconstructing an accurate 3D human shape by starting from clothed 3D scanned people (Yang et al. 2016; Zhang et al. 2014). In this case, the customers are not obliged to dress in tight clothes to get their correct measurements; however, designing a custom-fit garment also requires simulating the garment's wearability and comfort, considering the type of body shape (e.g. fat or slim) and customer's motion. This requires the use and integration within the clothing virtual environment of a motion capture system to gain the customer's movement and reproduce them when simulating garment behavior. The research work described in (Vitali and Rizzi 2018) presents a solution based on the combined use of a body scanner and a mocap system, both at low cost, to acquire the human body of the customer and his/her movement and their integration within a garment 3D computer aided design (CAD) modeler (Figure 7.7). This permits the creation of a kind of customized "animated

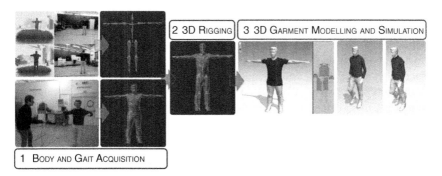

Figure 7.7 From customer body acquisition to customized "animated" avatar and garment design.

avatar" able to replicate the real movements of the subject and test the comfort of designed garment under real conditions.

The second example concerns the design of commercial refrigerated display units (Regazzoni and Rizzi 2013; Colombo et al. 2013b). This product is particularly interesting because the interaction with the different categories of people that may have to deal with it is crucial to provide an efficient and robust outcome. In fact, beside final user actions, maintaining, assembling, installing, or performing other activities related to product life cycle require a human intervention and, thus, a proper ergonomic design and check are necessary. We must consider several system variables, such as product configuration (e.g. different geometries or dimensions, with or without a door) as well as different human features (e.g. size, posture, position, orientation). In addition, various environmental conditions may occur. The methodology described in Colombo et al. (2013b) had the goal to increase repeatability and robustness of ergonomic analysis integrating virtual prototyping techniques and DHM tools, specifically those belonging to the second group, namely virtual humans for ergonomic analysis. Figure 7.8 shows the three categories of considered users (i.e. customers picking up goods, workers in charge of checking exposed products and filling out shelves with new ones, and maintenance technicians who need to access to specific components) and ergonomics issues analyzed for product design and assessment.

For each category some ergonomic aspects are more relevant than others. For example, visibility and reachability of goods are important for customers, while postures and stress for workers who have to load the display unit and repeat the task for hours and, therefore, may suffer of musculoskeletal disorders.

Ergonomic simulations have been carried out varying equipment type, display unit configuration (number and types of shelves, etc.), manikin size (5th, 50th, and 95th percentiles), and packed food (sizes and weight). The test campaign provided a large amount of data that has been organized in tables and histograms, allowing the designer to evaluate rapidly not only the

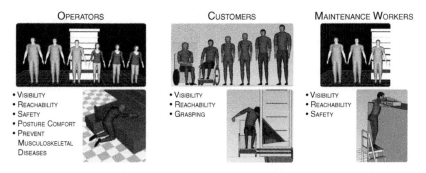

Figure 7.8 Refrigerator's users, role, and ergonomics issues.

AS-IS LOWEST SHELF				Female 5%	Female 50%	Male 50%	Male 95%
REACHABILITY			% reached	80	85	100	100
POSTURE			Owas class	2	2	4	4
F A T I G U E	Load 1 Kg	Lower Back analysis	L4L5 (Nm)	25	80	120	125
			Spinal Forces (N)	800	1500	2400	2600
			Muscle tension (N)	200	520	1100	1100
			Static force	390	390	450	430
		Fatigue-recovery	Cycles	30	30	26	25
			Recovery Time (s)	0,45	0,57	10,16	11,95
	Load 3 Kg	Lower Back analysis	L4L5 (Nm)	75	85	125	130
			Spinal Forces (N)	1500	1600	2600	2900
			Muscle tension (N)	520	600	1200	1250
			Static force	410	410	465	470
		Fatigue-recovery	Cycles	28	28	18	13
			Recovery Time (s)	4,48	4,536	30,28	37,89
	NIOSH		RWL[1]	3,88	3,65	3,37	3,65
			LI[2]	0,39	0,41	0,45	0,41
			CLI[3]	1,393	1,478	1,605	1,478

(1) Reccommended Weight Limit; (2) Lifting Index; (3) Composite Lifting Index

Figure 7.9 Results for the lowest shelf and postures assumed by the 5th female and 95th male percentiles.

specific machinery but also to compare different types of refrigerated units with respect to packed food loading and expository space. The most critical shelves were the highest and the lowest. As an example, Figure 7.9 shows the data acquired for the lowest shelf related to reachability and posture evaluation as well as the postures assumed by the 5th female and 95th male to load it. For example, we can observe that the females cannot reach the back with a posture that is barely acceptable (OWAS class 2), while males can reach 100% of the shelf but assuming a posture (OWAS class 4) that is risky and requires urgent correction to avoid musculoskeletal disorders.

On the basis of results analysis, three different lifting platforms and a new design concept have been designed to overcome the identified limits respectively for the highest and lowest shelves.

Two new simulation campaigns have been carried out to evaluate the lifting platform and the results have been compared with the previous situation. A design review process based on results comparison brought to the solution of creating a fourth platform (Figure 7.10a) that is able to raise and then to move toward the display unit getting the worker closer to the shelves especially for

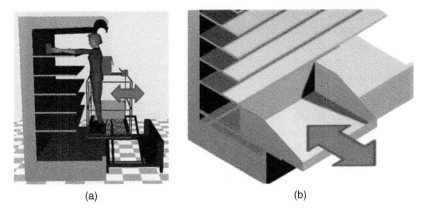

Figure 7.10 (a) Lifting platform; (b) sliding shelf.

the highest one. This solution permits 100% of the shelf to be reached for all percentiles as well as improving OWAS class of postures assumed by male percentiles. For the lowest shelf, it has been designed a sliding shelf (Figure 7.10b), which eases filling out, guarantees optimal reachability for everybody and, in most cases, decreases force needed.

Even if the company was aware of some drawbacks, the methodology permitted quantitative data to be acquired, alternative solutions to be evaluated, and, above all, design guidelines for future developments identified. For example, a reachability and visibility study carried out for a high temperature open display unit and low temperature closed display unit (with doors), being used by customers, has given significant insights to engineers for a better product design involving aspects such as shelf position, inclination, and lighting. In particular, the influence of frames and doors when taking a product out of the shelf and the visibility both from the lane and from a close position have been considered to redesign the display and guarantee a better buying experience.

Combining results from different human sizes and different postures and a graphical representation of results make it possible to identify guidelines for light positioning to augment visibility and reduce energy consumption, and to understand which parameters of the frame layout can easily and quickly increase or decrease visibility.

As mentioned, research and experience in different industrial contexts have proved the validity and potential of such tools; however, their usage is not widespread. One of the main problems is the lack of consolidated and systematic methodology that can guide the engineers to properly select and apply these tools along the whole design process. The adoption of a systematic approach to make virtual ergonomics a part of the product development process permits the occurrence of design mistakes to be decreased before realizing the physical prototypes (and therefore cost), enhances users' interaction, and improves workers' health and productivity.

Manufacturing

Ergonomic analysis of work tasks is a typical application of human models and may concern several types of products often manufactured by SMEs (Small-Medium Enterprises). Ergonomics in manufacturing is a combination of disciplines and the human factors come together with the equipment (machine/workplace design) and job requirements (task design). The use of DHMs is related to several aspects of the manufacturing in the era of the Digital Factory, from task design (e.g. management of time, ease of use, information overload, task scheduling) to workplace and manufacturing facilities to provide function and arrangement, assembly, servicing, reliability, manufacturability. It is becoming more and more important preventing or reducing occupational illness and injury by making changes to the design of work and workplace.

There are guidelines of "common sense" (e.g. use the right tool for any task, keep close to you objects you use frequently) but virtual humans are required to simulate a workers'/operators' population doing a set of tasks.

Simulations can be performed to:

- Optimize workers' efficiency (e.g. fix inefficiencies due to lack or excess of prevention).
- Optimize manual handling of material, goods, and tools.
- Prevent MSD.
- Decrease the risk of injuries.
- Improve operators' willingness to work by providing a more attractive workplace.

This means being able to guarantee occupational safety health and "fitting the work to the workers."

The potential of DHMs has been proven in various research works. As an example, Colombo and Cugini (2005) presented an application related to a workstation equipped with an automatic riveting system to assembe household products produced by a small manufacturer of machine tools. A DHM has been used to evaluate the comfort for the operator by varying the virtual workstation configuration (e.g. stool position and height or treadle drive orientation) and, therefore, to evaluate different working conditions and postures. The angles assumed by the human joints for each adjustment of the workstation have been analyzed to identify acceptable configurations. Figure 7.11 shows the acceptable values of the joints angles for a simulation carried out with a 50th percentile virtual human and a given configuration of the workstation.

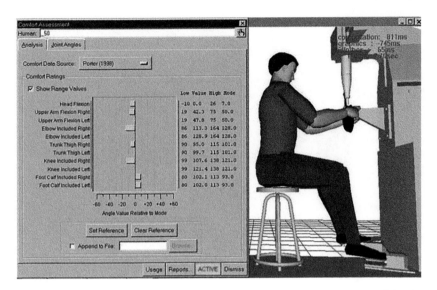

Figure 7.11 Acceptable values (green) for joints angles. Source: Image courtesy Colombo and Cugini 2005.

Similarly, Jovanovic et al. (2007) present an application where DHMs have been used to design an assembly system and plan assembly tasks. First, the design of a workplaces should fit the workers' population of the company. For example, available furniture (e.g. chairs and tables) and tools should be reconfigurable according to the worker's body sizes. Then, the assembly process should respect the ergonomic rules and this means that tasks should be evaluated considering postures, task repetitiveness, tool weight, etc. Figure 7.12 portrays an example of an assembly line ergonomically validated.

Another example, described in Regazzoni and Rizzi (2013), concerns maintenance. The case study refers to plants made of compressor units placed on the roof of a supermarket, connected to and serving a number of refrigerated display units containing fresh food. The case is particularly interesting because there are several issues concerning ergonomics for both maintenance workers and for the compressor unit, which must guarantee a continuous functioning for several years despite some well-known issues related to maintenance. The compressor unit comprises a modular steel frame provided with antivibration devices, 2–6 piston compressors of different sizes connected with the line to and from the display units with a complex system of copper pipes including filters, valves, instruments to monitor the functioning, and some other auxiliary devices. The authors propose a systematic approach based on a step-by-step procedure, which implies not only the use of DHM tools for ergonomics but also of a mocap system when data about maintenance operations are not available from the company and it is necessary to get real

Figure 7.12 Assembly line validated using DHMs. Source: Image courtesy Vukica et al. 2007.

postures assumed by the workers. In fact, the methodology foresees two initial steps to acquire knowledge acquisition/formalization and operators' behavior formalization (best practice). The analysis has been carried out considering input coming from the design department, the manufacturing experts, workers, and the team in charge of maintenance operations; the substitution of a filter for the refrigerant fluid has been considered. Even if this is not a frequent condition, it is extremely crucial because it implies the cut and the welding of copper pipes, which may be very difficult to access and operate. Virtual ergonomic analysis permitted evaluation of visibility, reachability and postures, stress, and fatigue. Figure 7.13 shows the compressor unit and an example of posture and visibility of the welding site, while Figure 7.14 shows two frames of the mocap acquisition with the operators in two different positions.

The formalization into a precise algorithm met the need of the enterprise for standardized procedures at any level: in the product development department as well as for the manufacturing and assembly process and for teams in charge of plant maintenance. The methodology and the results permitted the identification of new triggers for a better architecture of the compressor unit and quick and reliable tests of new ideas. The analysis highlighted known problems about posture or comfort, but also aspects never taken into consideration. In the specific case study, the pipe is on the bottom of the compressor unit and the operator was on his knees to reach the zone. However, there were also unexpected results and indications related to the operator's size. For instance, in

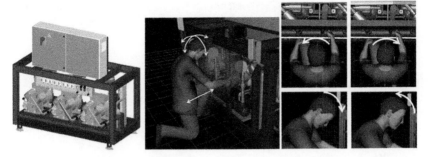

Figure 7.13 Compressor unit and operator's posture and visibility.

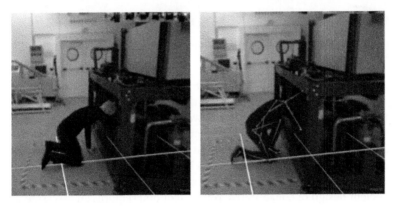

Figure 7.14 Frames of a mocap process with the worker in two different postures.

company maintenance teams tall men were preferred to perform repair activities, while simulations results showed that average men and women had better reachability performances.

Another interesting application concerns the configuration of an assembly line for automotive industry, integrating a Knowledge Base Editor (KBE) system with virtual reality (Rossoni et al. 2019) and DHM tools. The KBE system automatically configures the assembly plant and, starting from technical data coming from the KBE system, a virtual reality environment is automatically created, including the animation of the objects present in the scene, like robots and pallets. The DHM tool permits the ergonomics issues to be analyzed when the operators interact with the assembly line stations. Figure 7.15 portrays an example of an assembly line with the operator in front of a station with which has to interact.

The last example regards the interaction between humans and robots. During the last three decades, robots have been introduced in manufacturing industry to substitute humans in performing tasks that typically require speed

Figure 7.15 An image of the virtual reality environment. Source: Image courtesy Rossoni et al. 2019.

and high repeatability. On the other hand, human beings execute better tasks such as picking and carrying and have also the capacity to handle unexpected situations. Therefore, collaborative frameworks, in which robots share the workspace with humans and closely interact with them to perform tasks can lead to increased levels of productivity and decrease the number of musculoskeletal disorders. In such a context, safety is one of the primary challenges in human and robot interaction (Murashov et al. 2016; Chryssolouris et al. 2014) and an efficient design of collaborative robots (cobots) requires an assessment of the ergonomic benefits they can offer. DHMs can be used to simulate the worker, evaluate the robot-worker system and ensure that workers avoid injuries and have the best working conditions. Various researches have been carried to face these problems (Matsas and Vosniakos 2017; Morato et al. 2013). For instance, Morato et al. (2013) presented "a real-time precollision strategy that allows a human to operate in close proximity with the robot and pause the robot's motion whenever a collision between the human and any part of the robot is imminent." The presence of the human being is detected by multiple sensors (MS Kinects) located at various points of the work cell. Each sensor observes the human and outputs a 20 degrees of freedom human model. The robot and the human model are approximated by dynamic bounding spheres and the robot's motion is controlled by tracking the collisions between these spheres. This solution represented one of the first successful attempts to build an explicit human model online and use it to evaluate human-robot interference. Figure 7.16 portrays the experimental setup developed to test the human-robot interaction scenario.

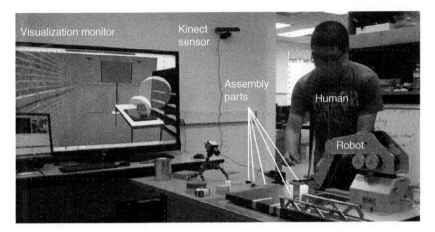

Figure 7.16 Experimental setup to test human-robot interaction. Source: Image courtesy Morato et al. 2013.

The previous examples demonstrate that the DHM tool can be used profitably to solve several problems related to occupational safety and health of workers, especially in era of the digital factory.

7.6 Summary

The benefits achievable by the introduction of digital humans in the digital factory are well known and consolidate concerns about ergonomics and anthropometric design of products and of production facilities. Digital manikins remove the need for costly physical prototypes and real human simulations and, by considering ergonomics from the beginning of the development process, cut the timeframe. Commercial solutions are generally increasing their usability and integration with other Industry 4.0 tools, fostering their application in everyday practices. As well as the commercially available solutions, there are also academic and research methodologies developed for ad hoc requests of simulation and assessment, which are not yet optimized, standardized, or integrated into the more extensive tools for digital factory planning.

The few commercial DHM solutions sharing most of the market were created for precise goals and present substantial differences. This leads to simulation outcomes that may seem to be remarkably different if they are obtained with two different systems having different constitution and parameter settings, and this is a major limitation in the acceptance of DHM tools in industry. Moreover, users are likely to use only one DHM tool. Also, portability of simulation models and results is almost impossible by means of ad hoc conversion modules and, thus, the upgrade to newest or better tools can be a hard task, too. Standardization of models, at least in terms of number and position and name of joints and anatomical parts, will ease the data exchange among different solutions and some more detailed tuning and convergence of conventions will make simulation tools more usable and reliable for technicians. Sharing a common base and best practices will facilitate the creation of a worldwide database of best practices for basic movements and tasks and improve the confidence in simulation results. To this aim, some researchers have been working to define, promote, and spread standards to make DHMs soar in industry. In Gunther and Wischniewski (2012) the reference to the existing ISO standards and the role of the standardization are reported. In particular, a database of parametric anthropometric measures (ISO 15535) and available data (ISO 7250) are the basis for the development of shared DHM format. In addition to this, the models describing the way human models interact with other human models and with the environment are crucial to gathering reliable results.

Designers know that using digital manikins to simulate work does not mean that real humans will act exactly as the manikins. The way any of us would perform a normal task on a shop floor is slightly different, and thus it is a nonsense to search for ultra-precise predictions. For specific cases, e.g. use of hands for precision tasks, it is difficult to model and simulate correct behavior and there are research groups that have been dealing with this issue for decades.

Rather than looking for unreachable forecasting of humans' activities, the aim of the DHM tool is to support engineers in making fast and correct decisions in the early phases of the product development process, either for product design or production planning. This will be more than enough to reach the required condition ensuring safety, comfort, and well-being of the humans involved in the process.

At present, research efforts are directed toward the incorporation of age-related parameters and to implement models capable of representing overweight or physically handicapped people. Concerning population aging, the idea of representing both anthropometric and mobility variability has been studied since the beginning of this century, mainly for healthcare applications and product design (Marshall et al. 2004; Goodman et al. 2007; Gonzale and Morer 2016). To determine a pragmatic approach to age modeling, in Högberg et al. (2008) the researchers propose a relatively small number of representative manikins (i.e. six per gender) and a pragmatic approach of employing them in design. The body mass index, when over a certain threshold, creates the need for a different modeling of male and female body shape, keeping into consideration reduced mobility and different behavior. In Ngo et al. (2016) an analysis was carried out to identify jobs that were challenging for older or obese workers and then create DHMs that represent heavier workers, so that their skills and condition can be considered. Analogously, considering specific human abilities or disabilities in DHMs provides the opportunity of better considering workers with specific conditions in regular work places. For able-bodied employees, ergonomic workplace design prevents the worsening of existing ailments and the development of new ones. Therefore, digital planning needs new workflows and interpretation of simulation results according to specific use cases.

A broader diffusion of DHMs in industry will stimulate the definition in a short–middle term of more specific and domain dependent best practices, standards, and use cases. Even if in some industrial contexts DHMs are quite common, there is still a lack of diffusion in SMEs, which could be due to several reasons, such as the *never used so far* syndrome, difficulties in estimating cost/benefit opportunity, inertia to change design paradigm, and cost of software solutions.

Nowadays, digital humans provide efficient support for product and product validation by means of rigid body mechanics. We can expect that more complex human models including soft tissue, already existing at academic level, will become a standard in industry in a near future. The most recent research lines are going in the direction of a more comprehensive approach in which the mechanical model of the body will be enriched not only with anatomical details but with mental features such as behaviors, abilities, and skills. In a long-term perspective, perceptual and cognitive models will be embedded into digital humans thanks to progress in artificial intelligence and virtual humans will get closer and closer to the digital twin in the digital factory.

References

Apeagyei, P.R. (2010). Application of 3D body scanning technology to human measurement for clothing fit. *International Journal of Digital Content Technology and its Applications* 4 (7): 58–68.

Bassily, D., Georgoulas, C., Guettler, J. et al. (2014). Intuitive and adaptive robotic arm manipulation using the leap motion controller. In: *ISR/Robotik 2014; 41st International Symposium on Robotics*, 1–7.

BCG (Boston Consulting Group) (2015) Report on Man and Machine in Industry 4.0: How Will Technology Transform the Industrial Workforce Through 2025? http://englishbulletin.adapt.it/wp-content/uploads/2015/10/BCG_Man_and_Machine_in_Industry_4_0_Sep_2015_tcm80-197250.pdf (Last accessed 9 June 2019).

Bernal, Y. et al. (2017). Development of underwater motion capture system for space suit mobility assessment. *Proceedings of the Human Factors and Ergonomics Society Annual Meeting* 6 (1): 945–949.

Bertetti, M., Bonacini, D., Colombo, G., and Magrassi, G. (2006). Reverse engineering and rapid prototyping techniques to innovate prosthesis socket design. In: *Proceedings of SPIE-IS&T, Electronic Imaging Three-Dimensional Image Capture and Applications VII*, San Jose, CA (15–19 January 2006), 1–11. Bellingham WA: SPIE.

Carulli, M., Bordegoni, M., and Cugini, U. (2016). Integrating scents simulation in virtual reality multisensory environment for industrial products evaluation. *Computer-Aided Design and Applications* 13 (3): 320–328.

Case, K., Marshall, R., and Summerskill, S. (2016). Digital human modelling over four decades. *International Journal of Digital Human* 1 (2): 112–131.

Chryssolouris, G. et al. (2014). ROBO-PARTNER: seamless human-robot cooperation for intelligent, flexible and safe operations in the assembly factories of the future. *Procedia CIRP* 23: 71–76.

Colombo, G. and Cugini, U. (2005). Virtual humans and prototypes to evaluate ergonomics and safety. *Journal of Engineering Design* 16 (2): 195–207.

Colombo, G., De Angelis, F., and Formentini, L. (2010). Integration of virtual reality and haptics to carry out ergonomic tests on virtual control boards. *International Journal of Product Development* 11 (1): 47–61.

Colombo, G., Facoetti, G., Regazzoni, D., and Rizzi, C. (2013). A full virtual approach to design and test lower limb prosthesis. *Virtual and Physical Prototyping* 8 (2): 97–111.

Colombo, G., Regazzoni, D., and Rizzi, C. (2013a). Markerless motion capture integrated with human modeling for virtual ergonomics. In: *Digital Human Modeling and Applications in Health, Safety, Ergonomics, and Risk Management. Human Body Modeling and Ergonomics*, Lecture Notes in Computer Science, vol. 8026 LNCS (PART 2) (ed. V.G. Duffy), 314–323. Springer.

Colombo, G., Regazzoni, D., and Rizzi, C. (2013b). Ergonomic design through virtual humans. *Computer-Aided Design and Applications* 10 (5): 745–755.

Colombo, G., Rizzi, C., Regazzoni, D., and Vitali, D. (2018). 3D interactive environment for the design of medical devices. *International Journal on Interactive Design and Manufacturing* 12 (2): 699–715.

Dassault Systemes, (2019) Enovia V5 DMU. https://www.3ds.com/products-services/enovia (Last accessed 8 June 2019).

Demirel, H.O., Zhang, L., and Duffy, V.G. (2016). Opportunities for meeting sustainability objectives. *International Journal of Industrial Ergonomics* 51: 73–81.

Duo3D, (2019) Duo3D Solutions. Available from https://duo3d.com (Last accessed 9 June 2019).

Dutta, T. (2012). Evaluation of the Kinect sensor for 3-D kinematic measurement in the workplace. *Applied Ergonomics* 43 (4): 645–649.

Gomez-Levi, W.N. and Kozak, W.J. (2013) Integration of motion capture with CAD for camera placement. 4th International Conference, Digital Human Modeling and Applications in Health, Safety, Ergonomics, and Risk Management, DHM 2013, Las Vegas, NV (21–26 July 2013).

Gonzale, I. and Morer, P. (2016). Ergonomics for the inclusion of older workers in the knowledge workforce and a guidance tool for designers. *Applied Ergonomics* 53: 131–142.

Goodman, J., Langdon, P., and Clarkson, P.J. (2007). Formats for user data in inclusive design. In: *Universal Access in Human Computer Interaction. Coping with Diversity. UAHCI 2007*, Lecture Notes in Computer Science, vol. 4554 (ed. C. Stephanidis), 117–126. Springer.

Green, R.F. and Hudson, J.A. (2011). A method for positioning digital human models in airplane passenger seats. In: *Advances in Applied Digital Human Modeling*. Boca Raton, FL: CRC Press.

Gunther, P. and Wischniewski, S. (2012). Standardisation of digital human models. *Ergonomics* 55 (9): 1115–1118.

Högberg, D., Hanson, L., Lundström, D., et al. (2008). Representing the elderly in digital human modeling. Proceedings of the 40th Annual Nordic Ergonomic Society Conference, Reykjavik, Iceland.

Hsiao, S.W. and Tsao, Y.C. (2016). Applying multiple kinect on the development of a rapid 3D mannequin scan platform. *International Journal of Mechanical and Mechatronics Engineering* 10: 1185–1189.

Human Solutions, (2019) Ramsis, https://www.human-solutions.com/en/index.html (Last accessed 4 September 2019).

Jiyeon, K. and Forsythe, S. (2008). Adoption of virtual try-on technology for online apparel shopping. *Journal of Interactive Marketing* 22 (2): 45–59.

Kim, M. and Cheeyong, K. (2015). Augmented reality fashion apparel simulation using a magic mirror. *International Journal of Smart Home* 9 (2): 169–178.

Kot, T. and Novák, P. (2014). Utilization of the oculus rift HMD in mobile robot teleoperation. *Applied Mechanics and Materials* 555: 199–208.

Kraft, M. et al. (2017). Toward evaluation of visual navigation algorithms on RGB-D data from the first- and second-generation kinect. *Machine Vision and Applications* 28 (1–2): 61–74.

Krzeszowski, T., Michalczuk, A., Kwolek, B. et al. (2013). Gait recognition based on marker-less 3D motion capture. In: *Proceedings of the 10th IEEE International Conference on Advanced Video and Signal Based Surveillance*, 232–237. IEEE.

Leap Motion, (2019) Reach into the future of virtual and augmented reality. https://www.leapmotion.com (Last accessed 4 September 2019).

Magnenat-Thalmann, N. and Thalmann, D. (1985). Computer animation. In: *Computer Animation: Theory and Practice*, Computer Science Workbench (eds. N. Magnenat-Thalmann and D. Thalmann), 13–17. Springer.

Magnenat-Thalmann, N. and Thalmann, D. (eds.) (2004). *Handbook of Virtual Humans*. Chichester, UK: Wiley.

Marshall, R., Cas, K., Porter, J.M. et al. (2004). Using HADRIAN for eliciting virtual user feedback in design for all. *Journal of Engineering Manufacture* 218 (B9): 1203–1210.

Matsas, E. and Vosniakos, G.C. (2017). Design of a virtual reality training system for human–robot collaboration in manufacturing tasks. *International Journal on Interactive Design and Manufacturing* 11: 139–153.

Morato, C., Kaipa, K., Zhao, B., and Gupta, S.K. (2013). Safe Human Robot Interaction By Using Exteroceptive Sensing Based Human Modeling. In: *Proceedings ASME 2013 International Design Engineering Technical Conferences & Computers and Information in Engineering Conference* , (IDETC/CIE 2013), Portland, OR (4–7 August 2013). ASME.

Murashov, V., Hearl, F., and Howard, J. (2016). Working safely with robot workers: recommendations for the new workplace. *Journal of Occupational Environment Hygiene* 13 (3): D61–D71.

NexGen Ergonomics, (2019) HumanCAD. http://www.nexgenergo.com/ergonomics/humancad.html (Last accessed 9 June 2019).

Ngo, S., Sommerich, C.M., and Luscher, A.F. (2016). Digital human Modeling of Obese & Aging Workers in automotive manufacturing. *Proceedings of the Human Factors and Ergonomics Society Annual Meeting* 60 (1): 1041–1045.

Oculus Rift. (2019) Play the next level of gaming. www.oculusrift.com (Last accessed 4 September 2019).

Ong, A., Harris, I.S., and Hamill, J. (2017). The efficacy of a video-based marker-less tracking system for gait analysis. *Computer Methods in Biomechanics and Biomedical Engineering* 20 (10): 1089–1095.

Ortengren, R. and Sundin, A. (2006). *Handbook of Human Factors and Ergonomics*. New York: Wiley.

Polášek, P., Bureš, M., and Šimon, M. (2015). Comparison of digital tools for ergonomics in practice. *Procedia Engineering* 100: 1277–1285.

Qualisys.(2019) Motion capture system. http://www.qualisys.com/ (Last accessed 4 September 2019).

Regazzoni, D. and Rizzi, C. (2013). Digital human models and virtual ergonomics to improve maintainability. *Computer-Aided Design and Applications* 11 (1): 10–19.

Romero, D. et al. (2016). Towards an operator 4.0 typology: a human-centric perspective on the fourth industrial revolution technologies. In: *Proceedings of the International Conference on Computers & Industrial Engineering (CIE46)*, , Tianjin, China (29–31 October 2016), 1–11.

Rossoni, M., Bergonzi, L., and Colombo, G. (2019). Integration of virtual reality in a knowledge-based engineering system for preliminary configuration and quotation of assembly lines. *Computer-Aided Design and Applications* 16 (2): 329–344.

Sharifi, A., Harati, A., and Vahedian, A. (2017). Marker-based human pose tracking using adaptive annealed particle swarm optimization with search space partitioning. *Image and Vision Computing* 62: 28–38.

Siemens (2019) Jack https://www.plm.automation.siemens.com/global/en/products/tecnomatix/human-modeling-simulation.html (Last accessed 4 September 2019).

Vicon (2019). http://www.vicon.com/ (Last accessed 4 September 2019).

Virtual Soldier (2019) The Virtual Soldier Research Program (VSR), The University of Iowa. https://www.ccad.uiowa.edu/vsr (Last accessed 9 June 2019).

Vitali, A. and Rizzi, C. (2018). Acquisition of customer's tailor measurements for 3D clothing design using virtual reality devices. *Virtual and Physical Prototyping* 13 (6): 1–15.

Jovanovic, V., Tomovic, M.M., Cosic, I., et al. (2007) Ergonomic Design of Manual Assembly Workplaces. Annual ASEE IL/IN Section Conference.

Wang, C.C.L. (2005). Parameterization and parametric design of mannequins. *Computer Aided Design* 37 (1): 83–98.

Wang, C.C.L., Wang, Y., Chang, T.K.K., and Yuen, M.M.F. (2003). Virtual human modeling from photographs for garment industry. *Computer-Aided Design* 35 (6): 577–589.

Wang, R., Choi, J., and Medioni, G. (2012). Accurate full body scanning from a single fixed 3D camera. In: *Proceedings of the 2nd Joint 3DIM/3DPVT Conference: 3D Imaging, Modeling, Processing, Visualization and Transmission*, 432–439.

Wuhrer, S. et al. (2014). Estimation of human body shape and posture under clothing. *Computer Vision and Image Understanding* 127: 31–42.

Yang, Y.I., Yang, C.K., Liao, X.L., and Chu, C.H. (2015). Virtual try-on of footwear in augmented reality using RGB-D cameras. In: *Proceedings of the 2015 IEEE International Conference on Industrial Engineering and Engineering Management*, 1072–1076. IEEE.

Yang, J., Franco, J.-S., Hétroy-Wheeler, F., and Wuhrer, S. (2016). Estimation of human body shape in motion with wide clothing. In: *Computer Vision – ECCV 2016* (eds. B. Leibe, J. Matas, N. Sebe and M. Welling), 439–454. Springer International Publishing AG.

Zhang, D., Wang, J., and Yang, Y. (2014). Design 3D garments for scanned human bodies. *Journal of Mechanical Science and Technology* 28 (7): 2479–2487.

8

The Dark Side of Using Augmented Reality (AR) Training Systems in Industry

Nirit Gavish

Synopsis

Augmented Reality (AR) systems, in which virtual information is superimposed on the real, physical environment, are becoming more and more used in industry. Some of their purpose is to serve as training systems. Surprisingly, when the purpose is training, some of the benefits that have been identified when the use of these systems is for on-the-job performance can turn into threats. Three of these threats were identified and studied in our lab in the recent years: physical fidelity vs. cognitive fidelity; the effect of feedback; and enhanced information channels. The findings, as well as practical implications, are discussed in this chapter.

8.1 The Variety of Options of AR Systems in Industry

Augmented Reality (AR) is a term defined as technologies in which virtual information is superimposed on the real, physical environment, and hence augments experience and interaction. The virtual information can be presented to the user by wearable devices, e.g. head mounted displays, or by small portable devices such as smartphones. The augmented information can take the form of visual objects, three dimensional sounds, vibrations, force feedback, smell, or any combination of these and more. AR is used in education, e.g. it enables learners to physically experience abstract material such as science rules (Klopfer 2008; Squire and Jan 2007), or to create virtual objects and connect them to specific places in the real world (Yuen et al. 2011). It is used to improve spatial ability and mental rotation skills (Dünser et al. 2006). AR is also used in many other fields such as design, the military field (Brown et al. 2006), and medical operations (Yeo et al. 2011). It seems that our world is moving toward

Systems Engineering in the Fourth Industrial Revolution: Big Data, Novel Technologies, and Modern Systems Engineering, First Edition. Edited by Ron S. Kenett, Robert S. Swarz, and Avigdor Zonnenshain.
© 2020 John Wiley & Sons, Inc. Published 2020 by John Wiley & Sons, Inc.

massive use of AR technologies, and these technologies have the potential to improve our life in many fields (Van Krevelen and Poelman 2010).

It is not surprising that in the recent years AR technologies have been implemented in industry. AR is used in picking and assembly production tasks to guide the user visually in new tasks in the spatial context where it is relevant (Paelke 2014). In the automotive industry, AR is used to compare real parts of a car with their associated construction data (Nolle and Klinker 2006). AR is highly useful in the construction industry for the purposes of virtual site visits, comparing as-built and as-planned status of projects, project planning, etc. (Rankohi and Waugh 2013). AR is used by shipyard operators to obtain information on their tasks and to interact with certain elements that surround them (Blanco-Novoa et al. 2018). Another important use of AR is in remote maintenance, because it enables cooperation between the on-the-spot technician and the manufacturer (Masoni et al. 2017).

8.2   Look Out! The Threats in Using AR Systems for Training Purposes

The use of AR in industry for the purpose of training has also been reported. AR was used for training of manual assembly tasks (Boud et al. 1999; Hořejší 2015), motherboard assembly (Westerfield et al. 2015), industrial maintenance (Besbes et al. 2012; Gavish et al. 2015; Schwald and De Laval 2003), ultralight and light sport aircraft maintenance (Christian et al. 2007), welding seams (Aiteanu et al. 2006), and aeronautical or aerospace maintenance (De Crescenzio et al. 2011; Haritos and Macchiarella 2005; Macchiarella and Vincenzi 2004; Rios et al. 2011).

However, while the benefits of using AR in industry are clear, using AR for training in industrial environments can be controversial. Several studies performed in recent years in our laboratory reveal that using AR in AR training systems can also have threats. Interestingly, some of the benefits of using AR can actually take the role of disadvantages when the AR system is being used for training purposes. It seems that on-the-job performance and training do not necessarily share the same system requirements. Designing the AR-based training systems should, therefore, consider the possible threats.

Three possible threats were identified in our laboratory. First, AR systems have the possible benefit that they may be highly correlated to the real world, because training tasks are performed in the real environment with the real device. However, it is possible that the so-called advantage of AR systems, i.e. of being closely related to the real world, may not be an advantage for training. Another threat is the ability to add feedback to the real-world experience when performing the task with the AR system. Feedback has many advantages but is it possible that it will decrease the effectiveness of learning? Lastly, the AR training system can enhance the real world with more information channels (visual, auditory, haptic, etc.). Nevertheless, do these added information channels really improve the training process?

The effects of each of these three threats on the effectiveness and efficiency of the industrial training are discussed here.

8.3 Threat #1: Physical Fidelity vs. Cognitive Fidelity

The Threat

Physical fidelity can be defined as the degree to which the simulation looks, sounds, and feels like the actual task (Alexander et al. 2005). The roots of preferring high physical fidelity systems for training can be found in Woodworth and Thorndike's (1901) paper, which claims that transfer from the first task to the second, real-world task occurs most efficiently when the two tasks have identical component activities. Similarly, Baldwin and Ford (1988) state that the more elements are shared between the two tasks, the better the transfer. In light of this notion, AR systems for training have the advantage of high physical fidelity, because the training takes place in the real world, in the real environment, and with the real devices and setting.

However, other researchers have questioned the importance of physical fidelity and state that the *cognitive fidelity* of the training system is much more important. Cognitive fidelity is the extent to which the training system engages the trainee in the types of cognitive activities involved in the real-world tasks (Kaiser and Schroeder 2003). Transfer of training, according to this notion, is obtained through training tasks and aids that maintain the correct stimulus–response relationships (Lathan et al. 2002). For example, in assembly tasks, high cognitive fidelity training system will involve training of the procedural skills needed for the task – the right steps in the right order. In tasks that require allocation of attention, as in driving, the training system should include practicing allocation of attention to several simultaneously tasks.

The benefit of high cognitive fidelity training systems was demonstrated in several fields. A simple computer game helped to train flight cadets in Israel Air Force to improve their attention management skills, showing a 30% increase in the actual flight scores after 10 hours in a training program (Gopher et al. 1994). Similarly, there are several examples of how training key perceptual–cognitive components of sports activities helps to improve performance in real-world tasks (Farrow et al. 1998; Williams et al. 2003).

Findings from Our Laboratory

Interestingly, physical fidelity and cognitive fidelity training systems were compared in our laboratory in a study of procedural tasks – assembling a LEGO®

helicopter model. The assembly task is similar to many industrial assembly and maintenance tasks in its procedural skill components, because it requires fulfillment of a series of steps in a defined order. The results demonstrated that although the physical fidelity training system caused trainees to perform the real-world task faster, the cognitive fidelity training system decreased the time they spent on error correction (Hochmitz and Yuviler-Gavish 2011). Hence, it can be assumed that the cognitive fidelity system helped trainees to improve their understanding and memory of the right steps included in the procedural tasks.

When considering the advantage of AR training systems in having high physical fidelity, one should take into account the cognitive aspects of the industrial tasks. Does the task require any attention management skills, procedural skills, perceptual skills, etc.? What are the benefits of training these skills in a training system that focuses on them with high cognitive fidelity? The answers to these questions may change the balance for preferring very technically simple training systems (e.g. computer games) to AR training systems.

8.4 Threat #2: The Effect of Feedback

The Threat

Feedback can be defined as information that the learner receives, in response to performance, about performance outcomes (Adams 1971, 1987; James 1890; Salmoni et al. 1984; Schmidt 1982; Schmidt et al. 1989). Feedback was found to be important for learning because the trainee can receive information that helps them to know whether their actions were accurate, to correct their errors, and, as a result, improve in performing the task (Adams 1987; Magill 2006; Trowbridge and Cason 1932). One of the advantages of AR training systems is the ability to add feedback to tasks in which the feedback is not inherent in the natural performance of the task or is very subtle and should be inferred. For example, in an assembly task, the operator night be unsure whether he assembled the parts in the right order and place. Adding virtual feedback to the real-world task has the potential to improve learning.

However, one of the main risks of feedback is in the phenomenon termed "*the guidance hypothesis*" (Salmoni et al. 1984; Schmidt et al. 1989). Under this hypothesis, providing the feedback during training might improve performance of the task when it is given, but when it is withdrawn, in the transfer phase, performance will deteriorate. The reason for this is that the learner might become dependent on the feedback, and will not develop capabilities such as detecting errors intrinsically or using proprioceptive information in motor tasks.

Findings from Our Laboratory

Several studies performed in our laboratory demonstrated the controversial effect of feedback. In Yuviler-Gavish and Krisher's (2016a) study, training visual temporal integration, a process that is crucial in reading, was done with a group that received feedback during training and a group which did not receive it. The performance of the feedback group was significantly poorer in the post-training transfer task, d2, Test of Attention (a standard test to evaluate the ability to identify patterns quickly). In a puzzle completion task, the group that received auditory feedback developed strategies that tryied to maximize short-term achievements (e.g. trial and error) compared to the no-feedback group, which developed preplanning strategies (e.g. separating out edge pieces) (Yuviler-Gavish et al. 2018).

Interestingly, another study implied a possible way to utilize the positive effects of feedback without risking the transfer. Yuviler-Gavish and Krisher (2016b) demonstrated that the mere availability of feedback in an executive task training improved performance both in the acquisition phase and in the transfer phase, when the feedback was withdrawn. The phenomenon happened even when trainees did not actually used the feedback but were only aware of the possibility of using it.

These studies join the long chain of studies on the effects of feedback on learning. When considering an AR training system as an opportunity to add virtual feedback to real-world industrial tasks, both the positive and negative influences of feedback should be considered, as well as more sophisticated ways to utilize feedback without taking the risk of guidance. For example, making the virtual feedback available to trainees, but making sure that they will use it only when it is necessary and it can improve their learning process.

8.5 Threat #3: Enhanced Information Channels

The Threat

In AR training systems, multiple virtual information channels can be used to enhance the real-world experience. Visual information can be presented to the trainee, auditory cues can be provided while performing the task, vibrations can be used to illustrate important points in the training process, etc. However, what is the benefit of adding these information channels? Moreover, do they have negative effects as well?

Several studies have demonstrated that using both visual and auditory channels for presenting information improves training compared to using a single channel (Brooks 1967; Frick 1984; Mayer and Anderson 1991, 1992; Mayer and Sims 1994) and that using three channels is even better (Hecht et al. 2006).

Explanation to these findings can be provided from three sources. According to the Baddeley's model of *Working Memory,* the working memory includes two processors – auditory and visual – which can be used in parallel (Baddeley 1997, 1992). The *Cognitive Theory of Multimedia Learning* (Mayer 2001) claims that having two modalities available during training enables the trainee to build two different mental representations, which enrich the stored memory. In addition, according to the *Cognitive Load Theory* (Sweller et al. 1998; Van Merrienboer and Sweller 2005), using multiple sensory channels reduces the cognitive load on working memory.

However, other studies have demonstrated the opposite. Some of them focus on task performance and not on training, but the patterns of behavior that were presented may imply a negative effect when learning these tasks. Cherubini and Dillenbourg (2007) showed that, in an online collaborative task, a referencing mechanism that presented visually the spoken words harmed performance when it distracted attention from the flow of the conversation. Cockburn and Brewster (2005) demonstrated that audio and haptic feedback in a menu-selection task impaired performance by overloading users with superfluous and distracting information. Similarly, auditory feedback in a virtual reality insertion task led participants to pay unnecessary attention to collisions (Lécuyer et al. 2002). Some studies demonstrated the negative consequences of adding a channel of information on transfer of learning. In training gait modification with a robotic exoskeleton, continuous complementary auditory feedback worsens the results of the post-tests, since it did not provide necessary information about the task (Zanotto et al. 2013).

Findings from Our Laboratory

Another negative effect of adding an information channel was demonstrated in our laboratory in Yuviler-Gavish et al.'s (2011) study of a 3D puzzle completion task. They found that incorporating an additional visual aid during training made the trainer's comments overly specific, and hence inhibited active exploration and deeper learning processes.

Multiple virtual information channels, as can be seen from the above overview, can be used in an AR training system for industry to enhance learning, but may have a negative affect on both the acquisition and transfer phase. When implementing the information channel in the training systems and the training program, the effect of them on the level of understanding and deep cognitive processes should be considered carefully.

8.6   Summary

AR systems are becoming widely used in industry (Blanco-Novoa et al. 2018; Masoni et al. 2017; Nolle and Klinker 2006; Paelke 2014; Rankohi and Waugh

2013) and, more specifically, several AR system for training industrial tasks have been developed (Aiteanu et al. 2006; Besbes et al. 2012; Boud et al. 1999; Christian et al. 2007; De Crescenzio et al. 2011; Haritos and Macchiarella 2005; Gavish et al. 2015; Hořejší 2015; Macchiarella and Vincenzi 2004; Rios et al. 2011; Schwald and De Laval 2003; Westerfield et al. 2015).

Several issues have been discussed in this chapter when considering the pros and cons of employing AR systems for industrial training. Some threats were identified and demonstrated in our laboratory and elsewhere. The high physical fidelity of these systems should be compared to the benefits of high cognitive fidelity. The ability to add virtual feedback to the real-world task should be considered carefully, given that feedback can also inhibit deeper learning processes. Finally, enhancing the real-world experience with more information channels has the potential of improving acquisition, but there is evidence that in some situations the additional information channels might interfere with essential processes such as active exploration and communication. In summary, the use of AR systems for training in industry should be done with consideration of the various aspects of these systems in the defined task and situation, and the various design options that address the training needs should be considered.

References

Adams, J.A. (1971). A closed-loop theory of motor learning. *Journal of Motor Behavior* 3: 111–149.

Adams, J.A. (1987). Historical review and appraisal of research learning retention and transfer of human motor skills. *Psychological Bulletin* 101: 41–74.

Aiteanu, D., Graeser, A., and Otto-Hahn-Allee, N.W. (2006). Generation and rendering of a virtual welding seam in an augmented reality training environment. In: *Proceedings of the Sixth IASTED International Conference on Visualization, Imaging, and Image Processing*, Palma De Mallorca, Spain (28–30) (ed. J.J. Villanueva). ACTA Press.

Alexander, A.L., Brunyé, T., Sidman, J., and Weil, S.A. (2005). From gaming to training: a review of studies on fidelity, immersion, presence, and buy-in and their effects on transfer in pc-based simulations and games. *DARWARS Training Impact Group* 5: 1–14.

Baddeley, A. (1992). Working memory. *Science* 255 (5044): 556–559.

Baddeley, A.D. (1997). *Human Memory: Theory and Practice*. Psychology Press.

Baldwin, T.T. and Ford, J.K. (1988). Transfer of training: a review and directions for future research. *Personnel Psychology* 41 (1): 63–105.

Besbes, B., Collette, S.N., Tamaazousti, M. et al. (2012). An interactive augmented reality system: a prototype for industrial maintenance training applications. In: *2012 IEEE International Symposium on Mixed and Augmented Reality (ISMAR)*, 269–270. IEEE.

Blanco-Novoa, Ó., Fernández-Caramés, T.M., Fraga-Lamas, P., and Vilar-Montesinos, M.A. (2018). A practical evaluation of commercial industrial augmented reality systems in an industry 4.0 shipyard. *IEEE Access* 6: 8201–8218.

Boud, A.C., Haniff, D.J., Baber, C., and Steiner, S.J. (1999). Virtual reality and augmented reality as a training tool for assembly tasks. In: *Proceedings of IEEE International Conference on Information Visualization 1999*, 32–36. IEEE.

Brooks, L.R. (1967). The suppression of visualization by reading. *The Quarterly Journal of Experimental Psychology* 19 (4): 289–299.

Brown, D., Stripling, R., and Coyne, J. (2006). Augmented reality for urban skills training. In: *Proceedings of IEEE Virtual Reality Conference*, 249–252. Los Alamitos, CA: IEEE Computer Society Press.

Cherubini, M. and Dillenbourg, P. (2007). The effects of explicit referencing in distance problem solving over shared maps. In: *Proceedings of the 2007 International ACM Conference on Supporting Group Work*, 331–340. ACM.

Christian, J., Krieger, H., Holzinger, A., and Behringer, R. (2007). Virtual and mixed reality interface for e-training: examples of applications in light aircraft maintenance. In: *Universal Access in Human-Computer Interaction. Applications and Services. UAHCI 2007*, Lecture Notes in Computer Science, vol. 4556 (ed. C. Stephanidis), 520–527. Springer.

Cockburn, A. and Brewster, S. (2005). Multimodal feedback for the acquisition of small targets. *Ergonomics* 48 (9): 1129–1150.

De Crescenzio, F., Fantini, M., Persiani, F. et al. (2011). Augmented reality for aircraft maintenance training and operations support. *Computer Graphics and Applications, IEEE* 31: 96–101.

Dünser, A., Steinbügl, K., Kaufmann, H., and Glück, J. (2006). Virtual and augmented reality as spatial ability training tools. In: *Proceedings of the 7th ACM SIGCHI New Zealand Chapter's International Conference on Computer-Human Interaction: Design Centered HCI*, International Conference Proceeding Series (ed. M. Billinghurst), 125–132). Christchurch, New Zealand. ACM.

Farrow, D., Chivers, P., Hardingham, C., and Sachse, S. (1998). The effect of video-based perceptual training on the tennis return of serve. *International Journal of Sport Psychology.* 29 (3): 231–242.

Frick, R.W. (1984). Using both an auditory and a visual short-term store to increase digit span. *Memory and Cognition* 12 (5): 507–514.

Gavish, N., Gutiérrez, T., Webel, S. et al. (2015). Evaluating virtual reality and augmented reality training for industrial maintenance and assembly tasks. *Interactive Learning Environments* 23 (6): 778–798.

Gopher, D., Well, M., and Bareket, T. (1994). Transfer of skill from a computer game trainer to flight. *Human Factors* 36 (3): 387–405.

Haritos, T. and Macchiarella, N.D. (2005). A mobile application for augmented reality for aerospace maintenance training. In: *Proceedings of the 24th Digital Avionics Systems Conference, Avionics in Changing Market Place: Safe and*

Secure, 5.B.3-1–5.B.3-9). Washington, DC (30 October– 3 November 2004). IEEE.

Hecht, D., Reiner, M., and Halevy, G. (2006). Multimodal virtual environments: response times, attention, and presence. *Presence: Teleoperators and Virtual Environments* 15 (5): 515–523.

Hochmitz, I. and Yuviler-Gavish, N. (2011). Physical fidelity versus cognitive fidelity training in procedural skills acquisition. *Human Factors* 53 (5): 489–501.

Hořejší, P. (2015). Augmented reality system for virtual training of parts assembly. *Procedia Engineering* 100: 699–706.

James, W. (1890). *The Principles of Psychology*, vol. 1. New York: Holt.

Kaiser, M.K. and Schroeder, J.A. (2003). Flights of fancy: the art and science of flight simulation. In: *Principles and Practice of Aviation Psychology* (eds. P.S. Tsang and M.A. Vidulich). Lawrence Erlbaum Associates, Incorporated.

Klopfer, E. (2008). *Augmented Learning*. Cambridge, MA: MIT Press.

Lathan, C.E., Tracey, M.R., Sebrechts, M.M. et al. (2002). Using virtual environments as training simulators: measuring transfer. In: *Handbook of Virtual Environments: Design, Implementation, and Applications* (eds. K.S. Hale and K.M. Stanney), 403–414. CRC Press.

Lécuyer, A., Mégard, C., Burkhardt, J.M. et al. (2002). The effect of haptic, visual and auditory feedback on an insertion task on a 2-screen workbench. In: *Proceedings of the Immersive Projection Technology Symposium*, vol. 2. Orlando, FL (24-25 March 2002).

Macchiarella, N.D. and Vincenzi, D. (2004). Augmented reality in a learning paradigm for flight aerospace maintenance training. In: *Proceedings of the 24th Digital Avionics Systems Conference, Avionics in Changing Market Place: Safe and Secure* (pp. 5.D.1–5.1–9 . IEEE.Washington, DC (30 October– 3 November 2004)

Magill, R.A. (2006). *Motor Learning and Control: Concepts and Applications*. New York: McGraw-Hill.

Masoni, R., Ferrise, F., Bordegoni, M. et al. (2017). Supporting remote maintenance in industry 4.0 through augmented reality. *Procedia Manufacturing* 11: 1296–1302.

Mayer, R.E. (2001). *Multimedia Learning*. New York, NY: Cambridge University Press.

Mayer, R.E. and Anderson, R.B. (1991). Animations need narrations: an experimental test of a dual-coding hypothesis. *Journal of Educational Psychology* 83 (4): 484.

Mayer, R.E. and Anderson, R.B. (1992). The instructive animation: helping students build connections between words and pictures in multimedia learning. *Journal of Educational Psychology* 84 (4): 444.

Mayer, R.E. and Sims, V.K. (1994). For whom is a picture worth a thousand words? Extensions of a dual-coding theory of multimedia learning. *Journal of Educational Psychology* 86 (3): 389.

Nolle, S. and Klinker, G. (2006). Augmented reality as a comparison tool in automotive industry. In: *Proceedings of the 5th IEEE and ACM International Symposium on Mixed and Augmented Reality*, 249–250. IEEE Computer Society.

Paelke, V. (2014). Augmented reality in the smart factory: supporting workers in an industry 4.0. environment. In: *Emerging Technology and Factory Automation (ETFA), 2014*, 1–4. IEEE.

Rankohi, S. and Waugh, L. (2013). Review and analysis of augmented reality literature for construction industry. *Visualization in Engineering* 1 (1): 9.

Rios, H., Hincapié, M., Caponio, A. et al. (2011). Augmented reality: an advantageous option for complex training and maintenance operations in aeronautic related processes. In: *Virtual and Mixed Reality - New Trends. VMR 2011*, Lecture Notes in Computer Science, vol. 6773 (ed. R. Shumaker), 87–96. Springer.

Salmoni, A.R., Schmidt, R.A., and andWalter, C.B. (1984). Knowledge of results and motor learning: a review and critical appraisal. *Psychological Bulletin* 5: 355–386.

Schmidt, R.A. (1982). *Motor Control and Learning: A Behavioral Emphasis*. Champaign, IL: Human Kinetics Press.

Schmidt, R.A., Young, D.E., Swinnen, S., and Shapiro, D.C. (1989). Summary of knowledge of results for skill acquisition: support for the guidance hypothesis. *Journal of Experimental Psychology: Learning, Memory and Cognition* 15: 352–359.

Schwald, B. and De Laval, B. (2003). An augmented reality system for training and assistance to maintenance in the industrial context. *Journal of WSCG* 11 (3).

Squire, K. and Jan, M. (2007). Mad city mystery: developing scientific argumentation skills with a place-based augmented reality game on handheld computers. *Journal of Science Education and Technology* 16: 5–29.

Sweller, J., Van Merrienboer, J.J., and Paas, F.G. (1998). Cognitive architecture and instructional design. *Educational Psychology Review* 10 (3): 251–296.

Trowbridge, M.H. and Cason, H. (1932). An experimental study of Thorndike's theory of learning. *The Journal of General Psychology* 7: 245–260.

Van Krevelen, D.W.F. and Poelman, R. (2010). A survey of augmented reality technologies, applications and limitations. *International Journal of Virtual Reality* 9 (2): 1.

Van Merrienboer, J.J. and Sweller, J. (2005). Cognitive load theory and complex learning: recent developments and future directions. *Educational Psychology Review* 17 (2): 147–177.

Westerfield, G., Mitrovic, A., and Billinghurst, M. (2015). Intelligent augmented reality training for motherboard assembly. *International Journal of Artificial Intelligence in Education* 25 (1): 157–172.

Williams, A.M., Ward, P., and Chapman, C. (2003). Training perceptual skill in field hockey: is there transfer from the laboratory to the field? *Research Quarterly for Exercise and Sport* 74 (1): 98–103.

Woodworth, R.S. and Thorndike, E.L. (1901). The influence of improvement in one mental function upon the efficiency of other functions.(I). *Psychological Review* 8 (3): 247.

Yeo, C.T., Ungi, T., U-Thainual, P. et al. (2011). The effect of augmented reality training on percutaneous needle placement in spinal facet joint injections. *IEEE Transactions on Biomedical Engineering* 58: 2031–2037.

Yuen, S.C.Y., Yaoyuneyong, G., and Johnson, E. (2011). Augmented reality: an overview and five directions for AR in education. *Journal of Educational Technology Development and Exchange (JETDE)* 4 (1): 11.

Yuviler-Gavish, N. and Krisher, H. (2016a). The effect of feedback during computerised system training for visual temporal integration. *International Journal of Learning Technology* 11 (1): 3–21.

Yuviler-Gavish, N. and Krisher, H. (2016b). The effect of computerized system feedback availability during executive function training. *Journal of Educational Computing Research* 54 (5): 701–716.

Yuviler-Gavish, N., Yechiam, E., and Kallai, A. (2011). Learning in multimodal training: visual guidance can be both appealing and disadvantageous in spatial tasks. *International Journal of Human-Computer Studies* 69 (3): 113–122.

Yuviler-Gavish, N., Madar, G., and Krisher, H. (2018). Effects on performance of adding simple complementary auditory feedback to a visual-spatial task. *Cognition, Technology and Work* 20 (2): 289–297.

Zanotto, D., Rosati, G., Spagnol, S. et al. (2013). Effects of complementary auditory feedback in robot-assisted lower extremity motor adaptation. *IEEE Transactions on Neural Systems and Rehabilitation Engineering* 21 (5): 775–786.

9

Condition-Based Maintenance via a Targeted Bayesian Network Meta-Model

Aviv Gruber, Shai Yanovski, and Irad Ben-Gal

Synopsis

Condition-based maintenance (CBM) methods are increasingly applied to operational systems and often lead to reduction of their life-cycle costs. Nonetheless, trying to evaluate a priori the cost reduction of various CBM policies under different scenarios and conditions is a challenging task addressed in this work. The chapter presents a CBM framework in which a targeted Bayesian network is constructed from data of a simulated system. Simulations of the system under different scenarios enable exploration of various CBM policies that could not have been considered by analyzing real operational data, thus leading to a robust CBM plan. The targeted Bayesian network is then learned from the simulation data and used as an explanatory and compact meta-model of failure prediction, conditioned on various realizations of system variables. The proposed framework exploits possible contingencies for determining a good CBM policy and is demonstrated through a real case study of a European operator of a freight rail fleet. This study demonstrates a significant profit improvement of the proposed approach compared to other CBM policies.

9.1 Introduction

Human-made systems are prone to deterioration over time and, therefore, require ongoing maintenance to avoid malfunctioning. Accordingly, it is essential to undertake an effective preventive maintenance (PM) policy that minimizes the life-cycle cost (LCC) of the system and maximizes its operational profit. A relatively simple PM approach is to use a time-based maintenance (TBM) policy, which implies, for example, an a priori scheduling of various

Systems Engineering in the Fourth Industrial Revolution: Big Data, Novel Technologies, and Modern Systems Engineering, First Edition. Edited by Ron S. Kenett, Robert S. Swarz, and Avigdor Zonnenshain.

PM tasks (based on elapsed time or system cycles). Over the years several extensions of the traditional TBM were suggested, such as the reliability-centered maintenance (RCM) approach that studies and analyzes the functionalities of different operational tasks and their effects on the system reliability (Nowlan et al. 1978; Moubray 1991). Yet, despite these developments, TBM is known to be suboptimal, as it does not account for the operational condition of the systems in real time (Dubi 2000). As an alternative to the prescheduled maintenance approach, condition-based maintenance (CBM) schedules the maintenance tasks according to the system's conditions and (partially) observed system states (Jardine et al. 2006). This chapter focuses on such a predictive approach, namely, addressing questions such as: When should a maintenance task be performed, to which system or a batch of systems in a fleet, and under which observed conditions?

In small systems, an optimal PM policy can be obtained by conducting an exhaustive search over all the feasible maintenance policies. In larger and more complex systems, simulation-based approaches are often used to obtain some estimation for satisfactory PM policies. Yet, even with powerful simulators the exhaustive search is typically too time consuming to evaluate each potential PM setting (Dubi 2000). Moreover, even if a powerful simulator exists, it is not clear how to use it efficiently to produce a good PM policy, as we shall propose in this chapter.

CBM methods attempt to define a PM policy according to the state of the system at various time periods. Previous works showed that the CBM approach can improve the PM plan considerably (Jardine et al. 2006; Peng et al. 2010). Yet, the implementation of CBM methods remains a challenging task, as it requires generating reliable prognosis models that analyze and predict system operational availability under various system conditions (Sheppard and Kaufman 2005). CBM models are often hard to formulate as closed-form analytical models without relying on physical models that are mostly feasible in simple mechanical systems. As an alternative to the exact modeling, it has been suggested to implement CBM in an automated fashion by using meta-models backed up by expert knowledge (Jardine et al. 2006). To address this challenge, model-free approaches were proposed for CBM optimization. For example, Marseguerra et al. (2002) optimized a CBM plan by means of genetic algorithms and simulation. Other surrogate models that could be used for PM have been suggested over the years (for example, Forrester and Keane 2009; Queipo et al. 2005; Keane and Nair 2005).

In this chapter, we follow these approaches and suggest a CBM-based policy that combines both a simulation model of the system and a predictive meta-model. The simulation model of the system is based on expert knowledge and historical data. Such a model can be generated by conventional simulation software, such as Arena or Matlab, and can be further enriched by data that are gathered from an operating system if such data are available. The simulator can be used to test various system settings, as well as to introduce variability and noise into the modeled system to evaluate the robustness of various CBM plans. The main required inputs into the simulation module are: (i) the state-transition distributions of individual system components, such as the components' failure-restoration distributions; (ii) an interaction scheme for these components that can be expressed, for example, by a reliability block diagram (RBD) or another system architecture schema; (iii) an initial PM policy by which the system is serviced and maintained – that will be improved by the proposed approach; and (iv) cost parameters that are related to the PM policy, for example costs of unscheduled failures, costs of PM operations, and costs of system downtime.

The simulation outputs are represented by system operational measures, such as the system reliability, availability, and maintainability, from which associated costs under various environmental conditions can be calculated. These measures, as well as the system settings, are then used as an input to the targeted Bayesian Network (BN) model. The BN is learned from the simulation scenarios, using the default parameters that are described in Gruber and Ben-Gal (2011). It represents, in a compact and descriptive manner, the effect of various system conditions and settings on the potential failures of components. The BN is then used as a meta-model for predicting system failures according to different system attributes and PM settings. These attributes are associated not only with the state of the physical system but can also be related to environmental, operational, or other factors that can affect the system availability (e.g. Prescott and Draper 2009). Note that system failures can be viewed as a stochastic process that cannot be anticipated precisely. Yet, often such a process can be characterized by the failure distributions of its individual components and by their interactions. Such information, together with a logistic support plan that includes the PM policy, spare parts considerations, and environmental conditions that affect the system availability, can be captured by the BN model.

Once the BN model is learned and fine-tuned, it is used as a prognosis meta-model for failure prediction of various system components. According to these predictions, different PM scenarios can be evaluated (e.g. triggering maintenance tasks to components with a failure probability higher than a certain threshold) and then an overall system performance can be evaluated to select the best PM plan. Using such an approach, the proposed CBM strategy does not require nor it is based on any underlying closed-form formula. Instead, it uses the targeted BN model designed to predict a system failure by automatically selecting the features that have the most influence for such a prediction. Using a BN as a meta-model is also appealing as it is general and can cover a wide range of systems. Systems that are suitable for the proposed framework are those in which the exact relationships between components' failure and the overall system maintenance requirement are undefined and unclear. Some examples for such systems include (but are not limited to) rail, aircraft, wind turbines, vessels, vehicles, nuclear reactors, and aerospace systems.

The rest of this chapter is organized into four basic sections. The next section overviews the challenges of PM optimization in reliability–availability–maintainability (RAM) models and refers to several methods for addressing these challenges; this section also describes targeted Bayesian Network learning (TBNL) and the motivation for using a BN as a meta-model for prediction. The following section provides a schematic framework for the implementation of the proposed CBM approach; this section demonstrates the implementation of the proposed approach on a freight rail fleet based on a real case study of a European operator. The penultimate section analyzes some key features of the proposed approach. Finally, the last section concludes the chapter with a short discussion on potential future directions.

9.2 Background to Condition-Based Maintenance and Bayesian Networks

This section is divided into three subsections. The first overviews key principles of maintenance plans and their implications on aging systems. The second subsection explains what a BN is, and emphasizes the properties of the targeted BN approach, which is implemented in the proposed framework. The third subsection briefly describes the advantages of using Monte Carlo simulation (MCS), particularly for modeling aging systems.

Maintenance Schemes

Maintenance is defined as a set of all activities and resources needed to uphold elements' specific performance and condition in a given time period. The precise definition of the term evolved over time from the simplistic "repair broken items" (Tsang 1995) to a more complete definition: "Combination of all technical, administrative, and managerial actions during the life cycle of an item intended to *retain it in, or restore it to,* a state in which it can perform the required function" (Bengtsson 2004).

When one examines the phrase "retain it in, or restore it to," it becomes clear that besides the activities that are focused on repairing broken items after breakdowns (restore), there is an additional approach of performing upkeep activities before the next breakdowns happen in order to prevent them (retain).

Bengtsson (2004) defined Corrective Maintenance (CM) and PM, as follows:

- *Corrective Maintenance (CM).* "Maintenance carried out after fault recognition and intended to put an item into a state in which it can perform a required function." CM can be divided into two cases. First, when the break-down is critical and affecting the functionality of the whole system – then it must be repaired immediately. In many situations it is characterized by sig-nificant costs caused by the breakdowns. Second, as long as the breakdown is not affecting the comprehensive function – then the repair can be delayed. In these situations, sometimes it is possible to defer the repair process to a more appropriate time, taking into account the production capacity. The first case is referred to as *Immediate* CM whereas the second case is referred to as *deferred* CM.

- *Preventive Maintenance (PM).* "Maintenance carried out at predetermined intervals or according to prescribed criteria and intended to reduce the prob-ability of failure or the degradation of the functioning of an item." PM can be obtained in two ways. The first is known as time-based maintenance (TBM), which is described as "Preventive maintenance carried out in accordance with established intervals of time or number of units of use but without previous condition investigation." The second is known as condition-based maintenance (CBM), which is described as "Preventive maintenance based on performance and/or parameter monitoring and the subsequent actions." Note that the difference between performance and parameter in this con-text is that the performance is often regarded as a target variable, whereas a parameter is considered to be any explanatory monitored variable. The described process of the CBM parameter monitoring can be either contin-uous, scheduled, or on request. When properly implemented, this approach enables one to combine the benefits of PM while suggesting a way to reduce the costs of unnecessary or excessive scheduled maintenance operations and still allowing one to keep the maintained equipment in healthy operational condition (Jardine et al. 2006).

For example, a general formulation of the LCC within a lifetime T is given as:

$$\text{LCC} = N_f \ \times \ C_f + N_{pm} \ \times \ C_{pm} + C_{T_d} \int_0^T (1 - A(t))dt \qquad (9.1)$$

where C_f represents the cost of an unscheduled failure, C_{pm} stands for the cost of a PM action, C_{T_d} is the downtime cost rate, N_f is the number of unsched-uled failures up to time T, N_{pm} is the number of PM actions undertaken within time T, and $A(t)$ represents the time-dependent availability, where the term *availability* stands for the probability that the system will be in an operation state.

Figure 9.1 depicts in a qualitative fashion the general behavior of the down-time of an aging system and of the LCC as functions of a PM policy. The figure illustrates that while a closed-form of these quantities is often not obtainable, a global optimal PM policy lies between two extreme policies. One extreme policy is to perform PMs too seldom (following the "less maintenance" direc-tion), resulting in an increasing failure rate, which could lead to unplanned

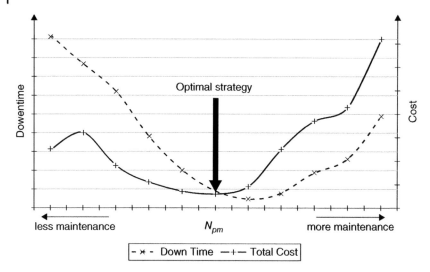

Figure 9.1 A qualitative illustration of the association between downtime and LCC vs. maintenance policy.

downtimes and high cost of repair and restoration. The second extreme policy is to perform PMs too often (following the "more maintenance" direction), resulting in an unavailable system due to PM and excessive PM costs, yet again, leading to high costs. Clearly, some other policy exists that minimize that cost, referred to as the "optimal policy," in a nonrigorous definition. A general closed-form equation for obtaining the optimal policy does not exist. The suggested framework, presented here, attempts to approximate this policy by detecting those cases where PM operation is truly required while, on the other hand, avoiding unnecessary PM operation.

A general survey and comparison of various maintenance policies of deteriorating systems is given by (Wang 2002).

A CBM program contains three key phases: (i) data acquisition – collection of the raw data that are considered relevant to system health; (ii) data processing – analysis of the collected raw data to allow better interpretation; (iii) maintenance decision making – recommendation of the efficient maintenance activity based on previous phases (Jardine et al. 2006).

It is worth noting that CBM practices occasionally distinguish between *Diagnostics* and *Prognostics*. The first term refers to detection, isolation, and identification of the failure when it occurs. The second refers to the prediction of failures based on some likelihood estimation, before they occur.

The literature on the various aspects of the CBM methodology is extensive and covers a variety of systems, components parts, and approaches. A review of different aspects of CBM implementation and methodology is provided by Jardine et al. (2006) and Peng et al. (2010).

This chapter presents to a class of studies that can be classified as an Artificial Intelligence (AI) based approach to CBM (Jardine et al. 2006). Among the AI related methods for prognosis models, there is a body of works using various types of Artificial Neural Networks (ANNs); examples are given in Paya et al. (1997), Samanta and Al-Balushi (2003), Spoerre (1997), and Li et al. (1999). Another approach is implementing Expert Systems (ESs), where the effort is concentrated on using domain expert knowledge in a computer program, allowing one to perform a computerized inference on an engine that is configured in advance for that task, such as demonstrated by Deuszkiewicz and Radkowski (2003) and Hansen et al. (1994). In addition, there are combinations of the above mentioned approaches, such as those of Brotherton et al. (2000), Li et al. (1999), and others, as extensively reviewed by Jardine et al. (2006) and by Peng et al. (2010).

There are several related studies dealing with applications of BNs for maintenance and reliability. Sheppard and Kaufman (2005) proposed a Bayesian framework for diagnosis and prognosis, providing an illustration for the approach on a stability augmentation system of a helicopter. Liu and Li (2007) discussed a decision support system for maintenance management using BNs. Arnaiz et al. (2010) proposed a decision support system based on operational risk assessment in order to improve aircraft operability. This work was later developed and discussed extensively by Arniaz et al. (2010). All these works, however, do not specifically use the targeted BN approach. The uniqueness of the proposed approach is in the use of a targeted BN as a surrogate model to support CBM prognosis and failure prediction.

The novelty of the proposed framework compared to the studies mentioned above is that the learned targeted BN model exploits merely the relevant information for predicting the failures as gathered from simulation data, and uses the predictions by the CBM model within the simulation, for assessing the actual effect on the LCC, as described in the next sections.

The Targeted Bayesian Network Model

A BN is a probabilistic graphical model that encodes the joint probability distribution of some domain in a compact and explanatory fashion. A BN is a directed acyclic graph (DAG), $\mathcal{G}$, containing vertices (or nodes) and edges, and a set of parameters, representing conditional probability tables (CPTs), denoted by $\boldsymbol{\Theta}$, of discrete or discretized random variables (Pearl 1988).

A BN $B(\mathcal{G}, \boldsymbol{\Theta})$ can often be used to represent the joint probability distribution of a vector of random variables $\mathbf{X} = (X_1, \ldots, X_N)$. The structure $\mathcal{G}(\mathbf{V}, \mathbf{E})$ is a DAG composed of $\mathbf{V}$, a vector of nodes representing the random vector $\mathbf{X}$, and $\mathbf{E}$, a set of directed edges connecting the nodes. An edge $(E_{ji} = V_j \rightarrow V_i)$ manifests conditional dependence between the variables X_j and X_i (given prior knowledge about X_i). E_{ji} connects the node V_j to the node V_i (Heckerman 1995), and thus V_j is also called the *parent* of V_i. We denote $\mathbf{Z}_i = \{X_i^1, \ldots, X_i^{L_i}\}$ as the set of "parent" variables of the random variable X_i represented by the set of parent nodes $\mathbf{V}_i = \{V_i^1, \ldots, V_i^{L_i}\}$ in $G(\mathbf{V}, \mathbf{E})$, where for any literal the superscript j stands for its index in the corresponding set and where $L_i = |\mathbf{Z}_i|$ is the size (cardinality) of the subset $\mathbf{Z}_i \subset \mathbf{X}$. The set of parameters $\mathbf{\Theta}$ represents the local conditional probabilities, $p(x_i \mid \mathbf{z}_i)$, over $\mathbf{X}$ that is estimated from observed data or given a priori by an expert (see also Gruber and Ben-Gal 2011).

One of the advantages of BN models is that they serve as an intuitive tool thanks to their qualitative graphical representation, while maintaining a rigorous well defined mathematical model that compactly and efficiently represents the domain (Ben-Gal 2007). A BN can be constructed manually, based on knowledge and hypotheses about the relationship among the domain's variables, or be obtained automatically through learning from data. The latter practice has grown remarkably in recent years, especially in the light of information technology, where data availability is growing massively in many industrial domains. Since BN learning is an NP-hard problem (Chickering et al. 1995), most BN learning methods split the learning procedure into structure learning (edges and their directions) and parameter learning (CPTs), given the learned structure of the BN (Claeskens and Hjort 2003). Moreover, most of the methods learn the BN model in a general fashion, namely encoding the joint probability distribution of the variables' set, irrespective of the application that will be used; hence, such methods that learn a general Bayesian network (GBN) are often referred to as "canonical."

While GBN learning methods attempt to best approximate the joint probability distribution, they address the trade-off between the model complexity versus the prediction error of the learned BN. Unlike the canonical approach of GBN learning methods, target-oriented methods learn the structure of the BN specifically for marginal prediction or classification purposes (Gruber and Ben-Gal 2011). These methods aim to be more effective in marginal focused purposes rather than approximating well the entire domain. The naïve-Bayes model (Duda and Hart 1973), for instance, is one of the simplest and most known target-based Bayesian classifier. It does not require structure learning. Instead, the structure is fixed a priori, where the node representing the class variable is predetermined as the common parent of all nodes that represent the attribute variables. The naïve-Bayes model is popular due to its simplicity but it bears a crude conditional independence assumption.

In the suggested framework, we employ the TBNL algorithm (Gruber and Ben-Gal 2011) as the meta-model to reflect the most influential failure causes as well as the interdependence among them. TBNL follows a targeted approach for accounting for the BN complexity by allowing for the final objective to be a target variable while learning. The target variable in the current application is an indicator of system failures in the subsequent time interval (in the next month, for example). Thus, rather than learning the joint probability distribution as a whole, the TBNL algorithm aims at learning only the monitored variables out of all the simulated ones, as detailed in the sequel, or those that best predict possible system failures.

In general, TBNL first attempts to obtain the most influential set of variables with respect to the target variable and then attempts to construct the connection among these variables. The influence of a variable or a set of variables on another variable is reflected by their dependencies, obtained by using mutual information measures that are well known in information theory (Gruber and Ben-Gal 2011).

One of the advantages of using TBNL for the current endeavor is that it enables managing the BN complexity and controlling it versus its prediction accuracy. TBNL refers to a common indicator of model complexity, defined as the number of free parameters that represent the network,

$$k = \sum_{i=1}^{N}(|X_i| - 1) \prod_{j=1}^{L_i} |X_i^j| \tag{9.2}$$

where N denotes the total number of variables, X_i represents the target variable, and accordingly $|X_i|$ denotes the number of entries it can obtain. Denote the parents' set of X_i by $\mathbf{Z}_i$, then $X_i^j \in \mathbf{Z}_i$ represents the jth parent of X_i. Similarly, $|X_i^j|$ represents the number of values that the jth parent of X_i can have (recall that the number of parents is L_i).

Two input parameters of TBNL that we shall use later on in the example demonstration are the MinPRIG and MaxPRIE. The MinPRIG stands for *minimum percentage relative information gain* (PRIG) and the MaxPRIE stands for *maximum percentage relative information exploitation* (PRIE). These parameters trigger the stopping conditions of the BN construction. The MinPRIG determines the minimal step for adding information while the MaxPRIE determines the total information to be cumulated about each variable. A rather detailed discussion on the properties of the MinPRIG and the MaxPRIE parameters can be found in Gruber and Ben-Gal (2011). These input parameters of TBNL enable selecting the most important and relevant information about the targeted variable, while controlling the model complexity. This "engineering approach" property of TBNL makes the resulting targeted BN an efficient surrogate model in the sense that it can be utilized particularly for predicting a system failure, given a specific condition of the system. The simulation model on the other hand, provides a holistic and generalized assessment of the system life cycle.

Modeling and Simulation in RAM Problems

MCS is an effective technique for modeling typical RAM problems, in that it is insensitive to their natural level of complexity (Wang 2002; Wang and

Pham 1997). The drawback of using MCS (and simulation in general), while attempting to optimize the PM policy of an aging system, is that the search space of possible policies is practically unmanageable, requiring a considerable number of computationally expensive simulations (Dubi 2000) of system availability and failures under various scenarios; (Gruber and Ben-Gal 2012). Barata et al. (2002) suggest a CBM optimization method based on a modeling and simulation approach, and propose an innovative way for addressing such a problem; however, their approach to the optimization requires the specific domain knowledge.

9.3 The Targeted Bayesian Network Learning Framework

In this section we describe a framework architecture and methodology, based on a combination of MCS and targeted BN-based decision engine for RAM optimization. The methodology consists of four main modules as shown in Figure 9.2. The framework can be applied to either a known operational system or to a new system in a design stage. The underlying assumption is that the state-transition (such as "operational to failed," "failed to repair," and "repair to storage") distributions of the components are based on expert opinion. Typically, the expert opinion is derived both from the equipment manufacturer manual and can be backed up by observed data analyses.

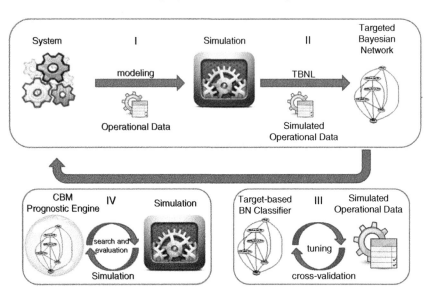

Figure 9.2 Targeted Bayesian Network Learning (TBNL) architecture.

- *System Modeling and Simulation*
  - o Modeling the system operational life cycle using a simulator, which is based on expert knowledge that can be supported by data.
  - o Subsequently conducting the necessary validation tests in order to ensure that the simulation reliably represents the system, in various operational scenarios.
- *Learning the Targeted Bayesian Network Model from Simulated Data.*
  - o Applying the TBNL algorithm to the simulated data to generate an initial targeted BN.
  - o The BN is used as a compact CBM prognosis model for failure prediction.
- *Refining the CBM Model by Tuning the Targeted BN Model Using Cross-Validation.*
  - o Performing iterative refinement of the targeted BN model to obtain a satisfactory failure prediction performance (not necessarily optimal) by tuning the parameters of the TBNL algorithm.
- *Searching for an Effective Maintenance Policy Based on the Developed CBM Model.*
  - o Using the BN model to generate a class of maintenance policies that are triggered by different thresholds on the predicted failure probability of the system.
  - o Evaluating each of these PM policies by the simulator and selecting the best one with respect to a desired objective function (e.g. min LCC, max profit).

Applying these four modules enables to select a satisfactory CBM policy can be adopted to the real system.

9.4 A Demonstration Case Study

The example simulation model is based on a real case study of a European operator of a freight rail fleet. For the sake of the CBM discussion, we demonstrate a reduced model, thus the demonstration does not account for all the original components, processes, and logistic considerations. The reason for the reduced model is not due to scalability issues, but merely to help the reader gather the concept and main principles of the proposed approach. Essentially, the simulation model can scale up to thousands of components of different types or thousands of systems in the deployed fleet.

The simulation model was validated by benchmarking it against the original model that was modeled by the SPAR™ modeling platform. SPAR is an advanced model development environment that has been designed to evaluate the life-cycle management of a system or fleet of systems. It is a discrete event simulator that is based on a Monte Carlo engine which incorporates a robust solution capability. The Monte Carlo engine samples the time points of all the modeled events, allowing for the modeled dependencies. For example, if a component is failed at a certain time point, the subsequent process (e.g. the restoration or removal) will start off by sampling its time point from the current time point. Also, any modeled dependence rule will apply; for example, a failure of a cooling subsystem can increase the failure rate of another subsystem because the latter is designed to work at some temperature boundaries. The simulation is based on a free-flight kernel, which means that between two subsequent events all the quantities, states, and responses do not change, hence the time step of the simulation is not fixed. The output resolution of the simulated case study is set to one day. SPAR simulates the system lifetime until the time reaches a predetermined service time. Since it is a Monte Carlo application, the entire simulation is repeated for as many iterations the user has requested and every numerical measure is averaged over the iterations throughout the service time. This enables a holistic prediction of asset operation and support, where it supports the evaluation of dynamic operations characteristics and supply scenarios, equipment aging, CBM, partial repair, and finite resource capacities. The following is a detailed description of the modeled case study.

The fleet comprises 104 wagons in total, of two types (78 of Type 1 and 26 of Type 2). Each type is deployed in a different field and can carry different loads over a period of 20 years. Each component is critical to the system (wagon) functionality; hence, the system function is serial. Upon failure, the failed component is repaired and inserted back – a process known as restoration. During the restoration, the wagon is unavailable.

The state-transition distribution is determined by an expert's opinion, based on preliminary assumptions and conventions. The common methodology is to use curve fitting on each of the state-transition mechanisms exclusively. A common practice for a failure process of an aging line replaceable unit (LRU) (e.g. mechanical assemblies) is to use a curve fitting of the Weibull distribution to obtain the scale and shape parameters of the distribution. A common practice for a nonaging process (e.g. foreign object damage or failure of electronic equipment) is to use a curve fitting of the exponential distribution to obtain the mean time between failures (MTBF). A common practice for a repair-restoration process of a failed LRU is to use either a curve fitting of the normal distribution or of the log-normal distribution, depending upon the logistic setup, to obtain the corresponding parameters (Balakrishnan and Varadan 1991; Saranga and Knezevich, 2000; Kelton and Law 2000). Using such input distributions allows one to introduce another level of uncertainty (e.g. by adding a noise component or increasing the distribution variance), to obtain a robust PM solution that accounts for input settings that were not observed in the data. The use of a simulation enables one not only to introduce variability into the system but also to map it through the TBNL meta-model to a component failures output. All the components in this example are considered to follow Weibull

Table 9.1 A list of the modeled components and their corresponding Weibull distribution parameters of failures.

Component/ Process	Failure dist. parameters (Years)		Restoration dist. parameters (Years)		Costs (Euros)				
	Scale	Shape	Mean	Std	Removal	Ship	Repair	Inspect	Replace/ build
Running gear	5.86	3.8	0.0014	0.0007	0	1000	2500	0	600
Wheel sets	11.05	3.8	0.3342	0.0002	0	670	1500	0	400
Draw gear	9.96	3.8	0.0854	0.0014	0	400	2000	0	600
Loading frame 1	9.00	3.8	0.0082	0.0001	0	500	5000	0	1200
Loading frame 2	9.05	1.9	0.5027	0.0001	0	500	5000	0	1200
Unloading	10.86	3.8	0.5027	0.0007	0	400	4000	0	1200
Headstock	9.05	1.9	0.5027	0.0001	0	0	0	0	0
Corrosion	10.14	1.9	n/a	n/a	n/a	n/a	n/a	n/a	n/a
PM frame 1	n/a	n/a	0.0274	0.0014	2000	0	0	2000	0
PM frame 2	n/a	n/a	0.0548	0.0027	2000	0	0	2000	0

time-to-failure distributions and normal time-to-restoration distributions. The modeled components are listed in Table 9.1 along with their properties.

The estimates are given by the operator as a coherent part of the model description. In addition, a RBD and other properties of the system are provided, together with its operational environment that is provided as a part of the model design. Corrosion is one of two failure modes attached to the headstock (the second, shown as headstock, is its wear and tear); therefore, corrosion is the only virtual process and is not being "restored," but this failure mode triggers and accelerates the wear and tear of the headstock. Thus, only upon a corrosion "failure" is the headstock failure process activated. After the headstock has completed a CM operation, the corrosion is "repaired" ad hoc. The "failure" process associated with it is reactivated and the headstock failure mode is set to passive.

The restoration efficiency is not perfect, thus a restored component is not as good as new one. We use the term probabilistic age (or simply age hereinafter) as the value of the cumulative distribution function (CDF) of any stochastic process apart from an exponential distribution (Dubi 2000). After restoration, the age of the repaired component is "reduced" by 60% relative to its age prior to the failure. Thus, after a component is repaired, its next failure time is shifted by a time interval that is equivalent to 60% off the failure CDF prior to the repair.

Apart from the physical components, the list includes some properties of the PM operations, for Type 1 frame and for Type 2 frame. The PM restoration

Profit Factor	Availability
30000	0.91 (desired)
20000	0.83 (satisfactory)
–30000(1–*A*)	Otherwise (a penalty)

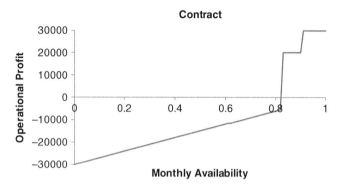

Figure 9.3 Contract figures of the operator.

designates its duration. The age reduction upon a PM operation is 100%, namely the component is considered as good as new.

For each component or process that is listed, a corresponding cost is shown, broken down to subtasks. The operator operates the fleet under a contract. The contract is summarized in Figure 9.3.

The penalty is not fixed, as it grows linearly with the average unavailability, should the availability go below the satisfactory value. This mechanism is considered in order to avoid cases where the availability drops dramatically merely due to maintenance considerations. The contract works as follows: at the end of every month, the availability over the past month is calculated and the operator profits the amount shown next to the corresponding figure, if the actual availability met that requirement.

As mentioned above, the example model was derived and reduced from a prototype model, characterizing a European operation. In order to enable integration with the BN models, the model was rebuilt in the VisualSPAR™ modeling platform (courtesy of Clockwork Solutions Ltd'), since the latter uses a .NET environment, which enables communication with the MATLAB libraries and tools by which the BN was learnt and used. These libraries and tools include the TBNL implementation, the Bayes Net Toolbox (BNT) package, which supports BN learning, and the structure learning package (SLP), which supports BN inference as well (Murphy 2004).

The simulation output was structured and aggregated, then used as an input to the TBNL algorithm. Operational data were aggregated on a monthly resolution, so that at the end of each month a new record of all operating systems is obtained. A data record includes features for each LRU, such as the number of failures and the time since the last restoration, as well as the following attributes: wagon ID, TIME, wagon Type, wagon STATUS, that is, whether the system is currently failed or available, PM_RECENTLY, that is, whether the wagon underwent PM in the preceding month, TIME_SINCE_PM, that is, the time since last PM, PM_TOTAL, that is, the number of PM operations in the wagon's history, accumulated MILEAGE and, as the target variable for the learning algorithm, FAILURE, which indicates whether the wagon has failed during the last month (even if it was restored by the end of the month). Note that some attributes do not necessarily cause a failure, but it is up to the BN to determine whether they are statistically dependent. It is also worth noting that the components' ages are not included in the model, as in practice it is not realistic to assume that they are known. The ages are monitored within the simulation and are often not obtained by the operator.

Since the focus of the meta-model is on failure prediction given the system conditions' vector (of any wagon out of the fleet), the BN model uses the wagon's condition over a month to predict the wagon's state in the next month (classifying it either as "failed" or "available"). Learning these sets of input–output vectors determines whether or not a given condition eventually leads to failure.

For compatibility with the BN form, continuous data, such as the TIME_SINCE_PM and MILEAGE are discretized using a supervised discretization algorithm (Ching et al. 1995). The supervised criterion used by this algorithm aims at maximizing the mutual information in the discretized variable about the target variable, normalized by their joint entropy. During the discretization, the number of symbols of each variable is bounded (in this case to 20). The scheduled PM policy from which the BN model was constructed was triggered for a five-year period. The main constraint with which the TBNL was executed was a MinPRIG of 2% for the target variable as well as for the rest of the attributes (MaxPRIE value of 100%). The graph of the learned BN following this learning process is shown in Figure 9.4 (a detailed view of a BN with CPT is given in Ben-Gal 2007).

The resulting BN includes five features (in addition to the target variable). The features that were selected by the TBNL are: the wagon status (referred to as STATUS); the time since the last PM of the wagon (referred to as TIME_SINCE_PM); the time elapsed since the last restoration of the running gear of Type 1 frame (referred to as Running_Gear_Uc_TSR) and of the wheel sets of Type 2 frame (referred to as Wheelsets_Upp_TSR); and the wagon mileage (referred to as Mileage). Recall that continuous variables are discretized using Ching's algorithm, the number of distinct states may vary accordingly. The PRIE of the failure expectancy was slightly above 32%, which means that the selected attributes provide nearly a third of the potential information about the failure prediction. All the rest of the features were not selected by the TBNL, and thus are not shown in the BN. If, for example the variables' realizations in particular states are: Mileage = 88 020 miles,

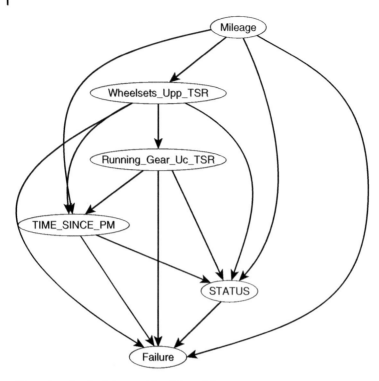

Figure 9.4 Graph of the resulting BN model.

Wheelsets_Upp_TSR = 15 years, Running_Gear_Uc_TSR = 1 year, TIME_ SINCE_PM = 4.89 years (about 4 years and 11 months), and the STATUS = 1 (namely, the wagon is currently available), then the conditional probability of a failure is 0.78, which will be classified as "failed" for a threshold value of 0.5. Thus, in this case, a PM task is performed on the observed wagon.

Note that the classification threshold in the above example can be treated as a PM parameter and modified by the user to trigger different PM policies, as shown in Figure 9.5.

At this stage one can further fine tune the BN model to obtain a better failure prediction performance by using, for example, a fivefold cross-validation test. Thus, dividing the data to train and test sets – learning the BN from the train set and testing the model performance on the test set, as commonly done with classification solutions (Maimon and Rokach 2005). However, in this case we did not test only the classification accuracy, but we generated also a receiver operating characteristic (ROC) curve (Figure 9.5) that draws the recall versus the false positive rate (FPR) of the model (Green and Swets 1966). The recall represents the ratio of correctly predicted failures when such occur, whereas the FPR is the ratio of falsely predicted failures in cases where the system did not fail.

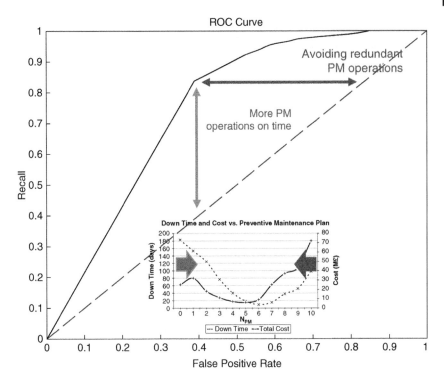

Figure 9.5 ROC curve of the learned BN and its CBM consequences.

Having tested the TBNL resulting model as a stand-alone element (Module III), we integrated it with the simulation. The integration was performed by plugging the CBM model into the simulation module. Each month, the condition vector of every single wagon was inputted to the CBM model, which returns the corresponding failure probability of the wagon for the following month. This probability was then used to classify whether the wagon should be sent to a PM operation according to a predetermined PM policy. This approach was found to yield the best performance (in terms of profit) out of a batch of evaluated policies. Each policy was predetermined by a different level of decision threshold as described in Module IV. The best performance was achieved by a decision threshold policy of 40% failure probability. Namely, if the failure probability of a wagon is equal 40% or more it is sent to PM (referred to as CBM-tbnl-Thresh40 policy). In this case study, all the components in a wagon undergo PM upon PM operation.

A performance benchmark of the resulting CBM policy was evaluated by comparing all the decision threshold policies with an upper bound result and with a lower bound result of the profit. The upper bound was obtained by simulating a scenario in which failures are perfectly predicted, and the

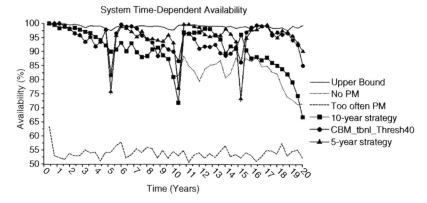

Figure 9.6 Wagon's availability across various PM management policies.

corresponding PMs are undertaken just beforehand. This can be artificially obtained by exploiting the technical details of the simulation, i.e. accessing the stack of future events.

There are actually two lower bounds of the PM policy or, more precisely, the lower bound is obtained by the higher profit out of two extreme PM policies. One extreme policy is to seldom perform PMs (denoted in Figure 9.6 as "No PM"), whereas the second is to perform PMs every time the system is inspected (denoted in Figure 9.6 as "Too often PM").

The system availability for all the above considered policies is presented in Figure 9.6.

Note the slight difference between minimizing the system downtime vs. minimizing the LCC. The profit is obtained by the simulation according to the contract detailed in Figure 9.3, while the rest of the cost contributors of the LCC as formulated in Eq. (9.1) are subtracted (the corrective and PM elements). This yields an objective function that can be positive or negative while the goal is to maximize it as can be seen in Figure 9.7.

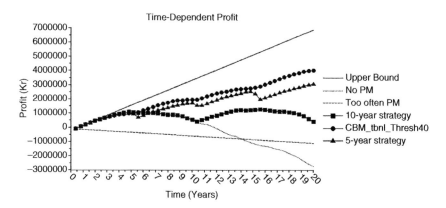

Figure 9.7 Operator profit across various PM management policies.

9.5 Summary

In this section we have discussed the architectural choices we made in the proposed solution.

In order to overcome previously mentioned difficulties in performing RAM optimization of real complex systems, we combine two types of modeling tool in the proposed methodology: MCS and targeted BN. Each of these tools has its pros and cons – leading to the conclusion to use each tool for a different purpose.

The MCS is used to create the most accurate and reliable model of the real-life system, expressing the existing couplings and complex relations among different system components, as well as key performance indicators. Apart from the ability to collect all the operational attributes that are required for the prognostics, this type of modeling tool serves as an "experimental laboratory," enabling how different settings might influence the system to be analyzed. These types of experiments would be very risky and difficult, sometimes impossible, if performed on the actual system.

The targeted BN model, built by the TBNL algorithm based on operational data, is a surrogate model in the suggested framework, hence it can be less accurate in assessing the behavior of the entire system, but it is a more efficient and descriptive modeling tool, and more accurate in predicting the failures using the selected variables (such as MILEAGE, TIME_SINCE_PM, etc.) as descriptors. The TBNL algorithm also takes into account the variables' complex interactions and provides an interpretable graphical representation, making the CBM model accessible for experts' inspection and validation.

The combination of the MCS and TBNL enables exploiting the advantages of each of these models and obtaining a robust framework for a CBM application. On the one hand, TBNL is efficient in predicting the failures, given the conditions vector of the wagon, thus, it is used mainly as a surrogate CBM prognosis model in this framework. On the other hand, the MCS is accurate in assessing the complex system behavior and provides the necessary framework for the examination of various key performance indicators under different operational policies.

The tuning is performed in two modules (modules III and IV) (Figure 9.2). In the first module (module III) the target BN model is tuned to provide good prediction performance, using cross-validation methods with the provided operational data. This module, which is performed outside of the simulation, is computationally cheaper than using the full simulation scheme, and it is used as a unit test for the prognosis model.

Module III is used for selecting the appropriate descriptors and tuning the parameters of the prognosis model for the best possible performance. In module IV, the prognosis model is plugged into the MCS to examine its overall influence on the system and to find the most suitable CBM-based operational policy that will provide the desired results.

The resulting BN clearly reflects the efficiency of TBNL for such purposes. Given a constraint of a minimum 2% of information gain, with respect to each variable, the network becomes very simple and compact. This constraint of 2% minPRIG was set after some parameter exploration, where it was mainly set in order to avoid noise. Although the resultant BN is very compact and exploits slightly over 32% of information regarding the target variable, it suffices to best predict failures under the specified constraints. The advantage of the network's compactness is reflected by the cheap computation complexity while using it for inference. The complexity of the problem, as defined in Eq. (9.2), is 7 M bits whereas the complexity of the resulted BN is 18 K bits (99.7% less). It can be also indicated that the selected features were a mixture of physical components and some other expected attributes. It would be extremely surprising if an expert could point out this exact mixture.

The resulting BN was benchmarked against two popular BN models: the Tree Augmented Naive Bayes (TAN) and naïve-Bayes algorithms. The accuracies of each type of BN models were 96%, 91%, and 90% respectively. However, the ROC curve shown in Figure 9.5 further emphasizes the contribution of using the TBNL as a decision support model for PM management. Each point on the ROC curve indicates a trade-off between the recall (a.k.a. sensitivity), which is the rate of detected failures out of those failures which did occur, versus the FPR, which are the cases in which failures were falsely anticipated. In Figure 9.5, it is illustrated how increasing the recall while maintaining limited FPR pushes the CBM model to the optimal PM policy. This stems from the fact that there are two extreme policies that provide a lower bound on the system's performance. One extreme PM policy is to maintain very poorly or not at all. This could be regarded as the CBM model underestimating failures, resulting in an increasing failure rate. The second extreme PM policy is to overmaintain. This could be regarded as the CBM model overestimating failures, resulting in a highly reliable, yet unavailable and expensive system. The more accurate the targeted BN model is, the better the trade-off achieved between recall and FPR, pursuing the optimal PM policy.

The optimal PM policy is reached when the profit is maximized. We pursued this upper bound with the policy which attempts to maximize the availability. More precisely, the upper bound was estimated by aborting anticipated failures as close to their failure time. Hypothetically, there might be situations in which the profit could be higher, because the components which put the profit together – namely the availability, the failures, and the PM actions – are all interdependent and are hard to break down. However, the upper bound was obtained via the simulation model, simply by aborting each ad hoc failure and maintaining the system instead. In between the upper bound and the two lower bounds, we ran two scheduled PMs, one every 10 years (referred

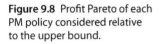

Figure 9.8 Profit Pareto of each PM policy considered relative to the upper bound.

to as 10-year policy) and one every five years (referred to as five-year policy). The TBNL follows the five-year policy, but exploits more information and manages better future analysis. Therefore, although it's overall availability is not as good as that of the original five-year policy, the profit related to it is boosted. Figure 9.8 shows the relative profit cumulated throughout 20 years of service time.

TBNL fulfills the contract in a more effective manner with respect to the profit, mainly since it spreads the PM actions slightly more than the five-year policy does. The five-year policy suffers from profit drops, as the penalty of maintaining a large portion of the fleet altogether is high. As a result, the TBNL improves the profit by 14% compared with the five-year policy, where the 100% line represents the profit that could have been gained by the upper bound policy. Note that the 10-year policy is considerably minor in profit and the lower bound of no-PM represents a loss of 40%.

This chapter proposes a CBM framework, based on a MCS model of the system and on a predictive analytics engine. The latter is a targeted BN that learns from data that are generated by the simulation, thus exploiting possible contingencies that are essential for determining the PM policy. On the other hand, the targeted BN model is used solely for the specific purpose of failure prediction and, hence, is used as a meta-model in the framework rather the model itself. As a result, the learned BN model is more effective while considering the failure causes.

Although the proposed architecture does not necessarily provide an optimal PM policy, this work provides a proof of concept of a tool that enables an effective design of a CBM policy. The added value of the proposed architecture is twofold: (i) since the CBM model is learned from simulation data of a modeled system, it enables exploring PM policies and scenarios that might not have been considered by analyzing real operational data; (ii) TBNL learns a predictive model from the simulation data that efficiently focuses on failure prediction conditioned on the system state, rather than learning a model of the entire operational and environmental system, and is used specifically as a CBM meta-model.

References

Arnaiz, A., Ferreiro, S., and Buderath, M. (2010). New decision support system based on operational risk assessment to improve aircraft operability. *Proceedings of the Institution of Mechanical Engineers, Part O: Journal of Risk and Reliability* 224 (3): 137–147.

Balakrishnan, N. and Varadan, J. (1991). Approximate MLEs for the location and scale parameters of the extreme value distribution with censoring. *IEEE Transactions on Reliability* 40 (2): 146–151.

Barata, J., Soares, C., Marseguerra, M., and Zio, E. (2002). Simulation modelling of repairable multi-component deteriorating systems for 'on condition' maintenance optimisation. *Reliability Engineering and System Safety* 76 (3): 255–264.

Ben-Gal, I. (2007). Bayesian networks. In: *Encyclopedia of Statistics in Quality and Reliability* (eds. F. Ruggeri, R. Kenett and F. Faltin). Wiley.

Bengtsson, M. (2004). Condition based maintenance system technology – where is development heading? In: *Proceedings of the 17th European Maintenance Congress*, Barcelona, Spain (11–13 May 2004). Asociación Española de mantenimiento.

Brotherton, T., Jahns, J., Jacobs, J., and Wroblewski, D. (2000). Prognosis of faults in gas turbine engines. In: *Proceedings of the IEEE Aerospace Conference*, Big Sky, MT (25 March 2000). IEEE.

Chickering, D.M., Geiger, D., and Heckerman, D. (1995). Learning Bayesian networks: the combination of knowledge and statistical data. *Machine Learning* 20: 197–243.

Ching, J.Y., Wong, A.K.C., and Chan, K.C.C. (1995). Class-dependent discretization for inductive learning from continuous and mixed mode data. *IEEE Transactions Pattern Analysis and Machine Intelligence* 17 (7): 641–651.

Claeskens, G. and Hjort, N. (2003). The focused information criterion. *Journal of the American Statistical Association* 98: 900–945.

Deuszkiewicz, P. and Radkowski, S. (2003). On-line condition monitoring of a power transmission unit of a rail vehicle. *Mechanical Systems and Signal Processing* 17: 1321–1334.

Dubi, A. (2000). *Monte Carlo Applications in Systems Engineering*. Wiley.

Duda, R. and Hart, P. (1973). *Pattern Classification and Scene Analysis*. New York: Wiley.

Forrester, A.I.J. and Keane, A.J. (2009). Recent advances in surrogate-based optimization. *Progress in Aerospace Sciences* 45 (1–3): 50–79.

Green, D.M. and Swets, J.A. (1966). *Signal Detection Theory and Psychophysics*. New York, NY: Wiley. ISBN: 0-471-32420-5.

Gruber, A. and Ben-Gal, I. (2011). Managing the trade-off between complexity and accuracy of bayesian classifiers. In: *Proceedings of the 5th Israeli Industrial*

Engineering Research Meeting (IIERM), Dead Sea Resort, Israel (17–18 March 2011).

Gruber, A. and Ben-Gal, I. (2012). Efficient Bayesian network learning for optimization in systems engineering. *Quality Technology and Quantitative Management* 9 (1): 97–114.

Hansen, C., Autar, R., and Pickles, J. (1994). Expert systems for machine fault diagnosis. *Acoustics* 22 (3): 85–90.

Heckerman, D., (1995). A Tutorial on Learning with Bayesian Networks. Microsoft Technical Report MS TR 95 06.

Jardine, A.K.S., Lin, D., and Banjevic, D. (2006). A review on machinery diagnostics and prognostics implementing condition-based maintenance. *Mechanical Systems and Signal Processing* 20: 1483–1510.

Keane, A. and Nair, P. (2005). *Computational Approaches for Aerospace Design: The Pursuit of Excellence*. Chichester, UK: Wiley.

Kelton, W.D. and Law, A.M. (2000). *Simulation Modeling and Analysis*. Boston, MA: McGraw Hill.

Li, Y., Billington, S., Zhang, C. et al. (1999). Adaptive prognostics for rolling element bearing condition. *Mechanical Systems and Signal Processing* 13: 103–113.

Liu, Y. and Li, S.Q. (2007). Decision support for maintenance management using Bayesian networks. In: *International Conference on Wireless Communications, Networking and Mobile Computing*, 5708–5711. IEEE.

Maimon, O. and Rokach, L. (2005). *The Data Mining and Knowledge Discovery Handbook*. Heidelberg: Springer.

Marseguerra, M., Zio, E., and Podofillini, L. (2002). Condition-based maintenance optimization by means of genetic algorithms and Monte Carlo simulation. *Reliability Engineering and System Safety* 77 (2): 151–165.

Moubray, J. (1991). *Reliability-Centered Maintenance*. Oxford: Butterworth-Heinemann.

Murphy K.P., (2004). Bayes Net Toolbox v5 for Matlab. Cambridge, MA: MIT Computer Science and Artificial Intelligence Laboratory. https://github.com/bayesnet/bnt (Last accessed 8 July 2019).

Nowlan, F. S., H. F. Heap, (1978). Reliability-centered maintenance. Report Number AD-A066-579. United States Department of Defense.

Paya, B., Esat, I., and Badi, M. (1997). Artificial neural network based fault diagnostics of rotating machinery using wavelet transforms as a preprocessor. *Mechanical Systems and Signal Processing* 11: 751–765.

Pearl, J. (1988). *Probabilistic Reasoning in Intelligent Systems: Networks of Plausible Inference*. San Francisco, CA: Morgan Kaufmann.

Peng, Y., Dong, M., and Zuo, M.J. (2010). Current status of machine prognostics in condition-based maintenance: a review. *International Journal of Advanced Manufacturing Technology* 50 (1–4): 297–313.

Prescott, P. and Draper, N.R. (2009). Balanced, nearly optimal mixture experiments for models with interactions with process variables. *Quality Technology and Quantitative Management* 6 (2): 67–86.

Queipo, N., Haftka, R., Shyy, W. et al. (2005). Surrogate-based analysis and optimization. *Progress in Aerospace Sciences* 41 (1): 1–28.

Samanta, B. and Al-Balushi, K. (2003). Artificial neural network based fault diagnostics of rolling element bearings using time-domain features. *Mechanical Systems and Signal Processing* 17: 317–328.

Saranga, H. and Knezevich, J. (2000). Reliability analysis using multiple relevant-condition parameters. *Journal of Quality in Maintenance Engineering* 6 (3): 165–176.

Sheppard, J. and Kaufman, A. (2005). Bayesian diagnosis and prognosis using instrument uncertainty. In: *Proceedings of AUTOTESTCON 2005*, Orlando, FL (26–29 September 2005), 417–423. IEEE.

Spoerre, J. (1997). Application of the cascade correlation algorithm (CCA) to bearing fault classification problems. *Computers in Industry* 32 (1997): 295–304.

Tsang, A.H.C. (1995). Condition-based maintenance: tools and decision making. *Journal of Quality in Maintenance Engineering* 1 (3): 3–17.

Wang, H.A. (2002). Survey of maintenance policies of deteriorating systems. *European Journal of Operational Research* 139 (3): 469–489.

Wang, H. and Pham, H. (1997). Survey of reliability and availability evaluation of complex networks using Monte Carlo techniques. *Microelectronics Reliability* 37 (2): 187–209.

10

Reliability-Based Hazard Analysis and Risk Assessment: A Mining Engineering Case Study
H. Sebnem Duzgun

Synopsis

The fourth industrial revolution has created a need for enhanced reliability-based hazard analysis and risk assessment for managing the uncertainties in earth resources industries. This chapter does not address directly the Industry 4.0 challenges but can serve as a basis for such work by providing an example from extractive industries. Quantification of risks due to failure of natural and engineered rock slopes is essential for supporting objective decision making in rock engineering. Although risk assessment is well developed in structural engineering, its adoption in rock engineering still involves various challenges. These challenges include modeling and quantification of uncertainties, implementation of probabilistic modeling frameworks for hazard assessment, prediction of losses in relation to potential instability, and evaluation of the assessed risk in terms of established acceptability/tolerability criteria. Such challenges also impact the development of engineering standards and, hence, the engineering practice. This chapter provides an overview of probabilistic approaches for rock slope stability analysis with emphasis on the first-order reliability method (FORM) in hazard assessment. The methods for quantification of uncertainties are also explained. Then, quantitative prediction of losses due to rock slope instabilities and associated risk assessment approaches are outlined considering acceptability and tolerability criteria. Finally, the chapter highlights future research directions and necessary standardization and adaptation strategies in rock engineering practice.

10.1 Introduction

The global population prediction by the United Nations is 9.8 billion in 2050 (UN 2017). It is also expected that, by 2050, 68% of the world's population will live in urban areas (UN 2018). It is obvious that the trends in population growth

Systems Engineering in the Fourth Industrial Revolution: Big Data, Novel Technologies, and Modern Systems Engineering,
First Edition. Edited by Ron S. Kenett, Robert S. Swarz, and Avigdor Zonnenshain.

and increased urbanization will lead to urban sprawl, which will bring about increased human activity with large exposure to risk of failure due to natural and engineered rock slopes. The design and analysis of natural/engineered rock slopes involve managing a large degree of uncertainty. Hence, a risk-based design and analysis of natural and engineered slopes is inevitable for effective decision making. In extractive industries like mining, the geotechnical stability of slopes plays a critical role, especially for mining the ore using surface mining methods as slopes constitute the open pit. Natural slopes around the mine facilities, like tailing dams, road, and power networks, also have equal importance. Any rock slope failure not only causes disruption of the mining systems but also has detrimental losses to the society.

Rock slopes are natural or engineered structures governed by intact rock and rock discontinuities/planes of weaknesses. Depending on the orientation, frequency, and persistence of these discontinuities various forms of rock slope stability problems emerges. Among the various failure types, plane, and wedge failure modes are the most common types of rock slope failures. For this reason they are taken into account for addressing the risk assessment in this chapter. When a single rock discontinuity daylights into a natural valley slope or an engineered cut at specific dip and strike conditions, failure of the plane geometry occurs. The intersection of two discontinuities and the slope in a specific orientation forms the conditions of the wedge failure geometry.

For rock slope design and analysis, a typical risk-based approach has four main steps (Figure 10.1), namely, data collection, hazard assessment, consequence assessment, and risk evaluation. Each step of the risk-based approach entails comprehensive data collection and rigorous data analysis. Due to the complex nature of the rock slopes, the data and the associated analyses exhibit various uncertainties; these uncertainties tend to propagate from the initial steps to the last ones. Hence, a thorough uncertainty analysis in the process is essential. Usually the uncertainty analysis is performed during the hazard assessment, as geomechanical properties of a rock slope have the highest degree of uncertainty. The hazard assessment (H) includes quantifying failure probability (Pf) of a rock slope for a given period of time (Figure 10.1). The computation of Pf necessitates assessing the structural reliability of the rock slope, which does not include a temporal component. As H requires computation of Pf for a specific time period, a time dependent analysis is needed. For this purpose, temporal probability for the occurrence of various rock slope failure triggers is usually computed (Duzgun et al. 2003).

The next step involves quantifying the potential consequences (C) if the rock slope fails (Figure 10.1). C is essentially predicting the losses due to a rock slope failure, which requires determination of elements at risk (E) and degree of damage (V) for a given element at risk. This step ideally requires identification of a damage probability for elements at risk, which is also called vulnerability/fragility assessment. Once H and C are quantified, computation

Figure 10.1 The basic steps of the risk-based approach for rock slope stability and design.

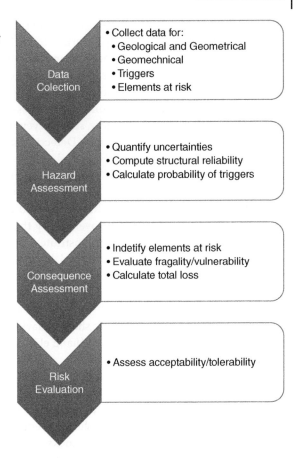

- Collect data for:
- Geological and Geometrical
- Geomechnical
- Triggers
- Elements at risk

Data Colection

- Quantify uncertainties
- Compute structural reliability
- Calculate probability of triggers

Hazard Assessment

- Indetify elements at risk
- Evaluate fragality/vulnerability
- Calculate total loss

Consequence Assessment

- Assess acceptability/tolerability

Risk Evaluation

of risk (R) is relatively straight forward as $R = H \times C$ or $R = H \times V \times E$. The last step in risk assessment is evaluating the significance of the calculated risk (Figure 10.1). This step is essential in communicating with the decision makers, which are composed of various stakeholders related to the elements at risk. Risk evaluation comprises an appraisal for the calculated risk in terms of its acceptability/tolerability.

10.2 Data Collection

The data collection for hazard assessment involves the data for assessing the structural reliability of the slope and the data for triggers. The data for assessing the structural reliability consist of the relevant data for geological and geometrical characteristics of the slope, geomechanical properties of intact rock, rock discontinuities and rock mass, type, frequency and magnitude of triggers. Geological data mainly consist of structural geology and lithology. As the rock slope geometry is mainly governed by rock discontinuities, collecting data on persistence, spacing, orientation, and roughness of rock discontinuities is essential. Such data collection is mainly conducted in the field by using scan line surveys in a conventional approach, where frequency, spacing, persistence, orientation, and roughness of discontinuities cross-cutting a specified scan line are measured (Priest 1993). Recently, such surveys have been conducted using image analysis and/or LiDAR (Light Detection and Ranging) scans (Kocal et al. 2007, Tiruneh et al. 2013, Han et al. 2017). The data for the geometry of the rock slope includes height, width, and orientation of the slope and the discontinuities forming the slope. The geomechanical properties connotes the strength of the intact rock, discontinuities, and the rock mass. The plane and wedge failure modes are mainly governed by the strength of rock discontinuities. Therefore, data collection constitutes obtaining strength properties of rock discontinuities. This is achieved by testing rock discontinuities in the field and/or in the laboratory. Due to the cost of field testing, laboratory tests on representative rock samples are conducted to predict the strength properties of the rock discontinuities. Recently, use of 3D printing methods for understanding the discontinuity shear behavior has been investigated and some promising results are obtained (Jiang et al. 2016; Woodman et al. 2017; Isleyen and Duzgun 2018).

The ideal way of obtaining the temporal component of the hazard, i.e. the probability of slope failure for a specified time period, is to model the time-dependent behavior of parameters used in structural reliability assessment (e.g. time-dependent shear strength of rock discontinuities, time-dependent shear deformation of rocks and rock discontinuities) and to incorporate them into structural reliability analysis. However, such models are quite insufficient in scaling up to actual rock slope structures; this indicates a large research gap. This research gap also highlights potentially new horizons in rock slope stability and design processes. Currently, due to lack of time-dependent slope reliability assessment models, the temporal component is mainly obtained by analyzing various phenomena that can trigger a rock slope failure. Rock slope failures are usually triggered by precipitation, seismic activity, groundwater, and weathering. Hence, data for determining the thresholds that will trigger rock slope failure are essential for the hazard assessment. The data for triggers, therefore, include frequency and magnitudes of triggers.

The data for elements at risk is mainly identifying, locating, and characterizing the elements at risks. The elements at risk could be number of buildings, infrastructure segments or components (e.g. length of road/railroad segments, pipeline nodes, etc.), amount of farmland or forest area, size of population or number of communities, traffic intensity, cultural heritage, etc.

The data related to hazard and consequence analyses should be analyzed in order to understand the links between the slope failure and the associated consequences. The hazard assessment focuses on understanding the failure mechanism and the main causes of slope instability. This allows one to predict the run-out area of the rock slope failure, velocity, depth, and volume of the slide, so that vulnerability of elements at risk can be estimated. Analysis of data for the triggers serves to characterize the long/short-term behavior of triggers so that direct and indirect consequences for various temporal scales can be identified. Usually direct consequences are losses of elements at risk – such as casualties, destroyed buildings and infrastructure, etc. – and relatively easy to quantify. Indirect consequences – disturbance of an economic activity, permanent disability, environmental degradation, losses due to service interruption of infrastructure, etc. – on the other hand, are challenging to predict.

10.3   Hazard Assessment

Hazard assessment for rock slopes constitutes computation of the probability of a slope failure (Pf) for a given time period. Natural/engineered slopes form when rock discontinuities and the slope cut (natural/engineered) satisfy certain geometrical and geotechnical conditions, which are checked by kinematic analyses. The probabilistic kinematic instability can be evaluated either using so called lumped models or stochastic discontinuity and geostatistical models. In lumped models, the discontinuity properties, such as spacing, trace length, and orientation, are fit to a statistical or empirical distribution. Then, Monte Carlo simulation is applied by using the fitted distributions for discontinuity parameters, and kinematic tests are performed to determine the probability of forming unstable blocks. The lumped models treat the discontinuity data, which are obtained from different locations in the field throughout the study region. Hence lumped models do not consider the spatial distribution and correlation of the discontinuity parameters (Nadim et al. 2005.) The stochastic discontinuity and geostatistical methods eliminate the shortcoming of the lumped models. Recently, various software (e.g. DipAnalyst, RocScience-Dips), which incorporate these approaches, have been good tools for identifying the possibility of a specific rock slope failure mode (e.g. plane, wedge) forming. Once the possibility of the formation of a rock slope failure is assessed, the next step is to calculate the kinetic (mechanical) stability of the slope. The multiplication of kinematic ($Pf_{kinematic}$) and kinetic ($Pf_{kinetic}$) stability probabilities is conditioned on the formation of failure modes (Eq. 10.1)

$$Pf = Pf_{kinematic} \times P\,(Pf_{kinetic} \mid Pf_{kinematic}) \qquad (10.1)$$

$Pf_{kinetic}$ can be evaluated either using reliability-based approaches or Monte Carlo simulations. However, the computed Pf value cannot be used as an estimate of hazard (H) as it does not contain information about the temporal occurrence of rock slope failure. There are three ways of computing H. In the first approach the time-dependent behavior of rock slide parameters are modeled in a probabilistic manner. Modeling of time-dependent behavior of rock slope failure parameters is possible for creeping rock slopes, where expected frequency of triggers for increasing the acceleration of the movement can be estimated The second method is based on simply recording the frequency of rock slope failures in a region where similar intact rock and discontinuity properties yield similar rock slopes. Duzgun (2008) outlines the use of this approach in quantitative risk assessment, where a Poisson distribution is adopted for the observed frequency of rockslides recorded for a century. In the third model, structural reliability of the slope is computed for the condition of failure and then the probability of a trigger (Pt) that would exceed the threshold (e.g. probability of precipitation that exceeds a certain amount for 100 years) to cause the slope failure is incorporated with the probability of failure computed from structural reliability approach. In this case H is expressed by Eq. 10.2:

$$H = Pt\,(Pf \mid Pt) \tag{10.2}$$

where Pf | Pt is the conditional probability of having a rock slope failure given that the triggers exceed a certain threshold value for a given time period.

Structural Reliability of a Rock Slope

For a set of basic variables $(x_1, \ldots, x_n)$ the performance of a slope is expressed by a limit state function $g(x) = g(x_1, x_2, \ldots, x_n)$, where the driving forces (Df) acting on the slope exceed the resisting forces (Rf) on the slope (Eq. 10.3)

$$g(x) = Rf - Df \tag{10.3}$$

The safe and failure states are expressed as $g(x) > 0$ and $g(x) < 0$, respectively. Hence, safety of a slope is measured by the reliability index, Ps (i.e. the probability of survival), or equivalently by probability of failure, Pf. If the joint probability density function (p.d.f.) of the basic variables, $x_1, \ldots, x_n$ is $f \times 1, \ldots, x_n$ $(x_1, \ldots, x_n)$, the Ps and Pf are defined as in Eqs. 10.4 and 10.5:

$$P_s = \int\!\!\int \cdots \int_{g(x)>0}^{n} f_{x_1 \ldots x_n}(x_1 \ldots x_n)dx_1 \ldots dx_n \tag{10.4}$$

$$P_f = \int\!\!\int \cdots \int_{g(x)<0}^{n} f_{x_1 \ldots x_n}(x_1 \ldots x_n)dx_1 \ldots dx_n \tag{10.5}$$

Due to the high level of nonlinearity in $g(x)$, close-form solutions to Eqs. 10.1 and 10.2 are not easily available. For this reason, Ps, and Pf are approximated by the first-order reliability method (FORM) or second order reliability method (SORM). Abdulai and Sharifzadeh (2018) provide a comprehensive review of

FORM approaches for the calculation of Pf in different modes of rock slope failures. Dadashzadeh et al. (2015) outline an example implementation of SORM and compares the results with FORM approximation. The estimation of Pf from SORM is computationally costly, however it is more accurate. Due to the high degree of uncertainty involved in the input parameters of g(x) in rock slope reliability analysis, the order of magnitude for the computed Pf value is between 10^{-1} and 10^{-3}. As SORM provides accurate Pf values for higher orders of magnitude, use of SORM for rock slope stability is quite rare.

The associated uncertainties in a rock slope stability analyses are high. For this reason, prior to any reliability analysis it is essential to quantify uncertainties. The shear strength of rock discontinuities plays the most critical role in slopes with plane and wedge failure modes. The forms of Df and Rf in Eq. (10.3) involve mainly geometrical parameters and shear strength properties of the rock discontinuities. Examples of typical g(x) for plane failure can be found from Duzgun et al. (2002), Duzgun and Grimstad (2007), Duzgun and Bhasin (2009). Similarly, examples of g(x) for wedge failure are given in Park and West (2001), Jimenez-Rodriguez and Sitar (2007), Low (2008), Fadlelmula et al. (2008).

Analysis of Uncertainties

Depending on the shear failure criteria used for constructing the g(x), the frictional parameters of discontinuities vary. Coulomb linear failure criteria (Eq. 10.6) and Baron–Bandis nonlinear failure criteria (Eq. 10.7) are among the most widely used failure criteria for rock slopes.

$$\tau = c + \sigma_n \tan(\phi_r) \tag{10.6}$$

$$\tau = \sigma_n tan \left[JRClog_{10} \left(\frac{JCS}{\sigma_n} \right) + \phi_r \right] \tag{10.7}$$

where τ is the shear strength, σ_n is the normal stress, JRC is the Joint roughness coefficient , JCS is the joint wall compressive strength, ϕ_r is the residual friction angle, and c is cohesion.

The shear strength of rock discontinuities is not only dependent on the shear behavior of the rock material forming the discontinuity surfaces but also on small scale roughness and large scale undulations, and filling materials with varying thickness. For this reason, they mainly exhibit a nonlinear behavior. On the other hand, assuming a linear failure criteria simplifies the form of the g(x). As residual friction angle (ϕ_r) is common in almost all the failure criteria, the quantification uncertainties in ϕ_r are mostly studied. The quantification of the uncertainties requires decomposing the sources of uncertainties, as illustrated in Figure 10.2.

The first source is the inherent variability (Figure 10.2), which is due to natural randomness in the rock properties. When measuring the friction angle

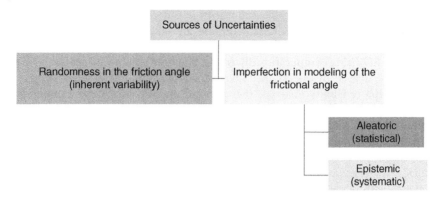

Figure 10.2 Sources of uncertainties in the friction angle of the rock discontinuities.

of a given discontinuity, a limited set of tests can be conducted. This causes a statistical (aleatoric) uncertainty (Figure 10.2) as the mean, variance, and the coefficient of variation change based on the number of samples tested. The statistical uncertainty decreases when the number of test samples increases. The tests conducted for measuring the fraction angle of rock discontinuities can be performed in situ or in the laboratory. However, it is almost impossible to mimic the conditions of the slope. Therefore, there is always a discrepancy between scale of the measurement and the real slope conditions, which results in systematic (epistemic) uncertainties. Duzgun et al. (2002) provide a comprehensive framework for quantification of the uncertainties in the friction angle of rock discontinuities. Park et al. (2005) demonstrate incorporation of uncertainties in the parameters of the g(x) for a case study.

Consequence Assessment

Consequence assessment is predicting the potential consequences (C) when a rock slope fails. For this purpose, analysis of elements at risk (E) and degree of damage (V) for a given element at risk are needed. For various structures under the risk of geohazards (e.g. earthquake, floods), fragility curves are well developed for a given hazard magnitude, where degree of loss for a given element at risk is expressed by a function defined in a range of 0 and 1, (0 being no damage and 1 being total destruction). However, in rock slope failure, there are basically two challenges for the identification of V. The first one is related to the magnitude, where the rock slope's destructive power should be predicted. The second one is linked to the spatial location of elements at risk when the slope fails.

In order to identify the elements at risk – those that can be damaged due to slope failure – it is essential to predict the run-out zone of the rock slope failure. The delineation of the run-out zone depends on volume, velocity, and topography of the region, which indirectly indicate the destructive power of the rock slope failure. The volume of a potential slope failure is a function of the slope geometry, which requires prediction of depth, height, and slope angle. Slope geometry is one of the least uncertain parameters of the rock slope stability, and hence is relatively easy to determine. Estimating the velocity of a potential slope failure relies on the failure mechanism and the phases of the failure. Usually a creeping rock slope has a limited run-out area but the amount and the direction of movement cause various damages to the structures on the slopes, like tilted posts, cracks on the walls of the buildings, failed infrastructure sections, etc.

Sudden rock slope failures, on the other hand, cause total loss of elements at risk, which are on the slope and in the run-out zone, (i.e. vulnerability is 1). For such failures, the main difficulty is to predict the vulnerability related to the existence of dynamic elements at risk in the run out or on the slide area, such as residents of the buildings, cars on the highway or trains on the railroad, etc., since the damage is related to the temporal occurrence of the slope failure. For example failure at midnight or in high season can have different vulnerabilities than slides during the daytime or in high season. Moreover, for rapid rock slope failures the elements at risk around the periphery of the run-out zone can have vulnerability between 0 and 1. In addition to that, some damages in elements at risk may be intangible, such as environmental pollution caused by a damaged pipeline due to a rock slope failure, disability of people due to an injury caused by the rock slide, etc., which makes it difficult to express the losses in monetary units. In these cases, life quality index (LQI) can be used for human life and multi-attribute utility methods can be implemented for combining losses in different units (Diamantidis et al. 2006; Uzielli et al. 2006).

Duzgun (2008) provides an example approach for predicting the vulnerability of a set of elements at risk from a potential rock slope failure whose major consequence is generation of a tsunami wave as the run-out area of the slope is a lake. In the proposed approach the temporal occurrence of the slope failure is considered to be time of day and season of the year. The area to be affected due to a rock slope failure is taken into account as the element at risk, which is composed of residents and visiting tourists, farmland, settlements, and property. Then, vulnerability of this area is predicted by using the damage probability matrix approach, where five categories of damage, namely, damages of no destruction, light destruction, moderate destruction, heavy destruction, and total destruction, are defined for the considered temporal scales.

For large rock slope cases, it is always possible to estimate the volume of rock in the case of failure (e.g. Duzgun and Bhasin 2009). Once the volume is identified, potential elements at risk downstream of the failure zone can be identified by using satellite image analysis or exiting maps. Erener and Duzgun (2011) demonstrated the use of satellite image analysis for identifying buildings and road sections that are in the downstream of a possible slope failure.

Once the identification of elements at risk and their vulnerabilities are calculated, the computation of the losses is straightforward. Duzgun (2008) shows a quantitative approach, where hazard and associated loss are calculated for computing the risk due to a rock slide.

Risk Evaluation

Risk evaluation refers to assessing the significance of the calculated risk in terms of its acceptability. It is essential for communicating with the decision makers so that they can take necessary actions for mitigating it if needed. Risk acceptability depends on various factors. Osei et al. (1997) overview these factors as:

- Voluntary vs. involuntary
- Controllability vs. uncontrollability
- Familiarity vs. unfamiliarity
- Short-term vs. long-term consequences
- Presence of existing alternatives
- Type and nature of consequences
- Derived benefits
- Presentation in the media
- Information availability
- Personal involvement
- Memory of consequences
- Degree of trust in regulatory bodies.

Among these factors, voluntary vs. involuntary, controllability vs. uncontrollability, familiarity vs. unfamiliarity, short-term vs. long-term consequences, and information availability are tightly linked with risk acceptability due to the risk of rock slope failures. When risks are voluntary, higher levels are accepted as compared to involuntary risks. Risks due to a natural rock slope and an engineered slope can be considered as voluntary and involuntary, respectively. An ability to control rock slope failure risk personally leads to a higher acceptability than risks controlled by other parties. Especially, slope stability controls taken by a trusted/untrusted party play a critical role in the level of acceptability. Societies experiencing frequent rock slope failures may have different risk acceptance than those experiencing rare situations. A short-term consequence of a sudden large rock slope failure is total destruction of the elements at risk, which may be directly linked with nature of elements at risk. For example, an irreplaceable cultural heritage site threatened by a rock slope failure may not be acceptable. On the other hand, when a rock slope failure cascades other hazards like a tsunami, which can have long-term consequences, acceptability of such risk is also quite low. Informed societies can have better preparedness for rock slope hazards. Societies experiencing frequent rock slope failures have fresh memories of consequences. Hence, societal differences determine the level of acceptability for the calculated risk.

When a calculated risk is accepted, it connotes no further action for its reduction. In many regulatory frameworks, a tolerable risk level is defined between acceptable and unacceptable risks. The tolerable risk level indicates a level that can be taken under certain risk control structures to obtain certain societal benefits. Risk below the tolerable level is considered to be unacceptable. The perception of a society for accepting the risk is called societal risk, where risks having low hazard and high consequence are taken into account when defining levels of acceptability/tolerability. The unit of societal risk is the loss of life/year, which is generally expressed by f-N or F-N curves. An F-N curve is obtained by plotting the frequency of rock slope failures causing at least N fatalities against N on a log-log scale. When the frequency scale is replaced by annual probability, the resultant plot is the f-N curve. The annual probability of N or more fatalities (f) plotted against N (log-log scale) gives the linear relation in Eq. (10.8).

$$\log f = a + b \log N \tag{10.8}$$

The F-N curves can be constructed for various geographical units such as country, province, state etc. Therefore, the number of rock slope failures and related fatalities within the considered geographical unit determines the acceptability and tolerability criteria. In Figure 10.3, f-N curves constructed by Duzgun and Lacasse (2005) for Canada, China, Colombia, Hong Kong, Italy, Japan, Nepal, and Norway are illustrated.

As it can be seen from Figure 10.3, acceptability of rock slope failure varies between the countries considered. When the spatial scale is reduced to provinces in a given country, the variations are also inevitable. Hence, establishing a database of rock slope failures and associated consequences is essential for well-established acceptability criteria.

10.4 Summary

Rock slope failures are detrimental to societies and quantification of risks related to them is challenging due to the uncertainties involved. However, a quantitative risk assessment for a rock slope provides an objective basis for mitigating the risks. The reliability-based models incorporated in hazard assessment for rock slopes are well established and excellent frameworks for quantification of hazard component. However, there is still a need for comprehensive research on generally accepted models for vulnerability assessment and criteria for risk acceptability. Recent developments in the collection of data from optical camera and LiDAR sensors from satellites, aerial platforms and drones, provide opportunities for analyzing damages due to rock slope failures, which can later be used for in depth understanding of vulnerabilities of elements at risk. Such methods also have good potential for automatic

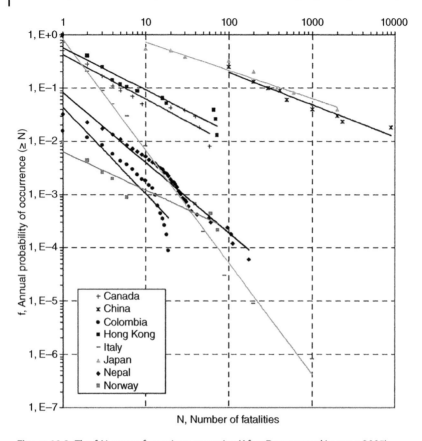

Figure 10.3 The f-N curves for various countries (After Duzgun and Lacasse 2005).

extraction of damage data for fast population of the existing databases. As the degree of damage is highly correlated with the behavior of the rock slope failure, numerical models for predicting the rock slope failures are needed to be incorporated in risk assessment frameworks. It is also equally important to develop acceptability and tolerability criteria not only for number of fatalities but also for damages to properties, infrastructures, lifelines, cultural heritage, and the environment. As rock slope failures can cascade formation of other hazards like tsunamis and environmental hazards, approaches of system safety has great potential to estimate the risks, realistically. The methodologies and practices of reliability-based hazard analysis and quantified risk assessment that are demonstrated in this chapter for mining engineering, can be applied for data driven and evidence-based risk management for systems engineering in the era of the fourth industrial revolution. Also, the technological revolution provides new technologies to collect and analysis of data for quantifying risks.

References

Abdulai, M. and Sharifzadeh, M. (2018). Uncertainty and reliability analysis of open pit rock slopes: a critical review of methods of analysis. *Geotechnical and Geological Engineering* 37 (3): 1223–1247. https://doi.org/10.1007/s10706-018-0680-y.

Dadashzadeh, N., Duzgun, H.S.B. , Gheibie, S. (2015). A second-order reliability analysis of rock slope stability in Amasya, Turkey, The 13th ISRM International Congress of Rock Mechanics, Montreal, Canada (10–13 May 2015)

Diamantidis, D., Duzgun H.S.B., Nadim, F. and Wöhrle, M. (2006). On the acceptable risk for structures subjected to geohazards, ECI Conference: Geohazards – Technical, Economical and Social Risk Evaluation, Lillehammer, Norway (18– 21 June 2006).

Duzgun, H.S.B., (2008). "A quantitative risk assessment framework for rock slides", 42nd US Rock Mechanics Symposium, San Francisco, CA (29 June – 2 July 2008).

Duzgun, H.S.B. and Bhasin, R.K. (2009). Probabilistic stability evaluation of Oppstadhornet rock slope Norway. *Rock Mechanics and Rock Engineering* 42: 729–749.

Duzgun, H.S.B. and Grimstad, S. (2007). Reliability-based stability analysis and risk assessment for rock slides in Ramnefjell. In: *Applications of Statistics and Probability in Civil Engineering: Proceedings of the 10th International Conference, ICASP10*, Tokyo, Japan (31 July – 3 August 2007) (eds. J. Kanda, T. Takada and H. Furuta), 189–198. CRC Press.

Duzgun, H.S.B. and Lacasse, S., (2005). Vulnerability and acceptable risk in integrated risk assessment framework, The International Conference on Landslide Risk Management and 18th Vancouver Geotechnical Society Symposium, Vancouver, Canada (31 May – 4 June 2005).

Duzgun, H.S.B., Yucemen, M.S., and Karpuz, C. (2002). A probabilistic model for the assessment of uncertainties in shear strength of rock discontinuities. *International Journal of Rock Mechanics Mining Sciences and Geomechanics Abstracts* 39: 743–754.

Duzgun, H.S.B., Yucemen, M.S., and Karpuz, C. (2003). A methodology for reliability-based design of rock slopes. *Rock Mechanics and Rock Engineering* 36: 95–120.

Erener, A. and Duzgun, HSB. (2011). A GIS and RS-based element at risk assessment for landslides. GeoInformation for Disaster Management, Gi4DM, Technical Session VI-B, Antalya, Turkey (3–8 May 2011).

Fadlelmula, M.M.F., Duzgun, H.S.B., and Karpuz, C. (2008). Reliability-based modeling of wedge failures. In: *Proceedings of the Asian Pacific Symposium on Structural Reliability and its Applications, APPSRA'08*, Hong Kong (18–20 June 2008), 341–348.

Han, X., Yang, S., Zhou, F. et al. (2017). An effective approach for rock mass discontinuity extraction based on terrestrial LiDAR scanning 3D point clouds. *IEEE Access*, (Special Section On Advanced Data Analytics For Large-Scale Complex Data Environments) 5: 26734–26742.

Isleyen E. and Duzgun, H.S. (2018). Evaluation of 3D printing in obtaining replicates of discontinuity roughness, 52nd US Rock Mechanics/Geomechanics Symposium, Seattle, WA (17–20 June 2018).

Jiang, Q., Feng, X., Gong, Y. et al. (2016). Reverse modeling of natural rock joints using 3D scanning and 3D printing. *Computers and Geotechnics* 73: 210–220.

Jimenez-Rodriguez, R. and Sitar, N. (2007). Rock wedge stability analysis using system reliability methods. *Rock Mechanics Rock Engineering* 40 (4): 419–427.

Kocal, A., Duzgun, H.S.B., and Karpuz, C. (2007). An accuracy assessment methodology for the remotely sensed discontinuities: a case study in andesite quarry area, Turkey. *International Journal of Remote Sensing* 28 (17): 3915–3936.

Low, B.K. (2008). Efficient probabilistic algorithm illustrated for a rock slope. *Rock Mechanics and Rock Engineering* 41: 715–734.

Nadim, F., Einstein, H., and Roberds, W. (2005). Probabilistic stability analysis for individual slopes in soil and rock. In: *Landslide Risk Management: Proceedings of the International Conference on Landslide Risk Management*, Vancouver, Canada (31 May – 3 June 2005) (eds. O. Hungr et al.), 199–236. London: Taylor and Francis.

Osei, E.K., Amoh, G.E.A., and Schandorf, C. (1997). Risk ranking by perception. *Health Physics* 72: 195–203.

Park, H. and West, T.R. (2001). Development of a probabilistic approach for rock wedge failure. *Engineering Geological* 59: 233–251.

Park, H.J., West, T.R., and Woo, I. (2005). Probabilistic analysis of rock slope stability and random properties of discontinuity parameters. Interstate Highway 40, Western North Carolina, USA. *Engineering Geological* 79: 230–250.

Priest, S.D. (1993). *Discontinuity Analysis for Rock Engineering*. London: Chapman & Hall.

Tiruneh, H.W., Oberling, Z.A., Morrison, D.R., Connolly, J.L, and Ryan, T.M., (2013). Discontinuity mapping using Ground Based LiDAR: Case study from an open pit mine. 47th US Rock Mechanics/Geomechanics Symposium, San Francisco, CA (23–26 June 2013).

UN (United Nations) (2017). World Population Prospects: The 2017 Revision, Key Findings and Advance Tables. ESA/P/WP/248, UN Department of Economic and Social Affairs, Population Division. https://population.un.org/wpp/Publications/Files/WPP2017_KeyFindings.pdf (Last accessed 14 June 2019).

UN (United Nations) (2018). World Urbanization Prospects: The 2018 Revision https://population.un.org/wup/Publications/Files/WUP2018-KeyFacts.pdf (Last accessed 14 June 2019).

Uzielli, M., Düzgün, H.S.B. and Vangelsten, B.V., (2006). A first-order second-moment framework for probabilistic estimation of vulnerability to landslides, ECI Conference: Geohazards – Technical, Economical and Social Risk Evaluation, Lillehammer, Norway (18–21 June 2006).

Woodman, J., Murphy, W, Thomas, M.E., Ougier-Simonin, A., Reeves, H.J., Berry, T. (2017). A novel approach to the laboratory testing of replica discontinuities: 3D printing representative morphologies. 51st US Rock Mechanics/Geomechanics Symposium. San Francisco, CA (25–28 June 2017).

11

OPCloud: An OPM Integrated Conceptual-Executable Modeling Environment for Industry 4.0

Dov Dori, Hanan Kohen, Ahmad Jbara, Niva Wengrowicz, Rea Lavi, Natali Levi Soskin, Kfir Bernstein, and Uri Shani

Synopsis

Modeling of software systems and hardware systems, along with their associated tools, has been evolving in parallel but with little, if any, effort to integrate software engineering with systems engineering. As we enter the era of Industry 4.0, and with the Internet of Things (IoT) quickly becoming a prevalent reality, the urgent need for a new paradigm that integrates systems engineering with software engineering into a coherent unifying framework is becoming ever more apparent. To address this challenge, we present MAXIM – Methodical Approach to Executable Integrated Modeling – and its implementation environment, OPCloud. The MAXIM framework enables concurrent modeling of the hardware and software system aspects, avoiding the need to make the painful and information-leaking transition from the abstract, qualitative conceptual system architecting stage to the concrete, detailed, quantitative design stage. Implementing the "holy grail" of model-based systems engineering (MBSE), MAXIM provides for involving systems engineering and software engineering professionals, as well as mechanical, electric, electronic, aerospace, and human factors engineers, all relating to the same conceptual-computational executable model as the most reliable, dynamic, up-to-date, and evolving source of authority. As the implementation platform for the MAXIM paradigm, we have developed OPCloud – a Cloud-based collaborative environment that uses OPM ISO 19450 to integrate, execute, and provide animated simulation of the combined hardware–software system model. The model is developed and presented from its top, abstract system level via increasingly operational and quantitative details, all the way to the nuts and bolts of the system's hardware and to the bit and bytes and the basic arithmetic operations of the system's software. Using a case study of improving Rolls

Systems Engineering in the Fourth Industrial Revolution: Big Data, Novel Technologies, and Modern Systems Engineering,
First Edition. Edited by Ron S. Kenett, Robert S. Swarz, and Avigdor Zonnenshain.
© 2020 John Wiley & Sons, Inc. Published 2020 by John Wiley & Sons, Inc.

Royce's response time to Request for Proposals for a new jet engine, we present OPCloud, the collaborative Cloud-based environment, which enables the application of MAXIM. With OPCloud, it is possible to model, execute, and present an animated simulation of the system at all levels of detail, naturally combining hardware, software, and humans in the loop. This environment is especially suitable to cope with the new challenges that Industry 4.0 presents, with focus on the design of cyber-physical IoT-intensive systems and products.

11.1 Background and Motivation

The term Industry 4.0, or Industrie 4.0, as originally coined by a 2011 initiative of the Federal German Ministry of Education and Research (BMBF 2016) as 1 of 10 future national projects, epitomizes the convergence of traditional manufacturing technologies with the exploding information and communication technologies, giving rise to the fourth industrial revolution we are currently experiencing. This convergence, which we are now witnessing, has been paving the way to emerging disruptive paradigms in the life cycles of systems, products, and services. Industry 4.0 has all the characteristics of the digital transformation, which, according to Gonzalez (2018), is concerned mainly with eliminating waste from the value chain, breaking down the silos between functional groups and information technology (IT) systems to improve collaboration and intelligence sharing, and integrating electronically with all partners in a Cloud- and network-based platform to reduce or eliminate inefficiencies along supply chains.

Industry 4.0 has four major digital transformation characteristics, elaborated from Deloitte (2015):

- Vertical integration of life-cycle support systems within an organization, characterized by the prefix "smart": smart manufacturing, smart logistics, smart services, smart marketing, and smart mass customization.
- Horizontal integration of businesses and business models, suppliers, partners, and customers across supply chains that transcend industry segments, political boundaries, and continents around the globe.
- End-to-end engineering throughout the entire value chain of the product or service, all the way from need or opportunity identification and ideation, through customer and engineering requirements, to architecting, detailed design, manufacturing, sales, use, service, all the way to end-of-life or evolution.
- Deployment of exponential technologies, such as ubiquitous low-cost connected sensors and computing power that exploits AI technologies, such as machine learning, natural language processing, and deep learning.

Figure 11.1 The hardware--software fusion of technologies involved in Industry 4.0. Software- and hardware-related concepts are in green and red, respectively.

SOFTWARE   HARDWARE

Smart **Robotics**
Smart **Manufacturing**
Data **Capture (sensors) and** analytics
Digital **Fabrication (3D Printing...)**
Cloud Computing **in Computer Farms**
Location & navigation services
Smart **Phones**
Internet **of Things**
Physical and Digital **Twins**
Autonomous **Transportation**

Clearly, such dramatic disruptive changes in almost any aspect of the industrial world as we know it, call for a commensurable development of engineering infrastructure that will match these changes. This newly required infrastructure must accommodate the fusion of hardware and software engineering, which traditionally have taken parallel, loosely connected evolution paths.

Figure 11.1 lists prevalent technologies that together make up the Industry 4.0 infrastructure, where the green and red colors designate software- and hardware-related concepts, respectively. Evidently, each technology in this list is an amalgam of the software and hardware aspects. Examples include digital fabrication, Internet of Things IoT, and physical and digital twins. Notably, the common term "digital twins" is a misnomer, as only one of the twins is digital, reflecting it physical twin sibling, so the correct term we use here is physical and digital twins.

As noted in BMBF (2016; translated from German), "The economy is on the threshold of the fourth industrial revolution. Driven by the Internet, *real and virtual worlds grow together into one Internet of Things.* With the project Industry 4.0 we want to support this process" (stress not in the origin).

Indeed, a major motto of Industry 4.0 is the coming together of the real and virtual worlds. In more practical terms, we refer to the fusion or amalgamation of hardware and software.

Figure 11.2 is yet another evidence of the hardware–software fusion in Industry 4.0, where the seamless link between the two is key. But here is the problem: While systems engineering and software engineering are complementary disciplines, their evolution has taken different paths. These two engineering disciplines have been developed by different approaches and schools of thought. This is a major source of the gap between them and consequent major problems in developing system, products, and services.

A key building block in Industry 4.0 is the "smart factory," where cyber-physical systems (CPSs) oversee physical processes and create a virtual copy of the physical world, enabling automated decentralized decision making.

Products, means of transport or tools are expected to "negotiate" within a virtual marketplace regarding which production elements could best accomplish the next production step. This would create a seamless link between the virtual world and the physical objects within the real world.

Figure 11.2 Citation of Dr. Dieter Wegener, Head of External Cooperation at Siemens AG, Munich, Germany. Software- and hardware-related concepts are underlined with green and red, respectively.

CPSs communicate over the IoT, enabling real-time collaboration with each other and with humans across the entire value chain (Hermann et al. 2016).

In Industry 4.0, software controls and governs hardware to the extent that the two are inseparable, and each one alone has almost no value without its counterpart: Hardware with no software is just a collection of physically connected brainless parts, while software without hardware is a bunch of source code lines or executable file bytes with no hardware to execute, command, and control it.

Properly designed and developed systems must be based on accurate formal modeling and design (Selic 2003). After the model is constructed and the system or product is developed according to the model, new insights are gained while operating the system and executing the software program that controls it with real-life data (Christie 1999; Gardner et al. 1994). Many important issues turn out to either be modeled incorrectly or not modeled at all, requiring significant rework, which delays the project's successful completion and makes the system or product more expensive.

Indeed, industrial product development projects experience many problems caused by a lack of a common language for creating and relating to a joint conceptual-computational model that serves engineers, especially system and software engineers (Böhm et al. 2014; Johanson and Hasselbring 2018). These problems are only going to worsen as we stride into the Industry 4.0 era, where the integration requirements between hardware and software are becoming ever more demanding, and as three tightly-coupled concepts of the IoT, CPSs, and physical and digital twins are getting more prevalent and weaved into all kinds of systems, spanning the range from small sensor arrays for shop-floor control all the way to large sociotechnical systems. These three concepts have one common thread: they involve tight coupling of hardware and software to the extent that it is no longer possible to consider and design them separately. Rather, hardware and software must be codesigned and coevolve in tandem. No longer can we afford to let these two critical aspects of any system, product or service be developed separately and combined only at certain late and remote points along their development path.

The systems engineering–software engineering gap is increasing, as Industry 4.0 is becoming part of our lives (Lasi et al. 2014), because hardware and software are becoming ever more intertwined and each hardware component has, or will soon have, its digital twin (Schleich et al. 2017) representation. In order to overcome this current hardware–software divide, a paradigm shift in our approach to systems engineering and software engineering must take place, so these two domains are perfectly aligned and "talk" freely to each other.

The commonly accepted approach to engineering of complex systems is model-based systems engineering (MBSE). MBSE is based on the principle that the conceptual model of the system-to-be is expressed as early as possible, ideally from the requirements stage. The model shall be formal and cover all the system's lifecycle stages and the prime source of reference for all the system stakeholders. Once the detailed design phase is reached, the conceptual model is communicated to the disciplinary engineers for further development, such as software development, at which point the common conceptual model usually ceases to evolve and loses its role as the up-to-date source of authority on the specification of the system under development. Usually, the disciplinary engineers lose the common big picture expressed in the original conceptual model, and critical information is lost in the transition. The result is lack of important information on intent, design considerations, constraints, delay in deadlines, and cost issues.

Enterprises and international organizations are starting to be aware of this major hardware–software divide problem. One example is INCOSE – the International Council on Systems Engineering, which has recently created SaSIWG – the Systems and Software Interface Working Group, with the mission to understand, clarify, and work to resolve issues with the systems-software interface that challenge our ability to engineer today's and tomorrow's systems. Another example is ISO – the International Organization for Standardization, whose Technical Committee TC184 Sub-Committee SC5 has initiated a study group on devising a new standard on mass customization, which is an emerging aspect of Industry 4.0. These initiatives and others indicate that industry's awareness of the need to coalesce hardware and software engineering is gaining momentum.

To be prepared for Industry 4.0, we must ask:

- What are the implications of the tightening hardware–software integration, which is at the heart of Industry 4.0, on the future of Systems Engineering (SE) in general and MBSE in particular?
- What does SE need in order to remain relevant and drive the Industry 4.0 vision forward?

The answer is that we must have the conceptual means – both language and methodology – to accommodate the mandatory change that will bring hardware and software engineering closer together. We must use a language and a methodology that cater to modeling and architecting systems with the Industry 4.0 hardware–software fusion paradigm mindset!

Traditionally, computational, software-oriented aspects of the system or product under development have been considered and incorporated into the system design in a variety of ways only at a late life-cycle stage, at which the conceptual model of the system's architecture, along with its design intents, have often been forgotten or abandoned. The research we present and the Methodical Approach to Executable Integrated Modeling MAXIM environment we developed aim to overcome the widening hardware–software modeling gap (Böhm et al. 2014; Pyster et al. 2015), stepping toward bringing systems engineering and software engineering closer together.

11.2 What Does MBSE Need to be Agile and Ready for Industry 4.0?

A basic requirement from MBSE in order for it to be agile and ready for Industry 4.0 is an underlying conceptual modeling language and methodology with the following characteristics:

- Capable of modeling complex systems of any kind:
  - ○ Technological systems, involving hardware and software.
  - ○ Natural systems.
  - ○ Sociotechnical systems, involving large-scale technological systems in the service of society at large and individuals within it.
  - ○ Systems-of-systems.
  - ○ Any combination of the above.
- Treats hardware and software on equal footing.
- Represents in the same and single kind of diagram the three system aspects:
  - ○ Function.
  - ○ Structure.
  - ○ Behavior.
- Includes humans in the loop.
- Based on a sound universal minimal ontology.
- Applies to both kinds of things (objects and processes):
  - ○ Physical, tangible, material, concrete.
  - ○ Informatical, abstract, logical, mental.
- Caters to humans' dual channel processing:
  - ○ Visual – use formal yet intuitive graphics.
  - ○ Textual – use any natural language that the modeler feels comfortable with.
- Enables managing system complexity in a methodical way.
- Seamlessly blends conceptual modeling with computations of any kind, integrating abstract, high-level model with detailed computations.

- Executable, amenable to simulation of the following kinds:
  - o Online, real-time animated simulation for human comprehension and testing.
  - o Offline, batch mode for generating data used for statistical analysis of system performance.
- Amenable to involving and communicating with nontechnical end users and other stakeholders through formal yet intuitive graphics along with text, which is a subset of any natural language.
- Easy to learn and apply for agile Industry 4.0-style MBSE.

The list above does seem excessively demanding, raising the question whether this is too good to be true, a pie in the sky that will never be achieved. In what follows, we present our MAXIM approach as an extension of OPCloud, which lays solid Object Process Methodology (OPM)-based foundations for achieving this ambitious goal.

11.3   OPCloud: The Industry 4.0-Ready OPM Modeling Framework

Motivated by the need to narrow the gap between system engineering and software engineering, we present OPCloud[1] – an ISO 19450:2015 OPM (Dori 2002; Dori 2016) MBSE environment, which we have developed and extended with MAXIM, adding quantitative and software engineering capabilities beyond ISO 19450. OPCloud enables the entire system to be modeled, including the software and quantitative aspects required to execute it. OPCloud is revolutionary in that it is the first and, currently, only modeling environment that enables modeling systems not just conceptually; the same environment also provides the modeler with the ability to proceed with detailed, quantitative design that is integrated into the qualitative model. The combined hardware–software–humans model constitutes a complete and accurate executable specification of the system from its abstract, qualitative top-level view all the way to modeling the most minute details of the system hardware and software elements and how they interact to produce the system's function.

Previous works (Johanson and Hasselbring 2018; Akkaya et al. 2016; Mellor and Balcer 2002; Seidewitz 2014; Object Management Group [OMG] 2018) have modeled complex systems by examining each domain separately, mostly using different languages and tools. No work to date has provided a unifying framework for qualitative and quantitative modeling over different domains governed by several engineering disciplines that facilitates visualizable execution capabilities. One way to view OPCloud is as a bridge between standard,

---

1 https://www.opcloud.tech

traditional textual programming and visual programming – two programming paradigms that have been competing for a long time. This bridge enables blending both programming approaches with smooth, seamless, bidirectional transitions between them. When building a qualitative model for describing a system, there is a certain point in which a transition to text-based program specification makes more sense, as demonstrated by Dori et al. (2016), and OPCloud offers an ideal facility for blending visual and textual programming.

In 2018, we started incorporating MAXIM into OPCloud. MAXIM, which stands for Methodical Approach to Executable Integrated Modeling, is a software–hardware integrative conceptual modeling approach that extends OPM ISO 19450:2015 with computational and executable capabilities via the introduction of computational OPM objects and software-executable OPM processes. An OPM computational thing (object or process) is a specialization (subclass) of an informatical thing – a thing that is intangible, nonphysical. In an OPM conceptual model, informatical things are intertwined with physical things, which are conceptual representations of physical objects and processes, in a manner that is common to all OPM things: Processes transform objects by consuming them, creating them, or changing their states (or values, in the case of a computational object). Thus, an OPM computational object is an informatical object that can be assigned a numeric or a character string value, while an OPM computational process an informatical process that can perform some mathematical computation or execute a piece of code. The code can be defined by the modeler (currently in Typescript, a superset of JavaScript) or obtained from an external software application.

Adopting OPM using OPCloud with MAXIM, corporations who must operate a collaborative environment for Industry 4.0 will get added value, as they will be able to fuse systems engineering with software engineering, bridging the "Grand Canyon" that currently separates the two disciplines. OPCloud executes and presents animated simulation of the combined hardware–software system model in the same framework. The model is developed and presented from its top, abstract system level, via increasingly operational and quantitative details, all the way to the most minute details that the modeler deems necessary to be included in the model.

In order to enable this kind of integration, we developed within OPCloud the ability to connect to other systems, be they primarily software systems or hardware systems. One connectivity method is implemented using OSLC for Open Services for Lifecycle Collaboration standard (OSLC 2010), which employs a model representation based on the Resource Description Framework (RDF) (Lassila and Swick 1999), and communication via a RESTful (REST is acronym for REpresentational State Transfer. It is architectural style for distributed hypermedia systems and was first presented by Roy Fielding in 2000 in his famous HYPERLINK "https://www.ics.uci.edu/~fielding/pubs/dissertation/rest_arch_style.htm" \t "_blank" dissertation (https://www.ics.uci.edu/~fielding/pubs/dissertation/rest_arch_style.htm)). (Richardson and Ruby 2008) protocol for direct connections with the OPCloud backend

OSLC service provider (a Web server). Another method is customized direct connection with external tools, such as the link between OPCloud and MATLAB. Here, OPCloud passes to MATLAB the parameters needed for its calculations. OPCloud receives back the computed results, presents them, and integrates them into computational objects that are needed as possible inputs to downstream calculations. Yet another example related to Industry 4.0 is connecting OPCloud to hardware equipment, such as robots or sensors of all kinds, including temperature, humidity, and air pressure, which are common in an IoT environment.

An example of OPCloud IoT capabilities is the design and development of an air conditioner remote control with the "I-feel" function of constantly checking the ambient room temperature, using its internal built-in temperature sensor, and comparing it to the user-defined desired temperature. If the remote control detects a temperature difference that exceeds a certain threshold, it sends a signal via another infrared sensor to the main air conditioning control unit to increase or decrease power in order to achieve the desired temperature.

In a common current practice of a development cycle of such a product, there are typically four main stages, each carried out separately: (i) building a conceptual model of the remote control, which specifies the structure and behavior of its hardware and software; (ii) developing, coding, and testing the remote control software; (iii) designing, building, and testing a prototype of the remote control hardware; and (iv) integrating the software into the hardware and testing them together.

Developing that same remote control in OPM using OPCloud with MAXIM, we can take a different, more agile approach, which is to collaboratively integrate the conceptual design of the hardware and the software and test the entire system via animated execution that simulates the hardware and executes what the software is supposed to do. This way, almost all the logical errors can be spotted and corrected before producing the actual product's hardware, loading the software on it, and only then experience failures due to logical design errors. This approach, which OPCloud enables, not only saves time and streamlines the product development process; it ensures that an up-to-date consistent and accurate conceptual model is always available for reference and consultations. This evolving model, which is sufficiently detailed to serve as the product's authoritative blueprint, is at the heart of the MBSE.

OPCloud supports seamless embedding of computations and software code into the conceptual model wherever these are needed, so all the functions required for the correct operation of the remote control can be incorporated into the model. Using the animated execution capabilities of OPCloud serves as a visual debugger at the system level, accounting for the interactions between the product's hardware and software. Once we are happy with the results, we can gradually replace the simulated hardware components – sensors in our case – with actual ones and continue performing animated exaction to find and resolve any new issues that were not detected when the hardware was still represented conceptually in the model. From this agile, model-based development process, we end up with a fully functional prototype of the remote control. OPCloud enables concurrent modeling of the hardware and software system aspects, avoiding the need to make the painful and information-leaking transition from the abstract conceptual system architecting stage to the detailed design stage.

11.4 Main OPCloud Features

OPCloud is a cloud-based software environment that utilizes the Web as a framework for creating OPM models according to OPM ISO 19450:2015. At its core, OPCloud is optimized to support system architects and modelers in the construction of correct OPM models. Figure 11.3 shows the current OPCloud graphic user interface, presenting a small part of the mRNA life cycle model (Somekh et al. 2014), developed originally in OPCAT (Dori et al. 2010).

OPCloud's correct-by-construction models is enabled, for example, by presenting to the modeler only the legal links between two entities that the modeler is about to connect, along with the corresponding Object-Process Language

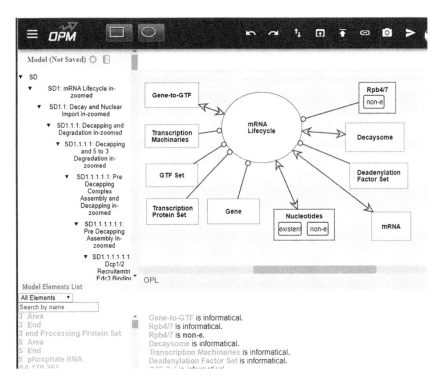

Figure 11.3 System diagram (SD) of the mRNA life cycle process, imported from OPCAT (Dori et al. 2010).

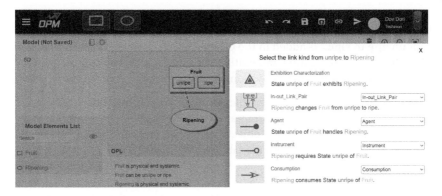

Figure 11.4 Example of correct-by-construction model building. Connecting the state **unripe** of the object **Fruit** to the process **Ripening** causes the menu on the right to pop-up, showing only the legal links for this context and the corresponding OPL sentences.

Figure 11.5 Example of correct-by-construction model building. Selecting the in-out link pair shown in Figure 11.4 yields the OPD above along with the OPL sentence in the second line.

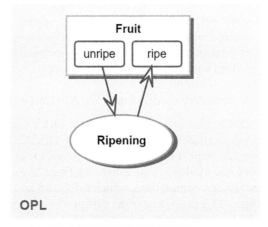

Ripening is physical and systemic.

Ripening changes Fruit from unripe to ripe.

(OPL) sentence – the textual counterpart of the graphical model – that would be added if the modeler selects that link. This feature is exemplified in Figure 11.4 and in Figure 11.5. In Figure 11.4, the modeler connects the state **unripe** of the object **Fruit** to the process **Ripening**. This causes the menu on the right to pop-up, showing only the legal links for this context and the corresponding OPL sentences. The modeler wishes to express the fact that ripening changes a fruit from unripe to ripe, so they find the OPL sentence that says so and clicks on that line.

Figure 11.5 shows the result of selecting the input-output (in-out) link pair: the in-out link pair is drawn and the OPL sentence "**Ripening** changes **Fruit** from **unripe** to **ripe**." is added in the OPL pane.

OPCloud enables creating and editing the graphical (visual) modality of an OPM model easily and effectively. Some prominent features are:

- Modern graphic user interface, which enables creating new things and duplicating existing ones easily by drag-and-drop operations.
- Automated layout of things (objects and processes) and states within an object for optimized diagram readability.
- Full control over styling of Object-Process Diagrams (OPDs) graphics and text.
- Performing editing operations, such as move, resize, align, cut, copy, and paste.
- Searching for elements within an OPM model.
- Easy creation of descendant OPDs via zooming in or unfolding of things.
- Navigation between model diagrams (OPDs).
- Support for online and offline collaboration, including:
  o model sharing with transferrable editing rights through a web browser;
  o creating groups within organizations for collaboration and security;
  o controlling viewing and editing authority for groups within an organization and individuals within a group;
  o commenting;
  o importing models from OPCAT (Dori et al. 2010) and exporting models.

OPCloud is backward compatible with its early versions and tool – OPCAT (Dori et al. 2010) – and can import OPCAT files. With this option, a modeler can import into OPCloud previous OPM models created in OPCAT and continue modeling and enhancing them to include new options and features, including computations, which OPCloud and its MAXIM extension provide. Figure 11.3 presents a small part of a very large model of the mRNA life cycle, which was imported from OPCAT (Somekh et al. 2014). OPCloud contains many features that enable it to function as the perfect tool for the MAXIM approach. In the following text, we review these features that together create the MAXIM framework of OPCloud.

A major power of a Cloud-based application is its ability to work anywhere and anytime, opening the door for real-time collaborative modeling. This is a key feature of OPCloud. Not only does collaboration enable a group of modelers to jointly construct a system model, it is also a great facilitator of knowledge sharing and integration. Collaboration skills can provide opportunities to experience multiple perspectives of other engineers and to develop critical thinking skills through argumentations and persuasion among the modelers

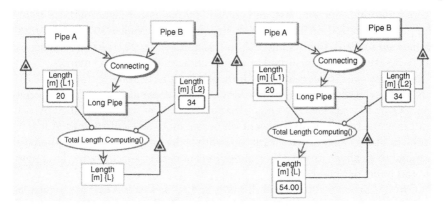

Figure 11.6 Pipe Connecting (a physical process) generates the object Long Pipe, and the Length of Long Pipe is computed (an informatical process) before (left) and after (right) model execution, with the result of 54.00 m.

in the group, while considering, evaluating, supporting, or opposing different viewpoints. Collaboration skills, with emphasis on distant collaboration, are key to Industry 4.0, as professionals in the twenty-first century are increasingly working remotely yet collaboratively across the globe, forming distributed virtual teams. OPCloud is designed to support this environment on a grand scale. For example, OPCloud includes a grouping mechanism, which provides for creating any group from a pool of users in a given organization and dynamically granting read or read/write authority to various participants.

Another major feature implemented in OPCloud, which is at the heart of MAXIM, is computation capabilities. These enable the modeler to augment the conceptual, qualitative model with quantitative aspects. OPCloud's computational capabilities range from simple, system-defined basic arithmetic calculations, as exemplified in Figure 11.6, to any mathematical function, such as trigonometric functions, which are built into OPCloud, all the way to user-defined functions that the modeler can program within OPM processes in OPCloud and connection to other systems for calculation, such as MATLAB/Simulink.

Software Architecture and Data Structure

The Industry 4.0 era, with its new hardware–software blend, has opened the door to further improvements in almost any conceivable area, and the aviation industry is a prominent example. The Boeing aerospace company is reported to have announced a blunt statement for its engine suppliers to think about a major improvement in jet engine characteristics to fit into an ever more economic, efficient, fast, and convenient passenger flight platform to be the next generation of Boeing's mid-size jets – the Boeing 797 (Reid 2018). Conceived in 2016, the Boeing 797, to be launched by 2025, has a forecast of 4000–5000 units to be absorbed into the middle of the air travel market.

Reviewing over 70 years of maturation of jet engine technology (Spittle 2003), it is hard to see where more efficiency, speed, and convenience can be squeezed from. Over 15 years later, jet engine manufacturers are still being called by aircraft manufacturing giants to invent yet better, more efficient engines. A major problem that jet engine producers cope with is the long period it takes them to respond to a call for proposals or quotations to deliver a new engine for a new aircraft being designed.

OPM has been used to model the as-is and to-be systems that are expected to deliver such response. Figure 11.7 presents a small part of an OPM model of a system aimed at shortening the time it takes the Rolls Royce company from getting a request for proposals (RFPs) from an aircraft manufacturer to supply a new kind of jet engine till having a complete technical and commercial response to that RFP from about 18 months to about 6 months.

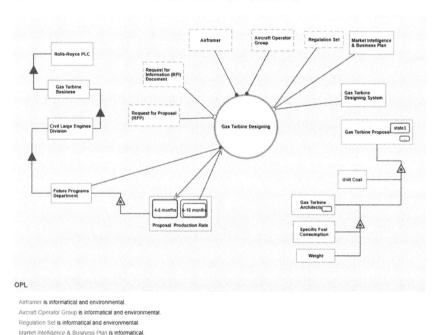

OPL

Airframer is informatical and environmental.
Aircraft Operator Group is informatical and environmental.
Regulation Set is informatical and environmental.
Market Intelligence & Business Plan is informatical.
Rolls-Royce PLC is informatical.
Request for Information (RFI) Document is informatical and environmental.
Gas Turbine Business is informatical.
Gas Turbine Designing System is informatical.

Figure 11.7 Part of an OPM model of a system for shortening the time it takes from getting an RFP to supply a new kind of jet engine to having a complete response to that RFP.

Collaboration

SE applied to large and complex systems requires collaboration among many engineers and interoperability among different expertise areas and tools common in these areas. We can distinguish collaboration among engineers from interoperability among different tools with the following considerations:

- Real-time collaboration, characterized as work done in the same time, although not necessarily in the same place, so engineers can share a view of a model and perform actions upon the model although seated far from each other.
- Sharing of model data across time and geography
  o using the same tool
  o across different tools

We first discuss the sharing of model data and then describe collaboration facilities as built into OPCloud. As a SE modeling tool, OPCloud must be able to share its models in a nonproprietary fashion; to this end, it must adopt some public standards that serve such needs. Broodney et al. (2013) compared MBSE with other engineering disciplines, where tools support collaboration and content sharing. Shani et al. (2016) presented semantic mediation – using ontologies to represent two languages as implemented and interpreted by the tool developers – to facilitate a transformational process, by which models of different kinds and from different modeling tools can exchange model data. This idea is further elaborated from a philosophical point of view in Shani (2017), arguing that adoption of such an approach would render conceptual models eternally meaningful. This claim should be taken with a grain of salt, as it requires that the ontologies used to perform semantic mediation are eternally valid. Semantic mediation relies on Web Ontology Languages (OWL), which portray first-order logic, so semantic mediation built on top of OWL has high likelihood to prevail.

Connectivity is essential for collaborative work, where engineers can share not only their work, but also work together on the same design. Seldom can a viable engineering project be carried out by a single engineer, having his limited expertise compared with the vast knowledge in an engineering enterprise, such as an aerospace conglomerate. Collaborating on such scales cannot be satisfied by merely sharing designs, even if this is done in real time. Engineers must have the facility to work together on the same diagrams and designs, and OPCloud is designed to facilitate such real-time design. Being a Cloud-based software environment, OPCloud is essentially collaborative almost by definition, as using the Web, resources in general, and OPM model files, can be made available effortlessly.

Working in a Cloud environment, engineers do not keep their work artifacts on their local machines as files. OPCloud leaves no traces on the work station being used; rather, it keeps its entire work spaces and generated artifacts on the

Web, using a Cloud storage service. However, one must design the appropriate security and authorization mechanism to ensure that the distribution of models in an enterprise environment is properly controlled. Given the proper authorization, a model being worked on by an engineer in any location on the Internet can be observed by other engineers anywhere else and at any time, including real time, on the global Internet space.

For real-time collaboration using OPCloud, the current modeler holds the permission ("token") to edit the model. Based on lessons learned about the perils of concurrent multiple model editing in Dori et al. (2004), OPCloud's real-time collaboration is a one-editor-at-a-time mode, which prevents the model from being edited by more than one editor at a time to avoid introducing logical errors in the model. At any point, the modeler can pass the token to another modeler, who can then edit the model immediately, and then pass on the token to any other engineer in the group.

Future OPCloud collaborative features might allow for expanding real-time observation to also be real-time intervention of one engineer in her or his peers' work. The web environment makes this possible, as demonstrated by real-time sharing of Google documents, but this will require strict real-time model checking and will be prone to race conditions that may render the model invalid, so research must precede any such facility, and the benefit is not obvious.

Connectivity and Model Interoperability: RDF, OWL, and OSLC

Model sharing is an essential part of a collaborative engineering space, but collaboration also mandates the ability to convert the OPM models into a representation that engineers employing other standards and tools can use. However, a complete and full conversion of a model in one tool, which abides by some modeling language and concepts, that makes it fully usable in another modeling tool, especially in a different modeling language, is not only impossible but also wrong. The common conceptual ground between the two languages governing the modeling capabilities of the two tools can be captured and translated. Even if the two tools work in the same modeling language standard, there are likely to be nomenclature differences in the interpretation of that standard or in the implementation of the standard in the tool, which cause differences between the capabilities of the corresponding tools.

Semantic mediation adopts the Open World Assumption, which does not require full coverage of facts, permitting holes to be left in the information fabric represented by the models being translated. This makes it possible to mediate useful language portions among tools and modeling concepts that differ to some extent, which may be quite large. A gap, such as the one between SysML (Friedenthal et al. 2014) and Modelica (Fritzson 2010), could be bridged through semantic mediation rules (Shani and Landau 2013; Shani et al. 2016;

Shani 2017) and coded in OWL (McGuinness and Van Harmelen 2004) – the ontological language in which both Modelica and the SysML were coded.

The promise of semantic mediation led us to develop an ontology for OPM (Jacobs et al. 2014), so we can use it to represent OPCloud models, making it possible to seek mediation rules to another widely used language – SysML. Unfortunately, SysML does not have a formal standard OWL language (Shani and Landau 2013; Shani et al. 2016; Shani 2017), so one has been developed for IBM's SysML tool Rhapsody (Harel and Kugler 2004). Similarly, the Modelica dialect has been implemented in the SystemModeler tool (Shani and Landau 2013), currently supported by Wolfram.

An approach complementary to semantic mediation over models that are managed by different tools is the W3 Consortium's Linked Data (Cyganiak et al. 2014), carried out through the OASIS standardization organization as the OSLC standard (OSLC 2010). To comply with this standard, a tool must provide a Web service access to its managed models. An element of an OPCloud model is represented as a subject node in an RDF structure, which requires it to include all the triples having it as an OWL Subject or OWL Object. This group of triples defines a small knowledge network that provides all there is to say about that element in the model. This structure is considered the full description of that element as an Internet resource, and it can be managed through RESTful API (an application program interface that uses HTTP requests to GET, PUT, POST, and DELETE data) to be obtained, added, modified or removed.

To further comply with OSLC, the contents of OPCloud models must be represented in RDF (Lassila and Swick 1999), a standard representation of knowledge. RDF represents knowledge as a graph of nodes and edges connecting them – triples consisting of two nodes and a link, so that <S, p, O> would read "p of S is O," where S is the subject, p is the predicate or property of S, and O is the object of that association, or the value of this predicate.

The same representation of triples is also used to represent the concepts in an OWL ontology. The classification of model elements is defined in the ontology. OPM examples of these terms are opm:Object, opm:Process, and opm:Thing, where "opm:" is the XML namespace of the OPM ontology. To be classifiers, each such term is defined in the OWL ontology as an "owl:class," with "owl:" being the namespace of terms in OWL. The dependencies we see here define a hierarchy of languages, with OWL being the base level, denoted as M0, in which ontologies define concepts in a modeling language such as OPM, denoted as M1. M1, in turn, is used to define concepts in an instance OPM model in OPCloud, which will be at level M2. This is the entire extent of the hierarchical depth we need.

An element of an OPCloud model is represented as a subject node in an RDF structure, which requires it to include all the triples having it as an OWL Subject or OWL Object. This group of triples define a small knowledge network, which provides all there is to say about that element in the model. This structure is considered the full description of that element as an Internet resource and it can be managed through REST API [RESTful] to be obtained, added, removed, or modified.

OSLC is all about implementing the API services for each tool, which can answer calls to that API. Each resource should be accessible as an Internet URL (Berners-Lee et al. 1994). When we denote a resource in the XML form, "namespace:resource," the namespace part of that format encompasses the long prefix that makes the combined string a legal URL. Typing that URL in the address slot of an Internet browser will answer with the RDF structure describing that resource.

OSLC is not just capable of obtaining the RDF description. Depending on the medium type requested, it can also render a graphical image of that element, or any other human-readable and comprehensible form that suits the context in which the URL is used. For OPCloud, we have developed the ontology in OWL terminology, and we use the concepts defined in it to generate the RDF responses to OSLC API calls using URL resource identifier for these model elements.

Complementary to that, OPCloud also implements links embedded in the model, so these elements can be linked with external OSLC resources of other tools, and perhaps entirely different kind of tools, which are part of the assembly of MBSE tools servicing the enterprise engineering activity, such as requirement tools, e.g. DOORS (Rational 2010), and problem tracking tools, e.g. Atlassian Jira, (Atlassian), to create a real web of knowledge among MBSE tools, including OPCloud, much the same as Web pages are linked to create the Web as we know it.

11.5 Software Architecture Data Structure

In this section, we present OPCloud under the hood. We describe its design criteria and implementation in order to portray a clear picture of the system. This exposure might help the modeler to better control the language and the methodology behind OPCloud. OPCloud is a Web-based single-page application (SPA). As such, it behaves much more like a desktop application rather than a classical website, where the user experience is interrupted by jumping back and forth between pages.

As Figure 11.8 shows, at its implementation level, an OPM model in OPCloud is composed of three layers: Drawn, Visual, and Logical. The Drawn layer is the part of the system with which the user interacts. This layer is package-dependent, and it currently uses the Rappid (Rappid 2018) diagramming package. An OPM object that the modeler draws and instantly sees on the screen is an instance of a JavaScript class that directly inherits from a Rappid

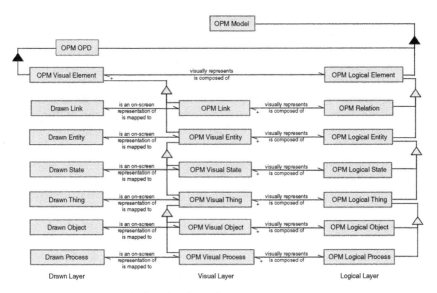

Figure 11.8 A meta-model of OPCloud's implementation.

class. Therefore, all technical aspects of Rappid are part of OPCloud Drawn elements.

With the separation of concerns principle in mind, and to enable future replacement of the Rappid package with other packages for achieving better functionality and/or performance, or adapting OPCloud to new platforms and interfaces, we create the Visual layer, which embodies a one-to-one mapping onto the Drawn layer and is decoupled from it. Thus, for every Drawn element (things or link), there is a corresponding Visual one that is package-independent (see Figure 11.8).

According to the *OPM model fact* principle, a modeler is allowed to have multiple Drawn (or Visual) instances of the same element. The same element can appear more than once in the same OPD or in different OPDs. The only difference between these instances are their graphical attributes, like position, size, and color. The rationale behind this principle is twofold. First, it is dictated from OPM's complexity management mechanisms, namely the refinement–abstraction options of in-zooming–out-zooming, unfolding–folding, and state expression–suppression. When the modeler refines a process by zooming into it, they create a new diagramming sheet ("canvas" or "paper"), where this process is refined to expose subprocesses and possible interim objects. Therefore, that process appears in both the current OPD and in the newly created OPD, where it is refined. Second, OPM caters to the human's limited channel capacity multimedia assumption (Mayer 2002), according to which people are capable of processing only a finite, relatively small amount of information in a (visual or verbal) channel at a time. Enabling multiple occurrences of the same element might be helpful when the modeler has a condensed diagram with many lines that cross each other, making the diagram harder to grasp and comprehend. In such cases, the modeler can visually clone a few of the model elements in order to enhance the model readability and avoid clutter.

Consequently, a given model might have many Drawn (and therefore also Visual) elements that relate to the same Logical one. To control this multiplicity of Drawn and Visual entities, OPCloud provides a third, Logical layer on top of the Drawn and Visual ones. Every Logical element can contain many Visual elements, each mapped onto a package-dependent Drawn element for drawing it on the model canvas.

To streamline data management, the Text attribute of a Drawn element (which is used to store the name of the object, process, or state) is defined in the logical layer, so that changing the name of an element occurs at this level without a need to propagate this change to its visual parts. When a visual element needs to know its name, it addresses its logical element. Similarly, other common attributes of a thing, such as its Essence (i.e. whether it is physical or informatical) and Affiliation (whether it is systemic or environmental), are also defined in the Logical layer.

The execution of an OPM model in OPCloud benefits from the layered approach, as it is divided into two parts: the computational part, which uses the Logical layer of the model, and the visual part, in which the execution is animated graphically using the Drawn layer.

Based on the description of OPCloud, we can provide two views of an OPM model (Figure 11.8): (i) a set of OPDs, where each OPD is a sheet that contains visual elements that are mapped onto drawn elements, and (ii) a set of logical elements. The former view depicts the way modelers perceive an OPM model, while the latter is a logical, developer-oriented view, which provides an overall picture that is useful for maintaining model-wide consistency and developing new features. For example, we use the Logical layer to develop an integrity check module for preserving our model consistency under any type of modification. The *Link Equivalence* consistency rule states that a link which connects two things is defined at the logical level of the model. An OPM model must not allow different types of links between pairs of entities that are logically the same.

11.6 Development Methodology and Software Testing

OPCloud has been developed following the Scrum agile software development methodology. In Scrum, the focus is on project management in situations where accurate planning is difficult. We keep the developments in increments, called *iterations*, and handle a backlog of issues to be resolved and features to be developed. We hold a weekly meeting where all the team members (about a dozen) share their progress, issues, and insights. The product owners can also decide what to work on from the backlog according to priorities set by the project management after meeting with the industrial partners and based on research needs.

OPCloud development faces challenges, such as tight, shifting schedules, changing of features and priorities, and different customer expectations. These challenges typically result from lack of process awareness (Flarup 2007). We have thus found the Scrum method to best fit our development needs. For example, we have used the weekly meetings to align all the parties involved in the development of OPCloud, including undergraduate and graduate students, researchers, and software developers, so they focus on goals prioritized by the project business needs.

To enable effective software project management, we have used the Scrum tool Zoho[2], which has enabled us to assign and track current tasks and backlog, issues, meetings, documents and milestones. Zoho can have two types of users: internal and external, e.g. the client. This option has enabled us to be transparent to our industrial partners and also get them involved, be aware of our milestones and tasks, or report issues found in early, Beta versions of OPCloud they had received as customer testers.

The testing of OPCloud has been aligned with the Scrum methodology. First, each developer designs and performs tests of her or his own developed feature, applying unit tests to verify that the requirements specified in the written "user story," which served as the basis for the feature development, have been met. Next, a QA person performs feature testing for each feature a developer inserted, and only if the QA testing is passed, the feature to be developed is marked as completed. Each iteration included integrated testing for all the features developed within that iteration. At the end of every version, which comprises about four iterations, we had a shorter, two-week iteration for "hardening," during which we do only testing and bug fixing.

11.7 Model Integrity

An OPM model is a source of knowledge of the underlying system. It is, therefore, of paramount importance that the facts that this model represents be consistent within and across OPDs. Some of the rules in OPM are derived from the language and others from the methodology principles. For example, a modeler is not allowed to connect two objects by an effect link. Another example is that any OPM fact must appear at least one time in the model, mandating that all the occurrences of the same construct (entity–relation–entity) must be consistent.

---

2 https://www.zoho.com

OPCloud follows the *validation by construction* approach, which supports error prevention rather than their later detection and correction. For example, when a modeler chooses to connect two objects, the system will present them with a list of links that can be valid between two objects, rather than letting them choose freely, and then attempt to detect and correct the error. In many cases, the consistency check must inspect more than one OPD for consistency, as the same model fact can appear many times, each time in a different OPD. In such cases, the system needs to find the instances of this fact and make sure that all of them are consistent. This process is feasible and relatively easy thanks to the tiered architecture of the model, as all the instances are visual facts that belong to the same logical one, hence we do not need to scan the entire model, just the visual elements of a specific logical one.

The implementation of the consistency checking module relies on the core data structure of OPCloud but it is highly decoupled from it. We define an abstract general consistency rule and its expected abstract functionality called check. Every specific rule inherits from the abstract rule and, as such, it must implement its checking method. This yields a hierarchy of all the rules in the system and enables us to add more in the future without impacting other modules. In addition, we define a consistency check module that constitutes a set of rules taken from the hierarchy. This module contains a check functionality that runs over the rules and executes its own check method for each rule. The consistency check method succeeds if all checks of the rules succeed.

11.8   Model Complexity Metric and Comprehension

Systems are complex by nature and are becoming ever more complex due to tighter integration between hardware and software, as the emergence of the IoT clearly demonstrates. This complexity is reflected in the conceptual models of these systems, whose importance is growing as a means to design, explain, and communicate systems' function, structure, and behavior. Conceptual models are recognized as the underlying source of knowledge about systems, and they are becoming a pillar in system engineering being the foundations of MBSE. Superimposed on systems' intrinsic complexity is their complicatedness – the human's difficulty to create and understand the conceptual models of systems due to improper use of languages or methods for modeling and visualizing complex systems.

Models evolve and serve as the authoritative resource of knowledge about systems. This basic premise of MBSE underscores the increasing centrality of model comprehension – improving the ways conceptual models are created and made comprehensible.

In natural languages, many of the reading comprehension approaches and metrics for text complexity have been adopted in software engineering for the active field of source code comprehension. Comprehensibility is highly related to complexity. The more complex the model, the harder it is to understand. Complexity metrics adopted from code comprehension are applicable also to models. For example, Halstead complexity measures provide a set of metrics for program vocabulary, length, volume, difficulty, and effort, based on operators and operands of a given source code. In OPM, the operators can be links, and the operands – objects and processes that these links connect. The Halstead measures are:

n_1 = *number of distinct operators*

n_2 = *number of distinct operands*

N_1 = *total number of operators*

N_2 = *total number of operands*

Program vocabulary: $n = n_1 + n_2$

Program length: $N = N_1 + N_2$

Volume: $V = N \times \log_2 n$

Difficulty: $D = \dfrac{n_1}{2} \times \dfrac{N_2}{n_2}$

Effort: $E = D \times V$

These metrics are based on the source code's *operands* and *operators*. Operands primarily represent data, while operators represent operations. This integration of the static and dynamic aspects is rare in conceptual models, as each aspect is represented separately, as is the case with the Universal Modeling Language UML and SysML. An OPM model, however, combines both in the same single kind of diagram – an OPD – and, as such, it highly resembles source code, making the adoption of code metrics straightforward and seamless. The first step toward applying these metrics on an OPM model is to map every OPM building block to an *operand* or *operator*. Having done this, the next step is merely counting and applying the aforementioned formulas.

Every OPM model is potentially composed of objects, processes, and links between them. Every computational object represents data, so it will be considered an operand. Every computational process is mapped to an operator, as it represents a mathematical operation or a piece of code. In an OPM model, a (noncomputational) process is presented to reflect its relation to other processes. For example, a process can inherit from another process or contain (zoom into) other process. In these cases, we capture the structural relations between processes. Links will be mapped to operators, as in OPL every link is translated to a verb in its gerund form (i.e. ending with ing, as in ending). Certain kinds of links should be considered distinct even though they belong to the same kind of relation. For example, tagged links are structural links that can be tagged. Every time we have a different tag, we actually have a different verb, which represents a different operation.

Conceptual models are not "write once" artifacts; they evolve as the system requirements change or defects are detected. It is, therefore, important to keep them highly maintainable. Maintainability is affected by, among other factors, comprehensibility, which is highly related to complexity. Complexity metrics can, therefore, be used to develop maintainability measures. For example, a maintainability index (*MI*) has been defined for measuring how maintainable a given source code is. This metric is defined by means of complexity metrics, including Halstead metrics, lines of code (LOC), and McCabe complexity metric. The *MI* original formula, which has several variants that have been defined over the years, is:

$$V = Halstead\ Volume$$

$$G = Cyclomatic\ Complexity$$

$$LOC = Lines\ of\ Code$$

$$MI = 171 - 5.2 \times \ln(V) - 0.23 \times (G) - 16.2 \times \ln(LOC)$$

In the *MI* formula, we use *V* as it was defined earlier in Halstead measures. The *G* component represents the cyclomatic complexity, which reflects the number of independent paths in the code – simply the number of conditions plus 1. For example, a code snippet without conditions has one path. In an OPM model, there are condition links that can change the flow of the model. Therefore, *G* will be based on this kind of links. The *LOC* component represents the number of lines in the source code, which is widely accepted as a measure of size. In an OPM model the size could be calculated as the sum of all components of the diagram. However, in source code, a line of code is generally a simple statement that is composed of many small expressions. Therefore, the most appropriate way to calculate a model size is by the number of sentences in the OPL, which is the counterpart of each OPD.

11.9   Educational Perspectives of OPCloud Through edX

Many courses in higher engineering education are taught in large lecture halls, where one teacher instructs many students using traditional pedagogical approaches, such as frontal teaching. This normally results in a learning experience that does not reflect engineering practice and does not induce the student engagement required for achieving meaningful learning (Dori and Belcher 2005).

Active learning involves the learner facilitating meaningful learning. It emphasizes skill development over direct transmission of information and has been found to improve learning outcomes for graduate students when compared with traditional teaching methods (Baepler et al. 2014; Dori et al. 2007; Kong 2014). This improvement applies to both low (Han et al. 2014) and high achievers (Dori and Belcher 2005). In courses with large numbers of participants, active learning has been found to be more effective than traditional teaching methods (Baepler et al. 2014; Dori et al. 2007), and the majority of students who have experienced active learning made overall positive reports about it (Baepler et al. 2014; Bower et al. 2015).

A *Massive Open Online Course* (MOOC) is a fully online distance learning course, taught over the Internet, open to the public, addressing obstacles of accessibility, availability, and space by providing traditional pedagogy, such as video lectures and readings, alongside active assignments, such as close-ended questions and forum discussions (Liyanagunawardena et al. 2013; Hone and El Said 2016).

The edX course *Model-Based SE with Object-Process Methodology* is a MOOC. It combines short video lectures with active learning through interactive exercises. Some of those exercises include the use of OPCloud integrated into the edX *Learning Tools Interoperability* (LTI), which is an application program interface that edX complies with. The LTI integration allows course learners to use OPCloud without having to go outside the edX platform. For the present course, we wished to evaluate OPM conceptual models constructed by numerous participants, requiring more elaborate assessment methods than what the edX platform has already made available. Since immediate, elaborate feedback is crucial to meaningful learning (Kay et al. 2013), especially in the context of collaboration (Arvaja et al. 2007), we developed an automated evaluation mechanism for assessing OPM models that learners had constructed. The assessment is based on a predefined scoring rubric, providing valuable feedback to both learners and instructors, and helping learners improve their conceptual modeling skills. This unique MBSE edX Professional Certificate program, endorsed by Whirlpool Corporation, is available at https://www.edx.org/professional-certificate/israelx-model-based-systems-engineering.

11.10   Summary

In this chapter, we presented OPCloud as a Could-resident software environment for conceptual modeling with OPM ISO 19450, which is especially suited for Industry 4.0 requirements of tight hardware–software integration. Further, we presented MAXIM – an extension of OPM ISO 19450:2015 that provides for incorporating computations and programming within the OPCloud framework. We also discussed issues of connectivity, interoperability, software development, and educational implications. We continue to develop the OPCloud environment, making it ever more user friendly, and adding to it useful functions, such as query capabilities using graph databases, and automatically creating design structure matrix (DSM) from an OPM model.

References

Akkaya, I., Derler, P., Emoto, S., and Lee, E.A. (2016). Systems engineering for industrial cyber-physical systems using aspects. *Proceedings of the IEEE* 104 (5): 997–1012.

Arvaja, M., Salovaara, H., Häkkinen, P., and Järvelä, S. (2007). Combining individual and group-level perspectives for studying collaborative knowledge construction in context. *Learning and Instruction* 17 (4): 448–459.

Atlassian (n.d.). Jira: Issue and Project Tracking Software. https://www.atlassian .com/software/jira (last accessed 15 June 2019).

Baepler, P., Walker, J.D., and Driessen, M. (2014). It's not about seat time: blending, flipping, and efficiency in active learning classrooms. *Computers and Education* 78: 227–236.

Berners-Lee, T., Masinter, L., and McCahill, M. (1994). Uniform resource locators (URL). RFC No. 1738.

BMBF (Bundesministerium für Bildung und Forschung.), 2016 Industrie 4.0. https://www.bmbf.de/de/zukunftsprojekt-industrie-4-0-848.html (Accessed 20 December 2018).

Böhm, W., Henkler, S., Houdek, F. et al. (2014). Bridging the gap between systems and software engineering by using the SPES modeling framework as a general systems engineering philosophy. *Procedia Computer Science* 28: 187–194.

Bower, M., Dalgarno, B., Kennedy, G.E. et al. (2015). Design and implementation factors in blended synchronous learning environments: outcomes from a cross-case analysis. *Computers and Education* 86: 1–17.

Broodney, H., Shani, U., and Sela, A. (2013). 4.5.2 Model integration–extracting value from MBSE. *INCOSE International Symposium* 23 (1): 1174–1186.

Christie, A.M. (1999). Simulation: an enabling technology in software engineering. *CROSSTALK – The Journal of Defense Software Engineering* 12 (4): 25–30.

Deloitte, Industry 4.0 – Challenges and solutions for the digital transformation and use of exponential technologies, 2015. https://www2.deloitte.com/content/ dam/Deloitte/ch/Documents/manufacturing/ch-en-manufacturing-industry-4-0-24102014.pdf (Last accessed 15 June 2019).

Dori, D. (2016). *Model-Based Systems Engineering With OPM and SysML*, 1–411. New York, NY: Springer.

Dori, Y.J. and Belcher, J. (2005). How does technology-enabled active learning affect undergraduate students' understanding of electromagnetism concepts? *The Journal of the Learning Sciences* 14 (2): 243–279.

Dori, D., Beimel, D., and Toch, E. (2004). OPCATeam – Collaborative business process modeling with OPM. In: *Business Process Management. BPM 2004*, Lecture Notes in Computer Science, vol. 3080 (eds. J. Desel, B. Pernici and M. Weske), 66–81. Berlin, Heidelberg: Springer.

Dori, Y.J., Hult, E., Breslow, L., and Belcher, J.W. (2007). How much have they retained? Making unseen concepts seen in a freshman electromagnetism course at MIT. *Journal of Science Education and Technology* 16 (4): 299–323.

Dori, D., Linchevski, C., and Manor, R. (2010). OPCAT – A software environment for object-process methodology based conceptual modelling of complex systems. In: *Proceedings of the 1st International Conference on Modelling and Management of Engineering Processes*, 147–151. Cambridge, UK: University of Cambridge.

Dori, D., Renick, A., and Wengrowicz, N. (2016). When quantitative meets qualitative: enhancing OPM conceptual systems modeling with MATLAB computational capabilities. *Research in Engineering Design* 27 (2): 141–164.

Dori, D. (2002). *Object-Process Methodology: A Holistic Systems Paradigm*. Springer.

Flarup, E. (2007). Best practices in software localization. In: *Internationalization and Unicode Conference (IUC32)*, San Jose, California (8–10 September 2007).

Friedenthal, S., Moore, A., and Steiner, R. (2014). *A Practical Guide to SysML: The Systems Modeling Language*. Morgan Kaufmann.

Fritzson, P. (2010). *Principles of Object-Oriented Modeling and Simulation With Modelica 2.1*. Wiley.

Gardner, L.L., Grant, M.E., and Rolston, L.J. (1994). Using simulation to benchmark traditional vs. activity-based costing in product mix decisions. In: *Proceedings of the 26th Conference on Winter Simulation*, 1050–1057. Society for Computer Simulation International.

Gonzalez, O. (2018). Digital Transformation: A Practical Definition. https://www.linkedin.com/pulse/digital-transformation-practical-definition-adrian-gonzalez (Last accessed 15 June 2019).

Han, S., Capraro, R., and Capraro, M.M. (2015). How science, technology, engineering, and mathematics (STEM) project-based learning (PBL) affects high, middle, and low achievers differently: the impact of student factors on achievement. *International Journal of Science and Mathematics Education* 13: 1089–1113. https://doi.org/10.1007/s10763-014-9526-0.

Harel, D. and Kugler, H. (2004). The rhapsody semantics of state charts (or, on the executable core of the UML). In: *Integration of Software Specification Techniques for Applications in Engineering*, Lecture Notes in Computer Science, vol. 3147 (eds. H. Ehrig, W. Damm, J. Desel, et al.), 325–354. Berlin, Heidelberg: Springer.

Hermann, M., Pentek, T., and Otto, B. (2016). Design principles for Industrie 4.0 scenarios. In: *2016 49th Hawaii International Conference on System Sciences*

(HICSS), 3928–3937. IEEE https://ieeexplore.ieee.org/xpl/mostRecentIssue.jsp? punumber=7426593 (Last accessed 15 June 2019).

Hone, K.S. and El Said, G.R. (2016). Exploring the factors affecting MOOC retention: a survey study. *Computers and Education* 98: 157–168.

Rappid Powerful visual tools at your fingertips. (n.d.). https://www.jointjs.com (last accessed 15 June 2019).

Jacobs, S., Wengrowicz, N., and Dori, D. (2014). Exporting object-process methodology system models to the semantic web. In: *2014 IEEE International Conference on Systems, Man and Cybernetics (SMC)*, 1014–1019. IEEE.

Johanson, A. and Hasselbring, W. (2018). Software engineering for computational science: past, present, future. *Computing in Science and Engineering* 20 (2): 90–109.

Kay, J., Reimann, P., Diebold, E., and Kummerfeld, B. (2013). MOOCs: so many learners, so much potential. *IEEE Intelligent Systems* 28 (3): 70–77.

Kong, S.C. (2014). Developing information literacy and critical thinking skills through domain knowledge learning in digital classrooms: an experience of practicing flipped classroom strategy. *Computers and Education* 78: 160–173.

Lasi, H., Fettke, P., Kemper, H.G. et al. (2014). Industry 4.0. *Business and Information Systems Engineering* 6 (4): 239–242.

Lassila, O., and Swick, R. R. (1999). Resource Description Framework (RDF) model and syntax specification. World Wide Web Consortium (W3C). https://www.w3.org/TR/1999/REC-rdf-syntax-19990222/ (Last accessed 15 June 2019).

Liyanagunawardena, T.R., Adams, A.A., and Williams, S.A. (2013). MOOCs: a systematic study of the published literature 2008–2012. *The International Review of Research in Open and Distributed Learning* 14 (3): 202–227.

Mayer, R.E. (2002). Multimedia learning. In: *Psychology of Learning and Motivation*, vol. 41 (ed. B.H. Ross), 85–139. Academic Press.

McGuinness, D. L., and Van Harmelen, F. (2004). OWL web ontology language overview. World Wide Web Consortium (W3C). https://www.w3.org/TR/owl-features/ (Last accessed 15 June 2019).

Mellor, S.J. and Balcer, M.J. (2002). *Executable UML: A Foundation for Model-Driven Architectures*, Object Technology Series. Addison-Wesley Professional.

Object Management Group (OMG). 2018. Semantics of a foundational subset for executable UML models (FUML), Version 1.4. https://www.omg.org/spec/FUML/1.4 (Last accessed 15 June 2019).

Pyster, A., Adcock, R., Ardis, M. et al. (2015). Exploring the relationship between systems engineering and software engineering. *Procedia Computer Science* 44: 708–717.

Rational, I. B. M. (2010). Rational DOORS. https://www.ibm.com/support/knowledgecenter/en/SSYQBZ_9.5.2/com.ibm.doors.requirements.doc/topics/c_welcome.html (Last accessed 8 July 2019).

Reid, D. (2018, June 27). Boeing reportedly tells engine makers to make bids for a new 797 plane https://www.cnbc.com/2018/06/27/boeing-797-ge-pratt-whitney-rolls-royce-told-to-bid-for-new-plane.html (Last accessed 15 June 2019).

Richardson, L. and Ruby, S. (2008). *RESTful web services*. O'Reilly Media, Inc.

Schleich, B., Anwer, N., Mathieu, L., and Wartzack, S. (2017). Shaping the digital twin for design and production engineering. *CIRP Annals* 66 (1): 141–144.

Seidewitz, E. (2014). UML with meaning: executable modeling in foundational UML and the Alf action language. *ACM SIGAda Ada Letters* 34 (3): 61–68.

Selic, B. (2003). The pragmatics of model-driven development. *IEEE Software* 20 (5): 19–25.

Shani, U. (2017). Can ontologies prevent MBSE models from becoming obsolete? In: *Systems Conference (SysCon), 2017 Annual IEEE International*, 1–8. IEEE.

Shani, U., and Landau, A. (2013). Tools interoperability platform for model-based systems-engineering. 8th Workshop on Model Based Software, Data, Process and Tool Integration, MBSDPTI 2013. Montpellier, France.

Shani, U., Jacobs, S., Wengrowicz, N., and Dori, D. (2016). Engaging ontologies to break MBSE tools boundaries through semantic mediation. 2016 Conference on Systems Engineering Research.

Somekh, J., Haimovich, G., Guterman, A. et al. (2014). Conceptual modeling of mRNA decay provokes new hypotheses. *PLoS One* 9 (9): e107085.

Spittle, P. (2003). Gas turbine technology. *Physics Education* 38 (6): 504.

OCSL (2010). OSLC core specification version 2.0. Open Services for Lifecycle Collaboration, https://archive.open-services.net/ (Last accessed 15 June 2019).

Cyganiak, R., Wood, D., Lanthaler, M. (2014). RDF 1.1 concepts and abstract syntax. World Wide Web Consortium (W3C). https://www.w3.org/TR/rdf11-concepts/ (Last accessed 15 June 2019).

12

Recent Advances Toward the Industrialization of Metal Additive Manufacturing

Federico Mazzucato, Oliver Avram, Anna Valente, and Emanuele Carpanzano

Synopsis

The times when additive manufacturing (AM) technologies were struggling against product complexity and manufacturability are over; additive manufacturing has indeed proven its benefits several industries and use cases, while research has widely invested in consolidating the opportunities envisaged years ago, also opening the door to bringing manufacturing into the digital era. Reports of successful industrial cases are increasing at a fast pace as an ever-growing number of companies are getting significant advantages from AM technologies by reducing production cost, material waste, and time-to-market. Nevertheless, open challenges and unresolved issues are still clouding the decision making process of a great number of industrial players in the favor of adopting additive technologies, regardless of their undeniable advantages and opportunities. An effective achievement of cutting-edge results in the field of AM will rely upon a larger scale pipeline aiming at comprehensively boosting the smart factory paradigm. Our research analysis on recent manufacturing trends motivates a framework that captures highly relevant activities to support AM technologies' move toward mass manufacturing. In this perspective this chapter has its onset on the quest of answers to the fundamental question of how to use digital technology to improve operational excellence, and to make AM mainstream at an industrial level.

First, the state of the art of AM is introduced followed by presentation of relevant industrial applications, recent advances, and open challenges, with a focus on metal AM. Starting from the premise that a convincing demonstration of innovative manufacturing processes, coupled with digital transformation of machinery, is a key factor for an accelerated adoption of AM technology, we introduce the latest developments on process design, control, and optimization

Systems Engineering in the Fourth Industrial Revolution: Big Data, Novel Technologies, and Modern Systems Engineering, First Edition. Edited by Ron S. Kenett, Robert S. Swarz, and Avigdor Zonnenshain.

and the roadmap set up by the ARM Laboratory at University of Applied Sciences and Arts of Southern Switzerland. We aim at the implementation of a digital platform capable of guaranteeing a consistent AM build across an extendable number of metal AM applications. We conclude the chapter with future work opportunities and conclusions.

12.1   State of the Art

Additive manufacturing (AM) is a new way to produce parts and components, defined by the American Society of Testing and Materials (ASTM) as "the process of joining materials to make objects from 3D data, usually layer upon layer, as opposed to subtractive manufacturing methodologies; Synonyms: 3D printing, additive fabrication, additive process, additive techniques, additive layer manufacturing, and freeform fabrication") ASTM 2012)

The main peculiarity that distinguishes AM from conventional manufacturing processes such as tuning, milling, casting, injection molding, or plunging is the capability to build parts without the employment of tools or molds, but by using a layer-by-layer material build-up approach directly from a digital model.

The term *"Additive Manufacturing"* is quite new. It was coined in 2010 by the International Committee F42 for AM Technologies to group all those technologies capable of managing and working a wide range of materials (from polymers to ceramics and metals) in an unconventional tool-less manufacturing method.

Before 2010, AM technologies were identified by various names, depending on their main industrial application. For instance, at the beginning of the 1980s, they were mainly known as *"Rapid Prototyping,"* a term used to group several techniques, mainly used to realize prototypes in thermoplastic polymers, in order to minimize the timing and the additional costs incurred in the process of developing a product.

In the 1990s, thanks to the patent of new additive technologies, AM found application in the manufacturing process to both realize sand cores with a higher geometrical complexity never reached before in the foundry and along the process chain to design and produce customized tools to support part manufacturing. Since the manufacture of cores and tools was faster than the traditional method then used, the involved AM technologies were known by the name of *"Rapid Casting"* (RC) and *"Rapid Tooling"* (RT), respectively.

Until that moment, industrial interest and attraction toward the AM world were of limited importance, since the technical and technological level of the systems and the employed material performances were low and the overall quality of the realized parts was coarse.

Only with the development of high performance engineering materials jointly with the emergence of new technologies and the technical improvement of preexisting AM systems was it possible to realize parts with enhanced mechanical properties and innovative design, highly appealing for the industrial sector.

The turning key that changed the role of AM in manufacturing, happened at the beginning of 2000s, when both research institutes and industries began to employ AM technologies to produce functional prototypes and parts having mechanical characteristics comparable with those realized through conventional technologies ("*Rapid Manufacturing*" [RM]). From that moment, AM powerfully entered in the industrial scenario and began to revolutionize the way to design and manufacture products.

AM Technologies: Industrial Impact

The industrial sectors mainly affected by the entrance of AM in the manufacturing world are: aerospace, naval, automotive, biomedical, fashion, jewelry, architecture, and electronics. Their interest in adopting this new way of manufacture is mainly motivated by the opportunities brought along by the AM technologies to address the changing needs and evolving requirements of the end-users in every specific market.

The aforementioned industrial sectors have to fulfill a demand that often requires a high degree of product customization, high operational performances and a low production volume. The conventional technologies usually employed in these fields have shown to be not cost effective and not capable of ensuring the required degree of part shape complexity. On the contrary, AM has the potential to overcome or support the technology traditionally employed, reducing the manufacturing timing and production costs drastically.

Year after year, the global additive market is increasing and it is expected to increase in the future too (Figure 12.1). In 2018, it was sized at $9.3 billion, marking an extraordinary increase compared to $5 billion in 2015 (www.smartechanalysis.com/). The preeminent 3D printer experts in the world, Wohler Associates (www.wohlersassociates.com), foresees a significant increase of the AM industry in 2018, which could be close to 21% (equal to $7.3 billion), while it estimates a sales growth of AM systems for metals equal to 80% (from 983 systems in 2016 to an estimated 1768 in 2017) (McCue 2018). Well established firms such as HP Inc, Optomec®, Renishaw, EOS, as well as other important industrial companies, are continuously investing in the development of new AM solutions and launching new AM systems (Metal AM 2018g). The aerospace, automotive, and medical industries are expected to account for 51% of the 3D printing market by 2025. In particular, between 2015 and 2025, AM revenue is expected grow at a compound annual growth rate (CAGR) of 34% in the automotive industry, 23% in medical devices, and 26% in the aerospace and defense industry.

The gradual and continuous climb of AM in the industrial world mainly stands on the following distinctive aspects:

- AM systems are "easy to use".
- The absence of cutting tools or molds.
- The possibility to realize Functionally Graded Material (FGM).

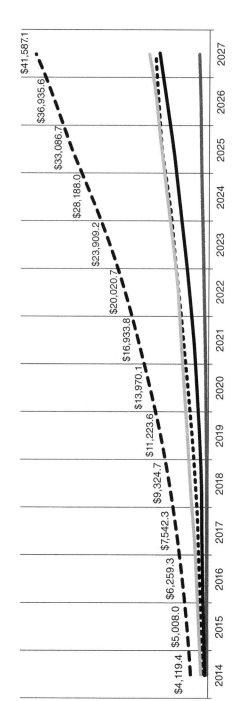

Figure 12.1 Projected global AM market size between 2016 and 2027 (billion US$). Source SmarTech Publishing.

When an AM process starts, the part takes shape adding material vertically, layer upon layer, from the bottom to the top. From the deposition of the first layer deposition to the last one, no external actions are required, but the process goes on until the complete realization of the part. Contrary to most conventional technologies, operations such as homing, reference definition, tool wear monitoring, or material supply during the part manufacturing are completely automated and do not require the external intervention of an operator. Moreover, most of the commercial AM solutions are "easy to use," that is user friendly, safe, and require low maintenance. From an industrial perspective, this translates into no need for specialized personnel to work with AM systems and in a low investment for employees' requalification. Therefore, the integration of such technologies in a preexisting process chain or in a new one is quite straightforward.

The absence of a cutting tool to machine and shape the part frees AM technologies from those constraints due to the presence of the tool itself. The collision risk between the part to be manufactured and the cutting tool is strongly reduced, enabling the possibility to realize near net shape or quasi-near net shape parts just starting from a nominal computer aided design (CAD). Very complex shapes, such as lattice structures and undercuts, are now possible. The final shape of the part is no more a compromise between functionality and the technological constraints of the manufacturing system, but it is only a matter of functionality. This allows complete design flexibility and the realization of customized geometry tailored for the specific application.

Moreover, a tool-less process is a process free from high dynamic forces, mechanical vibrations, and lubricants. This allows the construction of lighter, compact, clean, and noiseless systems that are easy to transport and move into more strategic locations for the manufacturing process. AM systems can be installed in designer offices as well in garages, workshops, or medical laboratories, bringing the manufacturing of the part to where it is immediately needed, cutting down the lead time and costs tied up with part transportation, and decreasing the time to market.

Thanks to the possibility to add material layer by layer, it is now possible to deposit the material only where it is strictly required to produce the part, decreasing the material waste drastically. In milling, for instance, the scrap can reach 90% of the starting raw material, heavily affecting the manufacturing costs. On the contrary, in AM technologies the waste material usually ranges between 10 and 50% depending on both the involved system and part complexity.

Another distinctive aspect of AM is the capability to realize FGM, managing different materials at the same time the process is running. This capability is unique and cannot be found in any other existing technology. This innovative aspect has become of great interest in industrial sectors, such as aerospace and biomedical, opening new ways to research and design parts.

Current Limitations and Improvement Opportunities

In almost 40 years, thanks to the effort of research institutes and private companies, AM technologies and systems have been developed strongly, increasing their performances in terms of process robustness and final quality of the part. Despite seeing great attention and awareness in the last few years, the widespread adoption and industrialization of AM are still held back by a number of compelling challenges that require immediate consideration. These drawbacks mainly regard low final dimensional accuracy, the presence of defects such as inclusions and pores, low production volume, anisotropic mechanical properties, residual stresses, lack of a standardization, and high production costs.

The geometrical and dimensional accuracy of a part realized through AM are, in general, lower than that realized through conventional technologies. This means that post-processing is often needed to fulfill the designed specification in terms of superficial roughness and tolerances. From an industrial point of view, this amounts to additional time and costs needed to reach the required final quality of the part. The post-machining of the part anyway could be difficult or impossible due to the complex geometry of the part.

Moreover, AM processes can cause the generation of uncontrolled residual stresses or the formation of internal defects (e.g. internal voids) during manufacturing of the part, which undermines the robustness of the AM process, making the mechanical behavior of the part unpredictable. In metal AM, to avoid distortions or undesired breakage, thermal treatments are often needed to relieve residual stresses and reach the desired mechanical properties of the realized part.

Even the production rate remains low if compared to milling or casting. Important efforts are being made to improve the system productivity, designing new AM systems (e.g. multilaser powder bed systems, hybrid systems) or working on the identification of the most suitable combination of process parameters through the execution of exhaustive experimental campaigns. Experimental tests are fundamental to understand the AM process in deep, but they are mainly based ona trial and error method that is time consuming and involves high material costs.

For this reason, new simulation and control tools are going to be developed and designed to increase the control on the additive process and improve the final quality of the part. Important companies, such as Materialise, Autodesk,

Simufact, and Ansys, are working to develop software tools capable of simulating and optimizing AM processes (Metal AM 2017b, 2018e).

Another important drawback of AM technologies is the strong sensitivity of the starting conditions of the raw material supplied during the process. To increase the robustness of the AM process, the raw material has to be certified and stored in controlled spaces to avoid changes of its temperature and residual humidity, which could introduce defects such as voids, lack of deposition or breakage of the part while the process is running. Companies planning to increase the usage of AM technologies in the future need to gain also a better understanding of bulk powder handling and containment techniques in order to guarantee consistent builds.

Another important critical point is the lack of standardization that guides the production though AM technologies and qualified AM parts. Despite these technological and processing challenges, the employment of AM technologies is growing up. Annual reports, specialized magazines, and documented user cases are showing that there is tremendous interest regarding all AM-related topics.

In 2017, in a hospital in Chongqing municipality a tantalum knee prosthesis realized by AM was successfully implanted to replace a large section of the knee bones damaged by severe arthritis (Metal AM, 2017a). From 2016, Yndetech (Fano, Italy) successfully produce dental implant for the Italian market (Metal AM 2018c). Surgeons at Morriston Hospital (Wales, UK) have successfully rebuilt part of a patient's chest using metal AM. The titanium implant was additively manufactured on a Renishaw system, as part of a collaborative innovation project between the company and the hospital (Metal AM 2018a). In 2018, Airbus Helicopters (Donauwörth, Germany) begun production of titanium latch shafts for the doors of the Airbus A350 passenger aircraft using metal AM. They are reported to save 45% of the weight and the 25% of cost compared to the traditional solution (Metal AM 2018d). In 2018, Porsche Classic (Stuttgart, Germany) declared production of rare, replacement parts by metal and plastic AM. In order to enable the economic manufacture of rare parts to a high quality in very short production runs, AM is now used (Metal AM 2018b).

12.2 Metal Additive Manufacturing

AM technologies are revolutionizing the way parts are manufactured. Year after year, more and more companies are employing AM to imagine, design, and realize end-use products having new functional shapes and enhanced operating performances. Parts with intricate internal features or embedded components, parts replacing an elaborate assembly with a single piece, and customized tooling produced quickly are all among the possibilities that additive manufacturing is bringing about.

The flexibility to design complex geometries paired with the freedom of employing multimaterials in integrated structures provides strong incentives for the growth of AM in the industrial sector

Relevant Industrial Applications

From the beginning of 2000s, industry is looking to make this technology more competitive, robust and fast, improving augmentation to standard production techniques (GKN). Academic institutes and industrial research and development continue to broaden the horizon of additive for the future.

Materials play a highly significant role in AM, particularly due to the different way AM they are handled with respect to the conventional manufacturing methods. Machines and specific AM technologies are linked to certain types, forms and states of materials. Recently metal AM technologies such as Powder Bed Fusion (PBF) and Direct Energy Deposition (DED) are drawing considerable attention from different industrial sectors.

The fulfillment of strong business cases will accelerate the learning curve and will bring these innovative technologies to mature level. Aerospace industries are demonstrating that the application of additive technologies in such critical and rigorous field is actual and real. Aircraft cabin interiors and UAVs' components are now manufactured additively thanks to a lightweight part design, reducing costs and meeting fuel consumption as well as CO_2 emission targets. For example, industrial giant GE Aviation has been realizing and successfully testing metal parts for commercial jet engines through AM (Metal AM 2015a); in 2018, it has offered improved turbine engine with metal AM parts to the US Army (Metal AM 2018c) (Figure 12.2).

In 2018, Airbus Helicopters (Donauwörth, Germany) has officially begun production of titanium-made latch shafts for the doors of the A350 passenger aircraft using metal AM, proving that AM components are cheaper to produce and weigh less than their counterparts made by conventional methods (Metal AM 2018f).

Safran, one of the world leading manufacturers of rotorcraft turbines, has been producing fuel injector nozzles for serial production of engine components using AM, simplifying the part geometry and reducing the number of components to one single piece with advanced injection and cooling functions (Metal AM 2015c). The US government space agency (NASA) is also following metal AM growth with interest, investing and employing AM technologies actively. In 2018, Engineers at NASA's Marshall Space Flight tested the first 3D-printed rocket engine prototype part made of two different metal alloys: copper alloy and Inconel, testing the part more than 30 times successfully (AM Additive Manufacturing 2018a).

Other important aerospace companies (e.g. Pratt and Whitney, Arconic) are testing structural parts realized with AM and integrating these additive

Latch shafts prior to depowdering   Knee tantalum implant   AM dental implants

BMW water pump wheel   Pressure and temperature sensor for engine
                                          control system

Figure 12.2 AM applications from various industries.

technologies on their process chain (Metal AM 2018h). The aerospace sector is not the only one taking advantages of AM. In the last 10 years, the automotive industry is another sector employing AM technologies widely. In the Formula 1 Championship, racing teams such as Alfa Romeo Sauber, Ferrari, and McLaren are successfully employing metal AM to design and realize metal parts to integrate in their sport cars. In the beginning of the 2010s, these car racing teams took advantage of AM for the rapid production of functional prototypes useful for car improvement and project choices, but now there are actually manufactured end-use parts being assembled into the racing car (AM Additive Manufacturing 2018b; Clarke 2017).

AM is also present in important automotive firms such as Volkswagen, Audi, and BMW. In 2016, Robert Hofmann GmbH (Lichtenfels, Germany) has additively manufactured a fully functional aluminum cylinder block for a Volkswagen VR6 engine. The AM cylinder block has undergone and passed a number of tests at Volkswagen, offering a significant weight reduction and showing a low porosity and significantly smaller distortions and deviations from the desired geometry when compared to the previous GJL250-made cast component (www .hofmann-innovation.com).

In 2015, BMW has celebrated its 500th water pump wheel made on a 3D printer. This component has been placed in the powertrain of Deutsche

Tourenwagen Masters race cars and Z4 GT3 customer vehicles. This AM component has replaced the previous one made through injection molding, achieving an ideal aerodynamics of the component and making the demand-oriented production more cost effective (BMW 2015).

Another important industrial field where AM is finding large application is the biomedical. In such a case, hospitals, medical companies, and laboratories are exploiting AM technologies to increase the quality of their services, customizing their parts depending on the patient needs and desires. The manufacture of prostheses is now tailored on the anatomy of the human body, increasing the comfort of the patient. In 2015 some successful realizations of graft and orthopedic implants were documented (www.anatomics.com) (Metal AM 2015d). In 2007, Lima Corporate, an Italian company involved in the development of innovative solutions to assist orthopedic surgeons, brought to market its first titanium cup of a hip implant component produced through AM. The cup surface has been carefully engineered to encourage the growth of the patient's bone into this surface (AM Additive Manufacturing 2017).

In 2015, 4WEB Medical (Texas, USA) has announced that surgeons have implanted over 3 000 of its 3D-printed orthopedic truss implants (Metal AM 2015b), whereas Dentsply Implants, one of the world's leading companies in implant dentistry, have started to manufacture dental products though AM.

Recent Advances

In the last 10 years, significant and important advancements have been carried out by AM machine manufacturers and research institutes to increase the performance and robustness of AM processes. In particular, this has concerned both PBF and DED technologies, with the definition of new standards, improvement of AM hardware, and development of new AM materials more suitable to be additively manufactured.

Recently new standards have been designed and published. For instance, ISO/ASTM52900 and ISO/ASTM52921 establish and define terms used in AM technology, classifying the terms involved into specific fields of application and including terms, definitions of terms, descriptions of terms, nomenclature, and acronyms associated with coordinate systems and testing methodologies for AM (ASTM 2015, 2013). ISO/ASTM52910 gives requirements, guidelines, and recommendations for using AM during the design of all types of products, devices, systems, components, or parts that are fabricated by any type of AM system, determining which design considerations can be utilized in a design project or to take advantage of the capabilities of an AM process (ASTM 2018). Moreover, ISO/ASTM52901 defines and specifies requirements for purchased parts made by AM, giving guidelines for the elements to be exchanged between the customer and the part provider at the time of the order, including the customer order information, part definition data, feedstock requirements, final part characteristics and properties, inspection requirements, and part acceptance methods (ASTM 2016).

Significant AM system improvements have been performed for both PBF and DED technologies. Machine manufactures have been focusing on two main technological aspects: increase of the system productivity and improvement of the final quality of the realized part, with the main goal to strongly reduce the production costs while increasing the production volume and limiting the post-processing operations required to fulfill the final part specifications. New PBF systems with improved laser controls have been produced. Nowadays it is possible to simultaneously employ up to four lasers to build metal parts, reducing the production time drastically.

Larger and inert building chambers have been designed to be able to accommodate a higher number of parts per print as well as to limit the oxygen content during the process. Moreover, these structures facilitate the processing of reactive materials such as titanium and aluminum and the reduction of oxidation phenomena. Preheated building platforms have been also implemented to control temperature distributions and to reduce the risk of crack formation during the metal part realization.

The manufacturing systems employing the DED technology were also subject to technological improvements. In addition to preheated building platforms, the kinematics and Computer Numerical Controls (CNCs) of DED systems have been improved to enable five-axis interpolation during the part realization. To increase the final quality of AM products, traditional subtractive tools (e.g. milling tools) have been integrated into preexisting DED systems, contributing to the enhancement of the surface roughness and the dimensional accuracy of the final metal part. Moreover, to improve the oxygen control during the printing, some DED solutions (such as Laserdyne 430 located at the Automation Robotics and Machine (ARM) Laboratory of SUPSI-DTI-ISTePS) employ a high temperature resistant polymer box to shield the deposition area and reduce the oxidation of the part.

In conjunction with both the establishment of new AM Standards and the enhancement of preexisting AM systems, new metal materials have been developed to fulfill AM process requirements. In particular, research has been carried out on:

- The form of the feedstock that should be compatible with the involved AM technology (e.g. shape of the raw material, quality, grain size distribution for metal powders, microstructure).
- AM metals showing "good behavior" during additive process (e.g. no material degradation, good flowability).
- AM metals exhibiting acceptable service properties (e.g. good mechanical behaviors, low deformations).

For the powder-based AM technologies the fusion, flowability, weldability, and castability of the supplied material are fundamental properties ensuring a successful additive process (Bourell et al. 2017). The range of usable materials is ever-extending thanks to ongoing studies aiming the implementation of high-entropy alloys (Tsai and Yeh 2014), magnetic alloys (Mikler et al. 2017), bulk metallic glasses (amorphous metals), high-strength alloys, and FGM (Ngo et al. 2018). For industrial sectors like aerospace, automotive, and biomedical, such materials present superior tensile strength, high hardness, wear resistance, and corrosion resistance in comparison with metal alloys traditionally used.

Open Challenges

All the aforementioned industrial cases of AM application and recent advances demonstrate that AM technologies have a strong presence in the manufacturing world and are getting a significant consideration in the industry. Nevertheless, the question naturally arises to why AM does not enjoy a wider acceptance and recognition in the manufacturing world when compared to other conventional technologies?

The main reasons beyond this question are mainly the following:

- *Steepness of the learning curve.* Despite the successful implementation of specific industrial cases through AM, the knowledge of the AM process in terms of capabilities, advantages, and applications is still unknown and linked with niche applications by the majority of manufacturing companies. Moreover, they still look at AM as a young and premature technique not capable of fulfilling their needs. The lack of confidence and the disillusion of too high and untimely expectations previously advertised have left the sense of a fascinating technology yet one which is difficult to be feasible in the manufacturing world. Today, feasibility is becoming real thanks to the development of new AM systems, new AM materials, and new tools (e.g. software, control devices) that are making AM processes more robust, precise, and consistent;
- *Resistance to change.* AM is an innovative and disruptive way to manufacture, which means it requires a new way to think, design, and produce parts. Many companies are afraid and strongly critical when they have to change their way of manufacturing or revolutionize their process chain. Well-known technologies (e.g. milling, tuning, casting) are preferred over AM since they are more widespread in industry and the "know how" is consistent, even if conventional technologies are shown to be less cost effective for some industrial cases;
- *Cost effectiveness.* AM is still a technology with high production costs in terms of facilities, materials, heat treatment and post-processing. The production volume is limited and the production speed for "simple" geometries is several times lower compared to traditional technologies such as milling or injection molding. Nevertheless, AM becomes impressively advantageous

when the complexity of the part increases. The manufacture of components having a high complex shape becomes slow, cost ineffective, and sometimes infeasible for the conventional technologies. Assemblies or free form parts required more process steps or the employment of several technologies to be realized. Structural lattice structure, hard-metal tools, and high-complex functional ceramic part are now possible through AM, reducing the material waste and collapsing the production time;

- *Part quality.* The maximum dimensional/geometrical accuracy and surface roughness of AM parts is worse than the maximum achievable by conventional technologies. Mirror surfaces or tight tolerances are still not feasible with the existing AM systems. Post-processing and post-treatments are often required to fully match the customer requirements;

- *Lack of standards.* The strength and reliability of a technology is measured by standards that define for instance their use, methodologies, performances, procedures to manufacture, and part characterization. The lack of a standardization to guide the implementation of AM technologies and equipment as well as the lack of performance criteria specification prevents the adoption of these techniques for several companies, which do not feel confident in making a fundamental shift in their manufacturing technology while being insecure about the return of such an investment.

Despite the efforts devoted to achieving hardware and material improvements, metal AM technologies are far from the exploitation of their full potential. Due to the dynamic behavior of the process, the control of the process is far from being straightforward. A static set of process parameters, employed along the entire deposition process, will very likely lead to an inconsistent build, which can be efficiently counteracted by an in-process adaptation of the most significant parameters. The common approach used to define proper recipes of process parameters for the production of defect-free parts is based on a trial-and-error experimental campaign, which usually involves high time and production costs that are not always sustainable by companies. This fact constitutes a critical barrier preventing the definitive adoption of metal AM technologies at the industrial level.

A key turning point in the transition from conventional toward AM technologies and their wide adoption in the industrial world is represented by the development of robust simulation and control tools. Specifically, for metal AM, the physics involved in the fabrication process are complex, not only from the perspective of the phenomena governing the process and their mutual interaction (e.g. heat exchanges, phase transformations, fluid dynamics, mass transfers, and microstructure evolutions are just some examples) but also due to the continuous variations in the boundary conditions that occur during part fabrication.

In order to cope with these challenges sophisticated and dynamic tools capable of both simulating the additive process during the phase of process design and controlling the manufacture of the part during its course are needed. Currently, AM simulation and control software do not have the technological level to fully assist the process. Consequently, the definition of process recipes for the manufacture of acceptable metal parts is still associated with huge trial-and-error experimental campaigns that involve the employment of significant human sources, material waste, and machine time.

Priority research efforts are directed toward the development and implementation of prediction tools making AM more robust, efficient, and sustainable. New mathematical theories are being applied to AM for a better management of multiphysics models, which are inherent in the characterization of almost all AM-based process–structure–property relationships. Predictive models are also being studied in the areas of powder properties, thermal distortion, and surface preparations and software developers (e.g. Autodesk, Simufact, Additive Works, Materialise, Ansys) are periodically improving and releasing to the market new AM simulation tools to improve final part quality and decrease process developing costs.

Nevertheless, most of the predictive software tools are mainly focused on PBF technologies, performing quite accurate predictions of both thermal gradient distributions and part deformations during the metal printing. These tools prove to be significantly useful in helping process engineers during the selection of a proper combination of process parameters, the placement of the part in the building chamber, and in the choice of the most suitable deposition strategy. Nevertheless, they are still limited and not applicable for all AM materials or PBF systems. Moreover, a real process control tool is still missing. Several solutions have been proposed, mainly based on laser power control through the monitoring of the melt pool shape and temperature, but, today, they are not sufficiently robust and fast to allow an efficient closed loop control of the process.

On the contrary, both significant simulation and control tools are currently missing for DED technology. Even if the development of process models and control algorithms are the subject of ongoing research, the current outcomes are not complete and far from providing a reliable and robust simulation tool. Existing solutions do not take into account the entire dynamics of the deposition process, ranging from the material–laser interaction to material growth models, from the kinematics and dynamics of the machines and feeding systems to the deposition strategy up to the possibility to compensate possible deposition deviations. This results in a myopic process control, partially reliable and hardly scalable to any piece of equipment.

12.3 Industrialization of Metal AM: Roadmap Setup at the ARM Laboratory

Mission

The Laboratory of Automation Robotics and Machines (ARM-Lab), at the Department of Innovative Technologies of the University of Applied Sciences and Arts of Southern Switzerland, is actively working to implement an innovative tool capable of simulating, monitoring, and controlling the process with the objective of transforming DED into an industrially available technology to be disruptively integrated into the manufacturing value chain (SUPSI 2019). Today, ARM-lab is participating in high industrial impact projects at both National level (i.e. DED-In718 (SUPSI 2019), Ground Control (SNSF 2018)) and European level (i.e.4DHybrid (http://4dhybrid.eu)), aiming to bring innovation on the following aspects: right-the-first-time parts (high quality final parts); improved efficiency (energy and material saving (Mazzucato et al. 2017b; Marchetti et al. 2017)); portability (i.e. solution initially developed for a specific machine setup but, at the same time, scalable to any DED machine architecture and setup).

One of the core research activities of the ARM-Lab is the development and implementation of an online control loop architecture capable of efficiently and effectively exploiting, on one side, the wealth of information gathered from the intensive offline experimentation and, on the other side, the field data from the ongoing deposition process. The data resulting from the offline experimentation will be employed to realize an offline module that, in conjunction with the final desired requirements of the part, will be capable of generating a data-driven process model endowed with machine learning techniques. Machine learning will be used to extract important findings from the data gathered by a consistent experimental campaign, determining what additive information and what measurements are truly necessary to the ongoing creation of this platform. The online control will be instead ensured by an online sensor-based elaboration module that will interpret the measurements captured during the deposition and will persistently close the loop on the laser power, spot feeding dynamics and kinematic parameters to ensure the matching of the optimal requirements.

The development of such an innovative platform for DED process simulation and control will involve strong collaborations between SUPSI institutes, departments, and research group such as the ARM-Lab, DynaMat Laboratory, and IDSIA. It will also contribute to establishing a solid knowledge ground on the DED process by sharing an open database on which researchers, students, and industrial experts can rely for benchmarking.

Infrastructure

In this section, an overview of ARM-Lab facilities related to AM activities is introduced. They mainly belong to two families:

1) DED/Hybrid systems for metal part construction.
2) Offline measurement instruments for metal part analysis and characterization.

Each facility will be dealt with in detail in the following sections, discussing their main capabilities, technological innovation, and applications.

DED Systems

ARM-Lab is equipped with two systems endowed with DED technology as depicted in (Figure 12.3):

- Laserdyne 430
- SYMBIONICA

Laserdyne 430 The system is the result of the adaptation and integration of a commercial deposition head solution (i.e. Optomec deposition head) in a preexisting laser cutting machine (i.e. Laserdyne 430 – Primapower®). The integration of both deposition head and powder feeding system in the machine CNC has been fully performed by the ARM, turning the previous commercial laser cutting system into a complete operating DED system. Table 12.1 shows the Laserdyne 430 datasheet.

In addition to the head and feeding system integration, the ARM group has performed other relevant hardware improvements by enriching the deposition system with innovative features, with the focus on the enhancement of process performances and final part quality. Each feature installed in Laserdyne 430 has been designed, manufactured, and tested by the ARM. It is worthwhile mentioning:

Laserdyne 430 DED   Symbionica Hybrid

Figure 12.3 DED machines available at ARM-Lab.

Table 12.1 Laserdyne 430 datasheet.

Max building volume	585 × 408 × 508 mm (X,Y,Z)
Axis number	3
Motion speed	12 000 mm min^{-1}
Position accuracy	12.5 µm
Table load capacity	250 kg
Laser spot	1 mm
Laser type	Fiber laser – 1070 nm (CF1000 Convergent$^{®}$)
Maximum laser power	1000 W
Powder feeder	Volumetric (Optomec$^{®}$ Lens, Print Engine)
RPM (powder feeder)	0–24
Deposition head	Multinozzle (Optomec)

- New clamping system, compact and functional, ensuring a stiff clamping of the baseplate on the deposition table within a limited envelope. Moreover, the clamping system has been designed to be flexible, allowing the arrangement of more than one baseplate or baseplates with different size.
- Preheating system, integrated with the motion table, allowing both the thermal gradients that could be detrimental for part integrity to be decreased during the process and temperature distribution to be controlled during and after the deposition process. The baseplate preheating temperature reaches temperature a higher than 400 °C, enabling the deposition and realization of high carbon steel components to be managed.
- Flexible high-temperature resistant glove box, improving the protection of the deposition area from oxidation; it targets the reduction of the oxygen content during deposition and contributes to an increase of the final quality of the part in terms of structural and surface integrity.
- CMOS (complementary metal-oxide semiconductor) camera installed coaxially to the laser beam to get a top view of the melt pool formation and evolution while the deposition is running. The recorded images of the melt pool morphology are employed to adjust laser power parameter to optimize the deposition process.

SYMBIONICA The SYMBIONICA machine is the final deliverable of the SYMBIONICA EU-funded project (www.symbionicaproject.eu), where the ARM group had the scientific and technical coordinator role and had successfully carried out the project activities pertaining to machine development and design, adaptive automation, system integration, DED process monitoring and optimization. As of today, SYMBIONICA is the direct reflection of the extensive competence and capabilities of ARM-Lab pertaining to metal AM process and machine construction. Unlike Laserdyne 430, SYMBIONICA embeds two different laser-based techniques and a 3D scanner to handle additive/subtractive manufacturing (i.e. DED and Ablation) and inspection operations, respectively, and defines itself as a new benchmark equipment for Hybrid Manufacturing. SYMBIONICA's design, construction, CNC integration and optimization have been completely managed by the ARM group. Fundamental and critical devices for the stability of the process, such as the powder feeder and deposition head, have been also totally designed, assembled, tested, and optimized within ARM-Lab. The final result was an innovative machine with improved performances capable of realizing metal parts with enhanced final quality while decreasing the production time and costs in comparison with Laserdyne 430 (Table 12.2 shows the SYMBIONICA datasheet).

Significant progress beyond the state of the art on the design and development of AM equipment has been achieved, in particular in the following system improvement areas:

- Machine architecture
- Machine control
- Process control.

Table 12.2 SYMBIONICA datasheet.

Maximum building volume	800 × 800 × 1000 mm (X,Y,Z)
Axis number	5
Translation axis speed	200 000 mm min^{-1}
Rotation axis speed	30 rpm
Tilting axis speed	60 rpm
Translation axis precision	± 5 µm
Rotation axis precision	± 1 arcsec
Table load capacity	800 kg
Continuous laser spot	0.2, 1 mm
Pulsed laser spot	0.1 mm
Laser type (continuous)	Fiber laser – 1070 nm (CF1000 Convergent)
Laser type (pulsed)	Fiber laser – 1064 nm (IPG laser, YLP-V2)
Maximum continuous laser power	1000 W
Maximum pulsed laser power	100 W
Powder feeder	ARM customized solution
RPM (powder feeder)	0–26
Deposition head	Multinozzle (ARM customized solution)

The main innovative outputs have been:

1) Increased building volume. The DED process is mainly employed to build massive metal parts thanks to its high building rate compared to powder bed technologies. Moreover, the process is flexible and can perform coating or repair operations on preexisting parts having massive volume and significant weight. To achieve this goal, SYMBIONICA features a working table with increased dimensions allowing the fabrication of parts with a maximum spatial size of $800 \times 800 \times 1000$ mm (X, Y, Z). Moreover, the table load capacity has been increased up to 800 kg, allowing the handling of heavy metal parts to be repaired or coated.

2) Clamping and preheating system. Laserdyne's clamping system solution allows high stiffness and flexibility to clamp baseplates with different sizes while its preheating system proved to be robust and highly functional. In consequence, it has been adopted for SYMBIONICA as well.

3) Five-axis CNC and motion control. Unlike the powder bed technologies, the design and realization of supports needed to correctly build metal parts displaying overhanging areas with respect to the build direction is not straightforward with DED technology. The capability to simultaneously interpolate up to five motion axes makes SYMBIONICA capable of managing parts with highly complex shapes without the employment of support structures. This translates into a faster process, enhanced capability to handle geometrically complex parts, and streamlines the design for the DED, allowing the manufacture of metal parts with reduced overmaterial and limiting post-machining operations.

4) Double laser source. It is well known that metal parts produced by AM technologies often required post-processing (e.g. milling, tuning) to fulfill part specifications in terms of surface roughness, dimensional and geometrical accuracy. In response to this requirement, SYMBIONICA has been designed as an hybrid systems capable of employing a continuous laser source to perform material deposition (i.e. DED) and a pulsed laser source to perform material removal (i.e. ablation). The double laser source allows an increase in both the DED system and process performance, shortening the process chain and increasing the final quality of the metal part. For DED deposition, two optics are available, one ensuring a laser spot of 1 mm and one of 0.2 mm, useful when a fine deposition is required. Moreover, the integration of a laser scanner can boost the deposition time even further reaching apparent spots up to 8 mm of diameter.

5) Deposition multinozzle head with a new concept design (Marchetti et al. 2017; Valente et al. 2018). Key points for a cost effective deposition process are the way the blown powder reaches the melt pool and the quality of the shielded curtain, to prevent the melt pool from oxidation phenomena. SYMBIONICA has been equipped with a new multinozzle deposition head,

designed and realized in SUPSI. Four nozzles have been integrated in a CuTe head body. The metal alloy chosen for the deposition head has been fundamental to ensure a proper thermal dissipation during the process and avoid component overheating. Moreover, additional channels have been manufactured close to the deposition nozzle to both improve the shielding of the melt pool, preventing oxidation, and to decrease the spread of the powder flow at the nozzle outlet (Mazzucato et al. 2017a). This innovative solution allows an increase in the deposition efficiency up to 17%, as more powder particles fall into the melt pool, enhancing the building rate, and decreasing the powder waste.

6) New flexible volumetric powder feeder. The SYMBIONICA machine has been integrated with an enhanced powder feeding system capable of accurately supplying the DED process with fine amounts of metal powders. To improve system performance, the feeder has been integrated with a retrofitted system control having the capability to reduce flux-powder oscillation and ensure constant powder feeding. More hoppers can be installed and managed, allowing different metal powders to be used at the same time.

7) Embedded 3D scanner. As previously mentioned, the SYMBIONICA is a hybrid machine capable of adding and removing material where it is strictly required by the process. For this reason, this system has been equipped with a 3D optical scanner employing structured light technology and scanning the part through the projection of precise fringe patterns. The integrated 3D scanner allows the size of the actual printed part to be measured and compared with the nominal desired CAD and triggers the adaptation of the process in case of undesired overdeposition or lack of material.

8) Closed loop control. The DED process is highly dynamic. The final quality of the part produced strongly depends on the involved process parameters, deposition strategies, and boundary conditions (e.g. thermal distributions, material properties, cross-section size of the part). As soon as the process starts, the printing conditions can change during fabrication of the part; to ensure a high final quality of the component an in-process control and adaptation of the manufacturing strategy is required. An innovative feature integrated in SYMBIONICA is the process adjustment based on melt pool monitoring (Vandone et al. 2018). Thanks to the installation of a thermo-camera and a Field-Programmable Gate Array (FPGA) camera for melt pool monitoring, it is possible to observe and record melt pool behavior during the deposition process in terms of temperature distribution and melt pool morphology. Such gathered information allows corrective actions to be taken during the deposition (i.e. parameter adaptations) by altering the values of the laser power, laser scan speed, and powder feeding changes either individually or simultaneously.

Measurement Instruments

ATOS Core 200 The ATOS core 200 is a 3D scanner employed by the ARM group to perform reverse engineering, 3D measurement of small and medium-size objects, and inspection. Measurements are mainly performed offline to characterize the metal printed parts in terms of dimensional and geometrical accuracy by identifying part distortions or deviations derived from the comparison between nominal CAD and the actual scanned mesh within the scanner's software.

The scanner uses a stereo camera setup, combined with the blue light technology, which makes the scanning process independent of the ambient light, resulting in better data output on complex and shiny surfaces. The sensor projects different fringe patterns onto the object's surface. These patterns are recorded by two cameras, forming a phase shift based upon sinusoidal intensity distributions on the camera chips.

Keyence VHX 6000 The Keyence VHX-6000 is a digital microscope that incorporates observation, image capture, and measurement capabilities, providing an on-screen interface for viewing objects. Real-time 2D measurements can be performed and profile, height, and volume data on 3D images can be obtained. Moreover, it is possible to extract a 3D point cloud of the detected area for further and external analysis. High-quality and fully-focused images can be generated in a very short time.

The Keyence VHX-6000 located at ARM-Lab of SUPSI is equipped with an adaptive multilighting function and a VH-ZST lens that features a 20–2000× magnification range. This measurement instrument is mainly employed for (Figure 12.4):

- External evaluation of surface.
- Preliminary estimation of the surface roughness.

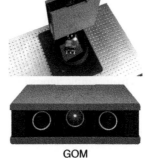

Keyence   GOM

Figure 12.4 Measurement instruments at ARM-Lab.

- Structural analysis in terms of porosity distribution or presence of inclusions for cross-sectioned samples;
- 2D and 3D profile detection and measurement.

DED Process Design

Process engineers working with metal AM systems have to deal with several process variables that have a strong influence on the final quality of the part. Due to their mutual interaction and dynamic behavior during a manufacturing process, a structured approach is required to adequately identify proper process recipes to fulfill the desired manufacturing requirements.

Generally, for metal AM technologies more than 15 process factors contribute heavily to part realization. In particular, for DED technology the number of significant factors can increase up to 20. Some of them pertain to system characteristics and machine capabilities, whereas others are related to material, ambient, and process properties. Table 12.3 summarizes the main variables that have to be taken into account during DED printing.

A common approach to designing an AM process is based on experimentation. Process variables linked with the DED system, powder shape (i.e. spherical) and ambient properties, are usually kept constant, while the effects of material and process properties on the final quality of the deposition are analyzed. System characteristics, powder properties, and ambient variables are usually defined by the machine in use for the deposition process and by the chosen metal powder supplier. The beam type, beam profile, and deposition head specifications have a strong influence on the final quality of the print, impacting mainly the powder distribution and melt pool formation mechanisms as well as the deposition efficiency. Nevertheless, they are specific to the employed DED machine and usually are not addressed during the AM process design. Material and process properties are two major groups of variables that engineers have to cope with during the AM process design.

The experimental investigation has to be carefully designed in order to efficiently characterize the influence of the process parameters on the final quality of the built, avoiding the risk of confusing or misinterpreting their main effects and mutual interactions.

The experimental campaign designed in ARM-Lab relies upon a structured approach aiming to investigate the effect of DED parameters on a step-by-step fashion. It starts from simple depositions, involving a limited number of process parameters, and moves on to the realization of massive builds that have to fulfill determined requirements. For a given deposition system and metal material, the evaluation and identification of the best process recipes is split into three main phases:

- Realization and characterization of single tracks;
- Realization and characterization of single layers (SLs);

Table 12.3 DED process variables.

DED system characteristics	Metal powder characteristics	Metal properties	Ambient properties	Process properties
Beam type	Particle shape	Chemical composition	Type of inert gas	Laser power
Beam profile	Particle size	Melting point	Not controlled/ controlled atmosphere	Laser spot
Type of deposition head	Particle size distribution	Absorptivity	Substrate shape	Axis speed (a.k.a. laser scan speed)
Axis relative velocity	Powder flowability	Reflectivity	Substrate size	Powder feed rate
Axis relative acceleration	Humidity content	Thermal conductivity		Hatching distance
Motion accuracy		Surface tensions		Layer thickness
Clamping system		Mechanical properties		Laser path
				Idle time
				Shielding gas flow
				Carrier gas flow
				Substrate preheating
				Stand-off distance

- Realization and characterization of 3D bulks.

The outcomes of every phase of the experimental campaign are characterized through offline measurements in order to analyze the influence of the process parameters on the final quality of the deposition.

Such an approach proved to work effectively, demonstrating to be an adequate tool for DED process characterization and design. Nevertheless, the execution of the entire experimentation is time consuming and requires several days to be performed and completed. Moreover, it will prove unreliable whenever a change related to the DED system, powder shape or environmental properties is made.

To conclusively overcome this limitation, ARM-Lab is currently exploiting the structured experimental approach for DED process design to support the realization of a dedicated process simulation and control platform (SUPSI 2019; SNSF 2018). This will support the reduction of the time required for process recipe identification and increase the performance of the technology while simultaneously improving deposition robustness and efficiency. The platform has the multifold ambition to:

- Simplify the DED process design, thanks to the development of a DED simulation model.
- Increase the final quality of the manufactured part, performing an active control and regenerating the manufacturing strategy by adapting significant process parameters or/and the tool path strategy during deposition.
- Promote the proliferation of AM machines endowed with integrated CAM–CNC software capable of guaranteeing consistent process and machine performance across a large number of DED applications and its further extension to hybrid equipment.

The final objective will be to design an intelligent, extendable, and fault tolerant platform to be integrated directly with a machine's CNC. This approach will provide a digitally enhanced user experience during the preparation, simulation, and execution of additive jobs.

The information needed for platform development and validation is gathered during the execution of the experimental phases (i.e. online data) and during the experimental characterization (i.e. offline data). In particular, online data collection is performed by melt pool and temperature monitoring and is correlated to the final quality of the deposition, allowing an active and robust closed-loop DED process control (Table 12.4).

Experimental data coming out from offline characterization are employed for the validation and optimization of an empirical model for DED simulation.

In the following sections, the structured approach designed by ARM-Lab is introduced and discussed in detail.

Single Track (ST) Deposition

The dynamics of melt pool generation is the heart of DED process. A proper melt pool formation and evolution is the key factor to get a successful metal deposition, free of pores and internal defects. Laser power, laser scan speed, laser spot size, and powder feed rate are the main process parameters responsible for laser/surface/powder interaction and are the main precursors for the generation of a proper melt pool. To investigate and characterize the effect of these four process parameters on the deposition process, single depositions are performed (commonly called STs).

STs are generally linear and designed to provide sufficient length for analysis under steady-state conditions, since the initial and final segments of the tracks are subject to uneven deposition associated with the transient behavior of the process parameters. They are usually realized on metal substrate having

Table 12.4 Offline and online variables.

Variable	Measurability
Track width	Offline
Track height	Offline
Local porosity	Offline
Yield strength	Offline
Surface roughness	Offline
Surface 3D scan	Offline
Thermal profile	Online
Melt pool image	Online
Melt pool main axes length	Online
Melt pool center of mass	Online
Melt pool area	Online
Melt pool intensity	Online
Melt pool tail length	Online
Sparks sparsity index	Online
Sparks mean direction	Online
Track profile	Online

the same chemical composition of the investigated powder to avoid crack formations induced by different thermal properties.

ST characterization consists of two interconnected analyses:

- External analysis.
- Structural analysis.

External characterization is performed to evaluate the superficial clad condition. For a given laser spot, inappropriate combinations of laser power, laser scan speed, and powder feed rate generate lack of deposition, inhomogeneity of growth, and delamination. In this step, metal tracks showing low or discontinuous clad formation are rejected, since they cannot ensure a proper and continuous melt pool formation. Moreover, metal tracks showing a massive clad growth, with significant spattering formation and unstable ST edges, are discarded as well, since they could bring to uncontrollable part growth, undesired formation of powder agglomerations, and lack of deposition between successive tracks (Figure 12.5).

STs having both a homogeneous and constant metal clad formation and regular edges are considered as acceptable for the deposition process (Figure 12.6). Nevertheless, STs with an acceptable external quality cannot guarantee a melt

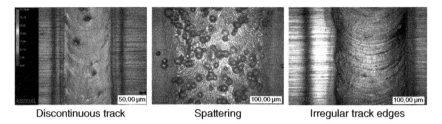

| Discontinuous track | Spattering | Irregular track edges |

Figure 12.5 ST deposition (AISI 316L) – common defects.

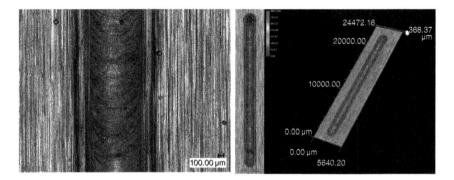

Figure 12.6 ST deposition (AISI 316L) – acceptable track.

pool formation free of internal defects. For this reason, a structural analysis of the acceptable tracks is also required.

In order to define a proper process window, the deposited samples are cross-sectioned, polished, and observed through optical instruments. The melt pool characteristics in terms of penetration depth and internal integrity are detected. Melt pools displaying either low penetration depth or internal pores (commonly called "key-holes") are rejected. Combinations of process parameters involving low penetration of the melt into the substrate can cause delamination or pores due to lacks of deposition during the ongoing process, deteriorating the structural integrity of the part.

From this second analysis, a process window capable of producing a proper and continuous melt pool formation is identified. Process window size can change depending on both the laser spot and metal material employed during the experimentation. Metal materials such as AISI 316L Stainless Steel can be considered as "easy-to-process" through AM technologies such as DED. In this case, the resulting process window is larger. On the contrary, for metal materials such as aluminum alloys where the majority of the laser energy is scattered due to material reflection, the process window becomes narrow.

Single Layer (SL) Deposition

If the generation of a consistent and constant melt pool is fundamental for a successful material deposition, the generation of a uniform and defect-free SL is the secret for a controllable and proper part growth. In this second step of the structured approach designed by ARM-Lab, the effect of the ST overlap between successive depositions is evaluated. Track overlap is also known as hatching distance in the AM world and it is defined as the distance between two consecutive tracks.

The definition of a proper hatching distance is fundamental for a homogeneous and defect-free layer filling. Well defined, controllable, and intact SLs are only possible if acceptable ST process parameters are employed in conjunction with proper track overlapping. Layer geometry can be different, depending on the size of the final part to be realized. Nevertheless, it is a good rule to choose SL size comparable with the cross-section of the part to be realized. Otherwise a significant over/undersizing of the SL could lead to wrong process considerations. A good practice to limit the number of experimental tests to be performed is to select a limited number of parameter combinations (i.e. 9) from the process window resulting from ST characterization.

The SL and ST characterizations are very much alike. After deposition, an external analysis is performed to detect SL growth profile. Waved or nonhomogeneous profiles are rejected, as they will not ensure a good control in the deposition process. Nonuniform layer heights could prevent good melt pool formations when successive layer depositions are conducted, as a constant layer/surface interaction can be prevented by an irregular metal surface previously created (Figure 12.7).

SLs showing homogeneous heights are subsequently analyzed through structural analysis. Layers showing internal pores are rejected. There are two main reasons for internal pore formations (Figure 12.8):

- An excessive material remelting. The temperature of the metal material increases with the number of tracks. Overheating can cause metal evaporation and the successive formation of internal pores.
- High ST overlapping combined with low laser power. The previously deposited metal clad can shield and prevent proper melting of the substrate, causing lacks of deposition.

Typical values for track overlapping range between 50 and 70% of the measured track width.

Nevertheless, the interaction between laser power, laser scan speed, laser spot, powder feed rate, and hatching distance is not straightforward. For instance, track overlapping of 50% can work well for some process parameters but not for others. This usually calls for an additional refinement of the process window resulting from ST characterization.

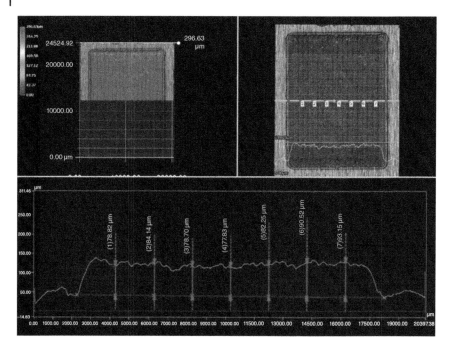

Figure 12.7 SL deposition (K890) profile measurement transversely to laser path.

3D Bulk Deposition

3D bulk is a term used to define sample parts with a relevant height and deposited volume. They could have different shapes, sizes, and geometries and they are realized to analyze the influence of layer thickness and deposition strategies on the realization of 3D metal parts.

3D bulk features should have the following characteristics:

- A deposited volume comparable with the final part to be realized.
- Geometrical characteristics equivalent to those present in the final part to be realized (e.g. fitting radius, corners, curved surfaces, overhang, thin wall features).

The reason is that 3D bulks should be representative of the final part complexity. Their manufacturing should provide to engineers all the information required for a good deposition and a controlled process.

In this last phase, the influence of the layer thickness and different laser paths are analyzed in conjunction with the combination of process parameters resulting from SL characterization.

Several methods can be employed to characterize the realized 3D bulks. Disruptive or nondisruptive techniques exist. Successive characterization methods have to be taken into account during 3D bulk design. Generally, an external

Single layer with good quality

Pores (external analysis)

Cracks (structural analysis)

Figure 12.8 SL deposition (K890).

analysis is performed first to detect external defects, such as superficial pores and external cracks, and to detect dimensional and geometrical accuracy of the deposited part. In this analysis, the accessibility of the external evaluation depends on both the size and shape of the feature with respect to the employed measurement instrument.

If the part that results is free of external defects, a structural analysis is performed. This characterization can be run with or without sample destruction, depending on both the analysis to be performed and the available characterization facility. Internal defects (such as internal porosity and cracks), microstructure, the presence of internal inclusions, and mechanical properties are usually the main investigated items.

Digital Platform for Consistent AM Build

Build failure is a common problem in AM. Intensive preparatory and experimental activities will obviously decrease the probability of its occurrence but this risk cannot be entirely eliminated due to extremely intricate phenomena associated with AM processes. In addition to the physical limitations of the

AM processes, which can be mainly considered the center of preoccupation of AM process engineers, a series of digital limitations related to AM value chain are now surfacing and drawing significant attention not only to the research community but also to the standard manufacturing software market and AM equipment manufacturers. This concerns mainly the lack of consistent quality and repeatability of additively manufactured products and calls for the definition and implementation of a set of sound digital initiatives.

For AM to scale at an industrial level the key is to use the power of digital services, customer experience, analytics, and automation. Digitally-enhanced machines present manufacturing companies building them with the opportunity to create not only their own smart manufacturing setup but also to help their customers to establish theirs. Digitization also makes output-based business models possible; that is to say that the equipment manufacturers must direct their attention toward providing to the market advanced solutions pairing hardware with intelligent software capable of guaranteeing the machine performance by continuously learning from previous experiences and applying knowledge for the implementation of future applications. Furthermore, the deployment of advanced digital solutions to address complex business and manufacturing challenges cannot be envisioned without giving full consideration to hybrid designs, materials, and processes. While additive and subtractive are generally perceived at opposite ends of the manufacturing spectrum the benefits of putting them together within the same system are endless and hybrid manufacturing will very much likely become a central building block of competitive manufacturing strategies in the future.

Approach

In order to capture and codify some of the best AM practices and to guarantee the avoidance of build failures ARM-Lab is fully committed and working toward the development and implementation of a new digital platform, empowering the creation of a seamless workflow between the digital process planning and the physical process execution (Avram et al. 2018). The planning and simulation of both DED and hybrid process workflows is a critical component of the platform, requiring the joint efforts of AM process and manufacturing strategy modelers, experts on process monitoring, form metrology, measurement, and characterization of part shape, CAM and controller software developers and AM equipment manufacturers, streamlined toward a smooth integration with the physical asset (i.e. AM machine).

Overall the successful deployment of such digital platforms requires the completion of the following major tasks:

1) The development of dynamic policies triggering an in-process adaptive mechanism of the hybrid manufacturing strategy based on process/machine simulations and online inspection feedback.

2) On-machine integration and implementation of bidirectional communication interfaces between the CAM and the CNC to drastically improve the efficiency of planning and control of the manufacturing of additive parts.
3) The development and integration of a Knowledge-Based System (KBS) and a Cyber-Physical System (CPS) capitalizing on existing manufacturing knowledge, sensing data and a set of control policies and rules to support the generation of the most adapted manufacturing strategies for future applications.
4) A new lean business model, suitable of lowering investment barriers adverse to the adoption of new AM applications; in order to thrive in this new digital ecosystem, machinery companies must define their digital manufacturing strategies, with one of the major decisions being to join/build a value-added platform supporting them to move away from exclusively selling hardware to selling services and guaranteeing outcomes (e.g. consistent AM part growth).

This digital platform concept exploits a number of advantages given by the integration between CAM and CNC directly on the machine environment: (i) supports the bidirectional flow of information between the planning and control of AM processes; (ii) enables the easy deployment of in-process adaptive recovery strategies; (iii) reduces tool path editing and programming time and boosts production efficiency; (iv) enables a realistic CAM simulation of what is expected to happen on the machine by leveraging the direct access to the parameter files (macros, subroutines, commands, etc.) used by the machine controller; and (v) enables an autonomous and intelligent CNC.

The work on the platform was initiated in past EU-funded projects (i.e. Borealis (2015) (www.borealisproject.eu), SYMBIONICA (2016)) and will be pursued within ongoing national and EU-funded projects (i.e. Ground Control (SNSF 2018), 4DHybrid), aiming toward the implementation of a high-performance software suite capable of automating the planning and simulation of additive-subtractive-inspection workflows and their transfer to execution.

No need for data conversion will be required amongst its main modules: Data preparation, Manufacturing Strategy (Operations Sequencing and Tool Path Generation), Simulation (Process and Machine) and KBS. A number of libraries will support the seamless data flow between the aforementioned modules as well as their future upgrade with new manufacturing processes and equipment, as can be seen in Figure 12.9.

HMI Interface
The workflow will be initiated with the support of an interface to be displayed on the machine's Human Machine Interaction (HMI) in order to collect user requirements and to be used sequentially by all modules of the platform. The

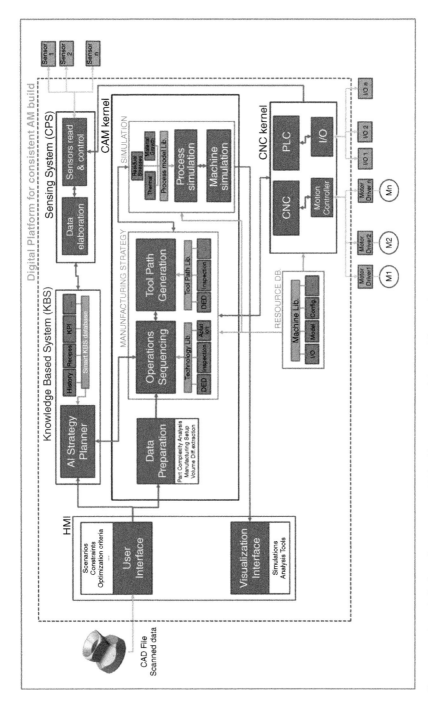

Figure 12.9 Digital platform for consistent AM build – overview of the workflow and main modules.

collected information pertains to scenario definition (i.e. new part or part repairing), the manufacturing setup and constraints, process switch type (i.e. manual or automatic), optimization criteria (e.g. deposition time, balance heat load, etc.), the geometry of the part as well as other nongeometric criteria (e.g. material type, powder features, technical performance considerations, etc.) organized in predefined templates. The HMI interface can also provide functions to visualize process and machine simulations with the possibility to directly edit parameter values if errors are detected. Furthermore, it will give access to the visualization of the analysis tools to support the decision making process when considering alternative manufacturing strategies.

Data Preparation

The aim of the data preparation module is to streamline the part geometric information toward the manufacturing planning strategy while preparing the manufacturing setup in function of the user's input. This module will provide support for the analysis of part complexity and will also handle the volume extraction activity when the part scanning data (e.g. cloud of points) need to be compared with a nominal CAD geometry. The input part file can be either a CAD (STL, IGES, STEP, etc.) solid file or a mesh resulting from an inspection activity performed internally by a 3D scanner (if available) or externally by other equipment. This module will be also responsible for a precise integration of solid geometry with mesh models in the scope of comparing a nominal CAD file with scanned data as well as detecting and extracting difference volumes between the two to be used for the subsequent modules. Furthermore, this module will collect a number of parameters required to compare the simulation results with the initial CAD (i.e. method of comparison, tolerances, etc.). The best orientation of the part will be decided based on the real time analysis of the critical areas of the part (e.g. overhang threshold, maximum allowable cusp height, etc.) while considering the user's preferences for the optimization criteria.

Manufacturing Strategy

The main objective of this module is to generate the most appropriate manufacturing strategy for a given part. For machines featuring exclusively DED technology, the generation of a complete strategy consists of the adequate sequencing of deposition and inspection steps as well as in the definition of an adapted tool path strategy for each one of them. For hybrid machines the aforementioned strategy will be complemented with the specifics of the available subtractive technologies. The sequence of operations and the process recipes will be automatically established with the support of the CPS sensing system and the AI Strategy Planner based on user's input and the manufacturing data, control policies, logics, and rules gathered together within the KBS. For instance, the selection of the process switch type is required as user input to support the definition of an adequate sequence of operations. If a manual process switch is selected then the manufacturing and inspection operations (i.e. deposition/ablation and 3D scanning) are alternated at specific checkpoints, defined manually by the user with respect to the geometry of the part. As a next step the tool path for each operation is automatically defined in conformity with the results of the process/machine simulation. If the automatic option is selected both operation sequencing and tool path generation are automatically defined based on the switch process constraints (collision constraints, physical machine constraints, continuity broken constraints) and process/machine simulation outcomes. For a part composed of a number of sections to be manufactured along different directions, the collision check can impose the deposition sequence of the sections in order to ensure a collision-free process between the nozzle/head and the part being built. The switch of the processes might be triggered also by the continuity constraint. If the continuity between two consecutive layers is broken (i.e. the surface area difference between layers is higher than a predefined threshold) then an inspection process might be required before starting the deposition of a new section. Additionally, the physical machine constraint can be linked with any limitation imposed by the axes of the machine with respect to the setup of the part defined in the previous step.

Generally, the tool path generation will be carried out as two-step process right after the slicing of the part. First, the tool path pattern ensuring a void-free deposition with respect to the geometry of every layer is selected. Second, the tool path is optimized (e.g. balance the heat load into the part) by enriching the manufacturing data defined in the previous steps with process simulation results (i.e. material growth, thermal history, melt pool size, residual stresses) updated in strict correlation with the machine kinematics (accessibility and collision avoidance), manufacturing rules and constraints (i.e. nozzle movement, laser scanner position, axis motion (Halevi et al. 2011)).

For a hybrid machine such as the SYMBIONICA, the tool path libraries to be considered fall under the following categories:

- 2D Layered Manufacturing – various tool path patterns (e.g. zig-zag, spiral, contour offset, etc.) can be generated for the deposition/ablation of consecutive planar layers by considering a build/remove direction driven by $2\frac{1}{2}$, 3 or $3 + 2$ axis kinematics.
- 3D Freeform Manufacturing – set a rapid and efficient way of morphing deposition/ablation tool paths for freeform surfaces by considering a build/remove direction driven by five-axis kinematics while maintaining the laser beam axis normal to the deposition surface.
- Inspection – generate adequate 3D part scanning strategies, in strict relation with the machine kinematics and the geometry of the part to be manufactured.

Inspection is an essential element in any closed loop manufacturing system. The offline inspection of a part presents a number of problems, mainly related to time, data compatibility, and data modeling. Inspection typically consists of three consecutives activities: programming, execution of the program and evaluation of the results. If these are carried out externally on separate systems it is likely to cause complex interfacing problems. On-machine inspection offers a better alternative by enabling measurement taking, data collection as well as data feedback and a fully automated and integrated process adjustment.

On the SYMBIONICA machine, two types of inspection operations can be considered:

1) Inspection carried out before any deposition/ablation operation in order to provide the positional information of the part being built to the CNC controller. Hence, the controller knows the current location of the part and the tool path can be planned/modified accordingly.
2) Inspections performed periodically to detect any possible deviation during the manufacturing of the part in order to enable the controller to make in-process changes.

Simulation
The simulation module is an invaluable tool to visualize and verify the deposition process while preventing costly collisions in the working area along with possible deposition errors, voids, or misplaced material. This module targets the solving of specific AM challenges (e.g. material growth, thermal effect, residual stress, etc.) through a realistic simulation to be performed both at process and machine level. The function of the process simulation is twofold: on one side, to provide reliable predictions of part geometry and mechanical performances of the deposited part based on the chosen process parameters; on the other side, to support the choice of some of these parameters by embedding automatic optimization functions that may have different targets (e.g. porosity, dimensional tolerance). The DED simulation model will deal with multiple physical aspects to reliably predict the material growth and keep track of where material has been added, the heat transfer and phase changes, as well as the evolution of material properties and residual stresses throughout the build time, by integrating empirical and precomputed partial models with rigorous multiphysics. Mass growth over the melt pool, coupled with the dynamic thermal distribution, will be computed numerically by using a combination of finite volume methods (Quarteroni 2009a,2009b) and the VOF (Volume of Fluid) approach (Mahady et al. 2015), which is an efficient method for managing computational domains with evolving boundary shapes.

The user will have the possibility of activating the simulation in optimization mode (receiving contextually an estimate of the increased simulation/optimization time). This version of process simulation will

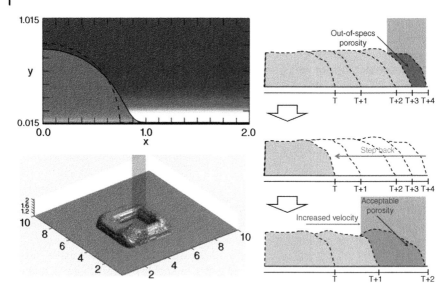

Figure 12.10 Example of mass growth (left) and process parameters optimization (right).

exploit concepts of optimal control for partial differential equations (PDEs) (Quarteroni 2009a, b) to provide optimal values of process parameters, combined with a "back-stepping strategy" for part program generation. Let us consider a simulation step where the model that relates porosity to energy density and cooling rate provides a predicted porosity value that is above the critical threshold for the part to produce (Figure 12.10). This suggests an increase of laser scanning speed, which can be obtained only with a machine-related acceleration ramp. A buffer of the last simulation steps allows going back in time and setting up a change in acceleration and final speed that satisfies the porosity requirement. This approach allows not only the generation of a tentative part program but also its updating and regeneration based on the results of the simulation and optimization run.

The process simulation will be complemented with machine kinematics and dynamics simulation to validate accessibility, collision avoidance, and machine axes performances (Halevi et al. 2011; Valente et al. 2017). The direct integration between the CAM and CNC will support not only local simulation of the axes motion but also real-time control of the virtual machine model by the real CNC. Moreover, the machine simulation will make sure that all the controlling additive/subtractive functions are employed correctly and respected all along the tool path defined to manufacture the part.

Knowledge-Based System and Sensing System

In the recent years data proliferation has made every business more data dependent and generated a growing interest for applying insight gained from past experiences to support decision making processes in the future. In order to completely define the manufacturing strategy, the automatic design of the tool path must be triggered with the support of an adapted set of process parameters. The entire amount of information required for the definition of a manufacturing strategy will serve not only the final scope of controlling the motion of the machine axes, the energy source, the powder feeders and gas supply systems but also to perform beforehand a realistic simulation of the deposition supported by process models.

In this context the importance of integration of both data- and analytics-driven functionalities to plan sound strategies for the manufacturing of new products in the future cannot be neglected (based on machine parameters and materials being used). In this perspective the new digital platform relies upon an AI strategy planner to provide strategy decisional support based on tested recipes (i.e. tables of process parameters, recorded melt-pool shapes, etc.) built on previous empirical experiences and stored in a smart KBS database.

To properly curate and retrieve knowledge objects from the database, the KBS interacts also with the CPS sensing system and deploys knowledge tools and constructs such as: if-then rules, logics, statistical, and machine learning models (trained offline by using sets of data previously acquired), Key Performance Indicators (KPIs), etc. In addition, the formulation of an adapted manufacturing strategy when confronted with process deviations, signaled with the support of the CPS, can be further improved by considering a number of adaptive control algorithms and optimization criteria, such as:

- Reduce deposition time and improve surface quality; accelerate the deposition time with automatic assignment of optimal deposition strategies to relevant feature of the build.
- Achieve better surface quality by assigning more accurate deposition strategies to specific zones; assign faster deposition strategies to internal volumes or zones that do not require high surface quality.
- Avoid under/overdepositions by defining adapted strategies for start/stop zones (e.g. gradient power or variable defocus).
- Achieve better control of the thermal gradient of the process by changing the traveling sequence of the deposition head across the generated tool path.

Regenerative Manufacturing Strategy

In the course of a deposition process, the geometry of the part being built will inevitably differ from the nominal CAD model. In order to guarantee a true physical representation of a CAD geometry and to avoid the irreversible accumulation of deviations, very close monitoring and adaptation of the initial manufacturing strategy must be performed. The initial strategy consists of the planning and sequencing of the operations to be deployed to complete a part as a function of the manufacturing scenario (i.e. new part manufacture or repair application) and the equipment available (i.e. DED or hybrid system). It will include also a number of predefined checkpoints for the general inspection of the part, defined during the part complexity analysis step.

Using the latest 3D measurement technology, sensors, and monitoring tools, deployed directly on an In-Envelope Hybrid Manufacturing System – i.e. additive, subtractive, and inspection systems are integrated within the same manufacturing system envelope – such as the SYMBIONICA machine, it's now easier than ever to quickly get accurate in-process geometrical and dimensional information on manufactured parts. Nevertheless, one of the biggest challenges lies in the interpretation of this information and the generation of adequate responses to be fed back into the ongoing manufacturing process to compensate for any deviation encountered. A fundamental issue linked with the regeneration of a hybrid manufacturing strategy resides on the fact that as soon as the initial part program was stopped for an inspection at the first checkpoint, the planning of the subsequent steps needs to be defined more with respect to the actual geometry of the part (i.e. manufactured up to this level) and less by referring to the nominal geometry. The adaptation of the strategy must match the actual geometry (inspected with the 3D scanner) since the consideration of the nominal geometry might lead to the creation of features that do not transition evenly in the previously manufactured section of the part. For instance, a misalignment with respect to the nominal CAD geometry might be noticed due to a slight warping effect of the deposited part. If this deviation from nominal is within predefined tolerances, then the manufacturing can continue but the use of the initial strategy to deposit the subsequent layers is not totally reliable anymore and must be adapted accordingly. If out of tolerance, then the process needs to be stopped and a morphed CAD reflecting the current status of the deposited part should be constructed and employed as the baseline for the generation of a deformed manufacturing strategy to be used for the future prints of the very same part.

It is straightforward that the adaptation must be specifically tuned on a part-by-part basis. To tackle this challenge, the digital platform will rely upon a dual-stage regenerative policy unifying a continuous output–feedback control (i.e. microscale regeneration) triggered by the data generated by embedded sensors (e.g. melt pool camera, pyrometer) with a periodic event-triggered strategy control (i.e. macroscale regeneration) performed at predefined checkpoints and generated based on the measurements provided by the inspection system.

The main purpose of the regenerative policy is to adjust the initial manufacturing planning of the build by compensating for the deviations caused by both intralayer and interlayer dynamics. The adjustment consists of either the individual or simultaneous alteration of the initial process parameters and the tool path strategy. For instance, the continuous monitoring of the melt pool will give valuable information about its shape, as predicted by the process model. If out of tolerance, the microscale regeneration loop will alter specific process parameters by overriding continuously their initially programmed values. As far as the macroscale regeneration is concerned, it will be triggered every time a checkpoint is encountered. The manufacturing process will be stopped and an inspection activity will be deployed. If an anomaly is detected, following the comparison between the nominal shape of the part and the current status of build, the initial manufacturing strategy jointly with the associated part program will be subject to either a partial or total regeneration. The output will be closely tuned as a function of the gravity of the deviation and generated as a result of running a new simulation interfaced with the KBS.

Microscale Regeneration

The microscale regeneration loop is responsible for undertaking the tasks of continuously monitoring and adapting the values of significant process parameters in order to ensure the actual additive process will fit the nominal one. Nominal conditions pertain to the prediction and control of the quality characteristics of an AM part by correlating melt pool dimensions, solidification microstructure, and thermal measurements with the process parameters used. This knowledge is defined based on a wealth of information (e.g. track width and height, local porosity, yield strength, surface roughness, etc.) gathered from intensive offline experimentation and structured within the platform's KBS.

To apply an effective integration of process knowledge into machine intelligence, it is necessary to both model the additive process and to define a closed-loop control model linking online monitoring signals with previous knowledge and trigger an adequate feedback response in case of misalignments. Due to the complexity of the phenomena and high number of involved parameters an analytical AM model is not available for prediction. Machine learning techniques are adopted instead, to model the additive process based on acquired data. Such a model is learned offline and it can be used online to implement an adaptive control loop, as shown in Figure 12.11.

The online control loop architecture will exploit effectively both the offline knowledge available within the KBS and the field data from the ongoing deposition process. We distinguish three sources of data within this control model: (i) the process parameters and control variables (represented by the subset of input data u); (ii) the real-time measurements from the field (represented by the subset of observation data y); and (iii) the quality parameters of the built parts (represented by the subset of target data x). A number of submodels will be defined in order to map the various types of data (i.e. map inputs to targets to estimate how the quality parameters of the build are affected by the process parameters; map observations to targets to estimate how the quality of the build are affected by the melt pool parameters measured in real time).

Online measurable variables (e.g. melt pool characteristics, sparks sparsity index) represent indirect process quality indicators that can be used for process

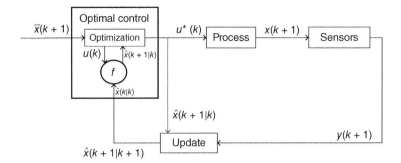

Figure 12.11 Microscale regeneration – adaptive control loop.

control, thus providing a feedback to tune the online control variables (e.g. laser power, spot size, standoff distance, powder flow, feed rate). In addition to the online control variables, there are other parameters with an influence on the final build attributes but they are not subject to monitoring/measuring activities and are considered as fixed known quantities (e.g. powder material, powder grain size, shielding gas speed, infill strategy, etc.).

The most suitable candidates for a quick adaptation in an AM process control system are the laser power and the feed rate while the melt pool is an attractive characteristic of the process to use for monitoring and control purposes. Real-time measurements from the melt pool include raw data from various sensors (camera or thermographic images) as well as quantitative features automatically extracted from the raw sensed data by means of image processing techniques (e.g. melt pool's area, center of mass, tail length, etc.).

Macroscale Regeneration

Nowadays any piece of information collected from an ongoing additive process cannot be directly interpreted and utilized by an offline CAM system in order to adapt the deposition strategy for the successive steps. Furthermore, the interpretation of the information and the elaboration of alternative adaptive strategies are strongly dependent on the experience and the knowledge of programmers and involve extensive manual interactions with the CAM system. The communication layer established by the integration of the CAM and CNC directly on the machine enables a fast and effective bidirectional dataflow supporting the automatic deployment of in-process adaptive manufacturing strategies and promoting a disruptive innovation for the AM equipment manufacturers and their customers by eliminating cumbersome post-processors and machine tool interpreters.

Figure 12.12 shows the digital workflow of activities contributing to the macroscale regeneration of the manufacturing strategy. In this setup we can consider that up to the first checkpoint the first section of the part

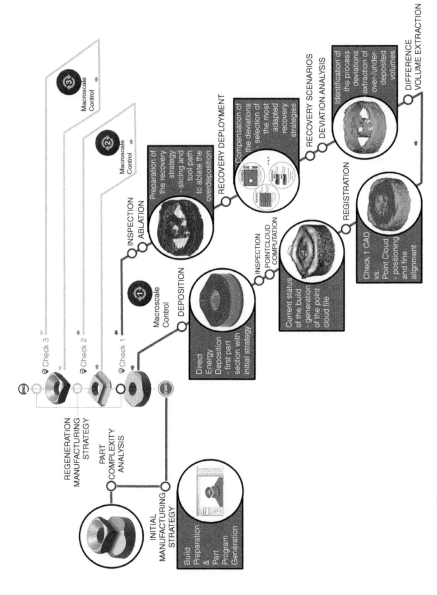

Figure 12.12 Digital workflow for macroscale regeneration loop.

was deposited according to the initial strategy. Meanwhile the microscale regeneration loop was active and performed an adaptation of the power and feed values in order to maintain a good deposition quality. Nevertheless this loop addresses exclusively deviations caused by intralayer dynamics and an inspection of the current status of the build is required in order to account for any possible errors that might have been caused by the process of stacking a specific number of layers one on top of another. The inspection is followed by the generation of a cloud of points describing the current shape of the part and by its registration with the corresponding CAD geometry. The difference volume extraction will reveal the possible deviations induced by the additive process, which will be divided in two separate files: overdeposited volume and underdeposited volume.

The KBS system of the digital platform embeds a Smart database that contain the full history of produced parts organized in KPIs and recipes, a collection of rules and logics, a machine learning model, and AI strategy planner module providing valuable information for the selection of the most adapted manufacturing strategy for a specific scenario. This functionality will be accessed in the next step, after a careful analysis of the deviations that will basically evaluate the extracted volumes against a predefined set of dimensional and geometrical thresholds and will provide decision making support for the selection of the most appropriate recovery scenario. Based on the type of deviation, an adapted recovery scenario can be selected. In the case of uniform overdeposition the most adapted recovery strategy might be to make the current build dimensionally and geometrically consistent with its nominal counterpart by deploying an ablation operation. Another alternative would be to regenerate the strategy for the subsequent layers by choosing parameters guaranteeing a better control of the layer height. The optimization criteria defined through the platform's HMI will be weighted according to the user's preferences and will guide the selection of the most adequate recovery strategy. The entire macrolevel regeneration workflow will be iterated in a similar fashion at every predefined checkpoint.

12.4 Future Work

A smart manufacturing strategy is not just about incremental change or cost savings; it is about innovating products and services and incorporating innovation at a much higher pace than ever before. A complete digital software infrastructure is extremely significant in connecting the individual links in the manufacturing value chain and must ensure that process engineering and machine intelligence are well integrated. Future work will be aligned with two research and development directions: increase of the process efficiency and improve the digital transformation.

Increase the Process Efficiency

The implementation of accurate simulation software modules will support the process design, giving indications on material growth, temperature distribution, and final quality of the part before running the additive process on the machine, thus avoiding having to perform huge trial-and-error experimental campaigns that involve extended periods of time and high resource consumption. Furthermore, the implementation of adapted control models will engender a higher robustness to the process while tracking and using real-time sensor information to optimize the ongoing strategy and reduce the process deviations during part realization.

Improvement of the Digital Transformation

The current market provides software solutions that are strictly connected with specific AM equipment manufacturers. The development and implementation of a hardware-independent, intelligent, and extendable digital platform to be integrated directly within the machine environment will be a game changer in the industry. The intelligence can be translated in the ability of the platform to handle in tandem with the CNC specific degrees of uncertainty and to increase the probability of success of a controlled AM or hybrid process. The architecture of the software also facilitates an easy integration of new knowledge and processes, enabling a smooth extendibility to other applications and a straightforward integration with various machines.

12.5 Summary

The adoption of Additive technology for metal parts opens new opportunities for a flexible and decentralized manufacturing of more complex and high added value components. Today, an increasing number of companies are taking profit from these disruptive technologies to manufacture prototypes, tools, spare parts, and components in a faster and cost effective manner, never done before with conventional solutions. Thanks to the highly achievable degree of geometrical complexity enabled by AM technologies, the realization of enhanced products with customized functional shapes becomes reality. The impact of these technologies on the process chain is also remarkable. Additive systems can be easily placed in loco (e.g. engineering offices, garages, workshops, laboratories, or medical hospital), cutting down the lead time and costs related to part production and optimizing the supply chain and logistic operations. The ability to produce on-demand new components and spare parts, directly from a digital model, reduces drastically the asset's downtime and necessary storage size, revolutionizing the way the products are designed and supporting the creation of additional revenue streams.

Furthermore, new data-driven simulation and control tools developed on the basis of real experimental results are needed to bring industrial progress in AM. The development and implementation of platforms capable of identifying in-process potential issues and providing consistent solutions to adapt the AM process and equipment response under different conditions will support companies in the creation of new business models and differentiated services while guaranteeing an improved product quality and performance.

Recent developments in AM have focused on building better, faster, larger, and more capable machines and on extending the range of new materials. While it is important to continue to innovate and increase manufacturing capabilities, the key of unlocking the next industrial transformation will be the true understanding of which applications benefit most from AM technology. Looking forward, Metal AM will become more and more of a necessity when solving specific manufacturing challenges and creating customized, complex end-use products.

In other words, machines and software platforms will become more and more application-driven to meet very specific customer and industry demands. The potential is huge. As the technologies evolve, AM equipment providers need to work closely with software providers to develop an integrated AM process. To take a leap on the technological ladder and to gain a competitive advantage, equipment providers must ensure that AM process and machine intelligence are very well and closely integrated, in a smart manufacturing fashion.

References

AM Additive Manufacturing (2017). What does it take to be safe? *AM Additive Manufacturing* 6 (1): 6.

AM Additive Manufacturing (2018a). Machine learning. How it's advancing AM. *AM Additive Manufacturing* 7 (1): 10.

AM Additive Manufacturing (2018b). Alfa Romeo Sauber F1 team extends partnership with Additive Industries. https://www.additivemanufacturing .media/news/alfa-romeo-sauber-f1-team-extends-partnership-with-additive-industries (Last accessed 17 June 2019).

ASTM (2012) ASTM F2792-12a, Standard Terminology for Additive Manufacturing Technologies, (Withdrawn 2015) ASTM International, West Conshohocken, PA.

ASTM (2015) ISO/ASTM 52900-15, Standard Terminology for Additive Manufacturing – General Principles – Terminology, ASTM International, West Conshohocken, PA. https://www.astm.org/Standards/ISOASTM52900.htm (Last accessed 17 June 2019).

ASTM (2013) ISO/ASTM 52921-13, Standard Terminology for Additive Manufacturing-Coordinate Systems and Test Methodologies, ASTM

International, West Conshohocken, PA. https://www.astm.org/Standards/ ISOASTM52921.htm (Last accessed 17 June 2019).

ASTM (2016) ISO/ASTM52901-16, Standard Guide for Additive Manufacturing – General Principles – Requirements for Purchased AM Parts, ASTM International, West Conshohocken, PA. https://www.astm.org/ Standards/ISOASTM52901.htm (Last accessed 17 June 2019).

ASTM (2018) ISO/ASTM52910-18, Additive manufacturing – Design – Requirements, guidelines and recommendations, ASTM International, West Conshohocken, PA. https://www.astm.org/ Standards/ISOASTM52910.htm (Last accessed 17 June 2019).

Avram, O., Valente, A., and Fellows, C. (2018). Adaptive CAx chain for hybrid manufacturing. In: *Proceedings of Direct Digital Manufacturing Conference*, 239–245. Stuttgart, Germany: Fraunhofer Verlag (14-15 March 2018P.

BMW (2015) Racing technology right from the 3D printer: BMW makes water pump wheel for DTM racecars using additive production method. https://www .press.bmwgroup.com/global/article/detail/T0215062EN/racing-technology- right-from-the-3d-printer:-bmw-makes-water-pump-wheel-for-dtm- racecars-using-additive-production-method?language=en (Last accessed 17 June 2019).

Borealis -the 3A energy class Flexible Machine for the new Additive and Subtractive Manufacturing on next generation of complex 3D metal parts (H2020), 2015, 2017. [Online]. Available: http://www.borealisproject.eu

Bourell, D., Kruth, J.P., Leu, M.C. et al. (2017). Materials for additive manufacturing. *CIRP Annals – Manufacturing Technology* 66: 659–681.

Clarke C. (2017). Ferrari uses 3D printing to develop new engine for 2017 F1 season. 3D Printing Industry. https://3dprintingindustry.com/news/ferrari- uses-3d-printing-develop-new-engine-2017-f1-season-104862 (Last accessed 17 June 2019).

GKN (n.d.) Our technology. GKN Aerospace. https://www.gknaerospace.com/en/ our-technology/?doccategory=71 (Last accessed 17 June 2019).

Halevi, Y., Carpanzano, E., Montalbano, G., and Koren, Y. (2011). Minimum energy control of redundant actuation machine tools. *CIRP Annals – Manufacturing Technology* 60 (1): 433–436.

Mahady, K., Afkhami, S., and Kondic, L. (2015). A volume of fluid method for simulating fluid/fluid interfaces in contact with solid boundaries. *Journal of Computational Physics* 294: 243–257.

Marchetti, A., Mazzucato, F., and Valente, A. (2017). Development and optimization of an innovative double chamber nozzle for highly efficient DLD. In: *Industrializing Additive Manufacturing –Proceedings of Additive Manufacturing in Products and Applications (AMPA 2017)*, 199–199. Springer.

Mazzucato, F., Marchetti, A., and Valente, A. (2017a). Analysis of the influence of shielding and carrier gases on the DED powder deposition efficiency for a new deposition nozzle design solution. In: *Industrializing Additive*

Manufacturing –Proceedings of Additive Manufacturing in Products and Applications (AMPA 2017), 59–69. Springer.

Mazzucato, F., Tusacciu, S., Lai, M. et al. (2017b). Monitoring approach to evaluate the performances of a new deposition nozzle solution for DED systems. *Technologies* 5 (2): 1–13.

Metal AM (2015a). GE Aviation's first AM part cleared to fly in commercial jet engines. *Metal Additive Manufacturing* 1 (1, Spring): 5.

Metal AM (2015b). Over 3,000 Additive Manufactured medical orthopaedic truss implants from 4WEB. *Metal Additive Manufacturing* 1 (1, Spring): 27.

Metal AM (2015c). Turbomeca uses metal Additive Manufacturing for helicopter engine components. *Metal Additive Manufacturing* 1 (1, Spring): 27, Vol. 1 No. 1, Spring, p. 27.

Metal AM (2015d). Arcam receives further orders from its customers in medical and aerospace sectors. *Metal Additive Manufacturing* 1 (3, Autumn): 13.

Metal AM (2017a). Chinese hospital implants additively manufactured tantalum knee joint. *Metal Additive Manufacturing* 3 (4, Winter): 39.

Metal AM (2017b). Simufact introduces the next generation of its metal AM simulation software. *Metal Additive Manufacturing* 3 (4, Winter): 56.

Metal AM (2018a). Hospital reconstructs patient's chest using titanium Additive Manufacturing. *Metal Additive Manufacturing* 4 (1, Spring): 28.

Metal AM (2018b). Porsche Classic supplies replacement parts using metal AM. *Metal Additive Manufacturing* 4 (1, Spring): 7.

Metal AM (2018c). Italian AM dental implant business Yndetech reports success with 3D Systems. *Metal Additive Manufacturing* 4 (1, Summer): 42.

Metal AM (2018d). Airbus Helicopters begins series production of A350 passenger aircraft components. *Metal Additive Manufacturing* 4 (3, Autumn): 36.

Metal AM (2018e). BASF invests $25 million in Materialise as companies seek to optimise AM materials and software. *Metal Additive Manufacturing* 4 (3, Autumn): 13.

Metal AM (2018f). GE offers improved turbine engine with metal AM parts to the U.S. Army. *Metal Additive Manufacturing* 4 (3, Autumn): 43.

Metal AM (2018g). HP Metal Jet: HP launches its first metal Additive Manufacturing system. *Metal Additive Manufacturing* 4 (3, Autumn): 7.

Metal AM (2018h). Pratt & Whitney and partners test additively manufactured turbomachinery components. *Metal Additive Manufacturing* 4 (3, Autumn): 43.

McCue T.J. (2018). Wohlers Report 2018: 3D printer industry tops $7 billion. Forbes Media. https://www.forbes.com/sites/tjmccue/2018/06/04/wohlers-report-2018-3d-printer-industry-rises-21-percent-to-over-7-billion/#6848c78b2d1a (last accessed 17 June 2019).

Mikler, C.V., Chaudhary, V., Borkar, T. et al. (2017). Laser additive manufacturing of magnetic materials. *JOM – The Journal of the Minerals, Metals & Materials Society* 69: 532–543.

Ngo, T.D., Kashani, A., Imbalzano, G. et al. (2018). Additive manufacturing (3D printing): a review of materials, methods, applications and challenges. *Composites Part B Engineering* 143: 172–196.

Quarteroni, A. (2009a). Optimal control of partial differential equations. In: *Numerical Models for Differential Problems*, 457–500. Springer-Verlag.

Quarteroni, A. (2009b). The finite volume method. In: *Numerical Models for Differential Problems*, 217–226. Springer-Verlag.

SNSF (Swiss National Science Foundation) (2018) Ground Control: closed-loop adaptive control for AM-DED process optimization. http://p3.snf.ch/project-179014 (Last accessed 8 July 2019).

SUPSI (2019) DED-In718 – Multi-physics platform for modelling, simulation and control of DED for structural Inconel 718 parts. University of Applied Sciences and Arts of Southern Switzerland.

SYMBIONICA-Reconfigurable Machine for the new Additive and Subtractive Manufacturing of next generation fully personalized bionics and smart prosthetics (H2020), 2016, 2018. [Online]. Available: http://www.symbionicaproject.eu

Tsai, M.H. and Yeh, J.W. (2014). High-entropy alloys: a critical review. *Materials Research Letters* 2 (3): 107–123.

Valente, A., Baraldo, S., and Carpanzano, E. (2017). Smooth trajectory generation for industrial robots performing high precision assembly processes. *CIRP Annals – Manufacturing Technology* 66 (1): 17–20.

Valente A, Brugnetti I, Marchetti A, Colla M (2018) Nozzle apparatus for direct energy deposition. Patent, Number: EP3330007A1.

Vandone, A., Baraldo, S., and Valente, A. (2018). Multi-sensor data fusion for additive manufacturing process control. *IEEE Robotics and Automation Letters (RA-L)* 3 (4): 3279–3284.

13

Analytics as an Enabler of Advanced Manufacturing

Ron S. Kenett, Inbal Yahav, and Avigdor Zonnenshain

Synopsis

Knowledge and information are critical assets for any manufacturing enterprise. They enable businesses to differentiate themselves from competitors and compete efficiently and effectively to the best of their abilities. At present, information technology, telecommunications, and manufacturing are merging as the means of production are becoming increasingly autonomous. Advanced manufacturing, or Industry 4.0, is based on three interconnected pillars: (i) Computerized Product Design and Smart Technology; (ii) Smart Sensors, Internet of Things, and Data Collectors integrated in Manufacturing Lines; and (iii) Analytics, Control Theory and Data Science.

This chapter consists of a critical review of the third pillar of data analytics, and how it is related to the two other pillars. Our objective is to present a context for a range of analytic challenges in the Industry 4.0 context. We first provide a general introduction to advanced manufacturing elements followed by a listing of trends in modern analytic tools and technology. We then list challenges in analytics supporting Industry 4.0. The information quality (InfoQ) framework serves here as a backbone for evaluating the analytics technology required for Industry 4.0. The eight InfoQ dimensions are: (i) Data Resolution; (ii) Data Structure; (iii) Data Integration; (iv) Temporal Relevance; (v) Chronology of Data and Goal; (vi) Generalizability; (vii) Operationalization; and (viii) Communication. These dimensions provide a classification of advanced manufacturing analytics domains and assist to identify gaps and new opportunities. The review provides a theoretical framework and suggestions for future research directions and implementations of analytics in Industry 4.0. It is designed to motivate researchers, practitioners, and industrialists and to facilitate Industry 4.0 deployments and expansions.

Systems Engineering in the Fourth Industrial Revolution: Big Data, Novel Technologies, and Modern Systems Engineering,
First Edition. Edited by Ron S. Kenett, Robert S. Swarz, and Avigdor Zonnenshain.
© 2020 John Wiley & Sons, Inc. Published 2020 by John Wiley & Sons, Inc.

13.1   Introduction

The first industrial revolution began with the introduction of the steam engine and the mechanization of manual work in the eighteenth century. Electricity drove mass production in the second industrial revolution during the early twentieth century. The third revolution in manufacturing was linked to the use of electronics and computer technology for manufacturing and production automation. At present, we are entering into a fourth phase, labeled smart manufacturing, Industry 4.0 or Manufacturing 4.0. In this context, the role of computer power has dramatically increased, and we shift from factory floors with one supporting computer to several computers supporting every stage of production.

Pai (2014) states that smart manufacturing is "a highly connected, knowledge enabled, industrial enterprise where all business and operating actions are optimized to achieve substantially enhanced productivity, energy, sustainability, and economic performance." Increasing digitalization and networking are changing the entire industrial production chain, and the volume of data worldwide is constantly expanding. Experts expect the volume of data in 2020 to reach 14 996 exabytes, roughly equivalent to about 15 trillion gigabytes (Siemens 2016). To properly analyze and use such huge volumes of data, one needs systems that enable us to understand their content. Part of this requirement is to know how devices and systems function in an integrated mode and which kind of sensor and measurement technology may be used to access the most useful data. According to the European Innovation Scoreboard (EIC 2017), for the third consecutive year South Korea is the most innovative country in the world and the second with the highest innovation growth (with China in the first place). Partly, it is due to the fact that many companies in South Korea have reached maturity in Industry 4.0 implementation. Chapter 1 provided more context, Chapter 21 presents an approach to assess maturity levels of organizations striving for advanced manufacturing implementations.

Information technology, telecommunications, and manufacturing are merging as the means of production are becoming increasingly autonomous. Yet it is impossible to predict exactly what smart factories will look like in the future. One possible scenario is that the factory of the future will have autonomous machines that will organize themselves. It will experience delivery chains that automatically assemble themselves and will witness how orders transform production information and flow into the production process. In this scenario, Industry 4.0 requires, more than ever and in advance, the conception of all the processes and procedures of operations and software that convey this information to machines. Industry 4.0 should be viewed via three interconnected pillars: (i) Computerized Product, Design, and Smart Technology; (ii) Smart Sensors, Internet of Things and Data Collectors; and (iii) Analytics, Control

Theory and Data Science. In this chapter, we focus on the third pillar of data analytics and its interconnectivity to the other two pillars.

To provide a perspective of data analysis in Industry 4.0, we review the literature from three different angles. The first consists of analytical advances in Industry 4.0 (Section 13.2). The aim of Section 13.2 is to provide a list of the most advanced analytic methods applied in advanced manufacturing. The second perspective (Section 13.3) addresses application areas in Industry 4.0, which are classified by advanced analytic technics. Lastly, we review the methodological challenges in Industry 4.0 and present some adapted approaches (Section 13.4).

Following this review, we propose a general framework based on information quality (InfoQ) dimensions to integrate the application domains in Industry 4.0, along with matching analytic methods and technologies.

13.2 A Literature Review

To systematically review the literature, we first searched for all the relevant papers from ScienceDirect (www.sciencedirect.com), using the search term [("advanced manufacturing") or ("industry 4.0")] (Figure 13.1). The search resulted in a list of approximately 9000 original articles. We then analyzed the papers both descriptively and content-wise, using text analysis of papers' titles and abstracts.

A similar approach was carried out for assessing analytical trends and applications by employing the keywords: [((("advanced manufacturing") or ("industry 4.0")) and analysis] and [("Industry 4.0" or "advanced manufacturing") and decision and management)], respectively. Our mining approach was followed by verification and modification of several experts and in-depth reading of top cited articles. Appendix 13.A extends on this methodology.

Advanced manufacturing requires analytics and operational capabilities to interface with devices in real time, at an individual level. Software development has become agile in what is called DevOps operations focused on providing continuous delivery as opposed to the traditional versioning approach (Kenett et al. 2017). Moreover, processing and analytic models evolve in order to provide a high level of agility as organizations realize data agility. The ability to understand data within a context and assume the right business action is a source of competitive advantage (Olavsrud 2017). The emergence of agile processing models enables the same data to support batch analytics, interactive analytics, global messaging, database, and file-based models. The result is an agile development and an application platform that supports the broadest range of processing and analytic models.

Another trend is the integration of machine learning (ML) and microservices. Previously, microservices' deployments have been focused on lightweight

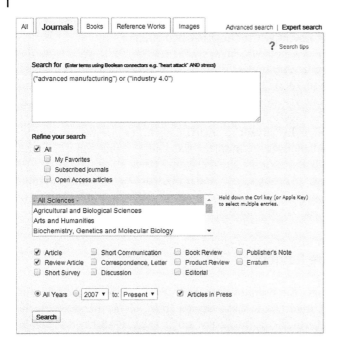

Figure 13.1 Article search in ScienceDirect.

services. Applications that incorporated machine learning have typically been limited to "fast data" integrations, applied to narrow bands of streaming data. Currently, one perceives the development of applications that leverage big data, and the incorporation of data analytics and machine learning approaches that use large amounts of historical data to better understand the context of newly arriving streaming data.

An IDC (2016) report states that during 2016 industries that invested most resources in advanced analytics systems were banking and retail, followed by healthcare and discrete manufacturing. Altogether, these four domains generated more than a half of the entire global analytics and artificial intelligence (AI) revenues. The predictions state that healthcare and discrete manufacturing present the greatest revenue growth over the period of 2016–2020, with compound annual growth rates (CAGR) of 69.3 and 61.4%, respectively. CAGR is the geometric progression ratio that provides a constant rate of return over a considered time period. This indicator dampens the effect of volatility of periodic returns that may render arithmetic means irrelevant and is particularly useful to compare growth rates from various data sets of common domains.

Nearly half of all analytics and artificial intelligence (AI) revenue, according to the IDC forecast, will be directed to software, including both cognitive applications (i.e. text and rich media analytics, tagging, searching, machine learning, categorization, clustering, hypothesis generation, question answering, visualization, filtering, alerting, and navigation) and cognitive software platforms, which facilitate the development of intelligent, advisory, and cognitively enabled solutions. As the largest and fastest growing category, the forecast for cognitive applications spending is $18.2 billion in 2020. Cognitive/AI-related services (business services and IT consulting) represent the second largest revenue category while hardware revenues (primarily from dedicated purchases of servers and storage) will grow nearly as fast as software with five-year CAGRs of more than 60%.

Decision management analytic supported tools have been identified as one of the most popular analytic technologies. According to the Forrester's analysis (https://go.forrester.com) and the IDC list of decision management tools, the top analytical technologies and their use in manufacturing 4.0 are:

Natural language generation, natural language processing, and text mining. Producing natural language from computer systems, processing the text, and extracting insights. These technologies are used in customer service, report generation, summarizing business intelligence (BI) insights. (Gatt and Krahmer 2017).

Speech recognition. Transcribe and transform human speech into recognizable formats. Used for example in interactive voice response systems and mobile applications (Policy Links 2017).

Virtual agents. Advanced virtual entity that can network with humans like Amazon Alexa. Commonly used in customer service and support and as a smart home manager (Rafaelli et al. 2017).

Machine learning platforms. Providing algorithms, development and training toolkits, data, as well as computing power to design, train, and deploy models into applications, processes, and other machines. Used in a wide range of enterprise applications, mostly involving predictive analytics or classification (Gatt and Krahmer 2017). These means include deep learning platforms that support very large data sets.

Artificial intelligence (AI) – optimized hardware. Graphics processing units (GPU) and appliances specifically designed and architected to efficiently run AI-oriented computational jobs.

Decision management. Entails all aspects of designing, building, and managing the automated decision making systems that are used to interact between organizations, users, and suppliers. Decision management leans on advanced analytics, including but not limited to (as extracted from our automated mining method) predictive analytics, risk assessment, network analysis, big data analytics, simulation, clustering, visualization, and data fusion. We will further discuss this point in the next section.

13.3   Analytic Tools in Advanced Manufacturing

This section focuses on applications of advanced analytics to Industry 4.0. Specifically, analytic methods that apply to several aspects of decision management in Industry 4.0 are reviewed. Following our review, we identified nine main application domains in Industry 4.0. These are:

1. Engineering design
2. Manufacturing systems
3. Decision support systems
4. Shop floor control and layout
5. Fault detection and quality improvement
6. Condition-based maintenance
7. Customer and supplier relationship management
8. Energy and infrastructure management
9. Cybersecurity and security related issues

Each one of these domains is described below with examples of the analytic support required to perform related tasks. These examples are not designed to be comprehensive, yet they demonstrate the role of analytics in Industry 4.0. The following section lists analytic challenges that need to be addressed in developing and implementing these methods. Identifying and addressing needs of industry in the context of Industry 4.0, in terms of analytic support, is an essential element in the fourth industrial revolution. The goals of this chapter are to stimulate such developments.

Engineering Design

System, subsystem, and parts modeling and simulation are an integral part of modern design and engineering. Advanced manufacturing opens new options for personalized production and low-volume high-mix processing. For example, Wang et al. (2016a) discuss simulations used to compare mathematical models for tissue mimicking with 3D printer processing. Recent advances in computer-aided design (CAD) and computer technologies have provided a rapid and low-cost method to generate patient-specific tissue-mimicking phantoms from computational models that are constructed from computed tomography (CT) or magnetic resonance imaging (MRI) results of individuals. Fast prototypes are now commonly based on 3D printing for testing design alternatives and enabling multidimensional optimization during the design and engineering process. 3D printing (or additive manufacturing (AM)) refers to the layer by layer fabrication of objects in an additive process from CAD models. It features a high ability for customization, high geometrical complexity, and cost effectiveness in some cases with low production volume, which is perfectly suited for biomedical applications like prosthetics, orthopedic implants, and tissue/organ printing. In their research, the authors present a "metamaterial" concept for constructing artificial models of tissue with heterogeneous microstructures with 3D printing. They designed and fabricated metamaterials with three microstructures and applied finite element analysis (FEA) to predict their mechanical behaviors under tensile loading (For more on such computer experiments see Kenett and Steinberg (2006).) Modern computer experiments and simulations provide the means for achieving robust designs with effectiveness and efficiency, which are far beyond the pioneering work of Taguchi (1978) (also, see Chapter 13 in Kenett et al. (2014)).

Manufacturing Systems

There is a growing tendency in the industry toward data-rich environments characterized by "intelligent" and autonomous machine tools, where several sources of information (i.e. sensors installed in production systems) are available for many purposes (e.g. monitoring, diagnostics, predictive maintenance, etc.). In this framework, many technological advances pave the way for a systematic and extended use of sensor data for industrial quality control. Furthermore, in many industrial applications, the process naturally switches from one operating mode to the following one, whilst producing streams of data from different distributions that follow one another over time. This kind of process is referred to as "multimode process." The monitored variables usually represent quantities that originate from one or multiple sensors, and each one of them requires dedicated preprocessing and raw signal elaboration steps (e.g. time-domain, frequency-domain, or more sophisticated kinds of analysis). Del Castillo et al. (2003), Colosimo et al. (2010a), Del Castillo and Colosimo (2011), Rato and Reis (2015a), Reis and Kenett (2018), among others, discuss manufacturing systems from an integrated view. For a review of data resolution and granularity issues see Reis (2019).

Decision Support Systems

Groger et al. (2012) present indication-based and pattern-based manufacturing process optimization as data mining approaches provided in advanced manufacturing. They discuss dashboards that are typically part of Manufacturing Execution Systems (MES) and custom BI applications based on Online Analytical Processing (OLAP). Their proposal relies on a Data Integration Layer that integrates process and operational manufacturing data in a holistic process within a centric data warehouse. The analytic tools they discuss include Bayesian classification, neural networks, support vector machines, and decision trees. For more on such tools see Yahav et al. (2016) and Kenett (2017).

Shop Floor Control and Layout

Shop floor control is an inherently multivariate problem. With modern computing power and advanced integration and visualization platforms, multivariate statistical process control (MSPC) is becoming operational and critical to advanced manufacturing. (For a background on MSPC see Kenett et al. (2014).) Ferrer et al. (2008) present partial least squares (PLS) as a tool for process improvement and optimization that handles multicollinearity in input variables. Wang et al. (2016b) evaluate the effect of control parameters on 3D printed meta-materials designed to mimic the strain-stiffening behavior of soft tissues. With simulation and physical experiments, they evaluate the effects of design parameters such as wavelength, amplitude, and radius of fiber on the sinusoidal wave design, pitch, radius of helix, and radius of fiber on the double helix design. Modern experimental design methods permit nonlinearities to be effectively accounted for in such responses (Goos and Jones 2011; Jones and Nachstheim 2011; Kenett and Nguyen 2017). In terms of factory layout, an early model of a process with a testing and repair stations has been proposed by Agni-hothri and Kenett (1985). In their research, the authors investigated the impact of the defect distribution on system performance measures such as yield, production lead time, and work-in-process inventory. They provided management guidelines for short-term control decisions such as identifying potential bottlenecks under increased workloads and allocating additional resources to release bottlenecks. They also discussed budget allocation method for process improvement projects initiated in order to meet the long-term goal of continuously decreasing defect levels. Shop floor control and layout is currently a very active area of research. For more on these topics see Chapters 6 and 12.

Fault Detection and Quality Improvement

Hybrid systems consist of continuous behavior and discrete states represented by modes. In each mode, the system is governed by continuous dynamics, and different modes correspond to different continuous models. System health monitoring is a key feature for early detection of faults, failure prevention, reliability, and condition-based maintenance. A health monitoring system integrates four main skills, namely: (i) fault detection; (ii) fault isolation; (iii) fault parameter estimation; and (iv) prediction of time to maintenance or replacement. In hybrid system diagnosis, two different types of faults are defined. The first is a parametric fault, where one or more model parameters are deviating from their nominal value to an unknown value. The second type is fault mode; in this case, the faulty state is recognized a priori and can be modeled by known parameters only. A health monitoring framework for hybrid system is presented in Arogeti et al. (2010) and in Low et al. (2010). Colosimo et al. (2005, 2008, 2010b), Colosimo and Pacella (2010),

Kenett and Pollak (1996), and Kenett et al. (2014) present various process monitoring methods such as the Shiraryev–Roberts (SR) detection procedure, MSPC, signal decomposition, and Empirical Mode Decomposition (EMD). Portillo-Dominguez et al. (2017) present an automation framework for the effective usage of diagnosis tools in the performance testing of clustered systems. Rato and Reis (2013a,2013b,2013c, 2015a,2015b,2015c), Reis and Saravia (2006a,2006b), Reis et al. (2008), and Reis and Bauer (2009) present advanced methods for time series analysis, such as autoregressive integrated moving average (ARIMA), vector ARIMA (VARIMA), compositional VARIMA (cVARIMA) models, dynamic principal component analysis (DPCA), dynamic data warping (DTW), and Wavelets. For a review of batch processing methods see Chiang and Reis (2019).

Condition-Based Maintenance

Man-made systems tend to deteriorate over time and, therefore, require ongoing maintenance to avoid malfunctioning. Accordingly, it is essential to undertake an effective preventive maintenance (PM) policy that minimizes the life-cycle cost (LCC) of the system and maximizes its operational profit. Saxena et al. (2008) present a damage propagation model for aircraft engine based on run-to-failure simulations. Gruber et al. (2013) present a condition-based maintenance method to define a PM policy per the state of the system at various time periods by combining both a simulation model of the system and a predictive meta-model. An up-to-date description of this methodology is provided in Chapter 9. Shalev and Tiran (2007) present a methodology for conditional based fault tree analysis. Meeker and Hong (2014), Lu and Anderson-Cook (2015), and Vining (2016) review challenges of big data to modern reliability engineering. See also Chapter 10 in this book.

Customer and Supplier Relationship Management

Supply chain management (SCM) is a systematic progression in which an organization manages the flows of products, services, currency, etc. Its purpose is to obtain maximum profits at minimum cost as well as fulfilling customers' demand. SCM includes the movement and storage of raw materials, work-in-process inventory, and finished goods, i.e. from raw material to point of consumption. Pal and Mahapatra (2017) reviewed this interconnected network. They developed a three-layer supply chain model with production-inventory model for reworkable items, including an inventory model for deteriorating items price and stock dependent demand. Their model is an economical order quantity one with imperfect quality and shortage backordering under inspection errors and deterioration.

Energy and Infrastructure Management

Nabati and Thoben (2017) focus on data gathered from operation and service of offshore wind turbines. They use and analyze field data or so-called Product Use Information (PUI) to improve maintenance activities and to reduce the costs. Their data are collected from sensors on the turbines, alarms information, signals from the condition monitoring, and supervisory control and data acquisition (SCADA) systems used in maintenance activities. In order to make the right decision, it is important to understand which PUI data source and which data analysis methods are suitable for various kinds of decision making task. The aim of their study is to discover in which way big data analytics of PUI may assist in maintaining processes of offshore wind power.

Cybersecurity and Security Related Issues

The European Union (EU) Cybersecurity Strategy defines Cybersecurity as "the safeguards and actions that can be used to protect the cyber domain, both in the civilian and military fields, from those threats that are associated with or that may harm its interdependent networks and information infrastructure. Cybersecurity strives to preserve the availability and integrity of the networks and infrastructure and the confidentiality of the information contained therein" (EU 2016). Cybersecurity is an important concern for Industry 4.0. Gruber and Ben Gal (2018) propose an application that characterizes the behavioral patterns of suspect users versus nonsuspect users based upon usage metadata. These are just a few examples of a growing area of research. See also Chapter 16 in this book.

13.4   Challenges of Big Data and Analytic Tools in Advanced Manufacturing

In this section, we offer a close examination of analytics and decision management tools that are applicable for advanced manufacturing. We review implemented tools and methods to verify if process behavior is consistent with normal operating conditions. These tool functionalities include:

- Detection – rapidly detect abnormalities in process operation.
- Diagnosis – examine the root cause of abnormal behavior.
- Fault criticality assessment – assess potential severity of the fault.
- Decision – stop the process and fix the problem or accommodate the fault and proceed.

Big data has been around for some time, e.g. in satellite imaging, genomics, particle physics, and the collection, provision, and analysis of data. The term *data science* is commonly used to describe these activities. Big data are considered to offer a high potential for learning, even though they are largely observational. In this context, many well-known issues recognized by statisticians should be addressed. For a comprehensive treatment of the real work of data science see Kenett and Redman 2019.

For example, data are subject to selection conditions that may determine their status as nonrepresentative for a target population or system (Yahav et al. 2016). This tendency may lead to biased decisions and overfitting in deployment applications. Another example relates to the fact that important variables or factors may not be included in the data and data may be uninformative; hence analyses could appear as misleading. These tendencies may reduce effectiveness of decision support systems. Additionally, data with large number of factors are susceptible to multiplicity and false discovery issues. The effects of this characteristic may be reduced specificity and poor robustness of methods. Lastly, theory-free analyses of association are fragile and difficult to generalize. Therefore, it implies that the generation consists of poor insights.

In the context of advanced manufacturing, several conditions relate to the challenges discussed previously: Data are of high dimension, hence there are many variables measured at a high sampling rate; observational data are available with no direct causal interpretation; process varies in lower dimensions leading to nonfull rank data; extensive missing data pattern is apparent, sometimes with some columns or rows missing 90% of the data; and low signal-to-noise ratio and little information in variables, which requires a multivariate analysis.

Additional challenges, specific to Industry 4.0, as reviewed by Reis and Gins (2017), are data integration under multiscale settings such as intelligent data fusion and interpretation toolboxes in process simulators and company information systems. This integration has to adapt to the needs and requirements of the company, but it must also include its supply chain and other networks of partners, making the integration challenge even more demanding. These data are often provided in real time. As a result, there is a need for a combination of real-time adaptive learning, as new data becomes available, with previous knowledge.

Manufacturing Analytics Best Practices

Harford (2014) states that "Big data do not solve the problem that has obsessed statisticians for centuries: the problem of insight, of inferring what is going on, and figuring out how we might intervene to change a system for the better."

The primary objective of analytics should be the development of "interpretable" models. Harding et al. (2006) review data mining methods in the context of manufacturing. They advocate for the use of a model entitled cross-industry standard process for data mining (CRISP-DM). In this framework, we provide a detailed review of the support provided by analytics to Industry 4.0 in a range of activities and application domain. The Sight Machine (2015) white paper lists five manufacturing analytics best practices for current competitive enterprises:

1. A direct connection to multiple data sources.
2. The ability to handle terabytes of data.
3. The capacity of integrating both structured and unstructured data.
4. Cloud-enabled functionality for both unlimited storage and anywhere/any time access.
5. Real-time dashboards along with deep-dive analytics for optimal insight, decision making.

The challenge is to put these practices into work in production enterprises.

Continuous and Batch Process Analytic Challenges

The principal analytical challenges in continuous and batch processes emerge from the need for transitioning: (i) from monitoring the mean to dispersion to correlation; (ii) from stationary, to dynamic, to nonstationary; (iii) from sensor data to higher-order profiles; and (iv) from detection, to diagnosis, to prognosis.

Continuous processes are characterized by a stable operation window. Classical approaches assume independent and identically distributed variables, which are rarely valid for industrial processes due the existence of inertial elements (big process units) coupled with fast acquisition rates. For dealing with autocorrelation, three types of approaches to follow exist:

1. Adjusting the control limits (this is feasible only for very simple systems).
2. Estimating a multivariate time series model and monitor the residuals with over a dozen variables.
3. Using a transformation in the time domain that decorrelates the autocorrelation of the series (namely the wavelet transform).

Batch processes monitoring is a greater challenging task than monitoring continuous processes. In this context, all variables may change in each cycle or batch. Therefore, the batch process is intrinsically multivariate and nonstationary. Moreover, in this case, nonlinearities are more noticeable than in continuous processes with more or less consistent cycles.

To characterize these challenges, we refer to a comprehensive framework designed to plan and assess the level of InfoQ provided by analytic tools and methods. After introducing the framework, some analytic methods mentioned in Section 13.3 are evaluated with InfoQ dimensions.

13.5   An Information Quality (InfoQ) Framework for Assessing Advanced Manufacturing

Kenett and Shmueli (2014, 2016) define InfoQ as the potential of a dataset to achieve a specific (scientific or practical) goal using a given empirical analysis method. InfoQ is different from data quality and analysis quality but is dependent on these components and on the relationship between them. Technically, the definition of InfoQ is the derived utility (U) from an application of a statistical or data analytic model (f), to a data set (X), given the research goal (g). This can be written algebraically as: $InfoQ(f, X, g, U) = U(f(X|g))$.

A key requirement for determining InfoQ is the nature of the study goal. We distinguish between explanatory, predictive, and descriptive goals. An explanatory goal is one that is based on causal hypotheses or seeks causal answers. A predictive goal is aimed at predicting future or new individual observations. A descriptive goal is aimed at quantifying an observed effect using a statistical or other approximation.

To assess the level of InfoQ in a study, Kenett and Shmueli (2014, 2016) propose eight dimensions of InfoQ:

1. *Data resolution.* The measurement scale and level of aggregation of the data relative to the task at hand must be adequate for the study. For example, consider data on daily purchases of over-the-counter medications at a large pharmacy. If the goal of the analysis is to forecast future inventory levels of different medications when restocking is done on a weekly basis, then weekly aggregated data is preferred to daily aggregate data.
2. *Data structure.* The data can combine structured, quantitative data with unstructured, semantic-based data. For example, in assessing the reputation of an organization one might combine data derived from the stock exchange with data mined from text such as newspaper archives or press reports. Doing it enhances InfoQ.
3. *Data integration.* Data are often spread out across multiple data sources. Hence, properly identifying the different relevant sources, collecting the relevant data, and integrating the data directly affect InfoQ.
4. *Temporal relevance.* A dataset contains information collected during a certain time framework. The degree of relevance of the data in that time framework to the current goal at hand must be assessed. For instance, in learning about current online shopping behaviors, a dataset with last year's records of online purchasing behavior might be irrelevant.
5. *Chronology of data and goal.* Depending on the nature of the goal, the chronology of the data can support the goal to different degrees. For example, in process control applications of discrete parts, we might collect data from previous processes that are relevant to a specific part. If the goal is to quantify the effect of previous manufacturing steps on the specific

Table 13.1 Relevance of Industry 4.0 domains by InfoQ dimensions.

Manufacturing 4.0 domains	Resolution	Structure	Integration	Temporal relevance	Chronology of data and goal	Generalization	Operationalization	Communication
1. Engineering design	X	X	X			X		
2. Manufacturing systems	X	X	X		X		X	
3. Decision support systems					X	X	X	X
4. Shop floor control and layout			X		X		X	X
5. Fault detection and quality improvement	X	X		X			X	
6. Conditioned-based maintenance		X	X		X		X	
7. Customer and supplier relationship management	X	X	X	X				X
8. Energy and infrastructure management	X	X		X		X		
9. Cybersecurity and security related issues	X		X		X	X		

part quality, then the chronology is fine. However, if the goal is to predict the final quality of a part, then the required information builds on data collected in future manufacturing steps and, hence, the chronology of data and goal is not met.

6. *Generalizability.* There are two types of generalizability: statistical and scientific. Statistical generalizability refers to inferring from a sample to a target population. Scientific generalizability refers to applying a model based on a particular target population to other populations. It may imply either generalizing an estimated population pattern/model to other populations or else applying it from one population to predict individual observations in other populations.

7. *Operationalization.* Observable data are an operationalization of underlying concepts. "Customer Satisfaction" may be measured via a questionnaire or by evaluating the impact of attributes that were assessed via conjoint analysis. Constructs play a key role in causal models and raise the question of what to measure and how. The sensitivity to what is measured versus the construct of interest depends on the study goal. Action operationalization is about deriving concrete actions from the information provided by a study.

8. *Communication.* In a case in which the information does not reach the right person at the right time in a clear and understandable way, the quality of information becomes poor. Data visualization is crucial for good communications, which is, therefore, directly related to the quality of information. Poor visualization of findings may lead to degradation of the InfoQ contained in the analysis performed on the data. Dashboards are about communication.

Formalizing the concept of InfoQ increases the value of statistical analysis and data mining, both methodologically and practically. In Table 13.1 the nine domains discussed in Section 13.3 are mapped in terms of the relevance of the eight InfoQ dimensions. The table also reflects the list of analytic challenges described in Section 13.4. (For more on InfoQ and InfoQ assessment see Kenett and Shmueli (2016).)

13.6 Summary

Eventually, analytics and statistical analysis are performed in order to generate information and to support decisions. In this chapter, we reviewed the role of analytics in the context of Industry 4.0 domains. We listed examples of analytic work and open challenges.

The framework of InfoQ is used here to evaluate analytic methods and tools. The chapter provides a detailed background for future research on analytics supporting Industry 4.0. The elements of a roadmap for establishing an infrastructure for developing and deploying advanced manufacturing consist of collaborative initiatives linking industry and academia. Test bed environments are needed where new technology and analytic algorithms are tested and information hubs need to be formed where knowledge is documented and shared. Education programs are now offered at different levels (schools, colleges, universities, workers, managers, and scientists) in order to prepare human resource infrastructures for data analytics developments. As more companies draw on analytics for their competitive edge, several complementary organizational trends are emerging around the emphasis on data. Businesses that consider data seriously are organized around data as an asset. These businesses are democratizing the access to data and "bring the right information to the right person at the right time." These businesses promote and support data sharing. Data sharing requires many parts of the organization to work together.

The objective of this chapter was to provide a context for research and development of analytic methods supporting advanced manufacturing initiatives. It was designed as an introductory review listing references as sources for additional information and case study examples.

References

Agnihothri, S. and Kenett, R.S. (1985). The impact of defects on a process with rework. *European Journal of Operational Research* 80: 308–327.

Arogati, S., Wang, D., and Low, C. (2010). Mode identification of hybrid Systems in the presence of fault. *IEEE Transaction on Industrial Electronics* 57 (4): 1452–1467.

Chiang, L.H. and Reis, M.S. (2019). Data-driven methods for batch data analysis Ð a critical overview and mapping on the complexity scale. *Computers and Chemical Engineering* https://doi.org/10.1016/j.compchemeng.2019.01.014.

Colosimo, B.M. and Pacella, M. (2010). A comparison study of control charts for statistical monitoring of functional data. *International Journal of Production Research* 48 (6): 1575–1601.

Colosimo, B.M., Pan, R., and Del Castillo, E. (2005). Setup adjustment for discrete-part manufacturing processes with asymmetric cost functions. *International Journal of Production Research* 43 (18): 3837–3854.

Colosimo, B.M., Semeraro, Q., and Pacella, M. (2008). Statistical process control for geometric specifications: on the monitoring of roundness profiles. *Journal of Quality Technology* 40 (1): 1.

Colosimo, B.M., Mammarella, F., and Petro, S. (2010a). Quality control of manufactured surfaces. *Frontiers in Statistical Quality Control* 9: 55–70.

Colosimo, B.M., Moroni, G., and Petrò, S. (2010b). A tolerance interval-based criterion for optimizing discrete point sampling strategies. *Precision Engineering* 34 (4): 745–754.

Del Castillo, E. and Colosimo, B.M. (2011). Statistical shape analysis of experiments for manufacturing processes. *Technometrics* 53 (1): 1–15.

Del Castillo, E., Pan, R., and Colosimo, B.M. (2003). A unifying view of some process adjustment methods. *Journal of Quality Technology* 35 (3): 286.

EIC (2017), European Innovation Scoreboard. http://ec.europa.eu/growth/industry/innovation/facts-figures/scoreboards_en (Last accessed 25 June 2019).

EU, (2016). Scientific Advice Mechanism: Cybersecurity, https://ec.europa.eu/research/sam/pdf/meetings/hlg_sam_012016_scoping_paper_cybersecurity.pdf (Last accessed 25 June 2019).

Ferrer, A., Aguado, D., Vidal-Puig, S., and Zarzo, M. (2008). PLS: a versatile tool for industrial process improvement and optimization. *Applied Stochastic Models in Business and Industry* 24: 551–567.

Gatt, A. and Krahmer, E. (2017). Survey of the state of the art in natural language generation: core tasks, applications and evaluation. *Journal of AI Research (JAIR)* 61: 75–170. arXiv:1703.09902 [cs.CL].

Goos, P. and Jones, B. (2011). *Optimal Design of Experiments: A Case Study Approach*. Wiley.

Groger, C., Niedermann, F., and Mitschang, B. (2012). Data mining-driven manufacturing process optimization. In: *Proceedings of the World Congress on Engineering, WEC 2012*, London (4–6 July 2012), vol. III (eds. S.I. Ao, L. Gelman, D.W.L. Hukins, et al.), 1475–1481. Newswood.

Gruber, A. and Ben Gal, I. (2018). Using targeted Bayesian network learning for suspect identification in communication networks. *International Journal of Information Security* 17 (2): 169–181.

Gruber, A., Yanovski, S., and Ben-Gal, I. (2013). Condition-based maintenance via simulation and a targeted Bayesian network metamodel. *Quality Engineering* 25 (4): 370–384.

Harding, J., Shahbaz, M., Srinivas, S., and Kusiak, K. (2006). Data mining in manufacturing: a review. *Journal of Manufacturing Science and Engineering* 128: 969–976.

Harford, T. (2014). Big data: are we making a big mistake? *Significance* 11 (5): 14–19.

IDC (2016). Worldwide Cognitive Systems and Artificial Intelligence Revenues Forecast to Surge Past $47 Billion in 2020, According to New IDC Spending Guide, https://www.businesswire.com/news/home/20161026005031/en/Worldwide-Cognitive-Systems-Artificial-Intelligence-Revenues-Forecast (Last accessed 25 June 2019).

Jones, B. and Nachtsheim, C.J. (2011). A class of three-level designs for definitive screening in the presence of second-order effects. *Journal of Quality Technology* 43: 1–15.

Kenett, R.S. (2017). Bayesian networks: theory, applications and sensitivity issues. *Encyclopedia with Semantic Computing and Robotic Intelligence* 1 (1): 1–13.

Kenett, R.S. and Nguyen, N.-K. (2017). Experimental learning: generate high information quality by comparing alternative experimental designs. *Quality Progress*: 40–47.

Kenett, R.S. and Pollak, M. (1996). Data-analytic aspects of the Shiryayev-Roberts control chart: surveillance of a non-homogeneous Poisson process. *Journal of Applied Statistics* 23 (1): 125–137.

Kenett, R.S. and Redman, T. (2019). *The Real Work of Data Science: Turning Data into Information, Better Decisions, and Stronger Organizations*. Wiley.

Kenett, R.S. and Shmueli, G. (2014). On information quality. *Journal of the Royal Statistical Society (Series A)* 177: 3–38.

Kenett, R.S. and Shmueli, G. (2016). *Information Quality: The Potential of Data and Analytics to Generate Knowledge*. Wiley.

Kenett, R.S. and Steinberg, D.M. (2006). New frontiers in the design of experiments. *Quality Progress* 39 (8): 61.

Kenett, R.S., Zacks, S., and Amberti, D. (2014). *Modern Industrial Statistics: With Applications in R, MINITAB and JMP*. Wiley.

Kenett, R.S., Ruggeri, G., and Faltin, F. (2017). *Analytic Methods in Systems and Software Testing*. Wiley.

Low, C., Wang, D., and Arogeti, S. (2010). Quantitative hybrid bond graph-based fault detection and isolation. *IEEE Transaction on Automation Science and Engineering* 7 (3): 558–569.

Lu, L. and Anderson-Cook, C.M. (2015). Improving reliability understanding through estimation and prediction with usage information. *Quality Engineering* 27 (3): 304–316.

Meeker, W. and Hong, Y. (2014). Reliability meets big data: opportunities and challenges. *Quality Engineering* 26 (1): 102–116.

Nabati, E. and Thoben, K. (2017). Data driven decision making in planning the maintenance activities of off-shore wind energy. *Procedia CIRP* 59: 160–165.

Olavsrud, T. (2017). 15 data and analytics trends that will dominate 2017, http:// www.cio.com/article/3166060/analytics/15-data-and-analytics-trends-that-will-dominate-2017.html (Last accessed 25 June 2019).

Pai, J.C. (2014). Industry 4.0: from the Internet of Things to Smart Factories, MSME, India, https://www.slideshare.net/jcspai/industry-40-pai (Last accessed 25 June 2019).

Pal, S. and Mahapatra, G.S. (2017). A manufacturing-oriented supply chain model for imperfect quality with inspection errors, stochastic demand under rework and shortages. *Computers and Industrial Engineering* 106 (c): 299–314.

Policy Links (2017). *Emerging Trends in Global Advanced Manufacturing*. University of Cambridge https://www.ifm.eng.cam.ac.uk/insights/national-innovation-policies/emerging-trends-in-global-advanced-manufacturing:-challenges,-opportunities-and-policy-responses/ (Last accessed 25 June 2019).

Portillo-Dominguez, A.O., Perry, P., Magoni, D., and Murphy, J. (2017). PHOEBE: an automation framework for the effective usage of diagnosis tools in the performance testing of clustered systems. *Journal of Software Practice and Experience* 47 (11): 1837–1874.

Rafaelli, A., Altman, D., Gremler, D.D. et al. (2017). Invited commentaries on the future of frontline research. *Journal of Service Research* 20 (1): 91–99.

Rato, T.J. and Reis, M.S. (2013a). Advantage of using decorrelated residuals in dynamic principal component analysis for monitoring large-scale systems. *Industrial and Engineering Chemistry Research* 52 (38): 13685–13698.

Rato, T.J. and Reis, M.S. (2013b). Defining the structure of DPCA models and its impact on process monitoring and prediction activities. *Chemometrics and Intelligent Laboratory Systems* 125: 74–86.

Rato, T.J. and Reis, M.S. (2013c). Fault detection in the Tennessee Eastman process using dynamic principal components analysis with decorrelated residuals (DPCA-DR). *Chemometrics and Intelligent Laboratory Systems* 125: 101–108.

Rato, T.J. and Reis, M.S. (2015a). Multiscale and megavariate monitoring of the process networked structure: M2NET. *Journal of Chemometrics* 29 (5): 309–332.

Rato, T.J. and Reis, M.S. (2015b). On-line process monitoring using local measures of association. Part I: detection performance. *Chemometrics and Intelligent Laboratory Systems* 142: 255–264.

Rato, T.J. and Reis, M.S. (2015c). On-line process monitoring using local measures of association. Part II: design issues and fault diagnosis. *Chemometrics and Intelligent Laboratory Systems* 142: 265–275.

Reis, M.S. (2019). Multiscale and multi-granularity process analytic: a review. *Processes* 7 (2): 61. https://doi.org/10.3390/pr7020061.

Reis, M.S. and Bauer, A. (2009). Wavelet texture analysis of on-line acquired images for paper formation assessment and monitoring. *Chemometrics and Intelligent Laboratory Systems* 95 (2): 129–137.

Reis, M.S. and Gins, G. (2017). Industrial process monitoring in the big data/industry 4.0 era: from detection, to diagnosis, to prognosis. *Processes* 5 (35): 2–16.

Reis, M.S. and Kenett, R.S. (2018). Assessing the value of information of data-centric activities in the chemical processing industry 4.0. *AIChE, Process Systems Engineering* 64 (11): 3868–3881.

Reis, M.S. and Saraiva, P.M. (2006a). Multiscale statistical process control of paper surface profiles. *Quality Technology and Quantitative Management* 3 (3): 263–282.

Reis, M.S. and Saraiva, P.M. (2006b). Multiscale statistical process control with multiresolution data. *AIChE Journal* 52 (6): 2107–2119.

Reis, M.S., Bakshi, B.R., and Saraiva, P.M. (2008). Multiscale statistical process control using wavelet packets. *AIChE Journal* 54 (9): 2366–2378.

Saxena, A., Goebel, K., Simon, D., and Eklund, N. (2008). Damage propagation modeling for aircraft engine run-to-failure simulation. In: *International Conference on Prognostics and Health Management (PHM08)*, Denver, CO (6–9 October 2008). IEEE.

Shalev, D. and Tiran, J. (2007). Condition-based fault tree analysis (CBFTA): a new method for improved fault tree analysis (FTA), reliability and safety calculations. *Reliability Engineering and System Safety* 92: 1231–1241.

Siemens (2016). https://www.siemens.com/innovation/en/home/pictures-of-the-future/industry-and-automation/digtial-factory-trends-industrie-4-0.html (accessed 14/4/2017).

Sight, M. (2015). Advanced Insights: 5 Manufacturing Analytics Best Practices for Today's Enterprise, http://sightmachine.com/advanced-insights-white-paper (Last accessed 25 June 2019).

Taguchi, G. (1978). *System of Experimental Design: Engineering Methods to Optimize Quality and Minimize Costs*. UNIPUB/Kraus International and American Supplier Institute.

Vining, G. (2016). Recent advances and future directions for quality engineering. *Quality and Reliability Engineering International* 32: 863–875.

Wang, K., Wu, C., Qian, Z. et al. (2016a). Dual-material 3D printed metamaterials with tunable mechanical properties for patient-specific tissue-mimicking phantoms. *Additive Manufacturing* 12: 31–37.

Wang, K., Zhao, Y., Chang, Y. et al. (2016b). Controlling the mechanical behavior of dual-material 3D printed meta-materials for patient-specific tissue-mimicking phantoms. *Materials and Design* 90: 704–712.

Yahav, I., Shmueli, G., and Mani, D. (2016). A tree-based approach for addressing self-selection in impact studies with big data. *MIS Quarterly* 40 (4): 819–848.

13.A Appendix

In this appendix we provide insights on the article mining from ScienceDirect (www.sciencedirect.com), using the search term [("advanced manufacturing") or ("industry 4.0")]. The analysis is based on approximately 9000 original articles.

In Figure 13.A.1 we examine the number of papers published under this search per year. The increased interest since 2011 onwards is apparent. Figure 13.A.2 presents the top terms[1] that appear in the titles and abstracts of articles published since and including 2011. Interestingly, analytics-related words do not float in this analysis. Rather the focus is on the engineering fashion of the discipline. A similar insight can be obtained from a nonsupervised topic modeling on the papers' title and abstract, as summarized in Table 13.A.1.

---

1 We preprocessed the term to remove stop words, punctuations, and numbers. We further normalized the terms using the well accepted TF-IDF weighting scheme.

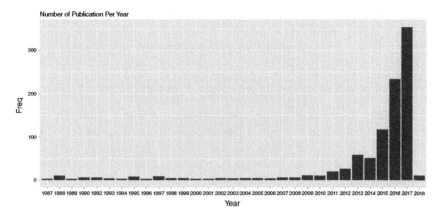

Figure 13.A.1 Article per year in ScienceDirect under the search term [("advanced manufacturing") or ("industry 4.0")].

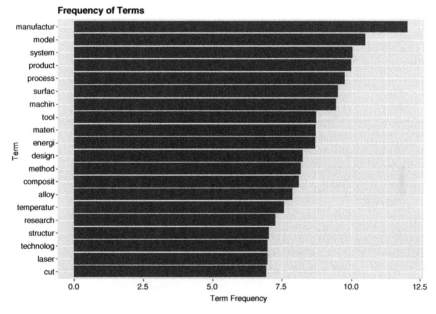

Figure 13.A.2 Top (normalized) terms in articles that meet the search phrase [("advanced manufacturing") or ("industry 4.0")].

The table presents the top keywords in a 10-topic analysis following the Latent Dirichlet allocation (LDA) model. Topics 3 and 6, which touch upon analytics (see keywords "data" and "analysis" in topics 3 and 6, respectively), consist of very few papers, compared with other topics.

Table 13.A.1 Topic modeling on articles that meet the search phrase [("advanced manufacturing") or ("industry 4.0")].

Topic 1	Topic 2	Topic 3	Topic 4	Topic 5	Topic 6	Topic 7	Topic 8	Topic 9	Topic 10
machin	composit	energi	system	structur	process	manufactur	applic	model	alloy
tool	mechan	data	propos	abstract	use	research	design	result	laser
surfac	phase	develop	product	high	method	industri	manufactur	simul	metal
process	sampl	effici	model	effect	measur	paper	technolog	distribut	process
cut	properti	industri	control	field	abstract	technolog	addit	valu	thermal
materi	increas	product	optim	perform	techniqu	develop	develop	effect	heat
paramet	temperatur	can	paper	materi	analysi	provid	fabric	experiment	use
wear	show	improv	design	magnet	can	product	structur	test	layer
forc	strength	use	problem	result	base	futur	materi	abstract	investig
use	addit	cost	base	film	present	literatur	advanc	deform	Weld
84	102	56	104	65	45	156	78	74	103

Table 13.A.2 Topic modeling on articles that meet the search phrase [(("advanced manufacturing") or ("industry 4.0")) and analysis)].

Topic 1	Topic 2	Topic 3	Topic 4	Topic 5	Topic 6	Topic 7	Topic 8	Topic 9	Topic 10
paramet	element	energi	model	manufactur	temperatur	analysi	system	perform	research
machin	structur	industri	method	product	weld	method	data	use	studi
surfac	stress	develop	base	process	composit	evalu	approach	result	develop
process	form	abstract	propos	design	laser	present	paper	factor	paper
tool	experiment	effici	analysi	requir	properti	use	propos	control	manag
cut	load	china	can	cost	show	effici	integr	oper	technolog
use	use	chang	valu	can	test	paper	provid	analyz	field
optim	numer	effect	predict	improv	thermal	problem	implement	measur	identifi
materi	finit	polici	simul	qualiti	mechan	select	applic	analysi	review
forc	deform	improv	result	servic	heat	techniqu	abstract	error	literatur
108	85	60	54	69	95	51	57	33	81

A similar analysis was carried on analytical papers under the search phrase [(("advanced manufacturing") or ("industry 4.0")) and analysis)]. Note that the search term returns words related to the term "analysis" as well (http://help .sciencedirect.com/Content/stadv_main_jnl.htm). Table 13.A.2 summarizes the topics found under this search criterion. Analytic-related terms are shaded.

The main analytic topics raised by the analysis are simulation, optimization, and predictive analytics. We further repeated this analysis with additional search terms (such as "data" and "analytics"), and a different number of topics, and compiled a list of common topics, as appear in the manuscript: predictive analytics, risk assessment, network analysis, big data analytics, simulation, clustering, visualization and data fusion.

An attempt to automatically mine decision management papers failed (the articles' topics are quite heterogeneous). Therefore, a manual study of papers under the search term [("Industry 4.0" or "advanced manufacturing") and decision and management] was carried out.

14

Hybrid Semiparametric Modeling: A Modular Process Systems Engineering Approach for the Integration of Available Knowledge Sources

Cristiana Rodrigues de Azevedo, Victor Grisales Díaz, Oscar Andrés Prado-Rubio, Mark J. Willis, Véronique Préat, Rui Oliveira, and Moritz von Stosch

Synopsis

The Quality by Design (QbD) paradigm and novel Process Analytic Technology (PAT) are reshaping the way that manufacturing processes are developed and plants are operated in the future, aiming in the pharmaceutical industry, for instance, at real-time release of a drug. The application of sensors that can provide real-time measurements of the process state and processed materials as well as risk assessment and process understanding are key topics of attention. Process understanding is relevant in particular, as it is at the core of process system engineering approaches, where the aim is to provide a better understanding of the involved risks and, therefore, their assessment. In addition, it can also support decisions as to what needs to be analyzed and where (in the process) to place sensors, or adaptive operating strategies. Given these requirements for process understanding, there seems to be a need to change the way we develop, handle, and maintain process understanding. Mathematical modeling is an attractive approach to distill, administrate, and conserve this knowledge in a compact manner.

For mathematical model development, data-driven and first-principle approaches can be regarded as the extremes of a systems-knowledge abstraction scale, the lowest versus the highest respectively. Moreover, the hypothesis-driven inference of higher abstract forms of knowledge from data in a manual or semi-automatic manner is a common objective across disciplines. The quality of the data, i.e. the information that is captured in the data, and the a priori available knowledge about the system play a critical role in this abstraction process and impact on the quality of the obtained system description – the model – and the "space" within which the description will

Systems Engineering in the Fourth Industrial Revolution: Big Data, Novel Technologies, and Modern Systems Engineering, First Edition. Edited by Ron S. Kenett, Robert S. Swarz, and Avigdor Zonnenshain.

be of satisfactory accuracy. In turn, the quality of the resulting model and the space in which faithful descriptions of a systems behavior can be obtained are of critical importance for (process) systems engineering, which normally exploits the developed models for systems design and operation.

In this chapter we review different methods of inferring system descriptions – models – from data, and we discuss their merits in light of data requirements and consideration of available a priori knowledge, which can describe either the structure of interactions and/or the way/form of the occurring interactions. We then demonstrate a hybrid semi-parametric framework that combines a priori available structural or forming knowledge about the system with nonparametric models where structure and parameters are inferred from the data. The framework is applied to a case study that considers the dynamic ultrafiltration of water, where a purely data-driven modeling approach is compared to a hybrid model in order to predict filtration performance. In a second case study, the framework is applied to controlled-drug-release modeling, showing that the method can describe drug release profiles based on the formulation characteristics, the synthesis parameters, and drug molecule descriptors. Finally, we outline how such a framework could shape the general approach of modeling across industry, the integration of different sensor signals (such as RAMAN, NIR) into one body of knowledge, and its application in process systems engineering (design and operation).

14.1 Introduction

The development, design, and operation of a process require process knowledge. Various strategies, such as the Quality by Design (QbD) paradigm or advanced process control approaches, further promote this, establishing process understanding that enables design and operation of a process so that it produces desired product quality. Process understanding is also a central element in the Process Analytic Technology (PAT) guidance, published by the US Department of Health and Human Services Food and Drug Administration (2004). In chemical and biochemical engineering, the generation of process understanding is a classic step, used for process operation/design and it allows the design of complex process flowsheets, model predictive control algorithms, plant-wide optimization strategies, and so on. In other words, process understanding/knowledge is the key to effective operation and design of process plants.

Process knowledge can have a number of different forms. Data are probably the least abstract form of knowledge, i.e. every data point captures a very local and potentially segregate snap-shot that taken alone does not help in understanding the bigger picture or overall dynamic behavior of the system.

In addition, more data do not automatically imply more knowledge and despite having "big data" it still may not be possible to understand much of the behavior of the system, as the information that is captured in the data is limited. Bühlmann and van de Geer (2018) refer to this as data from different "environments," $E = \{E_1, \ldots, E_n\}$.

The most abstract form, knowledge derived from first principles, provides a global comprehension of the behavior of the system. While first-principles models originate from theory, they will have been verified using a multitude of observations – data – generated under various settings over significant periods of time and under different environments. This means that first-principles models can provide consistent predictions, provided the assumption under which the model was developed holds (García-Muñoz et al. 2015), i.e. specific parts of "E" are defined. In contrast, the behavior of the system that is captured by a data point may not even be repeatable, as parts of "E" that are not measured may have varied, changing the outcome of the experiment. Between the two extremes several other forms of models may be used to represent knowledge e.g. mechanistic, phenomenological, heuristics, and empirical models. This is typically represented by a pyramid the bottom being constituted by data, the basis to all the other forms, and on the top knowledge from first-principles (Figure 14.1).

Much of the modeling exercise is an abstraction process that seeks to put the knowledge in a more general and usable form, while noting the assumptions/settings under which the model is applicable. The same mathematical representation can sometimes be attributed to different knowledge forms (e.g. the Monod and one form of the Michaelis–Menten kinetics are mathematically identical, yet the former is agreed to be an empirical model while the latter is considered a mechanistic model) and naming is not always consistent. Many times, the terms first-principles, mechanistic, phenomenological, or fundamental are used interchangeably whereas at other times these terms are seen

Figure 14.1 Pyramid of knowledge and the knowledge abstraction level. Between the extremes, knowledge from first principles and data, other forms of knowledge can be found, e.g. mechanistic, phenomenological, heuristics, and empirical models.

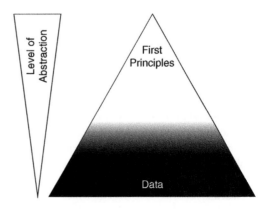

as distinct. Another nomenclature, commonly used in the field of statistics, is that of parametric and nonparametric models, differentiating between whether the structure of the model is fixed through knowledge or derived/inferred from data (both parametric and nonparametric models have parameters). This differentiation helps to contrast the two main ways to arrive at the model structure. It is possible though to combine parametric and nonparametric models, referred to as semiparametric models. One particular class of those models combines fundamental models with data-driven or heuristic-driven models, referred to as hybrid semiparametric models, or hybrid models. These models make use of fundamental knowledge and complement the description, where necessary, with nonparametric approaches. Artificial neural networks (ANNs) are often chosen as the nonparametric part, as they can approximate arbitrary complex nonlinear functions (Haykyn 1998) and because the research on hybrid modeling (HM) evolved from that of ANNs (Psichogios and Ungar 1992). For these reasons, they are also here the method of choice. However, other data-driven models can be incorporated, such as support vector machines, partial least squares, etc.

Since these hybrid models are based on a broader knowledge base they can typically describe the system's behavior more faithfully than models based on a single source of knowledge. The better representation of the system's behavior typically gives rise to an improved design and/or operation. Hence, in this chapter the focus is on hybrid semiparametric modeling.

14.2 A Hybrid Semiparametric Modeling Framework

Hybrid models combine fundamental models with data-driven or heuristic-driven models. The attractive traits of hybrid models are their good extrapolation capabilities, low data requirements, and transparent structure, which helps to develop process understanding. All of these properties stem from the incorporated fundamental knowledge. In fact, the original idea behind the development of hybrid models was that the structure of the fundamental knowledge would increase the prediction capabilities of neural networks (Psichogios and Ungar 1992). Why then would one not develop fundamental models in the first place? Though hybrid models benefit from the fundamental knowledge, due to the data-driven components, a model can also be easily established for parts that are fundamentally not understood. This increases the applicability of the fundamental knowledge to domains that are of interest but where it is difficult to describe all of the functional dependencies. The two applications described in the following sections demonstrate how hybrid models can be used to solve problems that are not easily solved using fundamental understanding alone, but where the fundamental backbone can be exploited by combining it with neural networks.

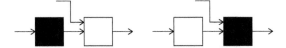

Figure 14.2 Serial hybrid modeling structures, the black box representing the parametric part, the white box representing the parametric one.

Serial Hybrid Models

Serial hybrid models are naturally evolving when only certain parts of the system can be described fundamentally, i.e. some part is not understood. The hybrid model, represented by a sequence of a white (representing the parametric/fundamental model) and black (representing the nonparametric/data-driven model) boxes in Figure 14.2, is adopted in situations where only a subpart of the behavior of the system is understood but the overall behavior is not understood. The model represented by a sequence of a black and white box, Figure 14.2, is applied in situations where the overall behavior of the system is described by a fundamental model and only parts of the mechanistic structure are unknown.

The development of these serial HM structures starts from gathering what is fundamentally understood about the system. In a second step, it should be assessed how certain one is about the correctness of the gathered knowledge, i.e. on which assumptions the gathered knowledge is based. Only those parts that are well understood and trusted should be incorporated in the beginning, to obtain an initial model that provides an "unbiased" reference against which subsequently developed/improved models can be benchmarked. Keeping in mind the objective of the modeling exercise – modeling should be a means to an end and not the final objective, i.e. modeling for a goal (Bonvin et al. 2016) – next, one needs to assess whether the available knowledge describes the overall behavior of the system one aims to describe. For the modeling of many systems, material, momentum, and energy balances provide a fundamental backbone and typically capture its overall behavior, containing small parts, such as transport, reaction, or other phenomena, yet undefined. Once one has decided upon which fundamental knowledge to incorporate, the "known" unknown parts are assessed, i.e. (i) What are their functional dependencies? and (ii) Which of those dependencies need to be captured to achieve the modeling objective? If no data are available, one should design a series of experiments in such a way as to decipher the impact of these components on the system's response. Statistical design of experiment methods seems to be the method of choice for this purpose. However, if data are available one should try to establish a first draft model and subsequently exploit it to design experiments in such a way as to improve the model (e.g. optimal experimental design), its applicability domain (e.g. focusing on a particular region of interest), or optimize the system (for more details see von Stosch (2018)).

Parallel Hybrid Models

In cases where a fundamental model is available, but its performance is inadequate for the desired application, a data-driven model can be used in parallel to adjust for the shortcomings of the fundamental model. The resulting parallel hybrid model, its schema is presented in Figure 14.3, typically shows an improved prediction performance.

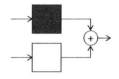

Figure 14.3 Parallel hybrid model, where the black box represents the data-driven (nonparametric) model and the white box represents the fundamental (parametric) model.

Each model can rely on a different source of information/inputs or it can describe different mechanisms. For instance, one could imagine using an empirical model for wastewater treatment in combination with a data-driven model that uses online data, such as oxygen or turbidity measurements, correcting or augmenting the prediction if the empirical model for monitoring purposes. Another example is the use of a thermodynamic model in combination with a partial least squares model for quantitative sequence-activity modeling, where each of the models accounts for a different mechanism (Portela et al. 2018).

The development of a parallel hybrid model is, therefore, very much driven by the plans for its application as well as the availability of a fundamental model. The data-driven model can easily be developed using standard techniques where the residual between the predictions of the fundamental model and the measured values are used as "measured" outputs of the data-driven model.

However, the intra/extrapolation properties of the parallel hybrid model stay limited to the domain in which one can rely on the data-driven model, i.e. its prediction performance will deteriorate for inputs that are far from those (in a multivariate sense) on which it was trained, referred to as the validity domain of the model. While methods can be used to restrict the application of the model based on the distance of a new input from the validity domain (Kahrs and Marquardt 2007; Teixeira et al. 2008; von Stosch 2018), one could alternatively use a weighting module (instead of the simple sum, shown in Figure 14.3) that gives more weight to the fundamental model in situations the hybrid model needs to extrapolate, whereas more weight to the data-driven model if one operates in the vicinity of the inputs the data-driven model had been trained on, e.g. see von Stosch et al. (2018).

Model Structure Discrimination and Parameter Identification

The development of the hybrid model comprises the determination of the data-driven model structure as well as the identification of its parameters. In addition, it may also be necessary to estimate the parameters contained in the fundamental part. As a rule of thumb, priority is given to the estimation of the parameters in the fundamental part (using standard techniques such as nonlinear least squares fitting) and only then structure discrimination and parameter identification for the data-driven model are performed. The parameters of the data-driven part can be identified in several ways, the two principle ones are the direct and indirect methods.

In the direct method, the parameters of the data-driven model are identified with the standard techniques, which requires the inputs and outputs of the data-driven model to be known/measured. Thus, where necessary the inputs and outputs need to be estimated from the available data, while accounting for the fundamental part of the hybrid model. For example, imagine the model: $y = z \cdot f(x, w)$ with y the output, z and x model inputs, w a set of parameters, and $f(\cdot)$ the functional dependence to be captured by the data-driven model. If, y, z, and x are measured (indicated by indices m in the following), one can compute $f_m = y_m/z_m$. As all inputs and outputs of the data-driven model are known, i.e. x and f_m, one can estimate w with standard techniques. However, it should be noted that the predictions of models trained with the direct method are not necessarily statistically optimal, as the mathematical transformation of the measured variables can affect the distribution of the variables.

The indirect method does not require computation of the inputs and/or outputs of the data-driven model, but instead makes direct use of the model structure for predicting the model output based on the measured inputs. Using a cost function and an optimization algorithm (such as nonlinear least squares), the parameters of the data-driven model are identified. For the example, one could e.g. minimize the cost function $\min_w \{(y_m - y(x_m, z_m, w))^2\}$. If a gradient-based optimization method is used to do this, the availability of analytical gradients (referred to as the sensitivity approach) will significantly speed up the solution of the optimization problem. For data-driven models for which special training algorithms have been developed, such as for Partial Least Squares models, a straightforward adoption of the indirect method is not always possible.

Other methods that are not only direct or indirect, have been developed for parameter identification of serial hybrid models (Kahrs and Marquardt 2008; von Stosch et al. 2014; Willis and von Stosch 2017; Yang et al. 2011).

Many data-driven models, for example neural networks, can be overfitted, meaning that structure is so flexible that from a certain moment during the optimization the model will learn the data-inherent noise. In order to avoid this, one can use methods such as early-stopping, regularization or cross-validation. Cross-validation is the most frequently applied; it uses a training dataset for the optimization, while checking at every optimization step whether the model starts learning the noise on a second, validation dataset. As it is assumed that the noise is different in all data, an increase in the model fitting error would be observed for the validation set, once the model starts overfitting. Final model performance is assessed using a test dataset, which contains data that have not been used for model development before.

Structure discrimination is typically accomplished by identifying the parameters of several models all with different structures. The performance of these structures in terms of the Akaike or Bayesian Information Criteria (see Burnham and Anderson (2004) for details) are then compared and the best performing structure is selected. Alternatively, certain models can also be aggregated, such as using the average of the predictions of the five best performing models. This strategy is recommended when the available dataset is small, reshuffling the data on which the models are developed in order to avoid bias stemming from certain data partitions.

Instead of separating parameter identification and structure identification, recently Willis and von Stosch (2017) proposed an algorithm for accomplishing these two tasks simultaneously; this is, however, limited to problems that can be transformed into a linear programming form. However, given the ever-ongoing increase in computational power as well as the development of novel optimization algorithms, one could imagine that in the future several alternative model structures (containing alternative structures in both the fundamental and data-driven part) are investigated in parallel.

14.3 Applications

In this section, two case studies are presented where the hybrid semiparametric framework has been exploited.

Case 1 : A Membrane Filtration Study

Ultrafiltration processes have demonstrated a potential for water treatment as a result of the separation performance (i.e. rejection and throughput), compact modular design, reasonable energy consumption, and the need for few chemical additives (Grisales Díaz et al. 2017). The main limitation for a wider industrial application of this technology is two coupled phenomena: concentration polarization and membrane fouling. As particles and solutes concentrate toward the membrane (concentration polarization), part of them are deposited on, or within, the membrane (fouling). As consequence, the membrane permeability and rejection are reduced over time. These phenomena are complex as they depend on the concentration and size of the particles/solutes, physicochemical interactions and operating conditions. Moreover, additional complexity rises in industrial processes as these concentrations change over time and their online monitoring is either expensive or unavailable. To reduce the adverse effect of polarization and fouling, chemical and physical cleaning strategies are usually performed using experience-based protocols. However, suboptimal scheduling of cleaning strategies represents higher investment and operational costs; therefore, there is a need to complement experimental efforts with mathematical model development. Herein, a hybrid semiparametric model of ultrafiltration is proposed to predict the time variant system performance. It is believed that this mathematical representation brings an alternative approach to determine operating conditions useful to optimize the design, operational strategy, and the process control.

Case 1: Hybrid Semiparametric Membrane Filtration Model
The membrane throughput (i.e. *Flux* [*m/s*]) in ultrafiltration is conventionally modeled using Darcy's law:

$$Flux = \frac{P \cdot TMP}{\sum Ri} = F \cdot P \cdot TMP \tag{14.1}$$

where, F (-) represents the fouling, P is the membrane permeability for the transported water divided by its viscosity (*m/KPa/s*), and *TMP* is the transmembrane pressure, which is the main driving force (*KPa*). The fouling in ultrafiltration is the inverse of the sum of all resistances ($\sum Ri$) such as the cake layer, boundary layer, gel layer, and membrane resistances (Wintgens et al. 2003). Currently, a completely knowledge-based approach that describes and predicts all the resistance dynamics in ultrafiltration is not available in the literature (Badrnezhad and Mirza 2014). However, several black box, hybrid models and neural networks have been proposed for the simulation of ultrafiltration processes (Chew et al. 2017; Prado-Rubio and von Stosch 2017). For example, Corbatón-Báguena et al. (2016) found that neural networks have a similar performance to a knowledge-based approach that partially describes the resistances in ultrafiltration (Hermia's model). Prado-Rubio and von Stosch (2017) proposed a methodology combining a data reconciliation strategy and system identification techniques to investigate an industrial ultrafiltration application. Their results demonstrated that a Multiple Input–Single Output (MISO) model can predict system behavior despite the high uncertainty in the input data. Grisales Díaz et al. (2017) developed a hybrid model combining Darcy's law and a neural network to describe the impact of the inputs on the membrane performance. Interestingly, the proposed model could predict the behavior of the ultrafiltration process with and without cleaning strategies; therefore, it is suitable to optimize ultrafiltration processes under several strategies of operation. Chew et al. (2017) proposed a model estimating two resistances in the Darcy's law, the specific cake resistance and the membrane resistance, also achieving high prediction accuracy whereas Chen and Kim (2006) found that a radial basis function neural network was the best model type for the simulation of ultrafiltration system.

Here, a hybrid model is proposed where only the parameter $F \cdot P$ is predicted by a neural network approach and the flux is calculated using the deterministic model represented by Eq. (14.1).

Case 1: Dataset

The dataset used in this application was reported by Chen and Kim (2006). It includes crossflow ultrafiltration experiments carried out considering different operational conditions such as: particle size (PS), ionic strength (IS), transmembrane pressures (TMP), operational time (t), and pH. This dataset is interesting since it has been studied in other applications, e.g. to test a genetic algorithm (GA) as the training method of for a neural network (Sahoo and Ray 2006).

Case 1: Model Structure, Training, and Validation

The inputs for the neural network approach were PS, IS, TMP, t, and pH. The data and the output of the neural network were normalized between 0 and 1 in order to have a more efficient training stage. The output of the neural network was the parameter $F \cdot P$ and, subsequently, the flux was estimated using Eq. (14.1), i.e. a serial hybrid model was developed.

Conventionally, 70% of the data are used for model training, 15% for testing, and 15% for validation. However, the neural networks studied previously in the literature (Chen and Kim 2006; Sahoo and Ray 2006) were developed using a small portion of the dataset (96 data points, corresponding to 17% of all data). Hence, to allow a fair comparison to be made with previous work, the same amount of data was used in the training (weight and bias estimation), while 33% (187 data points) was used for validation and the remaining 50% used for model testing. In order to compare the results with those reported by Chen and Kim (2006) and Sahoo and Ray (2006), the quality of the neural network predictions was assessed using the correlation coefficient (R) and the N10 index (the number of data points with an absolute deviation error higher than 10%).

The neural network structure used in this case study is the so called feed-forward ANN. In contrast with the neural networks reported by Chen and Kim (2006) and Sahoo and Ray (2006) which use two hidden layers, the model structure used in this work employs one hidden layer. The functions used in the inner and the output layers were the MATLAB® functions "tangsin" and "linear," respectively, as these functions had been reported as being optimal by Sahoo and Ray (2006). The default MATLAB function "train," the Levenberg–Marquardt optimization algorithm, was used to obtain the neural network weights.

The neural networks were trained 5000 times using random initial weight estimates in an attempt to achieve a globally optimum model. In addition, the minimum number of trained neural networks required to find an accurate predictive model was also studied. To do this, the 5000 trained neural networks were divided several times in different subgroups with the same number of trained neural networks. The number of trained neural networks in each subgroup was reduced several times from the greatest common divisor to the lowest common divisor of 5000. In this way, when the subgroups have 500 trained neural networks there are 10 subgroups or if each subgroup has 40 trained neural networks there are 125 subgroups, and so on. The neural network that gives the highest correlation coefficient was selected and the standard deviation of this correlation coefficient was calculated for the subgroups with the same number of trained neural networks.

Case 1: Results and Discussion
It has been reported that a GA is the best optimization method for neural network training using this dataset (Sahoo and Ray 2006). However, using this dataset the best correlation coefficient reported using GA to train a neural network (Sahoo and Ray 2006) is similar to that achieved by Chen and Kim (2006) using the Levenberg–Marquardt method (Table 14.1). In this work, the correlation coefficient reported by Chen and Kim (2006) was improved from 0.9936 (Chen and Kim 2006) to 0.9952 and the N10 index was reduced from 14 to 8 using the hybrid model and only one hidden layer with seven neurons (Table 14.1).

A parity diagram of the actual and predicted flux using the hybrid model is shown in Figure 14.4. Practically all the estimated membrane fluxes lie within a 10% prediction bound, demonstrating high prediction accuracy which is required for subsequent process optimization studies.

Table 14.1 Performance efficiency of hybrid model used for the ultrafiltration process case study.

Hidden-layer neurons		Performance efficiency		Hybrid	Reference
First layer	Second layer	R	N10		
4	—	0.9904	19	Yes	In this work
5	—	0.9934	11	Yes	
6	—	0.9947	8	Yes	
7	—	0.9952	8	Yes	
8	—	0.9958	9	Yes	
9	—	0.9950	11	Yes	
10	—	0.9955	12	Yes	
4	2	0.9947	7	Yes	
4	2	0.9940	14	No	Sahoo and Ray (2006)
5	5	0.9936	—	No	Chen and Kim (2006)

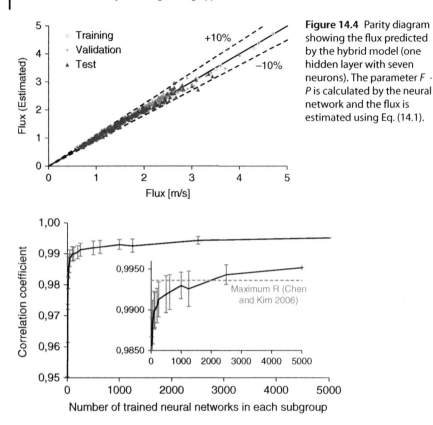

Figure 14.4 Parity diagram showing the flux predicted by the hybrid model (one hidden layer with seven neurons). The parameter $F \cdot P$ is calculated by the neural network and the flux is estimated using Eq. (14.1).

Figure 14.5 Correlation coefficient (R) as a function of the number of trained neural networks.

With regards to computational effort, at least 2500 trained neural networks were required to obtain a higher correlation coefficient than that reported by Chen and Kim (2006) (Figure 14.5). Although each optimization problem required a different number of trained neural networks to find the best structure and associated weights, in this case study at least 500 training iterations using random initial values of the weights were required in order to achieve the reported results (Figure 14.5).

Case 1: Summary
In this case study, the potential of using a hybrid model to predict membrane throughput in ultrafiltration has been investigated and compared to purely data-driven approaches reported in the literature. As a result, it has been shown that combining deterministic knowledge and a simple neural network structure gave a slightly better predictive performance than that previously reported by Chen and Kim (2006) and Sahoo and Ray (2006). It was found

that only one hidden layer (neurons between 6 and 8) was required to achieve accurate predictions. In addition, we found that the neural network models must be trained around 500 times to obtain the best possible structure and weights, and hence predictive performance. However, once trained, through exploiting the structure of the hybrid model it is possible to use the model to accurately predict *TMP*. This is particularly relevant in order to determine, for instance, optimal operating conditions and adaptive cleaning strategies and scheduling.

Case 2 : A Controlled Drug Release Study

One of the major challenges for achieving a controlled drug release (CDR) is the initial burst release. With the ultimate aim to control the amount of drug released during a burst from poly(lactic-co-glycolic) acid (PLGA) micro- and nanoparticles, in this section a hybrid modeling approach is presented that describes the drug release as function of the formulation characteristics, synthesis parameters, and drug molecule descriptors. The hybrid model combines the Corrigan model (Corrigan and Li 2009) with ANNs describing the changes in its parameters as functions of the designs parameters and drug. The model was developed using 132 release profiles and its predictive capability tested on a set of data not used for model development.

Case 2: Background

Substantial drug release from PLGA micro- and nanoparticles can occur during the first days of immersion, which is referred to as burst release (Batycky et al. 1997; Brazel 2000; Brazel and Peppas 1999; Esmaeili et al. 2008; Frank et al. 2004; Kim et al. 2002; Mao et al. 2007). Burst release has been attributed to a rapid desorption of the drug located on or near the surface of particles and to poorly encapsulated drug, which rapidly diffuses immediately upon immersion into a biological fluid. Typically the burst release is an unwanted event, since the high initial drug concentrations can have negative effects on the host (reviewed by Huang and Brazel (2001)). An unwanted burst release usually leads to a shorter drug release time and obstructs the cornerstones of CDR: predictability and reproducibility. Although a widely documented phenomenon (Guimaraes et al. 2015; Hasan et al. 2007; Hickey et al. 2002; Huang and Brazel 2001; Ito et al. 2008; Jeong et al. 2008; Li et al. 2001; Makino et al. 2000; Niwa et al. 1993; Xu et al. 2009) few of them report approaches how to prevent or control burst.

Mathematical modeling has been used to describe and predict the controlled release profile of specific drugs. Several researchers have derived modeling approaches based on a mechanistic or phenomenological understanding of drug release from micro- and nanoparticles (Cabrera and Grau 2007; Faisant et al. 2002; Goepferich 1996; Grassi and Grassi 2005; Grassi et al. 2011; Klose et al. 2008; Siepmann and Göpferich 2001; Siepmann and Siepmann 2008;

Juergen Siepmann et al. 2002). However, the experimental conditions and drug properties, which have a significant impact on the drug release, are rarely taken into account by these models. In some models, the mean particle size is included but the carrier formulation (e.g. monomers ratio, surfactants used during particle synthesis), which has a significant impact on the drug release, is not considered. Thus, the analysis of the impact of experimental conditions or drug properties on the drug release is not possible utilizing these models.

Recently, data-driven methodologies have been applied to model CDR from tablets (Matero et al. 2008) and from PLGA microparticles (Zawbaa et al. 2016; Szlek 2013) aiming at the correlation of the release profile with physicochemical properties of the carrier and/or drug. The utilized data-driven methods lack mechanistic interpretability and, hence, are often regarded as black-box methods. Thus, due to the black-box nature of the models little understanding is provided in relation to how the formulation characteristics or drug properties impact on the drug release. Also, these models completely disregard the knowledge formulated in the established drug release models.

In this work, the burst release from PLGA particles is modeled utilizing a HM approach, which combines the burst release model proposed by Corrigan and Li (2009) with ANNs describing the changes of parameters in function of synthesis, particle design, and the drug. PLGA micro- and nanoparticles were chosen as controlled delivery carrier in this work due to their widespread use (Danhier et al. 2012). Particular features are that: (i) PLGA is biodegradable and biocompatible; (ii) it allows for a fine-tuning of its properties and consequent degradation rate; and (iii) it is FDA/EMA (Food & Drug Administration/European Medicines Agency) approved in a range of several PLGA-based drug products (Wang et al. 2016). A thorough review on modeling of CDR from PLGA microparticles can be found in Versypt et al. (2013).

Case 2 : A Hybrid Semiparametric Drug Release Model

The model proposed by Corrigan and Li (2009) (Gallagher and Corrigan 2000; Langenbucher 1972) contains a term that explicitly describes the drug release during the burst phase. Due to its ability to describe a wide range of drug release profiles as well as its similarity to other models (Batycky et al. 1997; Donaldson et al. 2013; Langenbucher 1972), this model and, in particular, the term describing the drug release during the burst phase, was chosen as the parametric backbone of the hybrid model:

$$Q = F_{B,in} \cdot (1 - \exp(-k_b \cdot t)) \tag{14.2}$$

where Q is the total fraction of drug released at a given time t, (a value between 0 and 1), $F_{B,in}$ is the fractional amount of the drug released during the burst, and k_b is a first order rate constant associated with the kinetics of the burst release. The values of $F_{B,in}$ and k_b will change when varying the drug, the formulation characteristics or the experimental conditions; therefore, the aim is to use data-driven models to describe these dependencies.

Two ANNs are used to model $F_{B,in}$ and k_b as functions of the drug (represented by molecular descriptors), the formulation characteristics, and the experimental conditions. The structure of the ANNs has to be determined from data, therefore they belong to the class of nonparametric models. Here, ANNs with three layers – an input, output, and hidden layer – were used, as three layers typically suffice to model complex nonlinear functions. Linear transfer functions were chosen for the nodes in the input and output layers, whereas hyperbolic tangential transfer functions were chosen for the hidden layers.

Case 2: Dataset

The dataset from Szlek (2013) was curated and extended with data from literature, in total comprising data of 152 *in vitro* CDR experiments. The 152 experiments comprise 41 different active pharmaceutical ingredients (in this study for simplicity referred to as "drug") that have wide therapeutic applications, as described in more detail in Rodrigues de Azevedo et al. (2017). The dataset encompasses: (i) the cumulative drug release profiles over time (extracted from articles where possible either directly or using the image recognition software Plotdigitizer [version 2.6.8]); (ii) 50 molecular descriptors for each drug (calculated with the ChemAxon plugin from Marvin [v5.2.1]); and (iii) formulation characteristics of the drug-carrier synthesis, which were extracted from the protocols of the drug-loaded PLGA particles preparation methods (see Rodrigues de Azevedo et al. (2017) for more details).

Case 2: Input Selection

The input variables for the $F_{B,in}$ and k_b neural networks were selected from the 74 available variables, choosing the eight most significant variables by feature selection. For $F_{B,in}$ the inputs are: (i) aliphatic atom count of the drug molecule, (ii) PVA (poly(vinyl alcohol)) Mw, (iii) PVA concentration on the outer phase, (iv) encapsulation efficiency, (v) particle size, D, (vi) Balaban index, (vii) PLGA, and (viii) initial drug loading.

For k_b the inputs comprise (i) the encapsulation efficiency, (ii) PLGA Mw, (iii) number of oxygen atoms on the drug molecule, (iv) $1/D^2$, (v) PVA concentration on the inner phase, (vi) isoelectric point, (vii) PLGA to PEG (poly(ethylene glycol)) ratio, and (viii) PVA Mw.

In theory, more input variables could be used if more drug profile data were available. However, the eight selected inputs (in each case) yield a good performing model while not overparameterizing the system, i.e. with the maximum tested number of six nodes in the hidden layer, the parameter/data points ratio was greater than 0.5.

Case 2: Data Partitions

One part of the data, 20 profiles, was allocated as the test partition. Since the number of the remaining drug profiles was relatively low, a bootstrap aggregated modeling strategy was adopted (Breiman 1996). Thus, the remaining data were randomly partitioned ten times into a training (80% of the points) and a validation (20% of the points) partition, i.e. 10 training-validation datasets were prepared. For each set, the neural networks were trained on the training partition and the validation partition was used to stop the training, i.e. cross-validation. The test partition was then used to evaluate the performance of the final aggregated model, the development of which is described next.

Case 2: Parameter Identification and Structure Discrimination

Parameter Identification A two-level identification procedure was used for determining the neural networks weights.

Level 1. At first, the parameter values of $F_{B,in}$ and k_b were estimated for the drug profile contained in the training-validation set using the MATLAB function "lsqnonlin," which uses the Levenberg–Marquardt method for minimization of a nonlinear least squares objective function. In order to estimate the confidence intervals of the parameters, Monte Carlo sampling (100 repetitions) was used on the experimental data assuming standard error of 2.5% of the experimental value. It is considered that no drug was released at the beginning of the experiment, $Q(t = 0) = 0$.

Level 2. Once the $F_{B,in}$ and k_b values had been estimated they were used to determine the weights of the neural networks. For $F_{B,in}$ the values were auto-scaled, using the mean and standard deviation values of the training-validation partition (which are identical for all datasets). The auto-scaled values of the training partitions were then used to minimize a least square function employing a Levenberg–Marquardt method, i.e. the MATLAB function "lsqnonlin." For each training partition the weight values of one neural network with varying number of hidden layers were identified. For k_b the natural logarithm of the values was calculated first to account for the significant differences in magnitude. The log-treated values were then auto-scaled, again using the mean and standard deviation values of the training-validation partitions. Then, the same procedure as for $F_{B,in}$ was followed. The training was stopped in both cases when the mean-squared error in the validation partitions did not decrease further. The identification for both $F_{B,in}$ and k_b was started 400 times from random weight values to avoid getting stuck in local minima, and the best performing model (in terms of Mean Squared Error (MSE) calculated for the validation set) was chosen.

Model Structure Discrimination

For each of the 10 sets and for both neural networks the number of nodes in the hidden layer was varied between one and six, and for each structure the best performing parameter set was chosen. Thus, six networks with varying number of hidden nodes were obtained for each parameter $F_{B,in}$ and k_b and for each of the 10 sets. The best performing structure for each parameter and each set was chosen using the Akaike Information Criterion (*AI*) (Burnham and Anderson 2004), described below. For each set, the best performing neural networks for $F_{B,in}$ and k_b were then used together with the parametric model to calculate the drug profile Q and compare its fit with the experimental drug profiles. The HM calculated drug profiles Q for all 10 sets were then averaged to establish the final aggregated model.

Criterion for Model Performance

Model performance criteria are used to assess the performance of the neural networks and the aggregated hybrid model. In terms of cost function minimization, several authors have adopted the Akaike Information Criterion (*AIC*) for the selection of the hybrid model structure. The *AIC* is based on the Kullback–Leibler (K-L) information loss, which may be conceptualized as a "distance" between the true model and the approximating candidate model (Burnham and Anderson 2004). In practice, the *AIC* with second order bias correction (*AICc*) is preferred over *AIC*, as *AIC* tends to select models that overfit for small samples while *AICc* is valid for both small and high number of samples (*n*). The *AICc* takes the simplified form:

$$AIC_c = n \cdot \ln(MSE) + 2 \cdot k + \frac{2 \cdot k \cdot (k+1)}{n - k - 1} \tag{14.3}$$

where k is the number of estimated parameters in the model and n is the total number of data points (Zhao et al. 2008).

The mean squared error (*MSE*) is a widely used qualitative measure to evaluate the fit of the model predictions with the experimental data. Its calculation is based on the average squared distance between the modeled and the measured values, i.e.:

$$MSE = \frac{1}{n} \sum_{i=1}^{n} (y_i - y_{meas,i})^2 \tag{14.4}$$

where the index *meas* refers to the measured values and y_i represents either the values of $F_{B,in}$ or k_b in case of the individual ANNs, or the cumulative drug release Q when analyzing the regression quality of the HM.

The aim is to find the minimum value of *AICc*, i.e. the most parsimonious model structure.

Table 14.2 Selection of the best performing network structures for $F_{B,in}$ and k_b in terms of AICc calculated for the training-validation data for each of the 10 partitions. The MSE performance is also shown.

No. of Set	Number of hidden nodes	MSE, mean	AICc
1 $F_{B,in}$	5	0.0101	−2791
1 k_b	4	0.0045	−3320
2 $F_{B,in}$	5	0.0089	−2871
2 k_b	6	0.0022	−3821
3 $F_{B,in}$	6	0.0081	−2908
3 k_b	4	0.0027	−3623
4 $F_{B,in}$	3	0.0134	−2767
4 k_b	4	0.0038	−3447
5 $F_{B,in}$	4	0.0122	−2683
5 k_b	4	0.0028	−3619
6 $F_{B,in}$	4	0.0116	−2704
6 k_b	4	0.0036	−3461
7 $F_{B,in}$	4	0.0118	−2709
7 k_b	4	0.0029	−3623
8 $F_{B,in}$	4	0.0100	−2818
8 k_b	5	0.0033	−3576
9 $F_{B,in}$	5	0.0076	−2670
9 k_b	6	0.0027	−3766
10 $F_{B,in}$	4	0.0093	−2805
10 k_b	4	0.0034	−3487

Case 2: Results and Discussion

Model Structure Discrimination The neural networks for each ten datasets were individually trained for $F_{B,in}$ and k_b. The number of nodes in the hidden layer of the neural networks, which minimize the *AICc* for each set, are shown in Table 14.2. Across the ten sets the performance of the HMs were similar, i.e. the number of hidden nodes that minimize the *AICc* varies between three and six, with four being the most frequent. For both $F_{B,in}$ and k_b the performance of the networks in terms of MSE and *AICc* are comparable.

Bootstrap Aggregated Hybrid Model Performance

Using the identified ANNs for $F_{B,in}$ and k_b in conjunction with the Corrigan equation, a hybrid model is obtained. The values of Q obtained by each of the ten HMs (with the ANNs structure specified in Table 14.2) were aggregated.

Figure 14.6 Aggregated HM simulated versus measured values of cumulative drug release, Q. Red crosses: training set, green circles: test set.

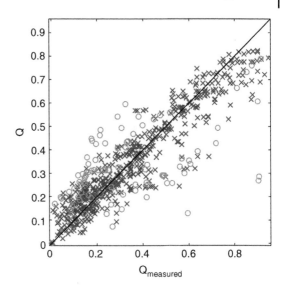

The modeled cumulative release values against the experimental are shown in Figure 14.6. It can be seen that the model generally fits the measured values well for both the training-validation and test set. The obtained MSE values for the training-validation and test sets are 0.005 and 0.037, respectively. The slightly greater MSE value of the test set is mostly due to some outlier points that can be seen at the lower right side of Figure 14.6. These points describe bovine serum albumin (BSA) release experiments and reasons for the inferior model performance might be that BSA is a relatively large molecule (66.5 kDa), wherefore different mechanisms might occur in the experiment than in the ones used for creating the models. In future, it would be useful to not only report findings in the form of articles but also to share the data and adopted experimental methods in a public database. This could significantly improve the quality and availability of data for the creation of predictive CDR models, as it would become e.g. possible to include more characteristics and descriptors in the models.

Analysis Example of the Impact of Particle Synthesis Parameters on Drug Release Profiles

A representative example of simulated and measured cumulative drug release over time is shown in Figure 14.7 for the case of ellagic acid (Bala 2006). The red profiles are from the training-validation set, while the green are from the test set. Different concentrations of PVA for particle synthesis, as well as nanoparticles of different size, were used in the three experiments (the inset table in Figure 14.7 gives details). It can be seen from Figure 14.7 that the agreement between simulated and experimental values is good. In particular, the variations

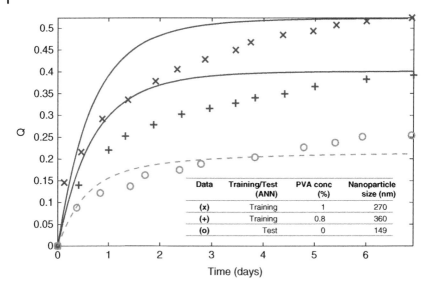

Figure 14.7 Comparison of predicted (continuous lines) and experimental values (markers) of cumulative release of ellagic acid over time (Bala 2006). Data shown in red are part of the training set and data shown in green are part of the test set.

in the release profiles, which are due to variations in nanoparticle size and PVA concentration, are well described by the model and consistent with the observations from the experiments. This is a good indicator that the application of the model for particle design and formulation is possible.

Case 2: Summary
In this section, drug release profiles are modeled via the Corrigan equation in which changes in its parameters associated with the total amount of drug released during burst and the burst kinetics are described by neural networks. The ANNs were derived as functions of the synthesis parameters and molecular descriptors of the drug. A bootstrap aggregating identification strategy was used for the development of the hybrid model. After development, the model was used to predict the cumulative drug release of an independent set of CDR experiments. Good agreement between the predicted and the experimentally measured cumulative drug release profiles was observed. On an example it was shown that the HM qualitatively describes the impact of changes in synthesis parameters on the release profile of ellagic acid. The use of more data and, in particular, more qualitative data could increase the performance of the model in the future. Sharing of the experimental data and accurate reporting of the associated methods in a database could, therefore, greatly help to exploit the full capabilities of modeling tools. All in all, the performance of the developed hybrid model is acceptable and it can be used for optimal drug carrier design by manipulation of synthesis parameters and formulation characteristics.

14.4 Summary

In this chapter, a HM concept for the integration of fundamental knowledge in terms of parametric models with nonparametric models that are developed from data was discussed. The concept was detailed for (i) a case study that considered process specific benefits and (ii) a case study that considered the benefits of using the same hybrid model across several processes with different products. In what follows we want to explore the implications, trying to envision what might become possible.

Process Focus

In the membrane filtration study, online data (short time interval measurements that become available fast enough to be used for process control) are available and the integration of fundamental knowledge allowed a better understanding of the process that may extrapolate more precisely the pressure beyond tested ranges. The fundamental knowledge may be the key to the described benefits. In addition, systematically exploring the fundamental knowledge and accounting for it in the design of experiments during process development studies, can reduce the number of experiments that are required for model development as compared to pure data-driven models. In the membrane filtration example, the impact of variations in initial conditions and transmembrane pressure are explicitly captured by the model and would not have to be tested experimentally. Explicitly accounting for the process dynamics, e.g. by using the dynamic material balances, offers the possibility to explore and exploit the temporal domain. From the design point of view, where design parameters and operating conditions should be estimated experimentally, this approach brings new perspectives to reduce experimental efforts. From operational point of view, the impact of process parameters that can be varied during operation can thus be assessed during the process run, rather than keeping them at a constant level as is conventional in the present design of experiment studies (Cruz Bournazou et al. 2016; von Stosch and Willis 2017). This is especially interesting as it would allow a move away from an offline step-wise unautomated optimization of the process (i.e. design of experiments, execution of experiments, analysis of the results, update of experimental design, proposal of optimal process conditions, experimental verification of the optimal conditions), to online, on the fly, self-optimizing process operation and control. Adopting such methodologies in a manufacturing process provides the means to adapt the process via operating conditions and cleaning strategies to changes in the raw materials, changes in the material/energy cost structure or changes on product quality/quantity demand, hence allowing for process adaptability as sought in Industry 4.0.

Another benefit, not yet mentioned and many times overlooked, is that the integration of fundamental knowledge can help to establish links between the data of different sensors, generating redundancies. These redundancies can be exploited to reduce the impact of measurement noise on the system identification and it can make the online application of the model more robust (if one sensor fails one still has the other one). Additionally, redundancy contributes to the implementation of diagnosis and fault detection strategies for a structured and systematic examination of the process operation contributing to risk assessment strategies. Although powerful, applying only data-driven techniques is not straightforward since the correlations between sensors are often nonlinear and upon detection one would have to investigate whether they are causal or just spurious. Another possibility would be to exploit the fundamental knowledge to reduce the number of sensors that are installed on the plant(s) or to reveal the best locations for their application. One could even keep a certain level of redundancy to account for potential sensor failures.

All in all, it should have become obvious that the integration of fundamental knowledge can bring significant advantages for its application in projects or manufacturing. While the fundamental knowledge could be transferred to a new project (new production process) it would be necessary to redevelop the data-driven model, which might also require the generation of data. A complete in silico design of a new plant or process would, therefore, not be possible using the model as presented. However, certain adaptations to the inputs of the data-driven model can be performed to extend its applicability, as discussed next.

Across Processes

The CDR study showed how fundamental knowledge about the drug release could be transferred from one drug to the next (i.e. across projects), the parameters in the fundamental model being given by neural networks that account for the drug, carrier, and formulation characteristics. In contrast to the first case study the hybrid model remains valid when changing the drug (product), i.e. tantamount to staying valid across "processes," and it can be updated/improved (both the parametric and nonparametric part) whenever new data become available. Thus, these modeling activities would greatly benefit from a central database, as mentioned before. Given that data management/storage solutions along with the recording of workflows (context) by use of robots and automation is at the base of Industry 4.0 activities, exploiting the accumulated data in alike modeling approaches has the potential to change the way we design complex process in the future.

Given the vast amount of data that will be produced one could ask why we would want to develop hybrid models and not stick to data-driven models only. There exist at least two good reasons in the (bio)pharma industry:

1) The number of factors that can be studied is so great that there will never be sufficient data to decipher the impact of each factor on the systems response

(also referred to as the curse of dimensionality). Integrating fundamental knowledge can structure the solution space, therefore reducing dimensionality, allowing to benefit better from the available data.

2) Experiments that are informative are those that have variations in the factors (which are typically those performed in R&D and not those of manufacturing). It is extremely expensive and time intensive to perform these experiments; thus, a potential reduction in its numbers might become decisive for the success of a company.

At this moment, the focus of Industry 4.0 concepts presented in public seems very data centric. In this chapter, the benefits of using more than one knowledge source (i.e. data) were detailed. These benefits seem to warrant a reflection on the current data-centric ideas, in particular it needs to be addressed how to bring in, integrate, manage, and query knowledge sources other than data and how to enrich/crystallize the learning/understanding in this knowledge types, therefore e.g. escaping from data storage limitations. In fact, we could even go so far as to ask us what the added value of storing all that data is, knowing that data have the lowest knowledge abstraction level and that more data do not imply more knowledge. Another point worth considering is that data-driven approaches to date only have inductive capabilities, implying that these models will not come up with completely new ways of designing or operating the plant, whereas this might be the case when considering fundamental knowledge.

Acknowledgments

The work on the membrane filtration study was financially supported by Newton Fund Institutional links program through the grant ID-216424051, a collaboration between Newcastle University and Universidad Nacional de Colombia. The work on the controlled drug release study was financially supported by FCT-MCES, Portugal, through grant SFRH/BD/17224/2004 and contracts POCTI/EQU/46715/2002 from the QCA III/FEDER and PTDC/EQU-EPR/119631/2010. CDA acknowledges financial support through the Wolf Schleinzer Stiftung.

References

Badrnezhad, R. and Mirza, B. (2014). Modeling and optimization of cross-flow ultrafiltration using hybrid neural network-genetic algorithm approach. *Journal of Industrial and Engineering Chemistry* 20 (2): 528–543. https://doi.org/10.1016/j.jiec.2013.05.012.

Bala, I. (2006). Sustained release nanoparticulate formulation containing antioxidant-ellagic acid as potential prophylaxis system for oral administration. *Journal of Drug Targeting* 14 (1): 27–34.

Batycky, R.P., Hanes, J., Langer, R., and Edwards, D.A. (1997). A theoretical model of erosion and macromolecular drug release from biodegrading microspheres. *Journal of Pharmaceutical Sciences* 86 (12): 1464–1477. https://doi.org/10.1021/js9604117.

Bonvin, D., Georgakis, C., Pantelides, C.C. et al. (2016). Linking models and experiments. *Industrial and Engineering Chemistry Research* 55 (25): 6891–6903.

Brazel, C. (2000). Modeling of drug release from swellable polymers. *European Journal of Pharmaceutics and Biopharmaceutics* 49: 47–58.

Brazel, C.S. and Peppas, N.A. (1999). Mechanisms of solute and drug transport in relaxing, swellable, hydrophilic glassy polymers. *Polymer* 40 (12): 3383–3398.

Breiman, L. (1996). Bagging predictors. *Machine Learning* 24 (2): 123–140. https://doi.org/10.1007/BF00058655.

Bühlmann, P. and van de Geer, S. (2018). Statistics for big data: a perspective. *Statistics and Probability Letters* 136: 37–41. https://doi.org/10.1016/j.spl.2018.02.016.

Burnham, K.P. and Anderson, D.R. (2004). Multimodel inference. *Sociological Methods and Research* 33 (2): 261–304. https://doi.org/10.1177/0049124104268644.

Cabrera, M.I. and Grau, R.J.A. (2007). A generalized integral method for solving the design equations of dissolution/diffusion-controlled drug release from planar, cylindrical and spherical matrix devices. *Journal of Membrane Science* 293: 1–14.

Chen, H. and Kim, A.S. (2006). Prediction of permeate flux decline in crossflow membrane filtration of colloidal suspension: a radial basis function neural network approach. *Desalination* 192 (1–3): 415–428. https://doi.org/10.1016/j.desal.2005.07.045.

Chew, C.M., Aroua, M.K., and Hussain, M.A. (2017). A practical hybrid modelling approach for the prediction of potential fouling parameters in ultrafiltration membrane water treatment plant. *Journal of Industrial and Engineering Chemistry* 45: 145–155. https://doi.org/10.1016/j.jiec.2016.09.017.

Corbatón-Báguena, M.J., Vincent-Vela, M.C., Gozálvez-Zafrilla, J.M. et al. (2016). Comparison between artificial neural networks and Hermia's models to assess ultrafiltration performance. *Separation and Purification Technology* 170: 434–444. https://doi.org/10.1016/j.seppur.2016.07.007.

Corrigan, O.I. and Li, X. (2009). Quantifying drug release from PLGA nanoparticulates. *European Journal of Pharmaceutical Sciences* 37 (3): 477–485. https://doi.org/10.1016/j.ejps.2009.04.004.

Cruz Bournazou, M.N., Barz, T., Nickel, D.B. et al. (2016). Online optimal experimental re-design in robotic parallel fed-batch cultivation facilities.

Biotechnology and Bioengineering 114 (3): 610–619. https://doi.org/10.1002/bit
.26192.

Danhier, F., Ansorena, E., Silva, J.M. et al. (2012). PLGA-based nanoparticles: an
overview of biomedical applications. *Journal of Controlled Release* 161 (2):
505–522.

Donaldson, O., Huang, Z.J., and Comolli, N. (2013). An integrated experimental
and modeling approach to propose biotinylated PLGA microparticles as
versatile targeting vehicles for drug delivery. *Progress in Biomaterials* https://
doi.org/10.1186/2194-0517-2-3.

Esmaeili, F., Atyabi, F., and Dinarvand, R. (2008). Preparation and characterization
of estradiol-loaded PLGA nanoparticles using homogenization-solvent
diffusion method. *DARU Journal of Pharmaceutical Sciences* 16 (4): 196–202.

Faisant, N., Siepmann, J., and Benoit, J.P. (2002). PLGA-based microparticles:
elucidation of mechanisms and a new, simple mathematical model quantifying
drug release. *European Journal of Pharmaceutical Sciences* 15 (4): 355–366.

Frank, A., Rath, S.K., Boey, F., and Venkatraman, S. (2004). Study of the initial
stages of drug release from a degradable matrix of poly(d,l-lactide-co-glycolide).
Biomaterials 25 (5): 813–821. https://doi.org/10.1016/S0142-9612(03)00597-0.

Gallagher, K.M. and Corrigan, O.I. (2000). Mechanistic aspects of the release of
levamisole hydrochloride from biodegradable polymers. *Journal of Controlled
Release* 69: 261–272.

García-Muñoz, S., Luciani, C.V., Vaidyaraman, S., and Seibert, K.D. (2015).
Definition of design spaces using mechanistic models and geometric
projections of probability maps. *Organic Process Research and Development* 19
(8): 1012–1023. https://doi.org/10.1021/acs.oprd.5b00158.

Goepferich, A. (1996). Mechanisms of polymer degradation and erosion.
Biomaterials 17 (2): 103–114. https://doi.org/10.1016/0142-9612(96)85755-3.

Grassi, M. and Grassi, G. (2005). Mathematical modelling and controlled drug
delivery: matrix systems. *Current Drug Delivery* 2: 97–116. Retrieved from
http://www.benthamscience.com/cdd/sample/cdd2-1/010AP.pdf.

Grassi, M., Lamberti, G., Cascone, S., and Grassi, G. (2011). Mathematical
modeling of simultaneous drug release and in vivo absorption. *International
Journal of Pharmaceutics* 418 (1): 130–141. Retrieved from http://www
.sciencedirect.com/science/article/pii/S0378517311000275.

Grisales Díaz, V.H., Prado-Rubio, O.A., Willis, M.J., and von Stosch, M. (2017).
Dynamic hybrid model for ultrafiltration membrane processes. In: *Proceedings
of the 27th European Symposium on Computer Aided Process Engineering*,
193–198. Elsevier https://doi.org/10.1016/B978-0-444-63965-3.50034-9.

Guimaraes, P.P.G., Oliveira, M.F., Gomes, A.D.M. et al. (2015). PLGA nanofibers
improves the antitumoral effect of daunorubicin. *Colloids and Surfaces B:
Biointerfaces* 136: 248–255.

Hasan, A.S., Socha, M., Lamprecht, A. et al. (2007). Effect of the microencapsulation of nanoparticles on the reduction of burst release. *International Journal of Pharmaceutics* 344 (1–2): 53–61.

Haykyn, S. (1998). *Neural Networks: A Comprehensive Foundation*, vol. 2. Prentice Hall.

Hickey, T., Kreutzer, D., Burgess, D.J., and Moussy, F. (2002). Dexamethasone/PLGA microspheres for continuous delivery of an anti-inflammatory drug for implantable medical devices. *Biomaterials* 23 (7): 1649–1656.

Zawbaa, H.M., Szlek, J., Grosan, C. et al. (2016). Computational intelligence modeling of the macromolecules release from PLGA microspheres – focus on feature selection. *PLoS One* https://doi.org/10.1371/journal.pone.0157610.

Huang, X. and Brazel, C.S. (2001). On the importance and mechanisms of burst release in matrix-controlled drug delivery systems. *Journal of Controlled Release* 73: 121–136.

Ito, F., Honnami, H., Kawakami, H. et al. (2008). Preparation and properties of PLGA microspheres containing hydrophilic drugs by the SPG (Shirasu porous glass) membrane emulsification technique. *Colloids and Surfaces B: Biointerfaces* 67 (1): 20–25. https://doi.org/10.1016/j.colsurfb.2008.07.008.

Jeong, Y.-I., Na, H.-S., Seo, D.-H. et al. (2008). Ciprofloxacin-encapsulated poly(dl-lactide-co-glycolide) nanoparticles and its antibacterial activity. *International Journal of Pharmaceutics* 352 (1–2): 317–323.

Kahrs, O. and Marquardt, W. (2007). The validity domain of hybrid models and its application in process optimization. *Chemical Engineering and Processing: Process Intensification* 46 (11): 1054–1066. https://doi.org/10.1016/j.cep.2007.02.031.

Kahrs, O. and Marquardt, W. (2008). Incremental identification of hybrid process models. *Computers and Chemical Engineering* 32 (4): 694–705. https://doi.org/10.1016/j.compchemeng.2007.02.014.

Kim, T.H., Lee, H., and Park, T.G. (2002). Pegylated recombinant human epidermal growth factor (rhEGF) for sustained release from biodegradable PLGA microspheres. *Biomaterials* 23 (11): 2311–2317. https://doi.org/10.1016/S0142-9612(01)00365-9.

Klose, D., Siepmann, F., Elkharraz, K., and Siepmann, J. (2008). PLGA-based drug delivery systems: importance of the type of drug and device geometry. *International Journal of Pharmaceutics* 354: 95–103. Retrieved from http://www.sciencedirect.com/science/article/pii/S0378517307008630.

Langenbucher, F. (1972). Letters to the editor: linearization of dissolution rate curves by the Weibull distribution. *Journal of Pharmacy and Pharmacology* 24 (12): 979–981. https://doi.org/10.1111/j.2042-7158.1972.tb08930.x.

Li, Y.-P., Pei, Y.-Y., Zhang, X.-Y. et al. (2001). PEGylated PLGA nanoparticles as protein carriers: synthesis, preparation and biodistribution in rats. *Journal of Controlled Release* 71 (2): 203–211.

Makino, K., Mogi, T., Ohtake, N. et al. (2000). Pulsatile drug release from poly (lactide-co-glycolide) microspheres: how does the composition of the polymer matrices affect the time interval between the initial burst and the pulsatile release of drugs? *Colloids and Surfaces B: Biointerfaces* 19 (2): 173–179.

Mao, S., Xu, J., Cai, C. et al. (2007). Effect of WOW process parameters on morphology and burst release of FITC-dextran loaded PLGA microspheres. *International Journal of Pharmaceutics* 334 (1–2): 137–148. https://doi.org/10.1016/j.ijpharm.2006.10.036.

Matero, S., Reinikainen, S.-P., Lahtela-Kakkonen, M. et al. (2008). Estimation of drug release profiles of a heterogeneous set of drugs from a hydrophobic matrix tablet using molecular descriptors. *Journal of Chemometrics* 22 (11–12): 653–660. https://doi.org/10.1002/cem.1148.

Niwa, T., Takeuchi, H., Hino, T. et al. (1993). Preparations of biodegradable nanospheres of water-soluble and insoluble drugs with D,L-lactide/glycolide copolymer by a novel spontaneous emulsification solvent diffusion method, and the drug release behavior. *Journal of Controlled Release* 25 (1): 89–98. https://doi.org/10.1016/0168-3659(93)90097-O.

Portela, R.M.C., von Stosch, M., and Oliveira, R. (2018). Hybrid semiparametric systems for quantitative sequence-activity modeling of synthetic biological parts. *Synthetic Biology* 3 (1): ysy010–ysy010. https://doi.org/10.1093/synbio/ysy010.

Prado-Rubio, O.A. and von Stosch, M. (2017). Towards sustainable flux determination for dynamic ultrafiltration through multivariable system identification. In: *Proceedings of the 27th European Symposium on Computer Aided Process Engineering*, 2719–2724. Elsevier https://doi.org/10.1016/B978-0-444-63965-3.50455-4.

Psichogios, D.C. and Ungar, L.H. (1992). A hybrid neural network-first principles approach to process modeling. *AIChE Journal* 38 (10): 1499–1511. https://doi.org/10.1002/aic.690381003.

Rodrigues de Azevedo, C., von Stosch, M., Costa, M.S. et al. (2017). Modeling of the burst release from PLGA micro- and nanoparticles as function of physicochemical parameters and formulation characteristics. *International Journal of Pharmaceutics* 532 (1): 229–240. https://doi.org/10.1016/j.ijpharm.2017.08.118.

Sahoo, G.B. and Ray, C. (2006). Predicting flux decline in crossflow membranes using artificial neural networks and genetic algorithms. *Journal of Membrane Science* 283 (1–2): 147–157. https://doi.org/10.1016/j.memsci.2006.06.019.

Siepmann, J. and Göpferich, A. (2001). Mathematical modeling of bioerodible, polymeric drug delivery systems. *Advanced Drug Delivery Reviews* https://doi.org/10.1016/S0169-409X(01)00116-8.

Siepmann, J. and Siepmann, F. (2008). Mathematical modeling of drug delivery. *International Journal of Pharmaceutics* 364 (2): 328–343.

Siepmann, J., Faisant, N., and Benoit, J.-P. (2002). A new mathematical model quantifying drug release from bioerodible microparticles using Monte Carlo simulations. *Pharmaceutical Research* 19 (12): 1885–1893. https://doi.org/10.1023/A:1021457911533.

von Stosch, M. (2018). Hybrid models and experimental design. In: *Hybrid Modeling in Process Industries* (eds. M. von Stosch and J. Glassey), 36–60. CRC Press.

von Stosch, M. and Willis, M.J. (2017). Intensified design of experiments for upstream bioreactors. *Engineering in Life Sciences* 17 (11): 1173–1184. https://doi.org/10.1002/elsc.201600037.

von Stosch, M., Oliveira, R., Peres, J., and de Azevedo, S. (2014). Hybrid semi-parametric modeling in process systems engineering: past, present and future. *Computers and Chemical Engineering* 60 (0): 86–101.

von Stosch, M., Portela, R.M.C., and Oliveira, R. (2018). Hybrid model structures for knowledge integration. In: *Hybrid Modeling in Process Industries* (eds. M. von Stosch and J. Glassey), 12–35. CRC Press.

Szlek, J. (2013). Heuristic modeling of macromolecule release from PLGA microspheres. *International Journal of Nanomedicine* 8: 4601–4611. https://doi.org/10.2147/IJN.S53364.

Teixeira, A.P., Clemente, J.J., Cunha, A.E. et al. (2008). Bioprocess iterative batch-to-batch optimization based on hybrid parametric/nonparametric models. *Biotechnology Progress* 22 (1): 247–258. https://doi.org/10.1021/bp0502328.

US Department of Health and Human Services Food and Drug Administration. (2004). Guidance for Industry. PAT – A Framework for Innovative Pharmaceutical Development, Manufacturing, and Quality Assurance.

Versypt, A.N.F., Pack, D.W., and Braatz, R.D. (2013). Mathematical modeling of drug delivery from autocatalytically degradable PLGA microspheres – a review. *Journal of Controlled Release* 165 (1): 29–37. https://doi.org/10.1016/j.jconrel.2012.10.015.

Wang, Y., Qu, W., and Choi, S. H. (2016). FDA's regulatory science program for generic PLA/PLGA-based drug products. https://www.americanpharmaceuticalreview.com/Featured-Articles/188841-FDA-s-Regulatory-Science-Program-for-Generic-PLA-PLGA-Based-Drug-Products/ (Last accessed 8 July 2019).

Willis, M.J. and von Stosch, M. (2017). Simultaneous parameter identification and discrimination of the nonparametric structure of hybrid semi-parametric models. *Computers and Chemical Engineering* 104: 366–376.

Wintgens, T., Rosen, J., Melin, T. et al. (2003). Modelling of a membrane bioreactor system for municipal wastewater treatment. *Journal of Membrane Science* 216 (1–2): 55–65. https://doi.org/10.1016/S0376-7388(03)00046-2.

Xu, Q., Crossley, A., and Czernuszka, J. (2009). Preparation and characterization of negatively charged poly(lactic-co-glycolic acid) microspheres. *Journal of*

Pharmaceutical Sciences 98 (7): 2377–2389. Retrieved from http://www
.sciencedirect.com/science/article/pii/S0022354916330052.

Yang, A., Martin, E., and Morris, J. (2011). Identification of semi-parametric
hybrid process models. *Computers and Chemical Engineering* 35 (1): 63–70.
https://doi.org/10.1016/j.compchemeng.2010.05.002.

Zhao, Z., Zhang, Y., and Liao, H. (2008). Design of ensemble neural network using
the Akaike information criterion. *Engineering Applications of Artificial
Intelligence* 21 (8): 1182–1188.

15

System Thinking Begins with Human Factors: Challenges for the 4th Industrial Revolution

Avi Harel

Synopsis

The 3rd industrial revolution (IR) brought us two new disciplines critical for system design: systems engineering (SE) and human factors (HF). In the 3rd IR, the two disciplines did not integrate very well, because both disciplines were system-centric. This is changing in the 4th IR, with the emergence of system thinking. System thinking has two aspects. The *internal* aspect is about the collaboration between system components; the *contextual* aspect is about the interaction with the real world, namely the customers and stakeholders, as well as the operational constraints. System thinking is a two-stage process, beginning with the contextual aspect, followed by the internal aspect. The role of HF is in the domain of contextual system thinking, namely of integrating people in complex systems. The framework recently established for relating the systems to the real world is human–machine interaction (HMI). This new transdisciplinary framework enables us to bridge the chasm between the two disciplines. A model of HMI design proposed here regards two distinct views of the user, corresponding to the two aspects of system thinking: as a system operator, the user corresponds to the contextual aspect, and as a system component, the user corresponds to the internal aspect. As a system operator, in contextual system thinking, we are concerned about the HMI. As a system component, in internal system thinking, we are concerned about the human capabilities, as well as about safety issues. In the 4th IR, we need to reengineer the HF, in order to integrate it better with the SE. Human–machine interaction engineering (HMIE) is a new discipline, intended to implement ideas and tenets of HMI for the sake of enabling effective system thinking. A new model of human–machine collaboration (HMC) enables representing formal definition of contextual information,

Systems Engineering in the Fourth Industrial Revolution: Big Data, Novel Technologies, and Modern Systems Engineering,
First Edition. Edited by Ron S. Kenett, Robert S. Swarz, and Avigdor Zonnenshain.

including the operational scenario, and the operator's goal. The model enables representing critical design dilemma, such as automation control and task management, in normal and exceptional operational conditions. New architectures based on the new model may result in better gain in terms of productivity and safety.

15.1   Introduction

A Case Study

A few years ago, I visited a friend of mine at his home in Toronto. He boasted to me about a new system that he bought, enabling him to control his house using several small, mobile devices. He could control the lights, the TV systems, the music systems, the windows, etc., all by these small devices. When he finished describing to me the features he had there, his wife mentioned to him politely that he was the only person in the house who could actually operate the system. She commented that she did not dare to touch the remote controls, because often, when she tried to close the lights in her bedroom, she ended up activating the TV or the stereo system in the living room, which was unpleasant and embarrassing.

The Legacy of the 3rd Industrial Revolution

Traditionally, systems engineers and software engineers disregard the ways people might operate the system. Typically, they believe that their intuition is good enough to design operational procedures that the operators can follow easily and reliably. Typically, they are not aware of the risks of letting the operators deviate from the intended operational procedures. Often, they overlook the possibility of choosing the wrong option from among the various features incorporated in the interface, as was the case of the computerized house in Toronto, described earlier. They might overlook the fact that more may be less. Typically, they do not bother to document the operational procedures, or to prevent diversions from the procedures by design. They are not ordinarily aware of the special education required to ensure that the interaction is efficient and reliable.

The Challenge of the 4th Industrial Revolution

The 4th industrial revolution (IR) is about a shift in our view of the effect of technology on our experience of using systems. In his keynote article on the significance of the fourth industrial revolution, Klaus Schwab (2016) sets the goal:

> All of these new technologies are first and foremost tools made by people for people.

He then explains that,

> [The] inexorable shift from simple digitization (the Third Industrial Revolution) to innovation based on combinations of technologies (the Fourth Industrial Revolution) is forcing companies to reexamine the way they do business.

Kenett et al. (2018) discuss different aspects of systems engineering (SE) in the context of the 4th IR and conclude that in the near future:

- Virtually all systems will have porous and ill-defined boundaries.
- Virtually all systems will have ill-defined requirements which are changed frequently.

According to Cooper (1998), technology consists largely of pieces that work, but not at all well. This is often the fault of poorly designed user interfaces. Too many devices ask too much of their users. Too many systems make their users feel stupid when they cannot get the job done. In the 4th IR, everything we regularly use in our home, work, transportation, is being equipped with new technology. Boy and Narkevicius (2013) observed that "requirements, solutions, and the world constantly evolve, and are very difficult to keep current". Can we use systems of the 4th IR safely?

Putting People First

A primary challenge of system design in the 4th IR is about what people experience in going through this change. In the words of Schwab: "In the end, it all comes down to people and values. We need to shape a future that works for all of us by putting people first and empowering them and constantly reminding ourselves that all of these new technologies are first and foremost tools made by people for people."

Crossing the Human Boundaries

In the first three industrial revolutions, the boundaries between technology and people were clear. Technology was a means to design products and systems.

The effect of technology on human behavior was indirect through these deliverables. The new technologies imply changes in the ways people interact with the systems. In the 4th IR, technology is gradually crossing the boundaries, affecting directly the body and soul of people.

15.2 Systems

Definitions of Systems

According to ISO/IEC/IEEE 2015, following Bertalanffy (1968), the term "system" is a short name for "system-of-interest." The general term "system" does not require any purpose or interest, and may refer to natural systems as well. According to the INCOSE System Engineering Body of Knowledge (SEBoK), the interest of systems engineering is in engineered systems. According to Bartolomei et al. (2006) an engineering system is

> An engineering system is a complex socio-technical system that is designed, developed, and actively managed by humans in order to deliver value to stakeholders.

The SEBoK definition of "sociotechnical system" is "an engineered system which includes a combination of technical and human or natural elements."

In this chapter, the system is a sociotechnical system, which means that it includes technical and human elements. The technical elements are called here "functional units." The human elements may be insiders, namely operators, or outsiders,(stakeholders), such as users.

System Thinking

Systems thinking is widely believed to be critical in handling the complexity facing the world in the coming decades. Richmond (1994), the originator of the term systems thinking, defines systems thinking as "the art and science of making reliable inferences about behavior by developing an increasingly deep understanding of underlying structure." With systems thinking, a systems engineer "can see both the forest and the trees; one eye on each." Senge (1990) defines systems thinking as a discipline for seeing wholes and a framework for seeing interrelationships rather than things, for seeing patterns of change rather than static snapshots. Sweeney and Sterman (2000) noticed that systems thinking involves the ability to represent and assess dynamic complexity (e.g. behavior that arises from the interaction of a system's agents over time), both textually and graphically.

Stave and Hopper (2007) observed that the term systems thinking is used in a variety of sometimes conflicting ways. Kopainsky et al. (2011) assert that systems thinking should include appreciation for long-term planning, feedback loops, nonlinear relationships between variables, and collaborative planning across areas of an organization.

Arnold and Wade (2015) compared the various definitions and came out with the definition that

> Systems thinking is a set of synergistic analytic skills used to improve the capability of identifying and understanding systems, predicting their behaviors, and devising modifications to them in order to produce desired effects. These skills work together as a system.

The authors conclude that "the use of systems thinking transcends many disciplines, supporting and connecting them in unintuitive but highly impactful ways."

Layers of System Thinking

Following the discussion above, we may consider two aspects of system thinking:

- The internal aspect is about the functional units integrated with the operators, collaboration between components of the engineered system.
- The contextual aspect is about the interaction of the engineered system with the real world, namely the customers and stakeholders, as well as the operational constraints.

The two layers of system thinking are illustrated in Figure 15.1.

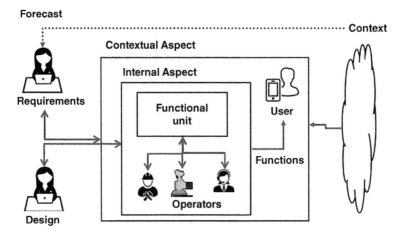

Figure 15.1 Layers of system thinking.

According to this model, the contextual aspect is defined based on requirements specifications, with respect to the user's tasks and capability, and considering forecast of the context. Also, the internal aspect is defined design considerations about the various roles of the operators, and their collaboration with the functional units.

This chapter is about ways people may integrate with systems, about necessary changes in the way engineers may implement the concept of system thinking, and about engineering activities enabling successful integration with its operators.

Agile System Thinking

In the early days of the 3rd IR, system development followed the waterfall model. According to this model, the system design is based on the requirement specifications, which remained unchanged until the version release. This model did not work very well, because during the system development new requirements emerge. Therefore, the waterfall model was replaced by other models, such as iterative development or agile development, which facilitated changing the requirements during the system development.

System thinking is a continuous process, integrated with agile development. The contextual aspect is the one that senses the need to change the requirements, and triggers the change. The internal aspect is the traditional response to changes, typical of agile development.

15.3 Human Factors

Traditional Interaction Definition

Traditionally, the engineers who define the interaction with the operators are systems engineers or software engineers. Typically, they are technology oriented, which means that they try their best to integrate state-of-the-art technological features. Often, they are feature oriented, which means that they include in the design as many features as the technology allows them to include, regardless of whether or how the operators will use them. Also, often, they are designer centric, which means that they optimize the interaction according to their knowledge about the operational procedures, and their own preferences.

In the 3rd IR, we were concerned of the effect of the human–machine interaction (HMI) on functional attributes: performance, production, etc. In the 4th IR we are also concerned about subjective attributes, such as customer satisfaction and operator's experience of making the system work as intended.

Technology-Driven Interaction Design

Technology-oriented engineers love to offer as many features as they can, and they try to assist the operators in as many ways as they can. In the early days of computing, when people just started to write software on the first mainframe computers, Weinberg (1971) demonstrated and analyzed the trouble with technology-driven software programming. Technology has developed much since 1971, but the troubles remain the same.

Typically, software engineers focus on providing user interfaces (UIs) that support the features specified in the requirement document. Often, they do not think how the operators will access these features, or if they use them correctly, in the proper situations (Cooper 1998). They feel responsible for preventing bugs but not for preventing operator's errors (Harel 2010). Infrequently used units may contain bugs that go unnoticed in normal operation. These bugs might hamper the interaction in exceptional conditions.

Standish (1995) analyzed reasons for the failure of software projects. The conclusion of this analysis was that there is a huge gap between software development practices and engineering disciplines. They demonstrated their finding by comparing the attitude to failure of software projects to those of building bridges. They concluded that beside 3000 years of experience, there is another difference between software failures and bridge failures. "When a bridge falls down, it is investigated and a report is written on the cause of the failure. This is not so in the computer industry where failures are covered up, ignored, and/or rationalized. As a result, we keep making the same mistakes over and over again" (Standish 1995). Figure 15.2 illustrates the effect of software bugs on the user experience.

Interaction design using the framework of HMI enables this problem to be solved. In the 4th IR, new failure models are available to the system designers,

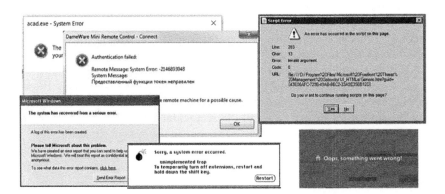

Figure 15.2 The inmates are running the asylum.

enabling then to handle also the unexpected. Interaction designers employ interaction protocols to ensure that the operators can perceive and handle rare events, and recognize and understand unexpected situation.

Feature-Oriented Design and Testing

Traditionally, interaction design is feature oriented, which means that by design the system offers as many features as the technology allows, regardless of whether the operators will need them, how they find them, or if they add noise to the operators' perception of the controls that they need to activate.

An example of the trouble caused by unnecessary features is the nuisance of beeps that many home appliances generate, just for the fun of the designer. Another example is of the delay option in home appliances. A user of an air conditioner or a drier that has this feature might confuse the delay control with other time-related controls, and activate it unintentionally. Then, the appliance would not work, because it is in delay mode. A user who just tried the basic features might not understand the meaning of the notifications on the panel or the remote control, and might not understand the meaning of the warning sound. The user might call a technician to see the appliance does not work, but when the technician arrives, the delay time is already over, and the appliance works perfectly. Somebody has to pay for the visit of the technician, which was in vain.

Designer-Centric Interaction Design

Software developers often optimize the user interface to suit their own needs, namely to facilitate the software development. For example, during the software debugging, the developers need to repeat the same action over and over again, each time testing the effect and outcome of a code or parameter value change. The activation of those actions which are tested frequently is tedious, because the access to the actuator may be through a sequence menu selections and searching. To facilitate such procedures, the software developers often define shortcut keys, such as those of Gmail, presented in Figure 15.3.

All the shortcuts in this figure were defined to facilitate the debugging of Gmail. A few, sophisticated users, may generate their own shortcuts. These shortcuts are sometimes problematic when they are activated accidentally. Typically, the operators are not aware of the accidental activation of the shortcut key and are confused by the unexpected change in the system situation.

The Usability Shift

In the beginning of the 3rd IR, system engineers were not aware of the role of human factors (HFs) and they did not consider them in the system design. The design was technology oriented. People judged the value of systems in

Figure 15.3 Shortcut keys in Gmail.

terms of functions. We did not care much about the operators. We thought about the ways the operators interact with the system but not about the effect of the interaction on the other operator's tasks. For example, we assumed that professional practitioners might be willing to spend time in order to learn how to use our system.

By and by, following investigation of many accidents, systems engineers realized that they need to consider human factors in the design. By the end of the 3rd IR, more and more HF practitioners became part of the design team. Their charter is to look at the ways the operators use the system, and how human errors contribute to accidents and to performance reduction. Still, human factors are of lower priority. The functions were evaluated primarily in terms of performance, productivity, and safety. Typically, the way to consider them is in two stages. First, the system is designed with the criteria of maximizing performance. Then, in a later stage, we review the system design, looking for design flaws, hampering the system operation.

From the preceding discussion we can conclude that HMI based on traditional practices might not be sufficiently good for the 4th IR.

The single most important factor of productivity assurance is employing usability assurance methodologies. The need to incorporate human factors in system design became obvious early when people started to use computers for designing office applications (Landauer 1996). According to a Gartner report (1995), system developers need to focus on business suitability and usability. Usability has a significant impact on the success of systems and products. It relates to the actual usage of a system and also to its effective design and development. According to Landauer, failing to build usable system may degrade a project's ability to deliver in time, budget, functionality, and quality.

The science of usability engineering started to develop even earlier. Shneiderman (1980) suggested that software programmers could do better to ensure that the user of their programs find them friendly and easy to use. Norman and Draper (1986) introduced the principle of User-Centered Design (UCD) and Shneiderman (1987) proposed guidelines to implement these principles in the software design. Later, Card et al. (1983) explained and demonstrated how software developers can incorporate human factors to augment the productivity of software products. Today, it is a common practice to assign the task of user interface design (UID) to usability professionals, who know and understand the operational needs.

Human Factors Engineering

Any large organization whose mission is to design and develop systems for humans needs a well-developed integration and process plan to deal with the challenges that arise from managing multiple subsystems. Human capabilities, skills, and needs must be considered early in the design and development process, and must be continuously considered throughout the development life cycle (Fitts et al. 1987). Human factors engineering (HFE) is a framework for describing how people may integrate with a human-made system.

People in prehistoric civilizations considered human factors whenever they had to design tools for their living. The need to assign an engineering term to natural behavior is due to accidents occurring during system operation (Meister 1999). The *Encyclopedia Britannica* defines human-factors engineering, a "science dealing with the application of information on physical and psychological characteristics to the design of devices and systems for human use."

Traditionally, software engineers and systems engineers did not bother to consider the human factors. They preferred to focus on functionality and technology.

They used to delegate the responsibility for interaction design to human factors engineers. After getting some experience with operating interactive systems, it became clear that user interfaces designed by systems engineers are often difficult to use, and sometimes disastrous. Harel and Weiss (2011) proposed that the methods for mitigating human risks should be integrated in the system engineering practices. Eventually, they proposed to extend the scope of systems engineering, adding methods and guidelines to help protect from unexpected events. The idea presented to systems engineers was that they could benefit from considering human factors in the system design (Jackson and Harel 2017).

For the purposes of systems engineering, it is helpful to consider two aspects of the HFE:

- The task view, in which we examine the ways people interact with the system.
- The capability view, in which we examine physical and mental limitation of the human operators, hampering successful operation.

The challenges of the 4th IR about the task view are primarily in system thinking, about methodologies for system development, and about the opportunities to implement these methodologies using new technologies. The challenges about the capability view are about leveraging the physical and mental capacity of people through technology. This chapter focuses on the task view, namely about methodologies for system development, enabling designers to improve the efficiency and safety in the HMI.

The Human–System Integration Engineering Framework

Unfortunately, systems engineers are not always aware of the benefits of considering human factors, and usability practitioners fail to explain their offer. There is a need to bridge this chasm from both sides. Systems engineers need to understand the benefits that they can get from incorporating human factors (Jackson and Harel 2017), and usability practitioners need to demonstrate and explain to systems engineers how to integrate the theories of cognitive sciences in the system development. The way systems engineers implement their part is by "systems thinking." The way usability practitioners implement their part is by "cognitive engineering."

Recently, usability practitioners have discussed the challenges of incorporating human factors in system development (McDermott et al. 2017). Also, Sillitto et al. (2018) have distinguished interdisciplinary from transdisciplinary. Interdisciplinary has to do with multiple disciplines to accomplish a task. Transdisciplinary, on the other hand, has to do with using multiple disciplines together to accomplish the task. Transdisciplinary addresses the cooperation and collaboration between the disciplines. Hence, using engineering, psychology, and human factors together constitutes a transdisciplinary science.

Unfortunately, these works have not yet matured to an engineering discipline. The bridge that will enable crossing the chasm between SE and HF should be built using new methodologies about the way we define the interaction between the human operator and the machine. This bridge is the transdisciplinary framework of Human–System Integration (HSI), being developed recently. The location of the new discipline is illustrated in Figure 15.4.

HSI Engineering (HSIE) addresses the cooperation and collaboration between the disciplines. Hence, using engineering, psychology, and human factors together constitutes a transdisciplinary science. Unfortunately, these works

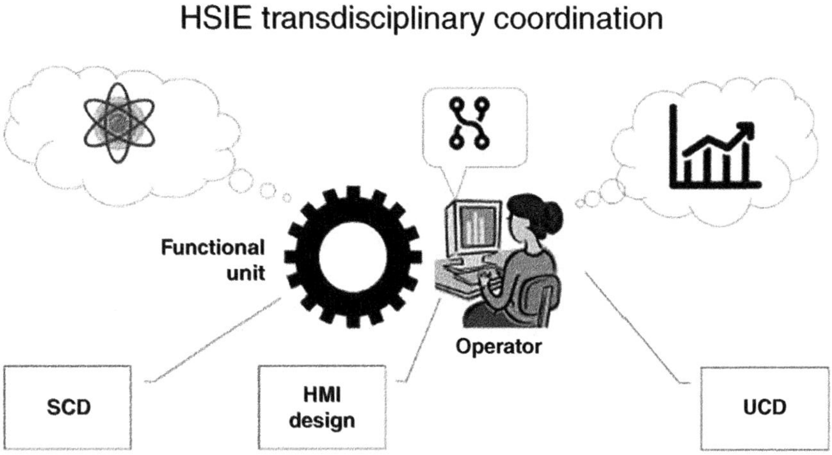

HSIE transdisciplinary coordination

Figure 15.4 HSI engineering discipline.

have not matured yet to an engineering discipline. The bridge that will enable crossing of the chasm should be built using new methodologies about the way we define the interaction between the human operator and the machine.

The transdisciplinary framework proposed here may enable the chasm between the SE and HF to be bridged.

Human–Machine Integration-Related Terminology

The concepts used in the Human–Machine Integration (HMI) framework apply to various kinds of systems, defined in different industry domains. However, the use of HMI terms in the various domains is not the same. Also, terms used in a particular domain are not always adequate to the other domains. This HMI framework proposes a generic, common terminology, which may apply to all underlying domains. Table 15.1 compares terms of various concepts (column 1) in the context of consumer product (column 2) with those used in the context of safety-critical systems (column 3), and the generic term used in this chapter (column 4):

The concepts of protection level and the protection analysis are specific to safety-critical system. The terms mentioned in this table apply specifically to the process industry. The generic terms used in this chapter are hereby explained:

Operators

The term User was adequate in the early years, when people used various electronic devices at home, and later, when they used computers for office application. Recently, the term Human replaced the original term, as it applies also to people integrated with the system, such as operators or passive people, who are not users.

Table 15.1 Human–machine integration (HMI) terms comparison.

Concept	Consumer products	Safety-critical systems	Generic term chosen for this chapter
The human part of the system	User	Operator	Operator
The interface between the system and the human operator	User Interface (UI)	Control panel	Manual control
Designing for facilitating the human part of the interaction	UCD	Procedure design	HCID
Disruption from normal operation	Exception	Hazard	Diversion
Situation awareness	User orientation	Situation awareness	Situation awareness
Recovery from a disruption	Resumption	Resilience	Adjustment
Protection level	N.A. in 3rd IR	System Integrity Level (SIL)	Protection level
Protection analysis	N.A. in 3rd IR	Layer Of Protection Analysis (LOPA)	Protection analysis

It is helpful to use two distinct views of the operator: as a system controller and as a system unit. As a system controller, we are interested in functions: production, performance, effect, etc. As a system unit, we are interested in the operator's ability to make the system work, and about safety. For example, we want to detect a situation of a pilot passed out due to G-LOC (g-force induced loss of consciousness) and activate an Auto-GCAS (Ground Collision Avoidance System) to stabilize the airplane and the pilot (Dockrill 2016).

As a system controller, the operator can have various roles: a user, motivated by functions and performance; a supervisor, motivated by the need to make sure that the system operates as intended; and a controller, who needs to manually make the system work. As a system unit, we are concerned about the operator's ability to function as a system controller, which is determined by qualification, motivation, vigilance, etc.

15.4 Human Factor Challenges Typical of the 3rd Industrial Revolution

Understanding the Designer's Responsibility

A common design mistake, typical of the 3rd IR, is of assuming that the human operators can learn and keep operating according to all operational rules imposed on them. In reality, system operation often fails due to failure of the operators or users to follow the operational procedures, or to obey any other operational rules.

According to prior studies (such as by Zonnenshain and Harel 2008) failure is mostly due to common flaws in the interaction design. Many operators have other tasks to perform and operating the system might hamper their primary tasks, which are not related to the system's operation.

The principle of HMI is that the system design should consider known limitations of the human operators and design the interaction such that the system operation is seamless and well protected from errors.

In the 4th IR, engineers are also concerned about the integration of the system operation with other operator's tasks, including operating complementary systems, as well as about the effect of the system on the environment.

Expanding the Scope of Quality Assurance

In the beginning of the 3rd IR systems, design focused on performance attributes, such as productivity and gain. Then, they realized that system failure is a key factor to performance assurance. They introduced the concept of system quality, which was about preventing system failure. The first stage in preventing system failure was to examine the failure of system components. At that stage, quality assurance was about the reliability of system components, namely reliability assurance.

By and by, toward the beginning of the 4th IR system scientists gathered statistics about the sources of system failure, and realized that operator's errors contribute to failure even more than component failure. For example, it has been reported that 70–80% of the aviation accidents are due to human errors (Wiegmann and Shappell 1997). Statistics about the productivity of text editing show that typing error result in extending the typing time by a factor of two, and that human errors are associated with most of the accidents. Consequently, in the 4th IR, the concept of quality should be extended to preventing human errors, beside component reliability (Zonnenshain and Harel 2008, 2015).

Similar to the traditional definition of quality, a way to evaluate the quality of a HMI is by measures of the rate of integration failure. We may regard the integration as being of high quality if the rate of failure is low. Accordingly, in order to evaluate the HMI, we need to assess the rate of various failure modes. A way to assess the failure modes was described by Zonnenshain and Harel (2008, 2015). According to their model, failure modes are associated with exceptional situations, namely due to distraction from normal operation. Accordingly, the quality of a HMI may be assessed in terms of the rate of operating in exceptional situations.

Traditionally, according to scientific disciplines (Popper 1968), quality assurance relies on verification testing. In the 4th IR, the most important concern about sustaining a required level of performance is the operator's capability to respond gracefully to extreme conditions. The challenge of quality assurance in the 4th IR is about defining ways for assuring the system resilience to faults, both of critical components and of the human operators.

In the 4th IR the role of human factors (HF) is changing and they lead the requirements. In the 4th IR, HF should be integrated in the design right from the beginning. HF should lead the requirements specifications, as well as critical design decisions, such as about the system architecture. The challenges of embedding HF in the 4th IR are to develop practices for:

- How to write usability-oriented requirements specifications.
- How to translate the usability considerations to design features.

Expanding the Value of Human–Machine Interaction

Originally, the primary goal of UCD was to facilitate the operation of customer products (Norman 1988). The vision was to design products that everybody can use. Therefore, UCD methodology targeted the novices. Accordingly, the focus in the 3rd IR was on basic, normal operation, in normal situations. Later, gradually, the principles of UCD spread to system design. Besides novices, the design principles also targeted experienced operators, which meant that the design also targeted advanced stages of the operation. However, the focus still remained on normal operation, in normal conditions. Deviations from the normal were considered errors.

In the 4th IR, we expect that the design will support the whole life cycle. A primary challenge of HMI in the 4th IR is to develop practices for supporting operation in auxiliary activities, such as maintenance and training, and also unusual conditions, such as testing and troubleshooting.

The Focus of Human–Machine Integration Design

In the 3rd IR (and also today), the primary concern of UCD is performance. New design practices are proposed for maximizing the performance in normal operational conditions, focusing on usability traits. For example, Boy (2016) suggested that tangible user interfaces may be easy and safe to use.

By and by, after gathering statistics about the sources of system failure, systems engineers realized that much of the operational time is spent operating in exceptional, nonproductive situations. Typically, exception management is of lower priority in the system design. It is an informal task, not included in the requirements documents. It is a leftover, for the software engineers. Consequently, software development is often a bottleneck in the system development, typically behind schedule. The challenge of the 4th IR is to reduce the effect of exceptions. A challenge about HMI in the 4th IR is to develop practices for:

- Preventing deviation of the system situation from normal to exceptional.
- Recovering from exceptional situations, and subsequently resuming normal operation.

Context-Dependent Interaction Design

In the 3rd IR, the architecture of HMI was based on concepts borrowed from system design. A common practice of describing the interaction between two hardware units was through a thin layer, an interface: each of the hardware units is capable of receiving input from the other unit through the interface, process the received data locally, and send output to the other unit, through the interface.

In a typical architecture used for the HMI design, one of the two units was the system and the other was the human operator. The interaction was through an interface, consisting of a control unit, enabling the sending of data and commands from the human operator to the system, and a display unit, enabling the sending of information about the system state from the system to the operator. A simple model commonly used in UCD proposes that the human activity is similar to that of the machine. However, in an HMI framework, the role of the two units is not symmetric. The human component is regarded as a black box. The interaction consists of combinations of the asymmetric architectures. The behavior of the human operator is subject to three mental activities, namely situation perception, decision making, and command execution. Its behavior depends on information not available to the system designer, such as operational context, intentions, state of mind, and personality. The designer cannot do anything to directly affect the operator's perception or decision making.

Using terms borrowed from hardware design, the human operator is a master and a client, and the system is a slave and a server. The master–slave model refers to the control aspect of the interaction, and the client–server model refers to the display aspect of the interaction. In the 3rd IR, it was the operators' sole duty to deal not only with the uncertainty, but also with unexpected behavior of the system part. This kind of architecture is described in Figure 15.5.

According to this model, the machine behavior is quite predictable, as long as all the components are reliable; besides the human operator, other external forces, defined as Context, also affect the machine behavior, through sensors, and algorithms. On the other hand, the behavior of the human operator is less predictable; the human perception is biased by improper vision, due to motivational, training, and vigilance factors. The decision making, based on the

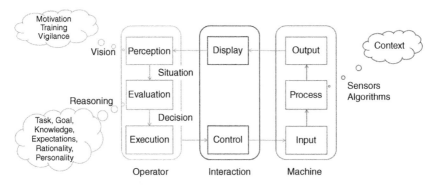

Figure 15.5 Classic HMI model.

situation evaluation, is biased by improper reasoning, due fuzzy task and goal setting, unsuited expectations, rationality, and other personality traits of the human operator, and also by organizational pressures (Reason 1990; Jackson and Harel 2018). According to these models, the primary goals of UCD are: facilitating reliable perception, decision support, and protection from execution errors. The challenge of HMI in the 4th IR is to enable and guide system developers in designing systems that consider these factors. Limitations of this model are:

- It is process oriented, therefore it does not highlight the value of the system, in terms of performance.
- Most significant factors affecting the system behavior are not in control. These factors are marked as clouds in Figure 15.5.
- Poor validity, inasmuch as there is no practical way to measure these, or to estimate the direction or magnitude of their effect (Popper 1968).

Human–Machine Task Allocation

The dilemma of task allocation is about the integration of automated processes with those controlled by the human operator. Automation is required when the system needs to respond quickly and accurately, and when the operators are not capable of fulfilling this need. Examples are when an airplane needs to respond quickly to a sudden gust of wind, or when a pilot becomes unconscious due to extreme G force (Learmount 2011).

Yet, automation might be a double-edged sword, when it disables the operator's control, as was the case the Air France AF 296 accident on 26 June 1988 (Casey 1993). Automation is applicable only for problems in which the system response is well defined, namely when the designer may assume that the data required for the response selection will be available when they are required, and that they are reliable. In the 3rd IR, this was not always the case. When the

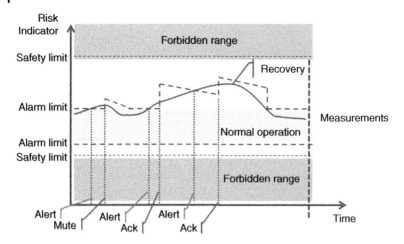

Figure 15.6 Automation control.

developers do not have all the information required to design resilient automation, they often let the operators take charge. The problem is that the operators often fail to recognize the situation (Norman 1990) and often are not trained to handle it (Bainbridge 1983).

In the 3rd IR, the task allocation was based on heuristics, namely on the intuition and experience of the designers. A systematic approach to human–machine task allocation, commonly applied in the process industry, is by statistical process control (SPC). Kenett et al. (2009) suggested that we may apply this method also for usability control by tracking usability changes. This method may be extended and applied for resilience control. The method is based on defining risk indicators. A risk indicator is a system parameter, such as operational temperature, with thresholds for warning, alarming, recovery time, and emergency. The automation control is based on tracing the values of these parameters, calculating the trends, and responding when crossing the thresholds. The method applied to alarm design is illustrated in Figure 15.6.

In the 4th IR, we may develop practices for optimal task allocation. Such practices should apply to both normal and exceptional situations.

Managing the Human–Machine Collaboration

Collaboration management is about adapting the operational procedures to scenario changes. The problem of collaboration management extends the problem of task allocation. The result of human–machine task allocation is a definition of the default behavior in normal and exceptional situations. Collaboration management is about situations when the default behavior is not appropriate. Specifically, when the operators need to take over a risky automated system behavior, as was the case of AF 296 accident (Casey 1993), or when the system needs to take over a risky manual control, as was the case of the Air France AF 447 accident (BEA 2012).

In the 3rd IR, the collaboration management was based on heuristics, namely on the intuition and experience of the designers. A key consideration was to assign tasks according to capability. Thus, routine tasks requiring intensive, reliable data processing, which were easy to formulate, were assigned to the machine, while other tasks were assigned to the human operator. An example from transportation is the cooperation between the driver/pilot/captain and the machine. The machine needs to keep track of the route, in order to detect potential threats and to inform the operator about them. It is the operator's job to take over the machine and cope with the threat.

Accordingly, the collaboration may be by assigning the task of bookkeeping the history of system activity to the machine, while the operator's tasks are about deciding on critical changes in the operational procedures. For example, the machine can validate the operation according to the procedures and notify the operators about distractions. The operator's tasks include responding to the notification, by changing the operational procedure, for example, to troubleshooting.

A challenge about collaboration management in the 4th IR is to develop practices, in the form of protocols, for collaboration management. The protocols may define the conditions in which the operators may override automation, as well as those in which the operator may enforce automated behavior, overriding the default manual control. The conditions may be stated in terms of priority, which may affect characteristics of the task termination, parameter preservation and reset, etc.

A New Human–Machine Collaboration Model

Adversity in the system operation is manifested by operating in exceptional situations and that mishaps are associated with difficulties in adapting to the exceptional situation. The traditional model of HMI is one dimensional, based on the concept of user interface (UI) comprising a display unit and a control unit: the system displays the situation through the display unit and the operators invoke their command using the control unit. This model does not represent well the need to deal with the exceptions. A new HMI model, emphasizing the role of human–machine collaboration that may be implemented in the 4th IR is illustrated in Figure 15.7.

The new model has several features, suited to comply with the opportunities of the 4th IR:

- An operator interface replaces the traditional user interface. The significance of the terminology change is that the interface serves all kinds of operators,

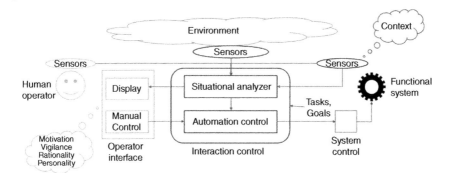

Figure 15.7 Human–machine collaboration.

including those who are not users of the system. The display to the operators comprises information received from a situation analyzer.

- A situation analysis, tracing the situation changes, based on sensory data received from the system, the operator, and the environment.
- An automation control unit, intended to control the automation according to the situation, enabling the operator to override the automated system control, and enabling the system to override improper operator commands.

The human–machine collaboration (HMC) design model assumes that the quality of the operators' decision depends on the perception of the situation in general, and of exceptional situations in particular. Accordingly, the model focuses on avoiding, rebounding, and recovery from exceptional situations. Critical factors not controlled in implementations the classical model (above) are controlled in implementations based on the new model. The system operation is subject to rules, derived from the operational tasks and goals specified in the requirements documents. Operator's errors are controlled by automation.

Scenario-Oriented Collaboration Design

Cooperation problems often emerge when the operators fail to comply with the operational procedure. This is often the result of a scenario change. Certain common barriers to successful integration may be attributed to the low level of interaction and about insufficient support for changes in the operational scenario. In the 3rd IR, the interaction was defined in using an event–response model: the system responses to single actions and the human response to information received from the system about its situation and activity. The specification of the high-level interaction was used as a framework for the interaction design. The design itself consisted of elaborating the specification in terms of the system response to events. What was missing is the operational context for the actions, namely the system situation and the step in the operational proce-dure (the history of the action).

In the 4th IR, we realize that the model of atomic event–responses is not sufficient for defining adequate integration because, besides the event, the ade-quate response should also depend on the operational situation and the oper-ational procedures. A challenging goal of scenario-based interaction design in the 4th IR is the development of means and practices for defining and storing task-level procedure protocols in a knowledge database, for tracing the inter-action, and for responding to deviations from the required procedures. Such protocols may be used for controlling the operational procedures, to fit into the STAMP (System-Theoretic Accident Model and Processes) paradigm proposed by Leveson (2004).

Behavior Management

When a scenario changes, all the system units, as well as the operators, should synchronize with the change, to enable recollaboration. Synchronization prob-lems are due to wrong termination of the first scenario, or wrong resetting of the second scenario. Another problem is of rollback, namely defining what happens when the second scenario ends, and how to resume the first scenario.

In the 3rd IR, the definition of the system behavior on scenario change was based on heuristics, namely on the intuition and experience of the designers. The requirement documents did not specify the protocols of the scenario transition. A significant challenge of behavior management in the 4th IR is to develop practices, including priorities and protocols, for changing an operational scenario. For example, a standard protocol may describe the rules about scenario changes as follows:

- Save the system state, including procedure step.
- Reset the system situation, to allow the next task.
- Execute the interrupt.
- Conditional (ask the operator) resume the interrupted procedure.

Interaction Styles

Interaction styles are attributes of the user interface affecting the user experience. The attributes mostly discussed are visual styles, such as screen background and layout design, shape, color and density of screen objects, appearance attributes, such as animation, salience, location, etc. Styles apply also to object selection, such as direct access vs. menu selection, and to audio attributes, such as melody, loudness, pitch, speed, etc.

Several guidelines about style definition were proposed during the 3rd IR. Shneiderman (1987) suggested employing eight golden rules for UID.

Nielsen (1993) proposed guidelines for designing web pages. Yet earlier, Shneiderman (1986) reported an experiment about interactive menu selection, in which interaction style optimized for novices was different from that of experienced user. Indeed, Zonnenshain and Harel (2008, 2015) have demonstrated that golden rules adequate for routine operation, with which the operators is very familiar, are not adequate for emergency operation, which the operators did not have any chance to experience beforehand.

In the 4th IR, the dilemmas of interaction styles should be examined and studied. Recently, Boy (2016) proposed the concept of tangible interaction as a means to improve the reliability of the human perception of the situation. However, different tangible interfaces may be adequate for different scenarios. A challenge about choosing the proper style is to define the boundaries of scenario validity of the various styles.

Beyond Root-Cause Analysis

In the 3rd IR (and even today) design for failure prevention was based on cause–effect fault analysis. The focus of common practices was on failure of hardware components. Common practices for such analysis include Fault Tree Analysis (FTA), Event Tree Analysis (ETA), Hazard and Operability (HAZOP), Failure Mode and Effect Analysis (FMEA), etc. The focus was on the fault triggering the sequence of events resulting in the accident. Other triggers, such as improper operator's action or a software bug, are typically regarded as unfortunate or unexpected.

In the 4th IR, new failure models are available to the system designers, enabling them to handle also the unexpected. Scenario-oriented interaction design enables detecting situations of deviation from the operational procedures, even if the source for the deviation is unknown. Special procedures may enable the operators to respond gracefully to the unexpected situation and to identify situations of operator's slip. Subsequently, system designers may explore the log of activity recorded during the operation and decide on means to prevent recurring events.

A challenge about root-cause analysis (RCA) in the 4th IR is to develop practices for coping with the unexpected, including:

- Develop database management systems (DBMSs) enabling the system designers to define the operational procedures, store them, trace them at run-time, compare the current activity with the expected, and notify about deviations.
- Develop guidelines for responding to unexpected events gracefully.
- Develop tools for analysis of recurring deviations from the expected, and for reporting on them.

Preventing Expected Diversions

Several methods for failure prevention were developed in the 3rd IR, based on cause–effect analysis. A well know example is the automatic shutdown of a system under high risk. Automatic emergency shutdown is part of collision avoidance systems (CASs), SCRAM at boiling water reactors, "reactor trip" at pressurized water reactors, etc. The common practice for deciding on activating the shutdown procedure is based on absolute criteria.

Unexpected events are often due to operating in exceptional situations. Zonnenshain and Harel (2008, 2015) analyzed 67 incidents and developed a model of operational failure. A resilience model, based on the failure model, attributed the failure primarily to operating in exceptional situations. In the resilience model, the role of trigger is secondary to the primary source, which is operating in exceptional situations. A primary challenge in the 4th IR is to cope with exceptional situations, namely to prevent them, and to facilitate the recovery, namely the resumption of operating in normal situations.

Preventing Unexpected Diversions

Common system engineering methodologies and practices do not mitigate the risks of unexpected events, such as operator errors or mode errors, typically attributed to "force majeure." Leveson (2004) proposed that the system should control its own behavior, by enforcing operation according to rules. Harel and Weiss (2011) proposed that we can mitigate such risks by design, by considering the human limitations in the interaction. Zonnenshain and Harel (2008, 2015) proposed that unexpected diversions are due to operating in exceptional situations. Therefore, to prevent unexpected diversions, the system needs to minimize the operation in exceptional situations. Also, they proposed that the system and the operators should collaborate in the troubleshooting and resuming a normal situation.

A challenge about preventing unexpected diversions in the 4th IR is to develop practices for balancing the need to enable operating in unexpected situations with the need to protect from the unexpected events while in an unexpected situation.

Responding to Diversions

Protecting from failure is costly in terms of time and budget. Typically, system engineers inform the UI designer about certain failure modes known to bother the stakeholders, but not about those that did not materialize yet to real costs. To illustrate, consider an example of basic control of a simple machine based on an On–Off switch. In a design typical of the 3rd IR, the switch enabled the basic functions of starting and stopping the machine. In a typical design scenario, a system engineer will analyze failure modes and require that the design will incorporate means to prevent these failures. However, it is rare that the systems engineer will require that the interaction design includes means to protect from operator's errors, or that it includes procedures for troubleshooting. In the 4th IR, HMI practices may include guidelines for supporting features for error prevention and for designing troubleshooting procedures.

Protecting from the Unexpected

The effectiveness of this approach is limited when a design needs to empower the human operator in order to cope with the unexpected. In the 3rd IR, unexpected events were nobody's business: typically, the design specifications did not mention them. Even today, it is the job of the software engineers to protect the system from unexpected events; typically, however, software engineers do not care much about how the operators may possibly become aware of such events and how they can possibly recognize and understand the exceptional situation.

A theoretical limitation of these methods is that they deal with the expected, while many of the accidents are regarded as unexpected, even with hindsight. Harel and Weiss (2011) studied the nature of unexpected event, employing observations from Taleb (2007), and suggested the need to integrate in the system design special means for detecting and protecting from unexpected events.

New methodologies developed in the 4th IR enable systems engineers to prevent the unexpected. For example, the STAMP methodology enforces the system to behave according to rules, thus avoiding unexpected situations (Leveson 2004). Robert et al. (1998) assumed that unexpected behavior is due to incomplete specifications and proposed a methodology for assuring that the requirements about the acceptable situations are complete.

The interaction between the operators' tasks is a major contributor to operational errors. The need to putting people first implies that we need to consider ways to prevent user errors.

Expanding the Concept of Fault-Tolerance

In the 3rd IR, the assurance of fault tolerance was based on heuristics, namely on the intuition and experience of the designers. Typically, the design for fault tolerance was about specific critical components. A fault of the operator, such as due to illness or vigilance problems, was regarded as misfortune. For example, in the history of combat aircraft, many of them crashed due to excessive G force that the human body could not tolerate.

The new technologies developed in the 4th IR enable systems developers to shift much of the processing from the operators to the machine. More than ever before, automation enables embarrassing limitations of the human operators to be overcome. For example, an autopilot can detect a situation of a pilot passed out due to G-LOC and alert colleagues, or activate safety measures such as Auto-GCAS to stabilize the airplane and the pilot (Dockrill 2016). Referring to the examples above, in the 3rd IR sensors for detecting sudden wind gusts or for identifying the pilot's mental state were not available, or were not reliable. In the 4th IR, such sensors may be available, enabling detection and proper response also in extreme operational conditions.

Challenges to achieving fault tolerance in the 4th IR are to develop practices for assuring extended fault tolerance, such as:

- Means to detect and protect from situations of disabled operators.
- Notifying the operators about all deviations from the supported procedures.
- Providing guidelines for designing resilient troubleshooting and recovery procedures.

Exception Handling

Much of the productivity is wasted when the system reaches exceptional situations. The reason for this is because the operators are not always trained well to cope with the exceptional situation, and the designers fail to facilitate the operator's mental activities in unfamiliar situations.

Software engineers often apply exception handlers to protect from expected exceptional situations. Even today, the unfortunate results of operating in exceptional situations is often regarded as an operator's error, and sometimes as unexpected.

A common limitation of traditional exception handlers is that they can capture only those exceptions which are expected, namely those which were identified in root-cause analysis. Another common limitation of traditional exception handlers is that they generate error messages with which the operators are not familiar.

In the 4th IR, exception handling may expand from normal to exceptional situations. A primary challenge about exception handling in the 4th IR is to develop practices, in the form of protocols, used to define the interaction in exceptional situations. These protocols may include definition of protocols for capturing unexpected events, as well as of the interaction following the exception. The practices should include specification of the way the system may notify the operators about the exception, as well as practices for minimizing the nuisance due to the notifications.

Decision Support

A decision is often defined as a conclusion or resolution reached after consideration. Decision making is sometimes defined as deciding on an action to execute, such as selecting a choice from available options.

By design, we can control the likelihood of selecting a desired choice (Thaler and Sunstein 2008). For example, Li et al. (2013) studied the effectiveness of alternative public policies targeted at increasing the rate of deceased donor organ donation. The experiment included treatments across different default choices and organ allocation rules. The results indicate that the opt-out with priority rule system generates the largest increase in organ donation relative to an opt-in only program.

Obviously, if this is the case, then the design should prefer the automated selection over asking the operator to make a decision. Therefore, we should assume that at design time we cannot know what will be the best choice at run time. Rather, we need to rely on the reasoning of the decision maker.

Another method for affecting the option selection is by marking the preferred option as recommended. Both the default choice and the recommendation mark methods were employed extensively in the 3rd IR. However, the impact of employing these methods is much higher when people started to use the Internet as a primary means for business. An interaction designer needs to decide which of the two methods may be better for each of the possible decision situations. This is a challenge for the 4th IR.

Norman (1990) has demonstrated that often the problem of decision making is actually a problem of missing critical information. An operator needs information in order to make the proper decisions. The challenge of decision support in the 4th IR is of providing the operators with the information they need. New guidelines should be developed, regarding dilemmas such as:

- How to make sure that the operators have all the information they need for decision making.
- How to make sure that the information required is available and that the operators know how to get it.
- How to make sure that the operators will notice the required information and that they perceive it correctly.
- How to avoid overwhelming the operator with distracting information.

Proactive HMI Terminology

In 3rd IR systems engineers realized that they need to provide the operators with means to prevent failure but they were not aware of the difficulties that the operators might experience in employing these means. If an operator failed to apply the means provided successfully, people attributed it to being irrational, or to making an error. The implication of this approach was that systems engineers did not feel responsible for assuring that the operators make the proper decisions, or for preventing errors. The approach to failure was reactive, namely looking for someone to blame for the error.

The term "human error" often refers to an unintentional action that triggered a failure. Such definition is commonly used in studies of organizational behavior (Frese and Keith 2014). The problem with this definition is that, in many cases, the loss cannot be attributed to any unintentional action, or even to a judgment error. In these cases, this term should rather be attributed to interactive complexity (Perrow 1984), namely to operating the system in exceptional situations (Hollnagel et al. 2006).

The new view of operators' errors is that the organization can and should prevent use errors. It does not make sense to demand that the operators avoid making errors, because they cannot. The operators follow the Human Factors version of Murphy's law: "If the system enables the operators to fail, eventually they will."

Operators' errors should be regarded as symptoms of the organizational deficiency that enables them, and not as the sources of the accident. The HFE approach to preventing user errors is by design, by considering the limitations of the users and the operators. This approach enables learning from incidents: instead of blaming the users; we focus on exploring why they failed in order to understand how to prevent similar mishaps in the future. Recently, a new methodology for safety culture has been proposed; this defines the investigation of the stakeholders in the organization such that safety considerations override personal interests (Reason 1990).

The term *Use Error* is a new term that is recently replacing the term *User Error*. The need to change the term was because of common malpractice of the stakeholders (the responsible organizations, the authorities, journalists) in cases of accidents: instead of investing in fixing the error-prone design, the management attributed the error to the users (Dekker 2007). User advocates (such as Reason, Hollnagel, and Dekker) have noted that the user action is classified as an error only if the results are painful, implying that it is not the user who should be considered responsible for the errors (Hollnagel 1983). The new term suggests that the incident should be attributed to the way the system is being used, rather than to the user. The term "*Use Error*" is used also in recent standards, such as IEC 62366: Application of Risk Management to Medical Devices. The standard defines *Use Error* as an:

> …act or omission of an act that results in a different medical device response than intended by the manufacturer or expected by the user.

The term *Use Error* suggests that the error is the result of temporal conditions. However, from the record of accident investigations, it is evident that use errors are enabled by poor design and other incessant operational conditions imposed by the responsible organization (Reason 1997). IEC 62366 includes an explanation (Annex A) [emphasis the authors]:

> This International Standard uses the concept of **use error**. This term was chosen over the more commonly used term of "**human error**" because not all errors associated with the use of medical devices are the result of oversight or carelessness on the part of the user of the medical device. Much more commonly, **use errors** are the direct result of "**poor user interface design.**"

Failure to prevent a mishap may be regarded as a design or implementation mistake. In the 4th IR, the preferred approach to handle failure is proactively, namely designing systems in which the operators cannot make errors. A problem in implementing the proactive approach is related to the terminology associated with failure. If the failure is associated with irrational behavior or with errors, then the system designer does not have sufficient incentive to prevent the failure. In the case of failure, the investigation focuses on the operators' mistakes instead of finding ways to prevent similar failures in the future.

A challenge of HMI terminology in the 4th IR is to force the stakeholders to use proactive terms, especially in accident investigation. When the operators make a decision that seems improper in hindsight, the reference to this decision should be in terms of the circumstances of the decision, such as "obscure information."

Dynamic Adaptivity

In the 3rd IR, the HMI was examined in usability testing, prior to the system delivery and deployment. The problem is that after getting experience with the system operator, the operator's behavior changes, and the demands for effective interaction change accordingly. Designers believed that adaptivity helps the operators. Accordingly, Microsoft designed its Office applications with adaptive menus, eliminating, hiding or disabling menu items that were not used frequently.

The problem is that the operational demands change with experience. After the operators learned how to use the menu, they change, which means that the learning was useless. The operators need to learn the new setting and to adapt the way the use the menus. This sequence repeats, as the system learns the new user's behavior, and responds by adapting the menu structure again and again.

Apparently, the straightforward adaptability described above is recursive, and to the user it appears as inconsistent. In the 4th IR, adaptivity should be by situation, not by the operator's experience. Items available to the operators are those relevant to the situation and, accordingly, supported by the design.

The challenge of adaptivity in the 4th IR is to propose tools for assessing the benefits vs. the drawbacks of adaptivity, as well as guidelines for when to prefer it over consistency.

Interaction Adjustability

During the system development and after deployment, it is almost always the case that some of the requirements need to change. Adjustability is about the changes in the ways we adjust the system parameters, such as alarm thresholds, to the changes of the operational context.

A simple, straightforward technique for supporting agile development is by replacing constants by parameters, which the developers could change easily.

This solution suited cases when the developer was not sure about the optimal value of a constant, for example when the optimal temperature in a container in the process industry was not known before the integration testing. This solution suited also parameters used for the interaction design, such as risk indicators, used for alarming and for switching the operational mode.

In the 3rd IR, and even today, the search for the optimal parameter is often by trial and error. This method is often expensive in terms of time and budget. Statistical methods, such as trend analyzers, are often employed to extrapolate the effect of parameter changes. A challenge about interaction design in the 4th IR is about the practices for deciding about the need to change operational parameters, such as switching to safe-mode operation, or alarm thresholds. The tools required to handle the optimization process are handy; however, the efforts and investment required to incorporate them in a project are remarkable. We may expect that in the 4th IR new methods and practices will be developed, enabling fast and easy adjustment of system parameters in general, and specifically of parameters used in the interaction.

In the 3rd IR the thresholds defining the system behavior in various situations were often defined at design time. Often, the designers enabled behavior to be customized by changing the thresholds. The typical way to decide on the need to change the settings was by experience. Typically, the way to decide on the rate of change was by trial. Typically, there were no guidelines for deciding on the optimal rate of S/N (signal-to-noise) in the variety of operational situations. Being unable to decide when and how to change the thresholds, system operators are reluctant to adjust the system behavior to the situation.

The adjustability dilemma, namely how to adjust the thresholds to the situation, has not yet been resolved even theoretically. A primary challenge for the 4th IR is to define guidelines for defining the optimal rate of S/N in various operational situations. Zonnenshain and Harel (2008, 2015) proposed an architecture enabling capturing situational changes, as well as means for getting the S/N ratio.

A challenge about interaction adjustment in the 4th IR is about the practices for deciding about the need to change operational parameters, such as switching to safe-mode operation, or alarm thresholds. The 4th IR may present opportunities for applying data mining techniques in order to define guidelines for adjusting the S/N ratio. We may expect that in the 4th IR new methods and practices will be developed, enabling fast and easy adjustment of system parameters used for the interaction.

Situation Awareness

Situation awareness is about the human operators understanding of system status and the actual system state (Woods 1988). Problems of situation awareness are key factors in accident development. Norman (1990) attributed this kind of problem to design flaws, of not providing the operators with the information about the situation, which is critical for deciding on the appropriate behavior.

Often, in the 3rd IR, the problem if situation awareness was due to missing information about the risks. For example, if a sensor designed to notify about a risk does not function, as was the case of the Pressure Operated Release Valve (PORV) in the Three Mile Island (TMI) accident (Perrow 1984). A primary barrier to solving this problem is primarily technological. The 4th IR offers a solution to this problem, as new sophisticated, reliable means and methods that fulfill future needs being developed. However, there is a problem of managing the high volumes of data and eliciting information relevant to the operator's perception of the situation.

Another problem is of nuisance due to excessive or nonspecific notifications. For example, Harel (2006) studied the sources of the failure of the public to respond properly to alarms about missile attacks during the Israeli war with Hezbollah, and concluded that the reason was that it was due to such nuisance. Situation awareness should be evidence-based. This means that the information used for the decision making should be elicited from a large body of data recorded during the operation. A challenge about ensuring situation awareness in the 4th IR is about the ways to manage the big data, and to mine it and how to present it to the operators. The tools required to handle the information elicitation for the big data are handy; however, the efforts and investment required to incorporate them in a project are remarkable. We may expect that in the 4th IR new methods and practices will be developed, enabling fast and easy information elicitation in general and, specifically, of the information required to facilitate the operator's situation awareness.

Another problem to solving this problem is design neglect. In today's sophisticated, complex systems it is quite probable that a designer might forget to include a sensor about a critical component in the design. Common design practices include ad hoc guidelines for testing critical components, but in a complex system even simple screws might become critical. Can we add a sensor for each screw in a complex system? This is a challenge for the 4th IR.

Information Overload

Let us assume that our system in the 4th IR has all sensors and algorithms to measure and report on all situational variables. The complementary problem is that the operators are overwhelmed with information, most of it irrelevant to their task. The problem is that they cannot find the needle in a haystack. The challenge in the 4th IR is to identify the information required for the decision that the operators need to make, and to present it to the operators, and nothing else. HMI in the 4th IR encourages interaction designers to present the data obtained from sensors in forms of information essential for proper decision making, and to attract the operator's attention to indication about critical situations.

Rebounding from Operator Slips

A method for preventing errors due to operator slips, commonly employed in the 3rd IR, is by verifying that the system can handle the operator's input at the particular situation. Theoretically, system designers can specify in the requirements documents the cases in which the system should reject the input. However, the amount of work involved in writing such requirements is huge, making it impractical. A practical method is to specify the rules for proper operation, namely the operational procedures, and to enforce the system to operate bound to these rules (Leveson 2004).

A challenge about rebounding from operator slips in the 4th IR is to develop practices for specifying the operational rules and for transferring them to a database that the system can use to verify that the operation complies with the rules.

Error Prevention

The best way to learn how to operate a system is by trial and error (Jones et al. 2010). In order to enable this kind of learning, the interaction should mitigate the risks of costly errors. In the 3rd IR, software providers did not invest in preventing costly errors. As a result, the operators hesitated before trying, and avoided trying features that they did not know. The term Error is used extensively in investigations, implying that the operator is accountable for the incident, in order to justify distracting the discussion from costly investment in resilience assurance to cheap and handy personnel changes (Harel 2011).

To prevent errors, we need to understand them. Norman (1980) and Rasmussen (1982) proposed methods for classifying human errors. Also, Norman (1983) proposed various design rules based on analysis of human errors. Hollnagel et al. (2006) suggested that errors are tightly connected to deficiencies in the system resilience. It is a challenge of the 4th IR to develop and enforce guidelines for preventing costly errors, in order to facilitate learning by trial and error.

Proactive Investigation

Traditionally, safety engineering in the 3rd IR used to focus on the system side of the integration. Practices of safety assurance include assuring the reliability of critical components, robustness and redundancy. Typically, safety engineers did not deal with the human side.

Investigators of many celebrated accidents, such as in transportation and in the process industry, attributed the source of the accident to the operators' behavior, explaining that it did not match the system situation. Often, they attributed the behavior–situation mismatch to ambiguity of the HMI (e.g. Perrow 1984). Traditionally, people expect the users to follow the operational instructions and avoid making errors. In case of a use error, the user is accountable. For example, people expect that nurses respond promptly to all medical alarms, even though most of them are irrelevant. In case of an operational error, the operator is to blame. People expect that operators understand the safety implications of each option that they choose during the operation, in any future operational situation, based on unknown designers' reasoning.

In practice, users often fail to identify exceptional operational situations, to recall the operational instructions, and to predict the system behavior in these situations. Typically, in the case of an accident, we accuse the user of negligence and we accuse the operator of unreasonable operation. We consider the user errors as the source of the accident. In fact, most accidents are attributed to user errors.

Typical reaction to accidents in the 3rd IR were, and are still today, emotion driven. Following an incident or an accident, the people involved typically focus on investigation issues rather than on improving the safety. In emotion-driven organizations, where the safety culture is biased, incident investigations often obey the "blame and punish" script (Zonnenshain and Harel 2008, 2015). Emotion-driven response to incidents prohibits improving resilience, because the investigations do not focus on the design changes needed to improving the resilience. On the other hand, when the organization adopts a safety culture, the investigations include recommendations for design change, and the management promotes implementing these recommendations (Dekker 2007). The guide proposes a procedure for continuous improvement of system resilience by learning from mishaps, preventing this bias (Weiler and Harel 2011). Following an incident or an accident, the stakeholders' reaction is typically emotion driven. In emotion-driven organizations, where the safety culture is biased by investigation, incident investigation often follows the "blame and punish" script, attributing the incident to the operators.

Emotion-driven response to incidents hampers the efforts to learn from them, because the investigations focus on the stakeholders instead of on the design changes needed to improve the resilience. Typically, they focus on investigation issues (looking for "bad apples," Dekker 2007) instead of on improving the safety. If a person blames the operators, then it may be the case that this person wants to distract the blame, or his/her impotence, to shift it to those who cannot protect themselves.

The interest of the stakeholders determines whether the decision is rational. In order to guarantee that learning from failure is effective, we need to avoid judging the decisions in terms of rationality. In case of a costly accident, the stakeholder often focus on looking for a person how may be nominated as responsible for the accident. This understandable behavior results in overlooking possible ways to prevent similar accidents by redesign.

Complying with the accountability bias is convenient for safety administrators, because if the operator is accountable for the accident, this implies that they are not. The accountability bias distracts the focus from the stakeholders, in charge of safety, to the operators, the victims of the design flaw (Jackson and Harel 2018). The problem with this approach is that it inhibits processes of safety improvements. The users' typical response is to think more about their own risks and less about the interests of the organization, or the public. The organization avoids acting to improve safety, because such actions are likely to manifest the accountability of the safety administrators (Dekker 2006). For example, admitting the design mistake that caused the Airbus 320 accident in Mulhouse Habsheim in 1988 could have prevented the accident in Bangalore, India in 1990 (Casey 1993). In this case, the safety administrators preferred to accuse the pilots instead of exploring the systemic circumstances. Also, accusing members of medical teams for accidents due to risky operational procedure is quite common.

The New View approach is often criticized for encouraging carelessness during the operation, which might result in accidents. Safety administrators often apply such reasoning to justify setting the system in ways that transfer their investigation to the users, which is risky to the public (Dekker 2007). For example, safety administrators are tempted to set alarm thresholds such that the users are overwhelmed with irrelevant alarms, in order to reduce the risks of missing alarms when needed.

In order to assure learning from incidents, we need to know the barriers to learning. The main obstacles to improving by learning from accidents are human biases by the system stakeholders. A common practice of organizations for coping with the investigation bias is by adopting a safety culture. This methodology encourages that investigations should include recommendations for design changes. However, this methodology contradicts the human nature of blaming people. People praise the concept of safety culture as long as everything goes according to the plans. In case of mishaps, people typically change their strategy and turn to their natural behavior of blaming others.

The IEC 60601-1-8 standard (IEC 2006) triggers awareness of the risks involved in using medical alarms, by warnings about what might go wrong. However, it does not provide sufficient guidance for how to avoid these risks. Insufficient guidance about maintaining "safety culture" results in organizational settings that overprotect the authorities, leaving "holes" (in terms of the "Swiss Cheese" model by Reason 1990) in patient safety (Harel 2011). Barriers to learning from incidents include:

- Reporting on incidents might suffer from the investigation bias.
- Information extraction might suffer from lack of means for data aggregation.
- Improving the operational procedures might be hampered by the responsible organization.

To enable learning from incidents we need to provide data about the circumstances of the mishap. The challenge of ***proactive*** investigation in the 4th IR is to develop tools and practices to capture, analyze, and report about incidents. These practices should consider the investigation bias and provide defenses against it.

Resilience Development

In the 3rd IR people believed that fixing a design flaw should always improve system safety. For example, Jackson (2009) discusses "a framework for implementation that both public and private organizations can use as a guide to establishing procedures for anticipating, surviving, and recovering from disruptions." In reality, this is not always the case. Indeed, Popper (1968) argues that "non-reproducible single occurrences are of no significance to science. Thus a few stray basic statements contradicting a theory will hardly induce us to reject it as falsified. We shall take it as falsified only if we discover a reproducible effect which refutes the theory."

The implication of this observation to engineering is that we can never be sure that adding a safety feature is always safe. Unfortunately, after fixing a safety problem, it may happen that the upgraded system suffers from a new, latent problem. A famous example is the reliability problem of the PORV indicator, which was fixed following a near miss in the Davis Hesse II nuclear power plant in 1978. The problem was fixed and the PORV indicator of the upgraded TMI nuclear power plant was more reliable. Because it was more reliable, the operators relied on it, even though it did not function properly.

In the 4th IR, usability trackers will be embedded in the software delivered, enabling system developers to learn the problems that the operators experience during the operation (Harel 1999).

15.5   Summary

This chapter addresses possible changes in the 4th IR relating to the role of human factors in systems engineering. It suggests that the 4th IR may involve various shifts toward HMI, associated with technology, methodology, and HMI thinking and practices. These shift may affect the people productivity, quality of life, and safety.

The main conclusion is that, in the 4th IR, HMI needs to evolve into a scientific discipline, offering degree programs in universities that teach students the essentials of HMI engineering and interaction design.

References

Arnold, R.D. and Wade, J.P. (2015). A definition of systems thinking: a systems approach. *Procedia Computer Science* 44: 669–678.

Bainbridge, L. (1983). Increasing levels of automation can increase, rather than decrease, the problems of supporting the human operator. In: *Automatica*, vol. 19, 775–779. Reprinted in: (1987) Rasmussen, J., Duncan, K. and Leplat, J. (eds.) *New Technology and Human Error*, Wiley, Chichester, pp. 276–283.

Bartolomei, J.E., Hastings, D.E., de Neufville, R., and Rhodes, D.H. (2006) Screening for real options. An Engineering System: A Step Towards Flexible Weapon System Development, 4th Conference on Systems Engineering Research. Los Angeles, CA (April 2006).

BEA (2012). Final Report On the Accident on 1st June 2009 to the Airbus A330-203 Registered F-GZCP Operated by Air France, Flight AF 447 Rio de Janeiro–Paris. https://www.bea.aero/docspa/2009/f-cp090601.en/pdf/f-cp090601.en.pdf (Last accessed 8 July 2019).

von Bertalanffy, L. (1968). *General System Theory: Foundations, Development, Applications*, rev. ed. New York: Braziller.

Boy, G. (2016). *Tangible Interactive Systems: Capturing the Real World with Computers*. New York: Springer. ISBN: 978-1-4471-4338-3.

Boy, G.A. and Narkevicius, J. (2013). Unifying human centered design and systems engineering for human systems integration. In: *Complex Systems Design and Management* (eds. M. Aiguier, F. Boulanger, D. Krob and C. Marchal), 151–162. London: Springer ISBN-13: 978-3-319 02811-8.

Card, S., Moran, T., and Newell, A. (1983). *The Psychology of Human- Computer Interaction*. Hillsdale, NJ: Lawrence Erlbaum Associates Inc.

Casey, S. (1993). A leap of faith. In: *Set Phasers on Stun, and Other True Tales of Design, Technology and Human Error* (ed. S. Casey), 92–108. Aegean Publishing.

Cooper, A. (1998). *The Inmates are Running the Asylum: Why High Tech Products Drive Us Crazy*. Sams – Pearson Education.

Dekker, S. (2006). *The Field Guide to Understanding Human Error*. Ashgate Publishing.

Dekker, S. (2007). *Just Culture: Balancing Safety and Investigation*. Ashgate Publishing.

Dockrill, P. (2016). WATCH: F-16 autopilot system saves the life of an unconscious fighter pilot. *Science Alert* 14.

Fitts, D.J, Sandor, A., Litaker, H.L. and Tillman, B. (1987) *Human Factors in Human Systems Integration*. NASA Space HFE Project.

Frese, M. and Keith, N. (2014). Action errors, error management, and learning in organizations. *Annual Review of Psychology* 66 (1).

Harel, A. (1999). Automatic operation logging and usability validation. In: *Human-Computer Interaction: Ergonomics and User Interfaces, Proceedings of HCI International '99 (the 8th International Conference on Human-Computer Interaction)*, vol. 1 (eds. H.-J. Bullinger and J. Ziegler), 1128–1133. Lawrence Erlbaum Munich, Germany, (22–26 August 1999).

Harel, A. (2006). Alarm reliability: what if an alarm goes off and no one hears it? *User Experience Magazine* 5 (3).

Harel, A. (2010) Whose error is this? Standards for preventing use errors, The 16th Conference of Industrial and Management Engineering, Tel-Aviv, Israel.

Harel, A. (2011) Comments on IEC 60601-1-8. Letter submitted to IEC/TC 62 working group. https://www.researchgate.net/publication/304523911_ Comments_on_IEC_60601-1-8 (Last accessed 8 July 2019).

Harel, A. and Weiss, M. (2011) Mitigating the Risks of Unexpected Events by Systems Engineering, Proceedings of the Sixth Conference of INCOSE-IL, Hertzelia, Israel.

Hollnagel, E. (1983) Human error. Position Paper for NATO Conference on Human Error, Bellagio, Italy (August 1983).

Hollnagel, E., Woods, D., and Leveson, N. (2006). *Resilience Engineering: Concepts and Precepts*. Aldershot, UK: Ashgate Publishing Limited.

IEC (2006) IEC 60601-1-8 Medical Electrical Equipment -- Part 1-8: General Requirements for Basic Safety and Essential Performance -- Collateral Standard: General Requirements, Tests and Guidance for Alarm Systems in Medical Electrical Equipment and Medical Electrical Systems. International Electrotechnical Commission.

Jackson, S. (2009). *Architecting resilient systems: accident avoidance and survival and recovery from disruptions*, Wiley Series in Systems Engineering and Management. Hoboken, NJ: Wiley.

Jackson, S. and Harel, A. (2017). Systems engineering decisions analysis can benefit from the added consideration of cognitive sciences. *Systems Engineering Magazine SyEN* 55: 6–22.

Jackson, S.; Harel, A. (2018) Improving Decisions to Mitigate the Risks of Organizational Accidents. *Preprints*, 2018030042 (doi: 10.20944/preprints201803.0042.v1).

Jones, R.S.P., Clare, L., MacPartlin, C., and Murphy, O. (2010). *The effectiveness of trial-and-error and errorless learning in promoting the transfer of training. European Journal of Behaviour Analysis* 11 (1): 29–36.

Kenett, R.S., Harel, A., and Ruggeri, F. (2009). Controlling the usability of web services. *International Journal of Software Engineering and Knowledge Engineering (IJSEKE)* 19 (5): 627–651.

Kenett, R.S., Zonnenshain, A. and Swarz, R.S. (2018). Systems Engineering, Data Analytics, and Systems Thinking: Moving Ahead to New and More Complex

Challenges, Proceedings of the 28th Annual INCOSE International Symposium, Washington, DC (7–12 July 2018).

Kopainsky, B., Alessi, S.M., and Davidsen, P.I. (2011). Measuring knowledge acquisition in dynamic decision making tasks. In: *The 29th International Conference of the System Dynamics Society* , (Washington, DC, 24–29 July 2011) (eds. J.M. Lyneis and G.P. Richardson), 1909–1939. Albany, NY: System Dynamics Society.

Landauer, T.K. (1996). *The Trouble with Computers: Usefulness, Usability, and Productivity*. Cambridge, MA: MIT Press.

Learmount, D. (2011). Global Airline Accident and Safety Review for 2010. FlightGlobal (18 January).

Leveson, N.G. (2004). A new accident model for engineering safer systems. *Safety Science* 42 (4): 237–270.

Li, D., Hawley, Z., and Schnier, K. (2013). Increasing organ donation via changes in the default choice or allocation rule. *Journal of Health Economics* 32 (6): 1117–1129. https://doi.org/10.1016/j.jhealeco.2013.09.007.Epub.

McDermott, P.L., Ryan, M., Bonaceto, C., Potter, S., Savage-Knepshield, P., Kosnik, W., Dominguez, C. (2017) Improving How We Weave Cognitive Engineering into Military Acquisition: Understand, Design, Validate. Proceedings of the Human Factors and Ergonomics Society Annual Meeting, Austin, Texas (9–13 October 2017).

Meister, D. (1999). *The History of Human Factors and Ergonomics*. CRC Press.

Nielsen, J. (1993). *Usability Engineering*. Boston, MA: Academic Press.

Norman, D.A. (1980) Errors in Human Performance. National Technical Information Service, US Department of Commerce, Springfield, VA.

Norman, D.A. (1983). Design rules based on analyses of human error. *Communication of the ACM* 26 (4): 254–258.

Norman, D.A. (1988). *The Design of Everyday Things*. New York: Basic Books. ISBN: 978-0-465-06710-7.

Norman, D.A. (1990). The "problem" of automation: inappropriate feedback and interaction, not "over-automation". In: *Human Factors in Hazardous Situations* (eds. D.E. Broadbent, A. Baddeley and J.T. Reason), 585–593. Oxford: Oxford University Press.

Norman, D.A. and Draper, S.W. (eds.) (1986). *User-Centered System Design: New Perspectives on Human-Computer Interaction*. Hillsdale, NJ: Lawrence Earlbaum Associates.

Perrow, C. (1984). *Normal Accidents: Living with High Risk Technologies*. New York: Basic Books.

Popper, K. (1968). *The Logic of Scientific Discovery*. New York: Harper and Row.

Rasmussen, J. (1982). Human errors. A taxonomy for describing human malfunction. *Industrial Installations, Journal of Occupational Accidents* 4, Elsevier Scientific Publishers: 311–335.

Reason, J.T. (1990). *Human Error*. Cambridge, UK: Cambridge University Press.

Reason, J.T. (1997). *Managing the Risks of Organizational Accidents*. Aldershot, UK: Ashgate Publishing.

Richmond, B. (1994). Systems Dynamics/Systems Thinking: Let's Just Get On With It. International Systems Dynamics Conference, Sterling, Scotland.

Robert, D., Berry, D., Isensee, S., and Mullaly, J. (1998). *Designing for the User with OVID: Bridging User Interface Design and Software Engineering*, Software Engineering Series. Macmillan Technical Publication.

Schwab, K. (2016). *The Fourth Industrial Revolution*. World Economic Forum. ISBN: 1944835008.

Senge, P. (1990). *The Fifth Discipline, the Art and Practice of the Learning Organization*. New York, NY: Doubleday/Currency.

Shneiderman, B. (1980). *Software Psychology: Human Factors in Computer and Information Systems*, Winthrop Computer Systems Series. Winthrop Publishers.

Shneiderman, B. (1986). Designing menu selection systems. *Journal of the American Society for Information Science* 37 (2): 57–70.

Shneiderman, B. (1987). *Designing the User Interface: Strategies for Effective Human-Computer Interaction*. Reading, MA: Addison-Wesley.

Sillitto, H.G., Martin J., Griego R., McKinney, D., Arnold, E., Godfrey, P., Dori, D., Krob, D. and Jackson, S. (2018). Envisioning Systems Engineering as a Transdisciplinary Venture, Proceedings of the 28th Annual INCOSE International Symposium, Washington, DC (7–12 July 2018).

Standish Group (1995) *The COMPASS report*, Forbes.

Stave, K.A. and Hopper, M. (2007). What Constitutes Systems Thinking? A Proposed Taxonomy. In: *Proceedings of the 25th International Conference of the System Dynamics Society*, (Boston, MA, 29 July – 3 August 2007) (ed. J. Sterman), 4508–4529. Albany, NY: System Dynamics Society.

Sweeney, L.B. and Sterman, J.D. (2000). Bathtub dynamics: initial results of a systems thinking inventory. *System Dynamics Review* 16 (4): 249–286. https://doi.org/10.1002/sdr.198.

Taleb, N. (2007). *The Black Swan: The Impact of the Highly Improbable*. Random House Trade Paperbacks.

Thaler, R.H. and Sunstein, C.R. (2008). *Nudge: Improving Decisions about Health, Wealth, and Happiness*. New Haven, CT: Yale University Press.

Weiler, M. and Harel A. (2011) Managing the Risks of Use Errors: The ITS Warning Systems Case Study. The 6th Conference of INCOSE-IL, Herzelia, Israel.

Weinberg, G. (1971). *The Psychology of Computer Programming*. Dorset House Books.

Wiegmann, D. and Shappell, S. (1997). Human factors analysis of post-accident data: applying theoretical taxonomies of human error. *The International Journal of Aviation Psychology* 7: 67–81.

Woods, D.D. (1988). Coping with complexity: the psychology of human behavior in complex systems. In: *Tasks, Errors, and Mental Models* (eds. L.P. Goodstein, H.B. Andersen and S.E. Olsen), 128–148. Bristol, PA: Taylor & Francis.

Zonnenshain, A. and Harel, A. (2008). Extended system engineering – ESE: integrating usability engineering in system engineering. *INCOSE International Symposium* 18 (1), (Utrecht, the Netherlands, 15–19 June)): 1542–1556.

Zonnenshain, A. and Harel, A. (2015). A practical guide to assuring the system resilience to operational errors. *INCOSE International Symposium*, (Seattle, WA, 13–16 July) 25 (1): 538–552.

16

Building More Resilient Cybersecurity Solutions for Infrastructure Systems
Daniel Wagner

Synopsis

Much of the world's global infrastructure was never designed to take cyber threats into account. Largely based on 1970s era SCADA technology, which is highly vulnerable to attack, many power plants, water control systems, and other critical infrastructures will remain vulnerable because the hundreds of billions of dollars required to replace them is not available. When funding does become available, what are some of the considerations that should be taken into account in the design of news systems to ensure that new types of vulnerabilities do not replace existing ones? This chapter will explore the topic broadly, elaborating on the scope of the problem, current types of solutions, and an orientation toward the future.

16.1 A Heightened State of Vulnerability

The world's infrastructure is generally vulnerable to cyberattack because so much of it is dependent upon software and so many of its systems are interconnected. Much of the world's critical infrastructure utilizes supervisory control and data acquisition (SCADA) systems, which automatically monitor and adjust switching, manufacturing, and other process control activities based on digitized feedback data gathered by sensors. These tend to be specialized, older computer systems that control physical pieces of equipment that do everything from routing trains along their tracks to distributing power throughout a country. SCADA systems have increasingly become connected to the Internet but were not designed with cybersecurity in mind.

This is a big problem and becomes more of a threat each year. A study carried out in 2014 concluded that almost 70% of critical infrastructure companies across multiple sectors in the United States had suffered at least

Systems Engineering in the Fourth Industrial Revolution: Big Data, Novel Technologies, and Modern Systems Engineering, First Edition. Edited by Ron S. Kenett, Robert S. Swarz, and Avigdor Zonnenshain.

one security breach that led to the loss of confidential information or the disruption of operations during the preceding 12 months (Vinton 2014). Virtual terrorists – those who perpetrate cyberattacks via the Dark Web anonymously and operate in a borderless and lawless world – could do tremendous damage if they wanted to, ranging from taking control of water treatment facilities to shutting down power generation plants to causing havoc with air traffic control systems.

Despite the fact that cyberattacks are occurring with greater frequency and intensity around the world, many either go unreported or are underreported, leaving the public with a false sense of security about the threat they pose and the lives and property they impact. While governments, businesses, and individuals are all being targeted on an exponential basis, infrastructure is becoming a target of choice among individual and state-sponsored cyberattackers who recognize the value of disrupting what were previously thought of as impenetrable security systems. This has served to demonstrate just how vulnerable cities, states, and countries have become.

16.2 The Threat Is Real

In December 2015, a presumed Russian cyberattacker successfully seized control of the Prykarpattyaoblenergo Control Center (PCC) in western Ukraine, leaving 230,000 without power for up to six hours. This marked the first publicly recognized time that a cyberweapon was successfully used against a nation's power grid. The attackers were skilled strategists who carefully planned their assault over many months, first doing reconnaissance to study the networks and siphon operator credentials, then launching a synchronized assault in a well-choreographed dance. The control systems in Ukraine were surprisingly more secure than some in the United States, since they were well-segmented from the control center business networks with robust firewalls, emphasizing just how vulnerable power systems are globally (Zetter 2016).

The PCC operated a SCADA system, which allows for remote controlling and monitoring of industrial processes. The attackers overwrote firmware on critical devices at 16 substations, leaving them unresponsive to any remote commands from operators, effectively leaving plant operators helpless (Zetter 2016). Given the degree of sophistication of the intrusion, the attackers could have rendered the system permanently inoperable; that they did not led some in Ukraine to speculate that the attack was a message from Russia not to pursue pending power plant nationalization legislation, since some of those plants were owned by a powerful Russian oligarch with close ties to top government officials.

The Ukraine example was hardly the first cyberattack on a SCADA system. Perhaps the best known previous example occurred in 2003, though at the time it was publicly attributed to a downed power line rather than a cyberattack. The northeastern US blackout that year caused 11 deaths and an estimated $6 billion in economic damages, having disrupted power over a wide area for at least two days. Never before (or since, for that matter) had a "downed power line" apparently resulted in such a devastating impact. Some cyber professionals maintain that it was a cyberattack, which makes a lot more sense (Leyden 2008). Subsequent to that attack, SCADA attacks occurred in the UK, Italy, and Malta, among others (Brenner 2013). Cyberattacks against SCADA systems doubled in 2014 to more than 160 000 (Zetter 2016).

In recent years, numerous forms of malware targeting SCADA systems have been identified, including Stuxnet, Havex, and BlackEnergy3 (Fortigard Labs 2015). What these three forms of malware have in common is their ability to sneak through Industrial Control Systems (ICSs) undetected by exploiting the weakest link in the cyber defense network (people) and posing as a legitimate e-mail or by finding a back door in the SCADA system (McAfee 2011). The power sector has already demonstrated itself to be particularly vulnerable and must dedicate substantially more resources to closing back doors and training employees to avoid clicking on malicious files.

The US energy grid connects more than 5800 power plants with more than 450 000 mi of transmission lines – the ultimate soft target. Yet some 70% of the grid's key components are more than 25 years old and most of them use older SCADA technologies that are readily hackable (Wagner 2017b). An investigation by the US House Energy and Commerce Committee revealed that more than a dozen US utility companies reported daily, constant, or frequent attempted cyberattacks, ranging from phishing to malware infection to unfriendly probes. One utility reported that it had been the target of more than 10 000 attempted cyberattacks each month. The US has more than 57 000 ICS connected to the Internet. In 2015, the US Department of Homeland Security (DHS) responded to 295 hacking incidents related to industrial controls, up from 245 in 2014 (Wagner 2017c, p. 234).

At the annual Chaos Computer Club (CCC) Conference, a gathering of global hackers held in Germany, analysts have demonstrated how to get full control of industrial infrastructure in the chemical, oil, gas, and energy industries. Hackers then shared this information with each other and created fully searchable public databases of known exploits that can be used to commandeer critical infrastructure. One well-known hacker database, Shodan, provides helpful tips on how to exploit everything from power plants to wind turbines and is searchable by country, company, or device. Since it is hosted on multiple servers across the world, Shodan cannot be easily shut down. Moreover, in many countries it is not illegal to publish such vulnerabilities (even though it clearly should be) (Wagner 2017c, p. 234).

Examples of other instances when SCADA systems have been breached abound, some of which did not involve sinister dark forces or governments, as far back as the 1990s. Another earlier example occurred in Maroochy Shire, in Queensland, Australia, in 2001, when a hacker gained control of a sewage treatment plant's ICS and caused millions of liters of raw sewage to spill out into local parks, rivers, and the grounds of the Hyatt Regency hotel.

Cybercriminals have also gotten into the act, as a means of extorting money from utilities and governments. A number of incidents were reported in Brazil between 2005 and 2007, when numerous cyberattacks were carried out in Rio de Janeiro and in the state of Espirito Santo, where nearly 3 million people were left in the dark because the local electricity provider had failed to meet the extortion demands of the hackers. The city of Vitoria, home of one of the world's largest iron ore producers, had numerous plants forced offline, costing the company almost $7 million (Goodman 2015).

Given the relative ease with which hackers can find a single system vulnerability, and the impossibility of plugging every conceivable security hole, cybersecurity professionals are in essence playing an endless game of cat and mouse, whereby a would-be attacker attempts to enter a system while security professionals attempt to defend a computer system from attack by applying continuous patches. The adversary then quickly moves to exploit the latest discovered vulnerability. That is why many computer security programs produce patches numerous times per day – even for home computers(Wagner 2016).

Systemic Vulnerabilities

Cyberattackers tend to view an infrastructure facility and its ICS in a holistic way, identifying physical vulnerabilities of the controllers and processes with options for exploiting its vulnerabilities via digital manipulations. Traditional cyberattacks focus on the Windows operating system using zero-day (previously unknown) vulnerabilities or other IT flaws to capture valuable data or cause a Distributed Denial of Service (DDoS) (loss of data) attack. Targeted ICS attacks such as Stuxnet and Aurora exploit system design features, with the IT system being typically focused on Advanced Persistent Threats (APTs) and traditional insider threats. Many ICS devices, including new ones, remain insecure because many legacy ICS cannot implement the latest security technologies. Even so, these devices tend not to be replaced because they still function and many infrastructure owners do not want to pay to upgrade their systems until the system no longer works, exacerbating the cyberthreats.

While the majority of the equipment that comprises a SCADA system resides in the control center network behind firewalls, localized SCADA communication equipment directly connected to the ICS can be as vulnerable as the ICS themselves. A digital attack or intrusion on these localized communication systems can have a greater effect on the overall system and allow the attacker access to all ICS connected to them. This gives the attacker the ability to operate all ICS, creating broader systemic impact.

Modern industrial systems operate with standard ICS produced by only a few vendors (about half are US-based) with similar architectures, training, and even the same default passwords. The control system designs generally lack the cybersecurity requirements and engineering (hardware and software) to be able to protect against the many failure modes related to attacks by hackers. Since the same ICS are used across multiple industries, a compromise of the ICS features in systems in one facility or industry can affect all facilities or industries that utilize the same systems and devices.

The BlackEnergy malware referenced previously was designed to attack the human–machine interface (HMI) of several major ICS vendors, allowing root access to HMIs used in multiple industries around the world. BlackEnergy was used in the 2015 Ukrainian hack and has infected many US electric grids since late 2014(Wagner 2017c, p. 236).

From the attacker's perspective, exploiting features rather than bugs has a significant advantage, as they cannot be expeditiously repaired by a vendor releasing a patch. The Stuxnet attack bypassed automated safety systems and prevented manual safety systems from being initiated. Aurora used the safety systems to produce the targeted attack. ICS with some protection against cyberattacks have been released but the installation and upgrade cycle is long and it cannot be guaranteed to be effective in new attack scenarios. Cyberthreats can damage equipment and simultaneously attack multiple locations, leading to extended long-term outages, with the need to replace or repair long-lead time equipment.

Given generally lax security standards, manufacturers are finding themselves potentially exposed at every point of their computer networks. Unlike IT networks, operational networks offer poor (or even nonexistent) visibility into ICS. Widely used industrial controllers are dedicated to making logic-based decisions to control industrial processes. Given the design of ICS networks and the absence of basic security protocols such as authentication and encryption, most ICS attacks need not exploit software vulnerabilities. Once the network is breached, an attacker typically gains free access to all the controllers and can alter their configuration to cause disruptions. In the absence of fully understanding which assets exist in a given ICS network, and who is accessing them at any specific point in time, they cannot be protected.

One of the biggest security challenges manufacturers face is the variety of communication protocols used in ICS networks. Standard data plane protocols such as Modbus and DNP3 are used by SCADA (and other) applications to communicate physical measurements and process parameters such as temperature, pressure, and valve status. By contrast, control plane protocols – which are used to configure automation controllers, update their logic, make code changes, and download firmware – tend to be proprietary and vendor-specific. Each vendor uses its own approach to standards for programmable controllers, which are rarely documented, making it very difficult to monitor critical activities.

Since the goal of most ICS cyberattacks is to prompt operational disruptions and/or physical damage, an attacker will typically try to modify the manner in which the process executes. While a predefined set of process parameters can be changed through SCADA applications, the logic maintained on the controller defines the process flow and its safety settings. Changing the controller logic is, therefore, the easiest way to cause such disruptions. Once inside the network, an attacker can easily download control logic to an industrial controller, or change its configuration. Since these actions are executed using proprietary vendor-specific protocols, there is no standard way to monitor control plane activities. As a result, changes made by an attacker often go unnoticed until damage starts to occur.

Gaining visibility into ICS networks is the first step in being able to protect them from cyberthreats. Discovering all assets, especially industrial controllers, is critical, as is maintaining reliable inventory of configurations, logic, code, and firmware versions for each controller. Once such baselines have been established, continuous monitoring of control-plane activities for unauthorized access or changes is required. This is especially true since the controllers are often modified by third-party integrators, which can result in changes that are not captured by network monitoring alone. However, new specialized monitoring and control technologies for ICS networks are available (Industry Week 2016).

Most ICS cybersecurity guidance and training is given to end-users and ICS vendors. By contrast, very little guidance is available to others (such as system integrators) who are often used to implement new designs and upgrade older legacy designs. This becomes an important issue with older legacy systems where the original vendor is no longer supporting its products and is, therefore, unaware of how its systems are being reconfigured. There have been many ICS cyber incidents that have occurred because of insiders creating or "exploiting" cyber vulnerabilities without being aware of it. Three examples that resulted in significant impacts were the 2007 Plant Hatch nuclear plant shutdown, the 2008 Florida outage, and the 2010 San Bruno, California natural gas pipeline rupture. Stuxnet was also a case where the system integrator is thought to have unintentionally inserted the malicious malware.

In July 2017, the DHS and US Federal Bureau of Investigation (FBI) jointly issued an urgent report stating that, since May of that year, hackers had been penetrating the computer networks of companies that operate nuclear power stations, other energy facilities, and manufacturing plants in the US and other countries (Wagner 2017c, p. 237). Among the companies targeted was the Wolf Creek Nuclear Operating Corporation, which runs a nuclear power plant near Burlington, Kansas. The hackers appeared determined to map out computer networks for future attacks, the report concluded, but investigators had not been able to analyze the malicious "payload" of the hackers' code, which would have provided more detail into what their objective was.

In the US, nuclear facilities are required to report cyberattacks that relate to their safety, security, or operations. None of the 99 nuclear plants in the US had reported that the security of their operations had been affected by the attacks but, in most cases, the attacks targeted people – industrial control engineers who have direct access to systems that, if damaged, could lead to an explosion, fire, or a spill of dangerous material. The report indicated that an "advanced persistent threat" actor was responsible, which points to government-backed hackers. The hackers' techniques mimicked those of the organization known as "Energetic Bear," a Russian group believed to have been responsible for attacks on the US energy sector since at least 2012.

Hackers wrote highly targeted e-mail messages containing fake résumés for control engineering jobs and sent them to the senior industrial control engineers who maintain broad access to critical ICS. The résumés were Microsoft Word documents laced with malicious code. Once the recipients clicked on those documents, attackers could steal their credentials and infect other machines on a network. In some cases, the hackers also compromised legitimate websites that they knew their victims frequented (watering hole attacks). In others, they deployed man-in-the-middle (MITM) attacks, in which they redirected their victims' Internet traffic through their own machines (Pelroth 2017).

16.3   A Particularly Menacing Piece of Malware

Hackers allied with the Russian government have devised a cyberweapon that has the potential to be the most disruptive yet against electric systems. The malware, believed to have been used in the Ukraine power grid hacks, has been dubbed CrashOverride and is the first malware framework ever designed and deployed to attack electric grids. With modifications, it could be deployed against US electric transmission and distribution systems with devastating impact. CrashOverride is the fourth ever piece of ICS-tailored malware (after Stuxnet, BlackEnergy 2, and Havex) used against targets, and the second ever to be designed and deployed for disrupting physical industrial processes. What makes CrashOverride particularly alarming is that it is part of a larger framework, designed like a Swiss Army knife that can be flipped open to extract the tool needed or to be added to achieve different effects. Theoretically, the malware can be modified to attack different types of ICS, such as water and gas.

CrashOverride manipulates the settings on electric power control systems, scans for critical components that operate circuit breakers, and opens them, which stops the flow of electricity. It continues to keep them open even if a grid operator tries to close them, creating a sustained power outage. The malware also has a "wiper" component that erases the software on the computer system that controls the circuit breakers, forcing the grid operator to revert to manual operations, which requires literally going to one or more substations to restore power. With this malware, the attacker can target multiple locations with a "time bomb" functionality and set the malware to trigger simultaneously. That could create outages in different areas at the same time (Nakashima 2017).

CrashOverride is not unique to any particular vendor or configuration but leverages knowledge of grid operations and network communications to cause impact. It can be immediately repurposed, and with a small amount of tailoring could be effective anywhere in the world. The malware could be leveraged at multiple sites simultaneously but would probably result in outages for hours or days rather than weeks or months. The functionality of the CrashOverride framework serves no espionage purpose; the only real feature of the malware is to lead to electric outages (Dragos 2017).

16.4 Anatomy of An Attack

On 17 December 2016, remote terminal units (RTUs) used to monitor and control circuit breakers in the Pivnichna (North) electrical substation in Ukraine went offline unexpectedly. The 330-kV transmission substation serves Kiev's power distribution system. The RTUs' failure resulted in power losses in the northern right bank section of Kiev but, a few minutes later, the power was restored. Ukrenergo's engineers were perplexed by the potential cause of the incident, which was eerily similar to the attack that occurred one-year prior, almost to the day. Ukrenergo's engineers wondered whether the event was another cyberattack, since the incident occurred amid a flurry of 6500 cyberattacks over two months.

Just as many professionals employ "best practices," so do elite hackers, otherwise referred to as the "cyber kill chain." There were primary similarities and differences between the attacks:

Similarity. BlackEnergy malware and KillDisk (a data deletion program) were used in both attacks. Hackers' direct interaction with the ICS caused the blackouts.

Difference. In 2015, hackers attacked multiple distribution substations (seven 110 kV and twenty-three 35 kV) while, in 2016, hackers attacked a single transmission substation.

Similarity. In both incidents, hackers targeted substation RTUs.

Difference. In 2015, hackers used malicious firmware to permanently disable the RTUs after opening breakers and then used KillDisk to damage operators' terminals, which prevented remote repair of the RTUs. In 2016, the hackers merely deactivated the RTUs, which made restoration easier (hence the short down time of just a few minutes).

Based on the facts associated with both breaches, it is believed that multiple groups were involved in the 2016 attack. The presumed narrative is that hackers would have first identified employees who hold key positions within and knowledge about Ukrenergo, such as system administrators and engineers. Hackers also would have identified target victims; detailed information could have been gathered about them online (from social media sites). Hackers would have gathered information such as devices, operating systems, and business applications used throughout Ukrenergo, which can be gleaned, for example, by looking at employee LinkedIn job descriptions and skill sets. Particularly valuable would be IT credentials, account details, weak passwords, and organizational processes and procedures. Hackers could have found some of this accidentally published online or cracked it with relative ease.

Hackers could have started learning about their ultimate targets by using the Shodan search engine to find Ukrenergo infrastructure exposed to the public Internet. Detailed information on the RTUs and ICS targeted in the 2015 attack was readily available online. In 2015, hackers weaponized Microsoft Word files with malicious macros. Hackers probably customized attacks toward targeted victims, such as recipients of spear phishing e-mails. Next, the hackers would have gained access to Ukrenergo's network, using clever coding techniques to obfuscate methods and evade signature-based attack detection. They would have also used macros to detect security technologies in Ukrenergo's environment, including intrusion prevention systems and sandboxes. The 2015 incident began with a spear phishing e-mail and the hackers gained access using native functionality in Microsoft Word to download BlackEnergy 3.

The tools used in the 2015 and 2016 attacks would undoubtedly have been similar, with BlackEnergy malware having been downloaded to a victim's computer via spear phishing, the hackers would have been "called back" to the hackers' remote command and control (C2) infrastructure. The C2 infrastructure would have allowed hackers to extract data from Ukrenergo's network, download additional tools onto its network, and issue commands remotely to its compromised systems. With C2 established, hackers most likely would have created multiple back doors into Ukrenergo's network to allow ongoing access. The hackers would then have begun to harvest administrator credentials for Ukrenergo's IT infrastructure. In the 2015 attack, hackers installed additional BlackEnergy plugins. This would have allowed hackers to map Ukrenergo's internal networks and move laterally across systems and subnets.

Failure to catch hackers early in an intrusion can be costly. According to a 2016 report (Scali and McBride 2016), hackers remained on victim networks in 2015 a median of 146 days prior to discovery. No one immediately noticed the hackers' network presence in the 2015 or 2016 Ukraine attacks. In the 2015 attack, hackers spent months exploring IT environments, eventually finding and accessing the virtual private network (VPN), which lacked two-factor authentication and, along with a misconfigured firewall, allowed hackers entry into, and ongoing access to, the ICS and SCADA system network. The hackers remained on Ukrenergo's network for months, gathering system logs, monitoring network traffic, and studying the behavior of system administrators. In both incidents, hackers skillfully hid their network presence. Investigators said the 2016 hackers "lived off the land," a technique whereby intruders use the same credentials and tools as system administrators to avoid detection.

Researchers investigating the 2016 attack have alluded to an ominous trend, suggesting it was merely "training." In early January 2017, as investigators were piecing together the 2016 incident, Russian cybersecurity firm Kaspersky Lab ICS-CERT reported a then-ongoing "targeted attack" against 500 organizations in 50 countries. Kaspersky wrote that the worst affected were companies in the smelting, electric power generation and transmission, construction, and engineering industries. Most of the organizations attacked were vendors of industrial automation solutions and system support contractors. In other words, the attack targeted organizations that design, build, and support industrial solutions for critical infrastructure. Kaspersky said the spear phishing campaign began in August 2016. As of mid-2016, 33% of 1552 publicly disclosed ICS vulnerabilities had no available security patch (Williams 2017).

16.5   The Evolving Landscape

In the US, the utility industry, NERC, the US Federal Energy Regulatory Commission (FERC), and the US Nuclear Regulatory Commission (NRC) have not required hardware mitigation to be implemented to prevent Stuxnet or Aurora-type attacks from occurring. The traditional mindset is that all industries rely on the grid, so the grid is presumed to have a high level of cyber protection. However, Aurora uses electric substations as the vehicle for launching attacks against any generator, AC motor, or transformer connected to a substation. Consequently, it is the grid that can be the source of the attack. Depending on the equipment, the damage from an Aurora attack can take months to repair or replace, assuming the equipment can be manufactured, transportation is available for delivery, and trained staff are available to install it. It would appear to be a matter of time until something similar happens in the US.

Many of the same control systems used in the electric industry are used in other industries. The vulnerabilities in general are the same, especially if they are based on the design features of the control systems. Both Stuxnet and Aurora affected a wide swath of industries and government entities in addition to the electric industry. Closely aligned to an electric utility is electrified rail, where operations depend on electric substations and AC equipment in electrified locomotives. Diesel-electric locomotives have their own vulnerability to Aurora. In the refining industry, pumps and compressors are critical infrastructure and their failure can shut down the entire refinery for days or weeks.

The burgeoning natural gas industry, which is providing some measure of energy independence to the US for the first time since the late 1960s, is vulnerable to Aurora because of the critical use of compressors. Pumps, motors, and compressors in the water and wastewater utilities are vulnerable as well. ICS honeypots (systems that look like real systems but are actually "test systems" used to identify who is attempting to attack them) are being attacked by Chinese, Russian, Iranian, North Korea, and other hacking groups. The chemicals, food, and pharmaceuticals industries are also vulnerable, to greater or lesser degrees. Far less well known for vulnerabilities are the users of rotating machinery, building automation systems, boiler control systems, and distributed control systems.

ICS cyber incidents have impacted electric grids, power plants, nuclear plants, hydroelectric facilities, pipelines, chemical plants, oil and gas facilities, manufacturing, and transportation around the world. Impacts have ranged from trivial to significant environmental releases, equipment damage, and widespread electric outages leading to injuries and deaths. While the risk is different depending on the industry and the application, the equipment is often the same (Weiss 2017). It seems clear that many vulnerable industries may not recognize their own vulnerabilities and are not in a position to do much about a cyberattack if and when it occurs.

16.6 The Growing Threat Posed by Nuclear Facilities

Notwithstanding important steps taken by the International Atomic Energy Agency (IAEA) to improve cybersecurity across the nuclear energy sector, the industry has less experience safeguarding against cyberattacks than some other areas of infrastructure. This is partly due to the nuclear industry's regulatory requirements, which have operating systems that were adopted later than other types of critical infrastructure. The industry's longstanding focus on physical protection and safety has also implied that while some aspects of nuclear risk response are relatively robust, less attention has been paid to developing cybersecurity readiness than one might expect. As a result, it makes sense that exploiting weaknesses in digital technology would be an attractive method of attacking nuclear facilities.

As nuclear facilities have become increasingly reliant on digital systems and commercial "off-the-shelf" software, the risk of cyberattack has naturally grown. Nuclear plant personnel may not realize the full extent of this cyber vulnerability and are often inadequately prepared to address potential attacks.

The notion that nuclear facilities are "air gapped" – or completely isolated from the public Internet – which protects them from cyberattack is nothing more than a myth. Not only can air gaps be breached with nothing more than a flash drive (as in the case of Stuxnet), but the commercial benefits of internet connectivity mean that nuclear facilities may now have VPNs and other connections installed by third party contractors or otherwise undocumented workers. At the same time, automatic cyberattack packages (targeted at known and newly discovered vulnerabilities) are now widely available for purchase, advanced hacking techniques are available on the Dark Web, and search engines can easily identify critical infrastructure components that are connected to the Internet.

Many ICS are "insecure by design," since cybersecurity measures were not designed in from the beginning. Supply chain vulnerabilities mean that equipment used at a nuclear facility risks compromise at any stage. The cybersecurity threat requires an organizational response by the civil nuclear sector, which includes, by necessity, knowledgeable leadership at the highest levels, and dynamic contributions by management, staff, and the wider community of stakeholders, including members of the security and safety communities. The nuclear sector as a whole needs to develop a more robust ambition to match or overtake its opponents in cyberspace, including an integrated approach to risk assessment that takes both security and safety measures into account. This includes engaging in robust dialogue with engineers to raise awareness of the cybersecurity risk (Baylon et al. 2015).

16.7 Not Even Close to Ready

A 2014 survey (Ponemon Institute and Unisys 2014) of critical infrastructure security preparedness among 600 global IT and IT security executives in 13 countries was conducted among individuals who were familiar with security and well-known standards and regulations regarding the protection of information assets and critical infrastructure. Among the key findings was that just 17% of the companies had most of their IT security program activities deployed, 50% of respondents said their IT security activities had not yet been defined or deployed, and only 28% believed that security was one of the top five strategic priorities across the enterprise.

This was particularly surprising since 67% of respondents said their companies had experienced at least one security compromise that had led to the loss of confidential information or disruption to operations over the previous 12 months. According to 34% of respondents, their companies do not get real-time alerts, threat analysis, or threat prioritization intelligence that could be used to stop or minimize the impact of a cyberattack. Despite recognition of the threats facing their companies, the majority of respondents believed they were not effective at managing security risks, they did not have state-of-the art technologies to minimize risk to SCADA, nor did they have sufficient resources to achieve compliance to existing regulations. Also, very few believed regulations and standards had decreased the risk of attacks.

Most companies in the research said they were stuck in the early or middle stage of maturity, meaning their IT security programs had not been defined or deployed, or were in the process of being deployed. Most organizations fell short in terms of security governance, with the majority saying important security governance activities were only partially implemented or not at all. More than half (57%) of respondents did not have fully implemented training and awareness programs about security requirements and 54% of C-level executives were not often or had never been briefed or made fully aware of corporate security initiatives.

Just 16% of the IT professionals were fully aware of the vulnerabilities that exist in their systems. The root cause of most of their breaches was most likely to be a negligent employee (noted 47%), with negligent insiders being recognized as the biggest single security threat. Nearly one-third of respondents said that more than a quarter of their network components were outside their control (Wagner 2017c, p. 247). All of this is evidence that critical infrastructure companies are not prepared to deal with a plethora of attacks against IT and ICS.

Many enterprises see implementation of new security measures or upgrades as a source of potential disruption to their existing operational systems. Further complicating their security posture, most companies do not receive actionable intelligence on APT, insider attacks, or any other alerts to contain or repel cyber threats. As a result, they must rely on reactive tactics, which by definition means the damage is often already done and remediation usually consists of an endless cycle of upgrades, patches, and maintenance. What these firms need to develop is a forward leaning security strategy that aligns security requirements with business strategies and long-term objectives.

A comprehensive, proactive strategy can eliminate many threats while remaining nondisruptive. Security systems must be agile enough to address the security requirements of a wide array of systems and be deployable with minimal effort. User credentials should be strictly enforced to shore-up existing network segmentation and security. Common security measures still present too many targets to external hackers and malicious code. Masking the number and types of targets can greatly reduce the number of incidents and hacking attempts. Employee and partner use of their own smart devices introduces yet another attack surface for hackers and an entry point for malicious code. Encrypting wireless data eliminates a vast number of potential vulnerabilities (Wagner 2017c, p. 247).

16.8 Focusing on Cyber Resiliency

Very few private sector firms can truly say they are cyber resilient. It is a very long road. Implementing sufficient IT countermeasures takes time, sucks up resources, and cuts into profit. Not factored into many organizations' thinking process on this subject is how to put a price on Virtual Terrorism or the inevitable loss of reputation that goes along with it when cyberattacks become public (which can of course be significant). Virtual Terrorism is defined as: "Any action taken by an individual, group, business, government or other entity – whether having direct or indirect impact on intended or unintended targets, whether it is within the confines of existing law or not, and whether visible to the intended target(s) or not – that has the objective of promoting its interests while either attacking, snooping, stealing, causing harm, or sowing fear through remote control or via the Internet." If corporate executives were thinking more along these lines, perhaps more companies would be taking the risk more seriously. One advantage individuals have is that they tend to upgrade their computers and other electronics every three to five years – more frequently than companies tend to upgrade theirs (which is every couple of decades in the case of control systems (Definition source: Daniel Wagner, Virtual Terror, p. 13.)).

So, what can be done, apart from raising awareness to the problem, devoting more resources, and making actions to counter Virtual Terrorism compulsory instead of voluntary? Creating a more holistic approach to the problem, becoming proactive (instead of reactive) in thinking about how to address the problem, implementing routine cybersecurity audits, and creating teams of individuals dedicated solely to the problem inside companies is a good place to start. Budgets therefore need to be adjusted to devote more resources to addressing the problem across the board. Security and privacy risk mapping, benchmarking, and scenario planning should become a standard component of a cyber risk management protocol.

Clearly, businesses, governments, and individuals must devote greater resources to becoming more cyber resilient, which means they must devote more resources toward anticipating and protecting against attacks. Governments and businesses need to also engage in more public–private partnerships (PPPs) in order to adequately address the issue. In 2013, President Obama issued Executive Order 13636 Improving Critical Infrastructure Cybersecurity (ICIC) which, among other things, called for the establishment of a voluntary risk-based cybersecurity framework between the private and public sectors.

ICIC was intended to enhance the security and resilience of America's critical infrastructure by encouraging efficiency, innovation, and economic prosperity. It was an attempt to set industry standards and best practices to help organizations manage cybersecurity risks. This framework allows for all US Government agencies, regardless of their size or cybersecurity capability, to apply the best possible risk management practices in improving the security of critical infrastructure. The primary importance of the framework is that it allows for all those who voluntarily participate to adequately communicate and understand the risks, which is vital to achieving a functioning national and international cybersecurity network.

Some 16 infrastructure sectors have been deemed "critical" by the US government, are obvious targets for Virtual Terrorism, and are now regulated under the National Institute of Standards and Technology (NIST) Cybersecurity Framework – including financial services, telecommunications, and food production and distribution. Both the NIST Cybersecurity Framework (NCSF) and ICIC were intended to be guidelines, rather than a mandate for corporate behavior – and herein lies a problem that is familiar to government attempts to address man-made risk more generally. As is the case with the COP(Conference of the Parties)21 guidelines, there is no law requiring compliance, nor any penalties for a failure to comply. The same may be said about the fight against terrorism. Both governments and companies are generally hesitant to implement strict security protocols to protect themselves without having experienced an attack.

The European Union (EU) finalized similar measures in 2016 as a critical first step in defending against cyberattack. The Network and Information Security Directive (the "NIS" Directive) is the first piece of EU-wide legislation on cybersecurity, designed to bring cybersecurity capabilities to the same level of development across all EU Member States while ensuring that exchanges of information and cooperation are efficient at the cross-border level. It forces member states to adopt more rigid cybersecurity standards and creates an avenue for all of its Member States – and the operators of essential services such as energy, transportation, and healthcare sectors – to communicate (Dragos 2017). Other nations are in the process of acting accordingly; however, no nation allocates sufficient resources to adequately respond to the increasing threat of a cyberattacks against critical infrastructure, nor does any nation have a truly comprehensive plan to prevent or meaningfully react to the outcome of such attacks.

The enormity of the challenge in determining the nature of the threat and monitoring vulnerability at the government level can be summarized by a 2015 report from the US Government Accountability Office, which noted that most federal agencies overseeing the security of America's critical infrastructure still lack formal methods for determining whether those essential networks are protected from hackers, and that of the 15 critical infrastructure industries examined – including banking, finance, energy, and telecommunications – 12 of them were overseen by agencies that did not have proper cybersecurity metrics. These sector-specific agencies had not developed metrics to measure and report on the effectiveness of all of their cyber risk mitigation activities, or their sectors' basic cybersecurity posture.

Although in 2016 the DHS then issued a report downplaying future cyberattacks against the US power grid, just a few months after issuing the report it joined forces with the FBI to institute a program warning utilities around the

United States of the dangers of future cyberattacks. A US Senate Committee on Homeland Security and Governmental Affairs hearing also discussed cybersecurity of the power sector and identified the most pressing concern as the need to create postattack plans to assist affected populations. Governments around the world have plans in place to deal with the consequences of natural disasters, yet none have disaster relief plans for a downed power grid. Clearly, this must change. Local and state governments must work together with their national counterparts to produce and quickly implement plans to address the future attacks that are coming (Wagner 2016).

In 2017, the US Department of Energy said in its Quadrennial Energy Review that America's electricity system "faced imminent danger" from cyberattacks, which were growing more frequent and sophisticated, and that a widespread power outage caused by a cyberattack could undermine "critical defense infrastructure" as well as much of the economy, and place at risk the health and safety of millions of citizens. The report added that the total amount of investment required to modernize the US electricity grid ranged between $350 billion and $500 billion (Natter and Chediak 2017). Governments, corporations and individuals are really only starting to understand what Virtual Terrorism is, and what its potential impact can be. If we fail to turn the tide against Virtual Terrorism quickly, it will soon be impacting us all in ways we had never imagined (Wagner 2017a).

16.9   Enter DARPA

The US Defense Advanced Research Projects Agency (DARPA) is working with a private sector firm to forge alternative communication networks that would come into use in case of a cyberattack on the US electrical power grid. Although the aim is to ensure safe connectivity among all civilian entities that depend on the power grid, the program is particularly focused on securing defense networks and operational combat activities. Entitled Rapid Attack Detection, Isolation and Characterization Systems (RADICS), the program consists of a variety of technical capabilities, including an ability to recognize or provide early warning of impending attacks, map conventional and ICS networks, and ad hoc network formation and analysis of control systems.

DARPA is interested, specifically, in early warning of impending attacks, situation awareness, network isolation, and threat characterization in response to a widespread and persistent cyberattack on the power grid and its dependent systems. The goal of the new protective technology is to detect and disconnect unauthorized internal and external users from local networks within minutes, and create a robust, hybrid network of data links secured by multiple layers of encryption and user authentication. The systems rely on advances in network traffic control and analysis to establish and maintain emergency communications. It also quickly isolates the attacked system and moves to an alternative Secure Emergency Network (SEN).

The purpose of the program is to provide a technology that quickly isolates both the enterprise IP network and the power infrastructure networks to disrupt malicious cyberattacks. The SEN can take the form of wireless Internet technology and radio communications or satellite systems to ensure the grid continues to function if under attack. The coordination that is essential to critical operational traffic necessary to restart and then facilitate the stable operation of the power grid is managed by the SEN. It is an indirect network control that can support a seamless switch-over and transport of critical communication. The SEN is designed to function essentially as a wireless network without an infrastructure that ensures end-to-end communication between power grid nodes to provide transport of critical real-time communication within the affected area.

Once activated, the private sector firm's technology detects and disconnects unauthorized internal and external users from local networks within minutes. It is designed to create a robust, hybrid network of data links secured by multiple layers of encryption and user authentication. Coordination of a SEN needs to be completed according to a sequential process, enabling relevant portions of the grid to connect with one another so a secure network can stand up. Threat analysis is a key element of the initiative because the RADICS effort seeks to anticipate and thwart both current and future attacks. The project is a three-phase, four-year program that ends in 2020. The first two phases are focused on research and development of technologies while the last phase is focused on technology transition. Potential recipients of technology transition include electric grid operators, DoD systems. and the DHS (Osborn 2017).

16.10 The Frightening Prospect of "Smart" Cities

Each year the US Federal Emergency Management Agency (FEMA) asks each US state to rank how prepared it is for various forms of disaster. In a report released in 2015, cybersecurity was ranked the weakest aspect for the fourth straight year, and it is taking its toll. San Diego, for example, is hit by an average of 60 000 cyberattacks per day. New York City faces about 80 million cyberthreats per year, ranging from phishing attacks to attacks meant to overwhelm websites. Hackers can tamper with traffic control systems, smart street lighting, city management systems that control work orders or other facilities, public transportation, cameras, smart grids, wireless sensors that control waste and water management, or mobile and cloud networks (Anand 2016).

A Smart City uses technology to automate and improve city services, and in the smart city of the future everything will be connected and automated. While this is not yet a reality, many cities around the world are committing significant funds to become "smarter." Saudi Arabia is investing US$70 billion into smarter cities, in Dubai 1000 government services are in the process of going smart, and Barcelona is already ranked as the world's smartest city. According to some estimates, by 2020 the potential market for smart cities could be more than $1 trillion.

In smart cities, city services deploy new technologies to automate as much as possible in an effort to make cities run more efficiently. Among the city services that would be automated in smart cities are smart traffic control (in which traffic lights and signals adapt based on volume and ongoing traffic conditions), smart parking (where citizens can use a parking application to find available parking slots and review pricing, including pricing changes based on time of day and availability), smart street lighting (by being managed centrally, street lights can adapt to weather conditions, report problems, or be automated by time of day), smart public transportation, smart energy management, smart water management, smart waste management, and even smart security (wherein traffic and surveillance cameras, gunshot detection sensors, and other security devices provide real-time information on what is happening and where). Sounds Utopian, right? From a cybersecurity perspective, it sounds like a real potential nightmare. Cities are (unsurprisingly) already implementing new technologies without first testing cybersecurity.

To put the potential for problems in perspective, consider these two examples in California that occurred because of "software glitches." In 2012, a traffic jam on Interstate 80 occurred when the Placer County court accidentally summoned 1200 people to jury duty on the same morning. And in 2013, San Francisco's Bay Area Rapid Transit was shut down due to a technical problem involving track switching, which began shortly after midnight, affected 19 trains and up to 1000 passengers who were trapped on board (Wagner 2017c, p. 254). If these two "minor" examples are anything to go by, a city-wide "software glitch" could easily become disastrous.

In 2003, a blackout affected an estimated 10 million people in Ontario (Canada) and 45 million people in eight US states. Its primary cause was a software bug in the alarm system at a control room of the FirstEnergy Corporation, located remotely in Ohio. The absence of alarms left operators unaware of the need to redistribute power after overloaded transmission lines hit unpruned foliage. This triggered a software bug known as a race condition in the control software, part of General Electric Energy's Unix-based XA/21 energy management system. Once triggered, the bug stalled FirstEnergy's control room alarm system for more than an hour. System operators were unaware of the malfunction, which deprived them of both audio and visual alerts.

What would have been a manageable local blackout cascaded into widespread distress across the electric grid. The impact included 508 generating units at 265 power plants that were shut down, water systems in several cities that lost pressure, and at least 10 deaths. New York City received 3000 fire calls, its 311 information hotline received more than 75 000 calls, mobile networks were overloaded and disrupted, and hundreds of flights were canceled (Wagner 2017c, p. 254). New York State was responsible for billions of dollars in associated costs.

Apart from halting the march toward smart cities (which of course will not happen), there are some basic precautions cities can take to reduce the risk of being attacked:

- Ensure that proper encryption, authentication, and authorization are in place, and that systems can be easily updated.
- Require all vendors to provide all security documentation. Make sure Service Level Agreements include on-time patching of vulnerabilities and 24/7 response in case of incidents.
- Fix security issues as soon as they are discovered (do not wait).
- Create specific city CERTs that address cybersecurity incidents, vulnerability reporting and patching, coordination, and information sharing.
- Implement and make known to city workers secondary services and procedures in case of cyberattacks and define formal communication channels.
- Implement fail safe and manual overrides on all system services.
- Restrict access to public data by requesting registration and approval to use it (track and monitor access and usage).
- Regularly run penetration tests on all city systems and networks.
- Create threat models and scenario testing for any conceivable event (Cerrudo 2015).

A program called Securing Smart Cities released guidelines in 2015 for smart city technology adoption, which included using strong encryption, designing systems that have strong protection against tampering, and ensuring systems can only be accessed by people authorized to use them. New York City's government created a Cyber Interagency Working Group with 90 members from 50 public and private organizations, including government offices, banks, ISPs, and local, state and federal law enforcement. It also wrote guidelines to address how it would inform the public, restore systems, and coordinate across agencies in the event of a city-wide attack.

In Michigan, a 50-person "cyber civilian corps" trains for emergencies as an IT version of a volunteer fire department. The group is working with other states to help develop similar response plans. To simulate the domino effect of a cyberattack on other city services, 10 states practice their response plans using Alphaville, a "cyber village" that includes a school, city hall, library, power grid, and more than 10 virtual machines that users can customize to simulate their own networks (Anand 2016). As smart cities become a reality, current "dumb" cities will need to become smarter about cybersecurity, but many of them will inevitably learn the hard way what must be done to foster greater cyber resiliency.

And that is the core issue where Virtual Terrorism collides with infrastructure: can any country (or city or state) afford to wait until a cyberattack or electromagnetic pulse (EMP) is successful to seriously address the problem, devote sufficient resources or act proactively? The short answer is of course "no," but that does not mean that much is likely to change in the near or even medium term. Regrettably, potential problems tend not to become a priority until they "have to" become a priority, which, in the case of infrastructure, is a serious – and potentially catastrophic – mistake.

Countries around the world should develop a separate, secure, Internet-like infrastructure to serve their critical infrastructure. A "smart grid consortium" is currently being explored by the Department of Energy by conducting research on ways to build a new, more secure national electricity infrastructure. The US Congress routinely creates economic incentives for corporations to pursue beneficial activities. It should create monetary incentives for companies to strengthen cybersecurity and become more aligned with government and industry standards. The Congress should enhance cybersecurity legislation and enforce cybersecurity regulations already in place. Going forward, perhaps a set percentage of spending on future infrastructure investment will automatically include a minimum amount of set spending for cybersecurity measures. This can only happen with enhanced PPPs – one more thing to be put on the laundry list of things that should be done to try to get ahead and stay ahead of the infrastructure cybersecurity curve (Visner 2017).

16.11   Lessons from Petya

An organization can do everything it is supposed to do to achieve cyber resiliency and still remain vulnerable, in part because software developers only know what it is they are supposed to be protecting against after a new form of malware has been deployed, identified, and a way is found to block it. The Petya malware that swept the world in 2017 is a useful case in point. The attack was different from the WannaCry virus unleashed a few weeks earlier in several ways. Although ostensibly referred to as a form of ransomware, Petya was ultimately more accurately described as a form of sabotage rather than an attempt to generate revenue. It is estimated that less than $10 000 in ransom was actually paid as a result of the attack, at least in part because the e-mail link that ransom was supposed to be paid to was taken offline shortly after the attack. Petya was ultimately intended to steal passwords and destroy data, and, unlike WannaCry, Petya had no kill switch, so it represented a much more insidious threat. With the arrival of Petya, more "blended" cyberattacks (combining ransomware with "wiper" or other forms of attack) should be expected, mixing elements of different threat vectors in new ways.

After the attack, it was learned that hackers penetrated the network of a small Ukrainian software firm (MeDoc) that sold a piece of accounting software used by approximately 80% of Ukraine's businesses. By injecting a tweaked version of a file into updates of that software, they were able to start spreading backdoor versions of it weeks before Petya was deployed throughout Ukraine, injecting Petya through the MeDoc system's entry points. In the process, it was learned that innocent software updates could be used to silently spread the malware.

Kaspersky Lab subsequently said it had seen two other examples in the year prior to the release of Petya of malware delivered via software updates to carry out sophisticated infections. In one case, perpetrators used updates for a popular piece of software to breach financial institutions. In another, hackers corrupted the update mechanism for a form of ATM software sold by an American company to be able to hack cash machines. Similar attacks should be expected in the future by infecting the supply chain that links dispersed computer systems together.

In the Petya case, hackers first breached another unnamed software firm and used its VPN connections to other companies to plant ransomware on a handful of targets. Only later did the hackers move on to MeDoc as a malware delivery tool. One reason hackers are turning to software updates as an inroad into vulnerable computers may be the growing use of "whitelisting" as a security measure, which, as noted above, strictly limits what can be installed on a computer to only "approved" programs. A basic security precaution that every modern developer should use to prevent their software updates from being corrupted is "codesigning," which requires that any new code added to an application be signed with an unforgeable cryptographic key. MeDoc did not implement codesigning, which would have allowed any hacker that can intercept software updates to act as a MITM and alter them to include a backdoor.

The hackers were deep enough into MeDoc's network that they likely could have stolen the cryptographic key and signed the malicious update themselves, or even added their backdoor directly into the source code before it was compiled into an executable program, signed, and distributed. None of this should dissuade anyone from updating and patching their software or using software that updates automatically.

Codesigning no doubt makes compromising software updates far more difficult, requiring much deeper access to a target company for hackers to corrupt its code. That means codesigned software that is downloaded or updated from Google's Play Store or the Apple App Store is far safer and, thus, significantly harder to compromise than a piece of software like MeDoc, distributed by a family-run company without codesigning. But even the App Store's security is not impenetrable. In 2015, hackers distributed infected developer software that inserted malicious code into hundreds of iPhone apps in the App Store that were likely installed on millions of devices, despite Apple's strict codesigning implementation.

What this all means is that on highly sensitive networks, such as the type of critical infrastructure disabled by Petya, even "trusted" applications cannot be fully trusted. Systems administrators need to segment and compartmentalize their networks, restrict the privileges of even whitelisted software, and retain backups in case of any ransomware outbreak (Greenberg 2017). In the era of Virtual Terrorism, no one can afford to believe they will not be targeted or that hackers will not be successful in attacking even the best protected system.

16.12   Best Practices

Fortunately, in this new and evolving world of Virtual Terrorism, there are a range of best practices that individuals, businesses, and governments can turn to in an attempt to protect themselves against virtual terrorists. Many of them are common sense and some are easy to implement, while others require sophisticated IT infrastructure. The point of going down the road of attempting to turn the tide is to embrace an ecosystem of protection that cascades from organizational culture to decision maker to the IT department and on down to the end user of data. Engaging in best practices implies comprehension of the scope of the problem and a willingness to do what it takes to fight back.

The abundance of affordable ransomware has made it easier than ever for cyberattackers and thieves to do what they do best. The number of threat vectors within organizations are growing exponentially, today's cybersecurity technology is capable of protecting organizations from the vast majority of attacks. Legacy firewalls may only prevent against known threats, while early next-generation firewalls (NGFWs) are limited to detecting rather than preventing zero-day attacks. However, some modern NGFWs are remarkably good at not only detecting but preventing ransomware, data theft, and the full assortment of advanced threats.

To be effective against the plethora of threats, firewalls must block files until they can be scanned and found to be innocuous, blocking zero-day threats at the gateway. Since most advanced threats enter organizations through email, protection should cover email security appliances. At the same time, organizations should leverage Secure Sockets Layer/Transport Layer Security (SSL/TLS) decryption technology to ensure that zero-day threats are not lurking in encrypted traffic arriving into the network. Among the things that can help achieve cyber resiliency are:

- Segmenting at-risk computers from critical data and services.
- Leveraging advanced network detection features, including a multiengine sandbox that supports automated prevention.
- Backing up and securely store our critical data offline.
- Performing tests to ensure that data can be retrieved if needed.
- Implementing and testing incident response plans.
- Applying whitelisting – only allowing approved programs to run on your networks.

- Limiting the ability for high-risk applications to run on your network.
- Conducting a thorough cybersecurity risk analysis for the organization.
- Determining which business operations are mission critical and how long your organization can operate without them.
- Attempting to hack your own systems and conducting a penetration test on public-facing portals (Sonic Wall 2017).

16.13 A Process Rather than a Product

Systems engineers are accustomed to designing technical approaches to cybersecurity utilizing network maps and schematics that detail layers of information and their interconnections. While obviously useful, future such approaches may be enhanced by integrating qualitative factors into the design process, encompassing an organization's strategy and objectives, and focusing on the people, processes, and technology required to achieve such objectives. Cybersecurity is, after all, a process rather than a product, which should ideally strive to be consistent with broader organizational risk management practices and objectives in order to be maximally effective.

Given that the vast majority of cyber breaches occur between the keyboard and the chair (i.e. as a result of human error), taking the human factor into consideration can be critical to a system's chance of success. Cybersecurity needs to be built in at every stage of system engineering design to be fully aligned to business requirements. This entails:

- Understanding the business goals and objectives.
- Determining the criticality of specific assets to a business.
- Understanding the threats, risks, and opportunities related to business operations and assets.
- Developing strategies, processes, mechanisms, standards, and tools that support the goals and take into account risks.
- Taking into account management roles and responsibilities that will underpin processes and mechanisms.
- Understanding the geographies, locations, and infrastructure where cybersecurity needs to be implemented.
- Understanding the business time aspects, such as processing schedules and sequences (Atkins Global n.d.).

Adopting a "passive defense" implies using technologies and products (for example, firewalls, cryptography, intrusion detection) and procedures (such as reconstitution and recovery) to protect IT assets. Some forms of passive defense may be dynamic (i.e. stopping an attack in progress) but, by definition, passive defense does not generally impose serious risk or penalty upon the attacker. By contrast, "active defense" imposes serious risk or penalty on the attacker because of the pursuit of aggressive identification, investigation and prosecution, or preemptive counter attacks.

By deploying passive measures, which are by nature reactive in orientation, attackers are free to continue the assault until they either succeed or decide to move on to the next target. Given the vulnerabilities of most cyber systems, the low cost of most attacks, and the ability of attackers to strike anonymously, a skilled and determined attacker is more likely to succeed than fail. Some defensive actions (such as stopping an attack in progress) can be pursued with both passive and active means – a defender might passively plug a vulnerability in real time while actively seeking to locate the source of an attack.

Defensive Systems. Multiple, overlapping, and mutually supportive defensive systems should be deployed to guard against single point failures in any specific technology or protection method. This should include regularly updated firewalls as well as gateway antivirus, intrusion detection or protection systems, website vulnerability with malware protection, and web security gateway solutions throughout the network. Implement smart data governance practices in your organization so that you know what business data are being stored on cloud services.

Encryption. Implement and enforce a security policy whereby any sensitive data are encrypted at rest and in transit. Ensure that customer data are encrypted as well and that passwords are strong. Important passwords, such as those for users with high privileges, should be at least 8–10 characters long (preferably longer) and include a mixture of letters, numbers, and special characters. Encourage users to avoid reusing the same passwords on multiple websites. Sharing passwords with others should be forbidden. Delete unused credentials and profiles and limit the number of administrative-level profiles created.

Software. Always keep security software up to date to protect against any new malware variants and keep your operating system and other software constantly updated (daily, if not multiple times per day, since multiple updates occur each day). Software updates will frequently include patches for newly discovered security vulnerabilities that could be exploited by attackers.

Websites. Regularly assess websites for any vulnerabilities – scan it daily to check for malware and secure websites against MITM attacks. Choose SSL certificates with extended validation to verify protection and display the green browser address bar to website users. Display recognized "trust" marks in highly visible locations on the website and be picky about displaying plugins. The software you use to manage your website may come with vulnerabilities included. The more third-party software you use, the greater your attack surface, so only deploy what is absolutely necessary.

The idea is to design a system to be secure from an attack from the beginning. If done so that it is visible to a would-be attacker, cyberattacks could be prevented because they would be perceived to be futile (or if launched, would cause little or no damage). The problem is that the vast majority of IT systems were not designed with security in mind. As a result, there is an enormous legacy of insecure systems that are in use and security protocols may conflict with organizational needs, which can be costly to introduce and may result in reduced efficiency and functionality. Most organizations are also not incentivized to redesign systems to be much more secure unless the equivalent of a digital 9/11 were to occur (Goodman 2007).

16.14   Building a Better Mousetrap

Astonishing as it may seem, Microsoft's Windows operating system still powers more than 95% of the world's desktop computers. Also, as of 2014, 95% of all ATMs in the US were still running the Windows XP system (which was the entry point for the WannaCry virus, and for which Microsoft had ceased providing security updates). As a result, with a single piece of malware, hackers have the ability to cause an entire software system to fail uniformly. Why else could more than one billion passwords be stolen at one time, or more than a dozen financial institutions be attacked at once? There is certainly an argument to be made that software systems should be able to self-detect defects and self-repair them. At a minimum, it should be made law that, upon first identification of a vulnerability or defect, the manufacturer becomes obligated to notify every single user of that system.

By the same token, we need an enhanced and more resilient means of protecting our information. Cyber resiliency implies adeptness in crafting a response and dexterity in determining which way is best to rapidly recover degraded technological capabilities. Part of the problem is that machines and systems are not designed to fail gracefully; when they fail, it is often total in scope. As artificial intelligence, robotics, nanotechnology, and other highly disruptive technologies become even more commonplace, the potential for cataclysmic failure and system-wide disruption will grow exponentially (Goodman 2015).

What we really need is a new form of operating system that matches the exponential changes occurring around us. One entity that is doing just that is the OS Fund, which partners with entrepreneurs who are rewriting the operating systems of life. Its founder notes that with the Fund's powerful tools of creation – including 3D printing, genomics, machine intelligence, robotics, software, and synthetic biology – they have the ability to create the kind of world people could previously only dream of (OS Fund 2017). That is exactly the type of thinking needed to use today's technology to beat virtual terrorists at their own game.

Our lackadaisical approach to cybersecurity and the profound technological vulnerabilities that lay before us have been akin to applying sunscreen and claiming it can protect us against nuclear meltdown (Silver 2015, p. 29). This is an apt analogy that should be like a cold slap in the face to everyone who opposes virtual terrorists. As others have also recommended, we need to create the cyber equivalent of the Manhattan Project, to put together the best minds from the private and public sectors, academia, and civil society to create a framework for combatting Virtual Terrorism in the twenty-first century. There may be some resistance to such an effort being spearheaded by a government so perhaps leading minds and cash-rich organizations from the private sector could take the lead in creating it. We need to collectively transcend away from linear thinking about cybersecurity and toward a multidimensional mindset to address a multifaceted problem.

16.15   Summary

The world's infrastructure is generally vulnerable to cyberattack because so much of it is dependent upon software and so many of its systems are interconnected. Despite the fact that cyberattacks are occurring with greater frequency and intensity around the world, many either go unreported or are underreported, leaving the public with a false sense of security about the threat they pose and the lives and property they impact. Many ICS are "insecure by design," since cybersecurity measures were not designed in from the beginning. ICS cyber incidents have impacted electric grids, power plants, nuclear plants, hydro facilities, pipelines, chemical plants, oil and gas facilities, manufacturing, and transportation around the world. Impacts have ranged from trivial to significant environmental releases, equipment damage, and widespread electric outages leading to injuries and deaths. While the risk is different depending on the industry and the application, the equipment is often the same. It seems clear that many vulnerable industries may not recognize their own vulnerabilities and may not be in a position to do much about a cyberattack if and when it occurs.

There is an enormous legacy of insecure systems that are in use and security protocols, which can be costly to introduce and may result in reduced efficiency and functionality, may conflict with organizational needs. Many enterprises see implementation of new security measures or upgrades as a source of potential disruption to their existing operational systems. Organizations need to develop a forward leaning security strategy that aligns security requirements with business strategies and long-term objectives. A comprehensive, proactive strategy can eliminate many threats while remaining nondisruptive. Security systems must be agile enough to address the security requirements of a wide array of systems and be deployable with minimal effort.

Creating a more holistic approach to the problem, becoming proactive (instead of reactive) in thinking about how to address the problem, implementing routine cybersecurity audits, and creating teams of individuals dedicated solely to the problem inside companies is a good place to start. Budgets therefore need to be adjusted to devote more resources to addressing the problem across the board. Security and privacy risk mapping, benchmarking, and scenario planning should become a standard component of a cyber risk management protocol. Clearly, businesses, governments, and individuals must devote greater resources to becoming more cyber resilient, which means they must devote more resources toward anticipating and protecting against attacks.

Systems engineers are accustomed to designing technical approaches to cybersecurity utilizing network maps and schematics that detail layers of information and their interconnections. While obviously useful, future such approaches may be enhanced by integrating qualitative factors into the design process, encompassing an organization's strategy and objectives, and focusing on the people, processes and technology required to achieve such objectives. What is needed is a new form of operating system that matches the exponential changes occurring around us. The objective should be to collectively transcend away from linear thinking about cybersecurity and toward a multidimensional mindset to address a multifaceted problem.

References

Anand, P., (2016) *The "Mind-Boggling" Risks Your City Faces From Cyber Attackers*, MarketWatch, CBS. https://www.marketwatch.com/story/the-mind-boggling-risks-your-city-faces-from-cyber-attackers-2016-01-04. (last accessed 8 July 2019).

Atkins Global (n.d.) *Cyber Resilient Infrastructure: Securing our National Infrastructure and Defense Capabilities*, 29. http://explore.atkinsglobal.com/cyber/Atkins_Cyber_Resilient_Infrastructure_Report.pdf (Last accessed 29 June 2019).

Brenner, B., (2013) *2003 Blackout: An Early Lesson in Planetary Scale?*, Akamai. https://blogs.akamai.com/2013/08/2003-blackout-an-early-lesson-in-planetary-scale.html (Last accessed 29 June 2019).

Cerrudo, C., (2015) *An Emerging US (and World) Threat: Cities Wide Open to Cyber Attacks*, IOActive. https://ioactive.com/pdfs/IOActive_HackingCitiesPaper_CesarCerrudo.pdf (Last accessed 29 June 2019).

Baylon, C., Brunt, R., and Livingstone, D., (2015), *Cybersecurity at Nuclear Facilities: Understanding the Risks*. A Chatham House Report. https://www.chathamhouse.org/sites/files/chathamhouse/field/field_document/20151005CyberSecurityNuclearBaylonBruntLivingstoneExecSumUpdate.pdf (Last accessed 29 June 2019).

Dragos, (2017) *Crashoverride: Analysis of the Threat of Electric Grid Operations*, Dragos. https://dragos.com/blog/crashoverride/CrashOverride-01.pdf (Last accessed 29 June 2019).

European Commission, (2016) *Your Guide to Policies, Information and Services*, European Commission.

FortiGuard Labs (2015) (Known) *SCADA Attacks Over the Years*. https://twitter.com/fortiguardlabs/status/565848138964279297 (Last accessed 8 July 2019).

Goodman, M. (2015). *Future Crimes: Inside the Digital Underground and the Battle for Our Connected World*, 2e, 476–481. New York: Anchor.

Greenberg, A., (2017, July) *The Petya Plague Exposes the Threat of Evil Software Updates*, Wired.

Industry Week, (2016, August) *Cyberthreats Targeting the Factory Floor*, Industry Week. Industry Week (18 August), http://www.industryweek.com/information-technology/cyberthreats-targeting-factory-floor (Last accessed 29 June 2019).

Leyden, J. (2008). *Chinese Crackers Blamed for US Power Blackouts*. The Register www.theregister.co.uk/2008/06/02/chinese_blamed_us_power_outage/ (Last accessed 29 June 2019).

McAfee, (2011) *Global Energy Cyberattacks: Night Dragon*. https://www.us-cert.gov/ics/advisories/ICSA-11-041-01A (Last accessed 8 July 2019).

Nakashima, E., (2017) Russia has Developed a Cyberweapon that Can Disrupt Power Grids, According to New Research, *The Washington Post* (12 June).

Goodman, S.E. (2007). *Cyberterrorism and Security Measures*. In: *Science and Technology to Counter Terrorism: Proceedings of an Indo-US Workshop*, 43–54. Washington, DC: The National Academies Press.

Natter, A., Chediak, M., (2017) *US Grid in "Imminent Danger" From Cyber-Attack, Study Says*, Bloomberg.

Osborn, K., (2017) *DARPA tasks BAE with workaround to secure the power grid in event of massive attack*, Defense Systems (13 April). https://defensesystems.com/articles/2017/04/13/grid.aspx (Last accessed 29 June 2019).

OS Fund, (2017) *About, OS Fund*. http://osfund.co/about (Last accessed 29 June 2019).

Pelroth, N., (2017) *Hackers Are Targeting Nuclear Facilities, Homeland Security Dept. and F.B.I. Say*, The New York Times (6 July). https://mobile.nytimes.com/2017/07/06/technology/nuclear-plant-hack-report.html?referer=http://drudgereport.com/ (Last accessed 29 June 2019).

Ponemon Institute and Unisys (2014). *Critical Infrastructure: Security Preparedness and Maturity*. https://www.huntonak.com/files/upload/Unisys_Report_Critical_Infrastructure_Cybersecurity.pdf (Last accessed 8 July 2019).

Scali, D. and McBride, S. (2016). *Inflection Point: Sandworm Team and the Ukrainian Power Outages*. https://www2.fireeye.com/WBNR-Inflection-Point.html?utm_source=webinar&utm_medium=blog&utm_campaign=ICS-iSIGHT (Last accessed 29 June 2019).

Silver, N. (2015). *The Signal and the Noise: Why So Many Predictions Fail, But Some Do not*. New York: Penguin.

Sonic Wall (2017). *2017 Annual Threat Report*. https://cdw-prod.adobecqms.net/content/dam/cdw/on-domain-cdw/brands/sonicwall/SonicWall-2017-Threat-Report.pdf (Last accessed 8 July 2018).

Vinton, K., (2014) *Hacking Gets Physical: Utilities at Risk for Cyber Attacks*, Forbes (10 July). https://www.forbes.com/sites/katevinton/2014/07/10/hacking-gets-physical-utilities-at-risk-for-cyber-attacks/#55db4ff87af9 (Last accessed 29 June 2019).

Visner, S.S., (2017) *The Cybersecurity of the Infrastructure – A Challenge and an Opportunity*, The Hill (2 March). http://thehill.com/blogs/congress-blog/technology/321987-the-cybersecurity-of-the-infrastructure-of-the-united-states (Last accessed 29 June 2019).

Wagner, D., (2016) *The Growing Threat of Cyber-Attacks on Critical Infrastructure*, The Huffington Post (24 May). http://www.huffingtonpost.com/daniel-wagner/the-growing-threat-of-cyb_b_10114374.html (Last accessed 29 June 2019).

Wagner, D., (2017a) *The Growing Threat of Virtual Terrorism*, International Policy Digest (20 March). https://intpolicydigest.org/2017/03/20/the-growing-threat-of-virtual-terrorism (Last accessed 29 June 2019).

Wagner, D., (2017b) *The Threat of Virtual Terrorism Against Infrastructure is Growing*, The Huffington Post (7 September). https://www.huffingtonpost.com/entry/the-threat-of-virtual-terrorism-against-infrastructure_us_59b13dbde4b0c50640cd64e6 (Last accessed 29 June 2019).

Wagner, D. (2017c). *Virtual Terror: 21st Century Cyber Warfare*, 234. CreateSpace Independent Publishing Platform.

Weiss, J., (2017) Industrial control systems: The holy grail of cyberwar, Christian Science Monitor (24 March). https://www.csmonitor.com/World/Passcode/Passcode-Voices/2017/0324/Industrial-control-systems-The-holy-grail-of-cyberwar (Last accessed 29 June 2019).

Williams, B. D. (2017). *Hackers' methods feel familiar in Ukraine power grid cyberattack*. Fifth Domain. https://www.fifthdomain.com/home/2017/01/29/how-a-power-grid-got-hacked (Last accessed 8 July 2019).

Zetter, K., (2016) *Inside the Cunning, Unprecedented Hack of Ukraine's Power Grid*, Wired (3 March). https://www.wired.com/2016/03/inside-cunning-unprecedented-hack-ukraines-power-grid/ (Last accessed 29 June 2019).

17

Closed-Loop Mission Assurance Based on Flexible Contracts: A Fourth Industrial Revolution Imperative

Azad M. Madni and Michael Sievers

Synopsis

The fourth industrial revolution blends physical, digital, and biological domains into what is commonly referred to as cyber-physical systems. This revolution is capitalizing on advances in technologies such as artificial intelligence, machine learning, biological sensing, human behavior modeling, nanotechnology, and the internet of things. As system complexity and integration challenges continue to grow, our ability to understand their behavior decreases, making them more vulnerable to disruptive events. Thus, fourth generation industrial systems must be able to detect, isolate, eliminate/withstand a continual barrage of unexpected, previously unseen operating conditions. Although traditional fault-tolerance and mission assurance (MA) methods are an essential component of the needed resilience, dealing with unknown-unknowns requires greater sophistication and vigilance than exists in today's systems. However, even with exquisite designs and novel resilience schemes, we cannot always ascertain a priori whether a complex cyber-physical system will operate as expected, slide into trouble, or face an imminent crash. Furthermore, we cannot always know a priori what to do even if we did know the state of the system. Therefore, what is needed is requisite flexibility in how a system can respond and a way for continually monitoring and evaluating system health. Partially Observable Markov Decision Processes (POMDPs) offer an effective means for continuously learning about system state and new/unexpected conditions while providing the requisite flexibility in actions to assure system safety and trustworthiness. Moreover, unlike traditional MA, the POMDP approach provides near real-time updates of system robustness metrics, while also determining trajectories that drive the system to nominal and safe operation.

Mission assurance (MA) is a system function that focuses on reducing uncertainty in the ability of systems to successfully complete missions despite disruptions. Current MA approaches tend to rely exclusively on information

Systems Engineering in the Fourth Industrial Revolution: Big Data, Novel Technologies, and Modern Systems Engineering,
First Edition. Edited by Ron S. Kenett, Robert S. Swarz, and Avigdor Zonnenshain.
© 2020 John Wiley & Sons, Inc. Published 2020 by John Wiley & Sons, Inc.

available at design time. In other words, they do not incorporate situation awareness data from mission execution in operational environments. More recent approaches attempt to combine risk management with situation awareness data from the operational environment. However, these approaches are not sufficiently flexible for systems that need to operate in partially observable environments. This paper presents a model-based approach to MA. The approach employs: probabilistic models to account for uncertainty and partial observability, and incremental information availability. The approach employs a flexible contract construct that introduces flexible assertions to enable adaptation to change in the operational environment, formal modeling to support system verification; and reinforcement learning to increase confidence in knowledge of system and environment states. The approach is discussed within the context of multi-UAV (unmanned aerial vehicle) swarm operations.

17.1 Introduction

MA is concerned with reducing uncertainty in the ability of systems to accomplish missions in the face of disruptions (Madni and Jackson 2009; Neches and Madni 2013; Goerger et al. 2014). MA has historically been a part of a variety of engineering subdisciplines including: high availability systems, failure analysis, performance engineering, quality assurance, reliability assessment, redundancy management, security engineering, hazard identification, software engineering, systems engineering, and safety engineering (Grimaila et al. 2010; Jabbour and Muccio 2011). Existing methods for MA are based on information available exclusively at design time (Jabbour and Muccio 2011). Recently, MA approaches employed in federated environments combine standard risk management with shared situation awareness among entities involved in executing operational missions (Grimaila et al. 2010). However, these approaches are not flexible enough when the system needs to operate in uncertain, partially observable environments. This recognition provided the motivation for the research reported in this paper.

This paper presents a model-based MA approach that accounts for uncertainties arising from partial observability of the operational environment and employs reinforcement learning based on incoming data (i.e. observations) from the operational environment to progressively determine system states.

This chapter is organized as follows. Section 17.2 presents the current approach to MA. Section 17.3 presents the flexible contract (FC) construct for real-time MA based on operational situation awareness data. Section 17.4 presents a closed-loop approach to MA based on FCs. Section 17.5 presents real-time MA concept of operations for a simple system. Section 17.6 discusses the technical underpinnings of the overall approach. Section 07

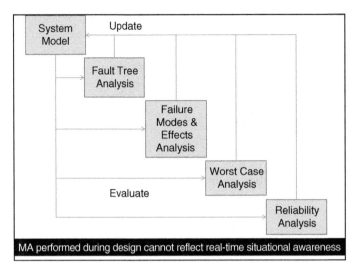

Figure 17.1 Current approach to mission assurance.

presents concluding comments and future developments to realize the proposed approach in operational settings.

17.2 Current MA Approach

MA today is a sequential, design time process (Figure 17.1). In other words, it does not reflect real-time *situational* awareness data that could potentially contribute to more accurate MA assessment. Figure 17.1 presents a simplified, high level view of the current approach to MA. Several key considerations, such as reliability, safety, performability, quality assurance, parts, materials, processes, and domain-specific considerations, such as radiation effects in space missions, which are all part of MA, are not explicitly represented to prevent cluttering the figure.

The traditional MA approach, based on the analysis of dynamic system characteristics, employs methods such as: bottom up evaluation of component faults (e.g. failure modes effects analysis); worst case circuit analysis; single event effects analysis; and reliability analysis. MA predicts how likely the system will provide the required level of service in its operational environment.

Thus, the key question is how valid are predictions that are based solely on information available at design time. While these predictions are useful for conducting trade studies and identifying potential risk areas, it is questionable whether they reflect the realities of the operational environment. To begin with, system behavior cannot be reliably predicted because the system's environment

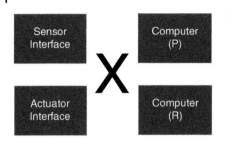

Figure 17.2 A simple system.

and operational context cannot be fully controlled. Furthermore, it is difficult to discern all the states that a system might be in. As a result, it is difficult to tell whether the system is performing satisfactorily (i.e. within safe states regime), is in trouble (i.e. in unsafe states), or heading into potential trouble (i.e. about to enter unsafe states). Thus, online assessment of MA becomes essential to understand a system's overall state and health status, and guide decisions about its operational use.

An illustrative example of MA for a simple system is presented next (Figure 17.2). The simple system comprises a prime computer (P) and a redundant computer (R) that are cross-strapped to sensor and actuator interfaces.

For this example, traditional MA typically involves analyzing a static block diagram (e.g. fault trees that are used to evaluate potential causes of mission failure in a top down fashion). A fault tree for the system in Figure 17.2 is shown in Figure 17.3.

With respect to this problem, reliability is defined by the equation:

$$R_{sys} = R_{sens} * R_{act} * (2 * R_{comp} - R^2_{comp})$$

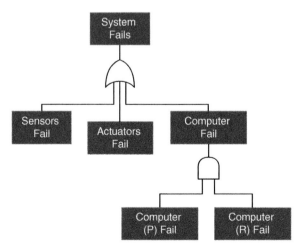

Figure 17.3 Fault tree for a simple system.

In this equation, R_{sys} = the reliability of the system, R_{sens} = the reliability of the sensors, R_{act} = the reliability of the actuators, and $(2*R_{comp} - R^2_{comp})$ = the reliability associated with one of two computers not working.

However, such traditional MA methods do not exploit sensed data from the operational environment during the conduct of the mission. The latter is needed for accurate, real-time, MA assessments. Specifically, what is needed is a flexible, closed-loop approach that reduces uncertainty by progressively learning system states from sensed data. This recognition led to the creation of the FC construct (Madni et al. 2017a, b; Sievers and Madni 2014, 2017; Madni and Sievers 2015, 2017;).

17.3 Flexible Contract Construct

A traditional (i.e. inflexible) design contract is a formal modeling construct defined by: assert–guarantee pairs and rigorous concepts such as composition, abstraction, and refinement. The assert–guarantee construct states that system/component properties are guaranteed under a set of assumptions about the environment. A traditional contract enables rigorous stepwise refinement and supports modularity and hierarchy. It is capable of representing conditions on continuous and discrete components. Design contracts are used for requirements validation, conflict detection and resolution, and identification of redundant or incomplete requirements.

A FC is a system modeling construct based on Markov Decision Process (Madni et al. 2017a, b; Sievers and Madni 2014, 2017; Madni and Sievers 2015). It extends the deterministic design contract through flexible assertions. It is suitable for probabilistic modeling and is capable of handling both observable and unobservable states. Implemented as a POMDP, a FC supports in-use learning, uncertainty handling, and pattern recognition. While developed at design time, the FC is trained during actual use ("learning").

More formally, FCs extend the concept of invariant contracts, which are defined by a pair of assertions, $C = (A, G)$, in which A is an assumption (precondition) made on the environment and G is the guarantee (postcondition) a system makes if the assumption is met. More precisely, invariant contracts describe a system that produces an output from set $o \in \{o_0, o_1, \ldots o_{o-1}\} \subseteq O$ when in the state $\sigma \in \{s_0, s_1, \ldots s_{s-1}\} \subseteq \Sigma$ for an input $i \in \{i_0, i_1, \ldots i_{i-1}\} \subseteq I$, where O is the set of all outputs, Σ is the set of all system states, and I is the set of all inputs. Systems defined by invariant contracts are compatible with formal analyses methods that enable rigorous design and validation. However, invariant constructs are not well matched with unknown and unexpected disruptions that might result from unpredictable swarm environments, internal faults, prolonged system usage, and previously undiscovered interactions with the operational environment. Flexibility is

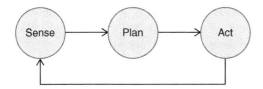

Figure 17.4 Flexibility is implemented with sense–plan–act cycle.

introduced within the "sense–plan–act" construct shown in Figure 17.4. This construct comprises iterations of: sensing the environment and system status (Sense ≡ assumption); planning action that maximizes the likelihood of achieving a goal (Plan); and executing those actions (Act ≡ guarantee). The environment and system health are sensed and assessed after each action. The planning function determines whether to continue with the current plan if the actions accomplish the desired outcome or make changes.

From a computational perspective, flexibility is introduced through a POMDP, which accommodates observable, unobservable, and unknown states. A POMDP models a decision process in which system dynamics are assumed to be a belief Markovian Decision Process (MDP), a memory-less decision process with transition rewards. A belief MDP comprises the 4-tuple:

β = infinite set of belief states
α = finite set of actions
$(b, a) = \sum_{s \in S} (s) (s, s'a)$. Expected reward at b (s) on transition from s to s' given a
$\Lambda (b'|b, a) = \sum_{o \in O} \Lambda (b'|b, a, o) \Lambda (o|a, b)$ Transition function

In the transition function, a belief represents an understanding of a system state, $s \in S$, with uncertainty. The MDP has a policy, π, which describes how to select actions for a belief state based on maximizing a goal defined by the reward function, ρ, within some time period, that is, $\pi: s \in S \to a \in \alpha$. We define the expected utility of executing π when started from s as $U^\pi(s)$. An optimal policy $\pi^* = argmax_\pi U^\pi(s)$ maximizes the expected *utility of an action*.

Expected utility may be computed iteratively using a number of methods, including a dynamic programming approach in which a discount factor, γ, where $0 \le \gamma < 1$, is used to penalize future rewards and is based on the "cost" for not taking immediate action. The k-th value of (s) is computed iteratively. The optimal policy is determined by finding the policy that maximizes $U^\pi(s)$ for each policy π_i.

$$U_0^\pi(s) = 0$$

$$U_1^\pi(s) = R(s, \pi(s))$$

$$U_k^\pi(s) = R(s, \pi(s)) + \gamma \sum_{s'} \Lambda (s'|s, \pi(s)) U_{k-1}^\pi (s')$$

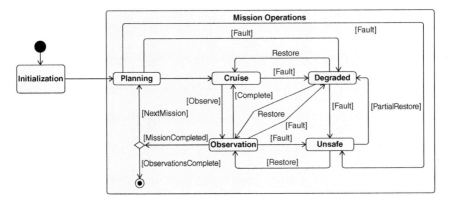

Figure 17.5 Mission meta-sates.

As noted earlier, in a POMDP model, some states are not observable (i.e. hidden) because of uncertainties about the current system state and the outcomes of actions due to imperfect information (e.g. noisy sensors). The FC approach begins with a naïve model of system behavior comprising known (designed) states and transitions, as well as predicted anomalous states and transitions. For the UAV swarm, the naïve model is refined over time by observing UAV swarm behavior as actions are taken.

Post deployment, swarms perform one or more missions defined by mission scenarios. Each scenario comprises a set of mission phases that are further refined into a collection of detailed task behaviors that are allocated to the vehicles in the UAV swarm to accomplish mission objectives. Mission scenarios are defined by instantiating meta-states and transitions, as shown in Figure 17.5. After initialization, the swarm transitions to a Planning state within mission operations. Planning takes stock of swarm health, mission objectives, risks, and prior knowledge of the operational environment to create an initial set of mission phases and individual vehicle tasks. Thereafter, the swarm transitions to Cruise, in which each vehicle positions itself for making observations. Vehicles typically make observations during Cruise and provide the information collected back to Planning. The swarm may encounter faults that require deciding how to best to accomplish the mission or select a secondary (less ambitious) mission when the primary mission becomes unachievable. The swarm enters the observation state when it arrives at its desired locations and begins intelligence gathering and operational mission data collection.

Figure 17.5 shows that the swarm may return to Cruise between observations or may transition to degraded or unsafe operation. Figure 17.5 includes restoration transitions that may return the swarm to a fully operational capability after a fault or may place the swarm into a degraded state by marshaling surviving swarm resources.

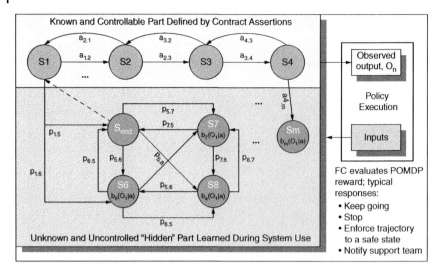

Figure 17.6 Generalized flexible contract.

The swarm may return to the Planning state after concluding the mission, if there is adequate time and resources remaining to prosecute another mission. Additionally, Planning may be needed when restarting a disrupted or an aborted mission depending on when the disruption occurs during task execution, and where the swarm is in relation to ongoing observations. The Planning function may also modify missions as a result of "interesting" observations. For example, a vehicle that detects a threat may request confirmation from another vehicle that has been tasked with observing a different geographic sector. Similarly, a vehicle might request support from another vehicle if a sensor needed for a particular observation has failed or is found to be untrustworthy.

Figure 17.6 shows the structure of a FC that implements the transitions shown in Figure 17.5. A belief state estimate is derived from environmental and health sensors and used to evaluate which actions to take as previously described. The actions are then mapped onto tasks.

A POMDP is a MDP in which the system state is not directly observable but probabilistically inferred from available sensory measurements. Thus, next action determination is not based on the current state but on the current probability distribution. Thus, computationally there is a major distinction between a MDP and a POMDP. A MDP assumes perfect knowledge of state, is relatively easy to specify, and is computationally tractable. A POMDP, on the other hand, does not require perfect knowledge of state, is a bit more difficult to specify, and is computationally intractable with respect to optimality. However, a POMDP treats all sources of uncertainty uniformly and allows for information gathering actions. Several approaches have been proposed in the literature to solve a POMDP (Monahan 1982; Sondik 1971; Cheng 1988; Cassandra et al. 1994). One class of solutions is locally optimal in Gaussian belief spaces (Patil 2015; Platt 2010).

In sum, with a POMDP, each belief is a probability distribution. Thus, each value in a POMDP is a function of an entire probability distribution. This property is problematic, because probability distributions are continuous, and belief spaces are large and complex. So, the real value of POMDPs is in *finite* worlds with *finite* state, action, and measurement spaces, and *finite* time horizons. In these cases, the value functions can be effectively represented by piecewise linear functions.

A POMDP transitions from one state to the next by evaluating a belief state and determining the "best" action to take in the most likely system state. It is initialized by invariant assertions that represent traditional design contracts for a system. As the system operates, the emission and transmission probabilities and actions are updated (get "trained") to reflect real world operational dynamics. Figure 17.6 presents a graphical depiction of a generalized FC.

As shown in this figure, the upper light blue portion represents the deterministic (i.e. known) parts of the system. These are defined by invariant (i.e. traditional) contracts such as "if X, then Y." The bottom light green portion depicts the unknown and uncontrolled "hidden" part of the system, which is represented probabilistically by a POMDP. This portion of the model, which is learned during actual system use, is akin to learning on the job. The initially postulated hidden states may or may not exist, and their probabilities are initially unknown. As observations are made during operation, these hidden states are "filled in" along with their emissions (outputs) that occur when the system is in those states. There are a few techniques to perform these calculations (Eddy 1996).

Intuitively, one can surmise how these techniques work. If the system is not in state S1 or S2 or S3 or S4, then it must be in one of the hidden states. The frequency with which the system goes into a hidden state determines the transition probability and the observations made in that state determine the emission probabilities.

The action policy determines the needed action at each state by evaluating a reward or penalty function. For example, the options might include: continue, stop, take an action that avoids an obstacle, put the system into a safe operational mode, notify the support team, and so forth.

17.4 Closed-Loop MA Approach

The closed-loop, FC-based, MA concept is depicted in Figure 17.7. A FC comprises a state estimator, a MA evaluation system (which evaluates factors such as reliability, availability, safety, and risk), and a response policy. The

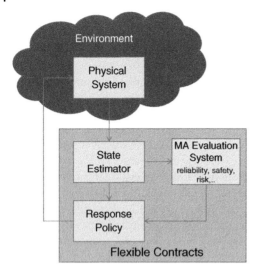

Figure 17.7 Flexible contract approach to mission assurance.

state estimator determines the belief state probability distribution from observations made and actions taken. MA is then evaluated from the belief distributions. The Response Policy determines the optimal action to take as a function of the state estimation and MA evaluation. Each belief estimate is associated with the system's MA metrics (e.g. reliability, safety, risk). The probabilities are then evaluated for each belief estimation, for example:

- P (system is sound) = Σ Prob (system is in a known "good" state).
- P (system has failed) = Σ Prob (system is in a known "failed" state).
- P (system is in a risky state) = Σ Prob (system is in an unknown or undesigned for state).

An exemplar hypothetical progression of belief states based on observations and actions taken is visually depicted in Figure 17.8.

As shown in this figure, initially all belief states are equally probable. When an observation "a" is made, the State Estimator updates the belief states making state S1 the most probable. When an observation "d" is subsequently made, the belief evaluation function makes state S4 the most likely. Finally, an observation "a" is made which makes S1 again the most likely. If state S4 = "failed," for example, then the reliability $R = 1 - P$ (S4).

17.5 POMDP Concept of Operations for Exemplar Problem

Returning to the simple system (Figure 17.2), we now examine the state model that is used for calculating reliability. The simple system comprises a prime

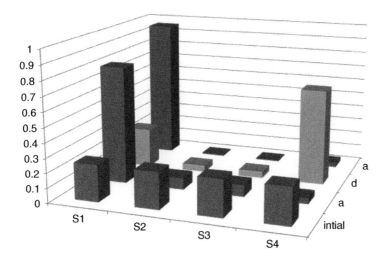

Figure 17.8 Graphical depiction of belief states progression.

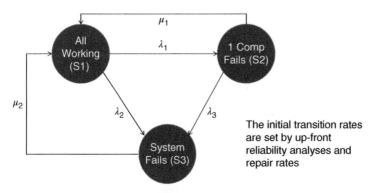

Figure 17.9 Markov model for exemplar problem.

computer (P) and a redundant computer (R) that are cross-strapped to sensor and actuator interfaces. An initial Markov model for a repairable system derived from invariant contracts is shown in Figure 17.9.

In this model, the system is initially in state S1 (i.e. all systems working). From this state, the system can transition to other states based on failure and repair rates, denoted by λ_i and μ_j. For example, the system transitions from state S1 to S2 if one of the computers fail with the failure rate λ_1. The system transitions from state S1 to state S3 at the rate $\lambda_2 = (1 - \lambda_1)$ if either the sensor, actuator, or both fail. The system transitions from state S2 to state S3, at the rate the remaining computer fails or the sensor, actuator, or both fail. The *reliability* of the system is the sum of the probabilities of being in S1 or S2, or equivalently,

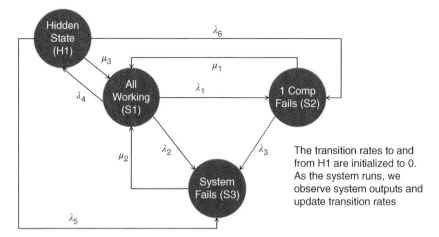

Figure 17.10 Introducing flexibility in a system model.

the probability of not being in state S3. *Availability* is the ratio of time when not in S3 to the total time.

We now address the problem of incorporating a hidden state into the Markov model. For simplicity, we add a single hidden state (H1) to the system model as shown in Figure 17.10. S1 now has a failure rate of λ_4 to H1 and H1 has a failure rate λ_6 to S2, and λ_5 to S3. Because H1 is hidden, the transitions rates into and from H1 are learned during system operation and/or test.

The "step-through" of states occurs as follows. Once the POMDP model is trained, we observe system outputs, evaluate the state we believe the system is most likely in, and take the action having the highest reward (or least penalty) for that state. For example, if the system is most likely in state S1, then it continues operating in that state in the absence of component/system failure. If the system is in state S2, it continues operating and requests a computer repair. An emergency repair is requested if system is in state S3. If the system is in the hidden state, H1, then it might continue monitoring its health for a period of time. If the system transitions to S1, S2, or S3 during the waiting period, then the above actions are taken. If the system stays in H1, then maintenance personnel are called to investigate during wait period. If maintenance personnel determine that H1 is "good," then the state diagram and actions are updated, otherwise the action in H1 is changed to "schedule repair." If the system state is ambiguous (i.e. two or more states are equally probable with the highest probability of occurrence), then the policy could either assume the worst, and take the associated action, or continue sampling outputs in the hope that the system state disambiguates. Table 17.1 summarizes the logic associated with stepping through the states.

Table 17.1 Stepping through states.

- After training, we observe system outputs, evaluate the state we most likely believe the system is in, and take appropriate action, e.g.:
  - if system is in S1, then continue operating
  - if system is in S2, then continue operating and schedule a computer repair
  - if system is in S3, then schedule an emergency repair
- If system is in H1, then monitor what occurs for a period of time:
  - if system transitions to S1, S2, or S3 during the waiting period, take above actions
  - if system stays in H1 during the waiting period, then schedule a maintenance check
  - if the check indicates H1 is "good," then update state diagram and actions; otherwise, change action in H1 to schedule repair
- If system state is ambiguous (i.e. two or more states are equally probable and have highest probability), then:
  - collect additional observations and retrain state model – which might require adding new states, and schedule a maintenance check as above, or
  - assume the worst and take the associated action

In sum, incorporating flexibility in MDP consists of: introducing hidden states in MDP representation, thereby making it a POMDP model; relaxing time-invariance restriction of the state space and action space; adding an evaluation metric to determine best action; and introducing the concept of time (e.g. temporal constraints/changes in probabilities that impact occurrence of events).

17.6 An Illustrative Example

We now discuss the different system modeling elements within the context of multi-UAV swarm control. These include: traditional contracts; FC, control flow model; swarm control architecture; and iterative Bayesian belief update.

Traditional (i.e. Inflexible) Contracts

These are "assert–guarantee" pairs derived from system requirements and expected system behavior. Table 17.2 shows two invariant contracts using linear temporal logic terminology.

Flexible Contracts

Invariant contracts can be made flexible by relaxing the assertions to account for uncertainty. The assumptions in a FC are represented by belief states, while the action policies replace guarantees in the traditional contract. Accommodating disruptions is accomplished by including belief states that represent environmental hazards and fault conditions and including

Table 17.2 Invariant contracts.

- Contract #1: At the next instant, if obstacle ahead, then turn left
- Contract #2: At the next instant, if no obstacle ahead, then continue path

---

Table 17.3 Penalty function calculation and decision.

- A simple penalty function:
  - P(leftThreat) * 10 + P(rightThreat) * (−100) + (1 − (P(rightThreat) + P(leftThreat))) * (−1)
- If P(leftThreat) = 0.95, P(rightThreat) = 0.01, and P(can't decide) = 0,04, then:
  - penalty = 0.95 (10) + 0.01 (−100) + 0.04 (−1) = 9.5 − 1 − 0.04 = 8.49, if turn left
  - penalty = 0.01 (10) + 0.95 (−100) + 0.04 (−1) = 1 + 0.95 − 0.04 = −95.94, if turn right
- Want to **turn right** because penalty for turning right is negative (i.e. **reward**), but penalty for turning left is positive

---

contingency or mitigation policies. For example, Table 17.3 shows a simple penalty function that is associated with a threat being possibly to the right or to the left of a UAV. If belief estimation determines that the probability of the threat on the left is 0.95 then the penalty for making a left turn is 8.5 but the penalty for making a right turn is −96, meaning there is a reward for turning right and a penalty for turning left.

Control Flow Model

System behavior is captured using the control flow model shown in Figure 17.11. Figure 17.11 shows transition "guards" or conditions that must be true for the transition to occur, e.g.:

b (failed) ≥ 0.95 is the threshold to transition from "normal motors" to "failed motors."

Belief state estimates are derived from observations about the system and used by the Auto Plan function that determines next actions to take.

An exemplar architecture that implements Figure 17.11 using software agents is shown in Figure 17.12. This figure shows that environment sensor outputs, the MDP model, and the policy feed the state estimator which produces an updated set of belief values and triggers an action for the UAV swarm.

Iterative Belief Update

The system's behavior reflects its beliefs based on latest observations. For example, suppose an UAV swarm finds itself in an environment in which it needs to avoid obstacles while being mindful of unknown threats. In this instance, the UAV swarm needs to veer left or veer right to avoid an obstacle which may be either to the left or the right of the swarm If the UAV swarm guesses correctly, it can avoid the threat. If the UAV swarm guesses wrong

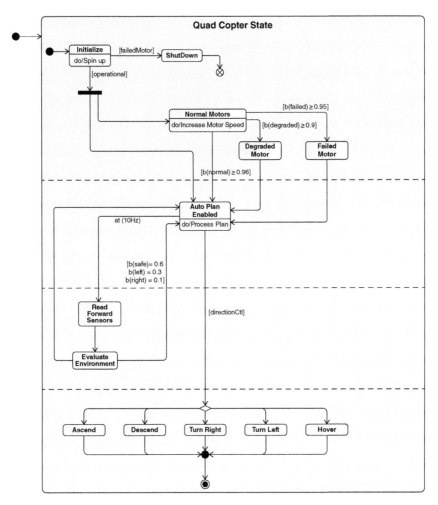

Figure 17.11 Control flow model.

(i.e. it decides to veer left and the threat is located on the left, or it decides to veer right and the threat is located on the right), then the swarm is likely to suffer damage. The potential actions that the swarm can take are: veer left, veer right, or continue on the current path making further observations of the threat. The algorithm for summing potential outcomes based on path taken is rooted in Bayesian Belief update with normalized rewards and penalties. Figure 17.13 presents a pictorial of iterative Bayesian Belief update and the formula for computing beliefs based on Bayesian Belief Network. The system starts with a 50–50 belief that the threat could be to the left or the right. The system makes an observation that suggests that a potential threat is to the

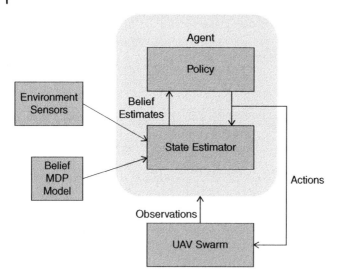

Figure 17.12 Multi-UAV swarm control architecture.

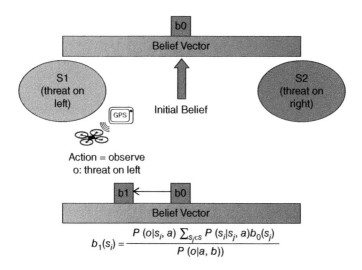

Figure 17.13 Iterative belief update.

left. So, the system's Bayesian engine moves the belief to the left as depicted in this diagram. This move indicates that there is greater belief that the threat is to the left. Eventually, the system concludes that the belief that the threat could not be to the right because it does not observe anything to the right. Belief updating occurs in accord with Bayesian analysis using observation and current state. The state history is contained in the current state. The "a" are actions, the "o" are observations, the "b" are beliefs, and the "S" are states. Belief state changes results from observations and actions taken.

17.7 Summary

The fourth industrial revolution blends physical, digital, and biological domains into what is commonly referred to as cyber-physical systems. This revolution is capitalizing on advances in technologies such as artificial intelligence, machine learning, biological sensing, human behavior modeling, nanotechnology, and the internet of things. As system complexity and integration challenges continue to grow, our ability to understand their behavior decreases making them more vulnerable to disruptive events. Thus, fourth generation systems must be able to detect, isolate, eliminate (or withstand) a continual barrage of unexpected attacks and previously unseen operating conditions. Although traditional fault-tolerance and MA methods are an essential component of the needed resilience, dealing with unknown-unknowns requires greater sophistication and vigilance than exists in today's systems. However, even with exquisite designs and novel resilience schemes, we cannot always ascertain a priori whether a complex cyber-physical system will operate as expected, slide into trouble, or face an imminent crash. Furthermore, we cannot always know a priori what to do even if we did know the state of the system. Therefore, what is needed is requisite flexibility in how a system can respond and a way for continually monitoring and evaluating system health. POMDPs offer effective means for continuously learning about system state and new and unexpected conditions while providing the requisite flexibility in actions taken to assure system safety, trustworthiness, and usefulness. Moreover, unlike traditional up-front MA analyses, the POMDP approach provides near real-time updates of system robustness metrics while simultaneously determining command trajectories that ideally drive the system back to nominal and safe operation.

The MA function is concerned with reducing uncertainty in the ability of a system (or System-of-Systems) to successfully complete its mission despite running into disruptions (Madni and Jackson 2009; Goerger et al. 2014; Madni et al. 2017b). Current approaches to MA tend to rely exclusively on information available at design time. This is a clear limitation of the traditional approach. This paper has presented a closed-loop, model-based approach based on FCs and reinforcement learning. Specifically, the approach employs probabilistic modeling to account for uncertainty arising from partial observability, and incremental availability of information. It employs a flexible extension of invariant contracts from contract-based design to accomplish some level of system verification and testing while having the ability to adapt to changes. It employs reinforcement learning to reduce uncertainty in the knowledge of system state and health status. The key concepts associated with this approach were presented in this paper within the context of resilient, multi-UAV swarm operations.

Acknowledgments

This work was supported by Department of Defense Systems Engineering Research Center (SERC), RT-180 contract No. HQ0034-13-D-0004.

References

Cassandra, A.R., Kaelbling, L.P., and Littman, M.L. (1994). Acting optimally in partially observable stochastic domains. In: *AAAI'94 Proceedings of the Twelfth National Conference on Artificial Intelligence*, vol. 2, Seattle, WA, 1023–1028. Menlo Park, CA: American Association for Artificial Intelligence.

Cheng, Hsien-Te. Algorithms for Partially Observable Markov Decision Processes. Doctoral dissertation. University of British Columbia. 1988.

Eddy, S.R. (1996). Hidden Markov models. *Current Opinion in Structural Biology* 6 (3): 361–365.

Goerger, S.R., Madni, A.M., and Eslinger, O.J. (2014). Engineered resilient systems: a DoD perspective. *Procedia Computer Science* 28: 865–872.

Grimaila, M.R., Mills, R.F., Haas, M., and Kelly, D. (2010). Mission assurance: issues and challenges. In: *Proceedings of the 2010 International Conference on Security and Management (SAM 10)*, Las Vegas Nevada (12–15 July), 651–650. CSREA Press.

Jabbour, K. and Muccio, S. (2011). The science of Mission assurance. *Journal of Strategic Security* 4 (1): 61–74.

Madni, A.M. and Jackson, S. (2009). Towards a conceptual framework for resilience engineering. *IEEE Systems Journal* 3 (2): 181–191.

Madni, A.M. and Sievers, M., A Flexible Contract-Based Design Framework for Evaluating System Resilience Approaches and Mechanisms, 2015 ISERC 2015, Nashville, TN.

Madni, A.M. and Sievers, M. Model Based Systems Engineering: Motivation, Current Status and Needed Advances, 2017, CSER 2017 (23–25 March), Redondo Beach, CA,

Madni, A.M., Sievers, M., Humann, J., Ordoukhanian, E., D'Ambrosio, J., and Sundaram, P. Model-Based Approach for Engineering Resilient System-of-Systems: Application to Autonomous Vehicle Networks, CSER 2017 (23–25 March), Redondo Beach, CA, 2017a.

Madni, A.M., Sievers, M., Humann, J., and Ordoukhanian, E., Model-Based Approach for Engineering Resilient System-of-Systems: Application to Multi-UAV Swarms, CSER 2017 (23–25 March), Redondo Beach, CA, 2017b.

Monahan, G.E. (1982). State of the art – a survey of partially observable Markov decision processes: theory, models, and algorithms. *Management Science* 28 (1): 1–16.

Neches, R. and Madni, A.M. (2013). Towards affordably adaptable and effective systems. *Systems Engineering* 16 (2): 224–234.

Patil, S., Khan, G., Laskey, N. et al. (2015). Scaling up Gaussian belief space planning through covariance-free trajectory optimization and automatic differentiation. In: *Algorithmic Foundations of Robotics XI*, Springer Tracts in Advanced Robotics, vol. 7 (eds. H.L. Akin, N.M. Amato, V. Isler and A.F. van der Stappen), 515–533. Springer.

Platt, R. Jr., Tedrake, R., Kaelbling, L., and Lozano-Perez, T. (2010). Belief space planning assuming maximum likelihood observations. In: *Proceedings of Robotics: Science and Systems (RSS) VI* (eds. Y. Matsuoka, H. Durrant-Whyte and J. José Neira) Zaragosa, Spain (27–30 June), 291–298. MIT Press.

Sievers, M., and Madni, A.M., A flexible contracts approach to system resiliency. 2014 IEEE International Conference on Systems, Man, and Cybernetics (SMC), San Diego, CA (5–8 October).

Sievers, M. and Madni, A.M. Contract-based Byzantine resilience for spacecraft swarm, 2017 55th AIAA Aerospace Sciences Meeting, Grapevine, TX (9–13 Janaury).

Sondik, Edward J. (1971). The Optimal Control of Partially Observable Markov Decision Processes. PhD thesis. Stanford University.

18

FlexTech: From Rigid to Flexible Human–Systems Integration
Guy A. Boy

Synopsis

Is human–systems integration (HSI) a necessary component of systems engineering (SE) or the opposite (i.e. is SE a necessary component of HSI)? It all depends on the perspective! If you are a technology-centered engineer, SE will provide you with methods and tools to develop technological systems, and you will need human factors specialists to develop user interfaces and test the usability of the end product. In this perspective, HSI is a necessary component of SE. However, if you are human-centered designer, you will need methods and tools to design and develop systems that enable the integration of human and machine requirements from the very beginning of design to dismantling of the system-of-systems at stake. This states the question of what we mean by "system." A system is simply a representation that helps to figure out physical and cognitive functions and structures of both people and machines. In this chapter, several areas related to HSI are covered, including task and activity analysis, cognitive engineering, organization design and management, function allocation, complexity analysis, modeling, and human-in-the-loop simulation (HITLS). Contemporary HSI design approaches are supported by virtual HITLS, which involves tangibility issues. Various kinds of data that should be collected and the tangibility indicators needed to develop appropriate HSI. An aeronautical example is provided to illustrate how HSI should be developed are discussed in the design and development of a system-of-systems. We conclude by discussing the necessary shift from rigid automation to flexible autonomy that prefigures the *FlexTech* (i.e. technological solutions, associated with organizational setups and human functions, that improve flexibility of operations).

Systems Engineering in the Fourth Industrial Revolution: Big Data, Novel Technologies, and Modern Systems Engineering,
First Edition. Edited by Ron S. Kenett, Robert S. Swarz, and Avigdor Zonnenshain.

18.1   Industry 4.0 and Human–Systems Integration

The fourth industrial revolution, also called Industry 4.0, results from the growing digitalization of industrial organizations, and more generally our sociotechnical society. Depending on approach and background, several new trends are popping up and developing: automation is leading toward cyber-physical systems; computer science to the Internet of Things; computer engineering to cloud computing; and human–computer interaction (HCI) and artificial intelligence to cognitive computing. The third industrial revolution was technology-centered, considering human–systems integration (HSI) as an adaptation of people to machines. After World War II, human factors and ergonomics (HFE) developed as a discipline attempting to make machines usable by people, either anybody (public use) or experts (mostly in life-critical systems). HFE was initially focused on physical ergonomics, as well as health and safety at work. HFE was handled by physicians until the beginning of the 1980s when microcomputers invaded our societies. Cognitive ergonomics started then to be prominent. At the same time, HCI became a requirement to enable people to commonly use computers in their everyday lives. As a matter of fact, HCI started to be developed as a discipline, involving computer scientists and cognitive scientists. HCI led to cognitive engineering, graphical user interfaces (GUI), and computer-supported cooperative work (CSCW), for example. At the same time, systems engineering (SE) developed to structure industrial organizations with technical processes. Several concepts started to be studied and operationalized, such as systems of systems, agile development (Schwaber, 1997, 2004; Sutherland, 2014), and model-based systems engineering. However, these concepts and approaches were mostly technology-centered and required to be revisited to include humans and organizations at the center. Human-centered design (HCD) then developed as a discipline. From the beginning of the twenty-first century, it became possible to effectively include human and organizational requirements into design and development processes. Why? Simply because human-in-the-loop simulations (HITLSs) could be taken seriously at design time, enable observing people's activity, virtually design and test possible solutions, and derive appropriate human and organizational requirements. HSI, which integrates HCD and systems engineering during the whole life cycle of a system (including people and machines), became a discipline.

Consequently, virtual prototyping has become a precious method supported by digital modeling and simulation tools that supports HSI. People can be included from the beginning of the design process to the end of the life cycle of a system. From the beginning of the twenty-first century, we have inverted the engineering approach. Instead of going from hardware to software (i.e. constructing structures and then refining functions), we now go from software to hardware (i.e. designing and testing functions and then producing structures) (Boy 2017). Nowadays, we can print 3D structures from digital models that have been functionally tested (in virtual environments). Products can be certified virtually before they are physically developed. This is very appealing, but the emerging concept of tangibility should be better mastered and tested. We will further describe and analyze tangibility in this chapter. Each system (e.g. a modular equipment) can be considered as a tangible interactive system (TIS) that can be easily manipulated, both structurally and functionally (Boy 2016). This TIS modeling and approach are very useful for managing rapid changes. Indeed, visualization of explicit knowledge of interconnected manufacturing systems, as virtual twins, provide useful and usable support for reconfiguration.

HSI will then be understood using a systemic approach. More specifically, we will specifically focus on real-time process control, data interoperability, consistency and integration based on a system-of-systems (SoS) approach. A SoS framework is very useful for mastering systems' structures and functions that lead to reducing risks, controlling costs, and improving function allocation. In other words, considering human factors from the beginning of design and all along the life cycle of systems improves safety, efficiency, and comfort. The SoS framework offers mediating concepts and tools that shareholders can use for collaborative work (offering a common frame of reference), distributed decision making and shared situation awareness.

The orchestration (Boy 2013) of various kinds of TISs requires developing an ontology of the manufacturing domain being investigated (i.e. music theory of we use the orchestra metaphor), making prescribed tasks explicit (i.e. analog to scores provided to the musicians that can be humans or machines), coordinating these tasks at design time (i.e. analog to the role of the composer), coordinating produced activities at operations time (i.e. analog to the role of the conductor), training the various TISs and human operators for improved performance and cooperation (i.e. analog to musicians learning), and seriously considering end-users of products being manufactured (i.e. analog to the audience of a symphony). This approach provides flexibility during the whole life cycle of systems being developed; we call it, *FlexTech*.

HSI applies to all stakeholders involved during the life cycle of a system. We should not restrict HSI to end users. Designers are the main actors of HSI. Think about designers of kitchen appliances that should be adapted to customers' needs and requirements. There are two extreme strategies: (i) procedural manufacturing of very well dichotomized generic systems; and (ii) expert-based crafting of customized systems. In the same way, fashion designers use the distinction: "ready to wear" versus "tailored." If we want to increase flexibility of production for customized products, we need to develop flexible technology and knowledgeable/skilled employees. Designing such FlexTech requires to investigate and use approaches such as systems of systems, teams of teams, function allocation, failure management, and creativity. In addition, we should not only develop systems in "normal situations" but also consider "abnormal and emergency situations" that may leads to "unexpected situations," where people have difficulty and even impossibility to find appropriate solutions in a reasonable amount of time. This is the reason why we propose to carry out an HSI study that considers tangibility of digital systems being developed in terms of complexity, stability, flexibility, maturity, and sustainability of the system being developed and its usages (Boy 2016).

How should we proceed? For example, we could define and develop several use cases, which could be derived from expert experience, accident and incident analyses and various possible projections based on current societal evolution. Two contradictory concurrent approaches could be mixed: (i) creativity and innovation (divergent processes); and (ii) use of experience (convergent process). Such use cases usually lead to the development of scenarios in terms of systems configurations and chronologies (i.e. storytelling, scripts like for a theater play). A prototype of the system is then developed and used to run HITLS in order to discover emerging patterns. The cognitive function analysis technique (Boy 1998) could be used to model the resulting SoS. The resulting model could then be used to support assessments and validation of the SoS being developed. Principle and criteria will be developed for such assessments.

Summarizing, HSI requires us to develop virtual prototypes, scenarios, and sociotechnical criteria that will support our HCD approach. It is by concurrently developing prototypes that incrementally become more tangible (physically – hardware part – and cognitively – software part) that we will reach these objectives. Current understanding of what tangibility of digital technology means is developed in this chapter.

18.2 HSI Evolution: From Interface to Interaction to Organizational Integration

It would be difficult to make sense of HSI without defining the distinction between task and activity. A task is what is prescribed to be done. An activity is what is effectively done once the task is executed. Task analysis has been extensively used to support the design of user interfaces. However, usability studies have shown that activity observation and subsequent analysis are crucial to improve interaction design (Kaptelinin and Nardi 2006). For a long time, the problem was that activity was only observable on existing systems, whether currently used systems prior to novel design or once a new system was fully developed. On the one hand, activity analysis on systems, which will become obsolete once a new system is developed, forces continuity or nondisruptive innovation. On the other hand, activity analysis once a new system is fully developed is likely to show design flaws that will or will not be possible to fix system-wise; only cosmetic patches can be brought at the user interface level. This has been done for a long time.

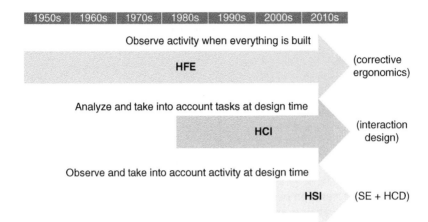

Figure 18.1 Evolution of human-centered technology approaches.

Historically speaking (Figure 18.1), HFE developed as based on analysis. However, activity analysis was only possible using an existing system (i.e. before the design of a new system using the old system when it existed, and after it was fully developed and usable). HFE then led to corrective ergonomics. In HCI, as a discipline, there is no, or very little, difference between task and activity in the manipulation of computers. This is the reason why task analysis was heavily used in HCI.

Interaction design was born in HCI. However, HCI has been very often limited to computing systems where complexity of multiphysical systems was almost absent. An aircraft, an air traffic control system or a nuclear power plant is such a complex system, where observed activity is generally different from related prescribed tasks, typically provided in the form of operations procedures. Related user interfaces, whether they are called cockpits or control rooms, involve deeper considerations than generic desktop GUI.

Since the beginning of the twenty-first century, systems engineering started to become aware of the importance of human factors in engineering design. More specifically, defining user interfaces when a complex system is developed is not satisfactory. Many user interfaces tend to hide engineering design flaws, and unfortunately adapt people to developed systems, creating situation awareness and decision-making problems in critical situations.

This is an important reason why HSI started to develop in engineering design. Which leads to systems engineering where systems include people and organizations. It is common to hear about sociotechnical systems to denote such systems. HSI attempts to concurrently consider technology, organizations, and people (TOP) during the whole life cycle of a system (Figure 18.2) and requires a more formal "system" definition.

People

Human
Centered
Design

Technology   Organizations

Figure 18.2 The TOP model for human-centered design.

18.3 What Does the Term "System" Mean?

A system is a representation of either a natural or an artificial entity. A natural entity can be a human being, an organ of a human being, a plant or an animal. An artificial entity can be an abstraction (e.g. a law, a legally-defined country, a method), an object (e.g. a chair) or a machine (e.g. a car or a washing machine) that was built by a human being to facilitate the execution of specific tasks.

A system can be either cognitive (or conceptual), physical or both (Boy 2017). It also has at least a structure and a function (Figure 18.3). Today, machines have cognitive functions (e.g. the cruise control function on a car enables the car to keep a set speed). Figure 18.3 presents a simple ontological definition of the "system" representation.

In addition, the conventional single-agent definition of a system function as something that transforms an input into an output (Figure 18.4) should be extended to a multi-agent perspective.

As a matter of fact, an agent can be defined as a society of agents (i.e. an agency is an agent itself) in Minsky's sense (Minsky 1986). For example, a postman is an agent of an agency that is commonly called "The Post." In this chapter,

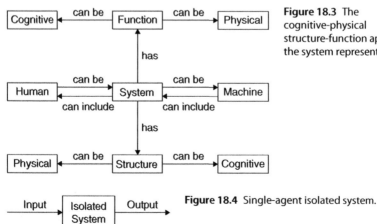

Figure 18.3 The cognitive-physical structure-function approach of the system representation.

Figure 18.4 Single-agent isolated system.

Figure 18.5 Interconnected
system-of-systems.

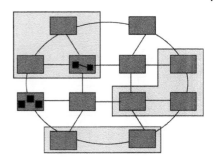

the term "system" will be a synonym of "agent." It can be said that a system is a representation of an agent, whether it is a human or a machine. In the same way an agent is a society of agents, a system is a SoS. Therefore, system's structures and functions can be defined as structures of structures and functions of functions. More generally, it is now common to use the system-of-systems concept to denote sociotechnical interconnected systems (Figure 18.5).

A typical postman's function can still be defined as having a (prescribed) task of "delivering letters" (i.e. function input). His/her activity (i.e. function output) may not always reflect such a prescribed task because the environment may change, his/her capacity may change (e.g. the postman is tired or starts to be sick) or other contextual factors may change (e.g. heavy rain or excessive traffic jam). This is the reason why a system function, in the multi-agent sense, should be defined by three attributes:

- a role;
- a context of validity;
- a set of possible and necessary resources.

Context of validity of postman's role (i.e. delivering letters) can be defined by a time context (e.g. from 8:00 a.m. to noon and from 2:00 p.m. to 5:00 p.m.) and a space context (e.g. the neighborhood where he/she has to deliver letters). Context can be normal (i.e. every day the same) or abnormal (e.g. some other postmen are absent and he/she needs to expand his/her time and/or space context). A postman's resources can be physical (e.g. a bag and a bicycle) or cognitive (e.g. a pattern-matching cognitive function that enables him/her to put each letter in the right box). At this point, it becomes clear that a function is a function of functions (e.g. postman's function to deliver letters is a function of another function, the pattern-matching function).

More generically, function's resources are systems. They can be either physical, cognitive or both. In many cases, systems may be represented by their structures. This representation is very convenient for function allocation in a SoS (i.e. among systems [or structures] in a network of systems).

In the same way, contexts can be represented as contexts of contexts. In aviation, for example, the overall flight context can be decomposed into smaller

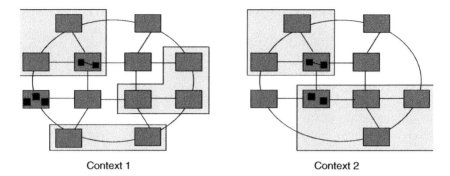

Context 1   Context 2

Figure 18.6 Evolution of system-of-systems from Context 1 to Context 2.

contexts that include taxiing, takeoff, after-takeoff climb, cruise, descent, approach and landing, and so on. Each of these contexts can be decomposed into even smaller contexts. A system may evolve from Context 1 to Context 2 (Figure 18.6), where Context 1 may represent a normal situation and Context 2 an abnormal situation, for example.

18.4 HSI as Function Allocation

Fitts and his colleagues were the first to be technically interested in function allocation among people and safety-critical machines (Fitts 1951). In this chapter, function allocation will be brought a step further. Instead of allocating functions statically only using a task-based approach, the system-of-systems representation enables systems being incrementally designed by redefining their functions using an agile activity-based approach. Since system-of-systems models can be simulated (using explicit system representation), emergent properties, and emerging functions, can be reallocated into the SoS (DeLaurentis 2005). Let's better explain how this approach works and can be effectively used. Here is a procedure:

1. A SoS is defined from innovative concepts, provided by creative people, consolidated by experience, provided by domain experts. A first set of functions is defined and allocated to the various human and machine systems. Systems should be defined using the ontology provided in Figure 18.3.
2. The SoS is simulated, producing an activity that should be observed and analyzed. Activity observation should be guided by appropriate metrics to be defined with respect to tangibility issues.
3. Effectivity of original functions should be analyzed and emerging functions induced from activity analysis. A new function allocation should be performed.

4. If there are still open questions on current function allocation, another sequence 1-2-3 should be executed, otherwise the allocation process is finished.

It is clear that such an approach can be very sensitive to chosen levels of granularity of systems (structures) and functions. The finer the grain, the longer and more difficult the process will be. Conversely, the bigger the grain, the easier but less informative the process will be. As in all processes of that type, expertise, and experience matter. This is the reason why one has to practice before becoming effective and efficient. It is also clear that such a function allocation process is never finished because the world in which the SoS will work is open, and therefore unexpected events will always create new emergent properties that need to be discovered, analyzed, and considered in a new allocation process. However, it is crucial to define principles and criteria that will help stop momentarily the allocation process to accept a "good-enough" solution validated by experienced people.

18.5 The Tangibility Issue in Human-Centered Design

HITLS and, therefore, HCD are based on software-based models that can be considered as digital twins of the system being developed and incrementally optimized. Tangibility is inversely proportional to the distance between such digital twins, also called virtual prototypes in design and development phases, and real systems that they represent.

Tangibility can be defined from two complementary perspectives: physical and figurative (Boy 2016). Physical tangibility denotes the capacity of an object or a system to be grasped, held, and manipulated correctly. Figurative (or cognitive) tangibility denotes the capacity of an argument, an abstraction or a concept to be grasped, held, or manipulated correctly. Tangibility is related to both realism and meaning. When something makes sense, it is tangible. It may make sense sensitively (physically) and/or cognitively (figuratively).

Tangibility should be assessed using appropriate properties and metrics. It can be decomposed into five considerations that lead to such metrics: complexity, maturity, flexibility, stability, and sustainability. Just a few properties and metrics will be provided in this chapter as illustrations. A deeper account will be available in a forthcoming book (Boy n.d.).

Considering the complexity of a system seriously requires finding out what parts are separable, in the sense of being investigated separately without disturbing the overall structure and function of the system. For example, an organ of a biological system is separable when it can be studied in isolation of the overall system it belongs to. Some organs are simply not separable (e.g. the brain of a human being should be investigated and treated connected with the rest of the body; it is not separable). Consequently, complex systems have important metrics that cannot be measured through independent variables. For example, workload of a person cannot be measured by simple variables such as heart rate or electroencephalographic signals. Indeed, workload is not only an output that people produce, it is also an input that people need to regulate their performance. Workload is an emergent property. Therefore, workload deserves to be modeled in order to qualify as a human-factors metrics. For example, time-wise workload could be expressed as the ratio of required time on the available time to perform a task. Workload is one of the metrics that can be used to assess tangibility of a complex human–machine system.

In the same way, maturity should be continuously assessed in order to make sure that we correctly ensure both physical and figurative tangibility when the system will be delivered. Maturity can be explored in three ways. Technological maturity expresses the amount of confidence one could have in using the system being developed without needing expert technical knowledge. For example, very few people considered computers as tangible during the 1970s because computing technology was not mature. Today, almost everybody tangibly uses a computer every day. Maturity of practice expresses the amount of virtuosity one could have in performing any task using the system. For example, there are various kinds of computing systems for specific usages such as smart phones usable for communication purposes and digital vision systems to detect intruders in specific venues. Organizational or societal maturity expresses the amount of lack of resistance an organization or a society has to effectively use the system. For example, our society was not ready (i.e. not societally mature) in the beginning of the 1990s to recognize Apple's Newton tablet as a tangible tool, even if it was a great system. It took about 15 years to make tablets (e.g. iPad) acceptable for our societal needs.

Flexibility is always needed when things go wrong. Designing a system that provides rigid alternatives in critical situations requires to be thought through twice before committing. Indeed, people are always more flexible than machines. This means that people can provide solutions that machines could never find if they are not programmed to do so. Conversely, even if people are unique creators of ideas, they may make mistakes. Consequently, in life-critical systems, flexibility should be bounded within reasonable limits, and people should be equipped with appropriate tools that support them in problem-solving tasks. Various kinds of safety solutions can be developed going from fail-safe systems to fault-tolerant systems.

Stability can be passive or active depending on the autonomy of the system. Passively stable systems are able to recover from various kinds of disturbances. Actively stable systems require external assistance to keep safe performance and sound activities. The more a system is autonomous, the more it can go back to a stable state when disturbed or faced with an unexpected event. Unstable systems are generally not considered as tangible without considering stabilization systems that go with them. Autonomy is further defined later in this chapter.

Finally, sustainability is also related to autonomy, but in the sense of the relationship between the system being developed and its environment. A system is totally sustainable when it is self-sufficient in a natural context and does not destroy its environment. At this point, tangibility deals with philosophical models where a choice should be made between whether ecology leads economy or the opposite.

18.6 Automation as Function Transfer

Automating a machine is delegating a human function to a machine. Obviously, the resulting artificial function is not exactly doing what a human can do, but it is close enough to say that the machine owns a cognitive function similar as what humans used to have (Boy 1998). For example, the function "heading control" in an aircraft cockpit was transferred to the machine in the early 1930s (e.g. Boeing 247), in the form of an autopilot. Autopilots have been used since then on aircraft. During the 1980s, commercial aviation introduced integrated and digital autopilot and autothrottle that defined a new control loop, the guidance loop based on high level modes. This was a clear evolution. At the same time, the flight management function was introduced on commercial aircraft; this integrated guidance and flight management on a computer, called the flight management system (FMS). This was a drastic revolution, moving flying activities from control to management, that is from controlling basic flight variables to managing systems, which themselves are controlling basic flight variables. This was a delegation shift. From then, pilots had to collaborate with the FMS. However, these changes were very much between pilots and aircraft systems, where each pilot was considered as a single agent. The aviation community learned over the years that technology introduced another factor, that is the organization among involved agents.

Indeed, the traffic-alert and collision avoidance system (TCAS), introduced during the 1980s, had a significant impact on the reduction of midair collisions (Kuchar and Drumm, 2007). It also introduced a new concept of authority sharing between air traffic controllers (ATCOs) and pilots. The question was no longer installing new systems onboard without taking care of other agents or systems outside the cockpit, but considering air and ground agents or systems. TOP (Figure 18.2) had to be "designed" concurrently. Using TCAS is not a trivial thing to do. The TCAS system has three internal functions: a surveillance function (and system), a trajectory extrapolation function (and system), and a threat detection function (and system). The surveillance function, F_S, monitors the presence of other aircraft in the vicinity. The trajectory extrapolation function, F_E, calculates, with respect to range, bearing,

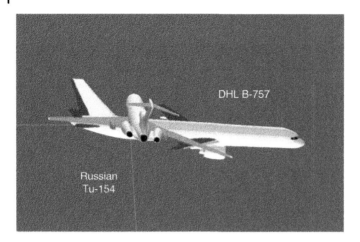

Figure 18.7 Überlingen accident. Source: Picture from the German investigation report BFU 2004.

and altitude, the possibility of conflict with another aircraft if detected by F_S. The threat detection function decides if there is a threat with respect to the F_E result. However, technological functions of the TCAS are not sufficient to solve the entire problem of collision avoidance. Indeed, the pilot is also involved. Once a possible conflict is detected, TCAS provides a "Traffic Alert" to the pilot (aircraft's traffic alert function F_{TA}), in the form of two possible resolution advisories (aircraft's resolution advisory function F_{RA}: "Climb" or "Descent"). Then, the pilot needs to quickly decide (pilot's decision function F_D), inform the ATCO (pilot's response selection function F_{RS}), and maneuver accordingly (pilot's maneuver function F_M). Normally, the pilot should obey the TCAS and not the ATCO (in the case the ATCO asks the pilot "Climb" or "Descent"). The 2002 Überlingen (Germany) disaster, which killed 71 people, is an unfortunate example of such confusion using TCAS (Figure 18.7).

Orders coming from TCAS and ATCO were contradictory, and one of the pilots obeyed the ATCO. Consequently, both conflicting aircraft maneuvered a "Descent"... They collided into each other. An HSI study, using a cognitive function analysis (Boy 1998, 2011), would have proposed that a technological link between TCAS and ground control should be implemented in order to provide ATCO with correct information of TCAS resolution advisory to pilots. In such a SoS, coordination is key. Without an HSI approach, technology-induced human errors remain quite possible (e.g. the Überlingen disaster).

This authority sharing trend is now developing as air traffic density is increasing (4.5% per year for the last 35 years). The growing number of aircraft in the sky, more specifically on top of large airports, deserves new kinds of studies on air traffic management complexity. The way air traffic should be handled in high density zones is not the same as what controllers used to be doing when the number of aircraft was reasonable to control. SESAR and NextGen programs are developing possible solutions for this kind of problem. 4D trajectories for example are planning-based solutions that still deserve investigations since planning and flexibility are contradictory. Indeed, air traffic is planned, the more it can be stabilized and handled more effectively, but also the more planning rigidifies and does not provide enough flexibility when failures or unexpected events occur. Consequently, current airspace system has to be considered as a SoS.

18.7 From Rigid Automation to Flexible Autonomy

When everything is "normal," well-done automation works perfectly without human assistance. However, as already stated previously, in abnormal and emergency situations automation tends to introduce rigidity, due to the fact that it executes procedures that cannot be changed easily. In case of machine failure, human error or, more generally, unexpected events, people require flexibility to handle corresponding situations. We have been focusing on the human error syndrome for a long time, considering that people were "the problem." It is time to consider that people can be "the solution," when human presence is necessary. This is the reason why not only HSI is essential, but also people's competence and skills are crucial. In other words, the question of autonomy needs to be considered on the human side first during the whole life cycle of a human–machine system.

At this point, it is crucial to better define what "autonomy" means. A human or machine system is autonomous when he/she is able to handle (almost) any situations without external help. An autonomous system is equipped with sensors that enable appropriate information leading to appropriate decision making and action to be performed. Situation awareness is a typical cognitive function that includes perception of the situation, comprehension, and projection to anticipate and perform correct actions (Endsley 1995a,1995b, 1998; Endsley and Garland 2000; Boy 2015). There is still a long way to go for identifying appropriate variables that will be implemented in situation awareness support on increasingly-autonomous systems. For example, birds flying in flocks have super TCAS systems that zoologists typically model in the form of three types of functions that can be simulated using differential equations: separation, alignment, and cohesion (Potts 1984; Reynold 1987; Ballerini et al. 2007; Pemmaraju 2013). These functions could be very purposeful to be adapted and implemented on aircraft for handling self-separation for example in highly congested environments. Of course, it goes without question that such solutions should be heavily tested in HITLS to look for emerging properties and functions that need to be further considered. In any case, the more aircraft become autonomous, in the sense of providing more

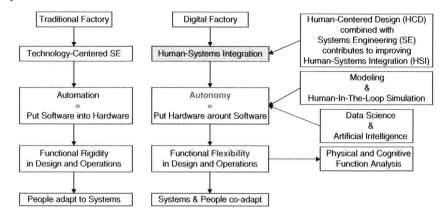

Figure 18.8 How could a digital world provide more autonomy and flexibility?

autonomy to pilots and also to aircraft themselves, the more coordination rules will be necessary. Again, without HITLS experimentations, such rules will not be discovered.

Let's summarize what the shift from rigid automation to flexibility means (Figure 18.8). The traditional factory typically led to technology-centered systems engineering (TCSE), as current digital factory is progressively leading to HSI because HCD combined with systems engineering contributes to improve HSI. Automation contributed to put software into hardware, as autonomy should contribute to put hardware around software using modeling and HITLS. Traditional functional rigidity in engineering design and operations can nowadays leave the floor to flexibility using HSI methods and tools, and more specifically physical and cognitive function analysis. Our evolving digital world can provide more autonomy and flexibility where systems and people coadapt instead of adapting people to systems. As already defined, systems include people and machines.

18.8 Concluding Remarks

HSI is both a process and a product; it should take place as early as possible during the design process and evolve during the overall life cycle of the product. While human factors have been considered in the past by HFE and HCI specialists, being able to observe activity at design time is a brand new capability, because modeling means are now available and realistic enough to support HITLSs.

HCD combined with systems engineering approaches, such as agile development and systems-of-systems, effectively supports HSI. Indeed, TOP must be considered concurrently during the whole life cycle of a system in which functions can be incrementally allocated to more appropriate agents with respect to principles and criteria. This means that HSI is no longer a matter of adapting people to machines by crafting user interfaces and operations procedures but an integrated approach based on the concept of SoS that integrates people and machines into articulated sociotechnical systems. In fact, HSI theory is far from being fully developed, and requires more formal investigations and heuristic experimentations. More specifically, HSI currently leads to investigating autonomous complex systems that could be better designed, where autonomy should be thought for all agents, whether people or machines.

Finally, HSI developed using digital twins should be based on strong tangibility principles and assessment criteria, properly defined along five directions: complexity, maturity, flexibility, stability, and sustainability. Number and content of these dimensions will be probably extended with respect to ongoing research and innovation developments. Instead of twentieth century's automation (i.e. incorporation of software into hardware) leading to rigid operations, especially in critical situations, we promote a twenty-first century approach based on coordinated autonomy of technology, organization, and people (i.e. deducing appropriate hardware structures from software functionalities) leading to more flexible operations. This is the birth of FlexTech.

18.9 Summary

HSI denotes an evolution of the conventional HFE discipline that focuses on evaluation of existing systems and usages, as well as HCI that provides methods and tools for interaction design. HSI evolution consists in considering the TOP model that supports symbiotic integration of TOP from the beginning of design to the end of the life cycle of a system. The resulting technology is called FlexTech, which enables more flexibility in systems operations.

HSI can be seen as the association of HCD and TCSE where multi-agent modeling and HITLS are used from the beginning of the design process to increase knowledge included in the corresponding TOP model, keep enough flexibility during the life cycle of the overall system being developed and maintained, and do not commit too early on necessary resources required to support the overall system.

HSI should guaranty that technology be adapted to people and not the opposite from a systemic point of view. Since HSI is strongly based on modeling and HITLS, using virtual prototypes, tangibility has become a key concept that requires deeper analysis. Mastering system's complexity, flexibility, stability, maturity, and sustainability contributes to improving system's tangibility. Making tangible things requires both innovation and experience (two concepts, often considered as antagonist, that should be combined to make sense of what is unknown and known), and systematic testing. HSI is a goal in industry that requires more scientific attention.

This chapter proposes a system-of-systems structure–function approach combined with a cognitive-physical distinction, and more specifically on cognitive-physical function allocation among a society of agents.

References

Ballerini, M., Cabibbo, N., Candelier, R. et al. (2007). Interaction ruling animal collective behavior depends on topological rather than metric distance: evidence from a field study. *Proceedings of the National Academy of Sciences of the United States of America* 105 (4): 1232–1237.

Boy, G.A. (1998). *Cognitive Function Analysis*. Praeger/Ablex. ISBN: 9781567503777.

Boy, G.A. (ed.) (2011). *Handbook of Human-Machine Interaction: A Human-Centered Design Approach*. Ashgate Publishing.

Boy, G.A. (2013). *Orchestrating Human-Centered Design*. Springer. ISBN: 978-1-4471-4338-3.

Boy, G.A. (2015). On the complexity of situation awareness. In: *Proceedings 19th Triennial Congress of the IEA*, Melbourne, Australia (9–14 August) (eds. G. Lindgaard and D. Moore), 9–14. International Ergonomics Association (IEA).

Boy, G.A. (2016). *Tangible Interactive Systems: Grasping the Real World with Computers*. Springer. ISBN: 978-3-319-30270-6.

Boy, G.A. (2017). Human-centered design of complex systems: an experience-based approach. *Design Science Journal*. Cambridge University Press, U.K. 3: e8. https://www.cambridge.org/core/journals/design-science/article/humancentered-design-of-complex-systems-an-experiencebased-approach/F726100E2E7D0C825CB526FD7412C48E.

Boy, G.A. (in press). *Human-Centered Design for Tangible Operations of Complex Systems*. FL: Taylor & Francis.

DeLaurentis, D. (2005). Understanding transportation as a system of systems design problem. In: *Proceeding of the 43rd AIAA Aerospace Sciences Meeting*, Reno, Nevada (10–13 January). Reston, VA: American Institute of Aeronautics and Astronautics AIAA-2005-0123.

Endsley, M.R. (1995a). Measurement of situation awareness in dynamic systems. *Human Factors* 37 (1): 65–84.

Endsley, M.R. (1995b). Toward a theory of situation awareness in dynamic systems. *Human Factors: The Journal of the Human Factors and Ergonomics Society* 37 (1): 32–64.

Endsley, M.R. (1998). A comparative analysis of SAGAT and SART for evaluations of situation awareness. In: *Proceedings of the Human Factors and Ergonomics Society 42nd Annual Meeting*, Chicago, IL (5–9 October), 82–86. Santa Monica, CA: The Human Factors and Ergonomics Society.

Endsley, M.R. and Garland, D.J. (eds.) (2000). *Situation Awareness Analysis and Measurement*. Mahwah, NJ: Lawrence Erlbaum Associates.

Fitts, P.M. (ed.) (1951). *Human Engineering for an Effective Air Navigation and Traffic Control System*. Washington, DC.: National Research Council.

BFU (2004). *Investigation Report AX001-1-2/02: Bundesstelle für Flugunfalluntersuchung (BFU, German Federal Bureau of Aircraft Accident Investigation)*. Braunschweig, Germany.

Kaptelinin, V. and Nardi, B. (2006). *Acting with Technology: Activity Theory and Interaction Design*. Cambridge, MA: MIT Press.

Kuchar, J.K. and Drumm, A.C. (2007). The traffic alert and collision avoidance system. *Lincoln Laboratory Journal* 16 (2): 277–296.

Minsky, M. (1986). *The Society of Mind*. New York: Simon & Schuster. ISBN: 0-671-60740-5.

Pemmaraju, V. (2013). 3 Simple Rules of Flocking Behaviors: Alignment, Cohesion, and Separation. Envatotuts game development tutorials. https://gamedevelopment.tutsplus.com/tutorials/3-simple-rules-of-flocking-behaviors-alignment-cohesion-and-separation--gamedev-3444 (Last accessed 2 July 2019).

Potts, W.K. (1984). The chorus-line hypothesis of coordination in avian flocks. *Nature* 309: 344–345.

Reynold, C.W. (1987). Flocks, herds, and schools: a distributed behavioural model. *Computer Graphics* 21: 25–34.

Schwaber, K. (1997). Scrum development process. In: *OOPSLA Business Objects Design and Implementation Workshop Proceedings* (eds. J. Sutherland, D. Patel, C. Casanave, et al.), 117–134. London: Springer.

Schwaber, K. (2004). *Agile Project Management with Scrum*. Microsoft Press. ISBN: 978-0-7356-1993-7.

Sutherland, J. (2014). *Scrum: The Art of Doing Twice the Work in Half the Time*. Crown Business. ISBN: 13: 978-0385346450.

19

Transdisciplinary Engineering Systems

Nel Wognum, John Mo, and Josip Stjepandić

Synopsis

Transdisciplinary processes are aimed at solving problems that cannot be solved by one person nor one discipline, like urban planning, waste treatment facility, or a disruptive innovation. A system view is a good way to describe transdisciplinary processes. Transdisciplinary systems are complex systems. This means that goals of the system may conflict due to the people that act in the system. Transdisciplinary systems are also organizational systems. Inherently, system processes are performed by people, possibly with the help of information systems and other tools. All kinds of arrangements exist to make the processes manageable, like organizational structures, teams, or social norms known as culture. Different levels of transdisciplinary systems exist, while in each level different kinds of people interact. For example, a system aimed at creating a complex service system is different from the complex service system itself. In this chapter, we mainly focus on transdisciplinary processes in the engineering domain. The concept of a transdisciplinary system will be explored and defined. Two examples of a two-layer system will be described to illustrate the concept.

19.1 Introduction

The concept of transdisciplinary processes has been the subject of discourse in the past few decades in the context of large, complex, ill-defined problems, also called wicked problems. Solutions to such problems are not obvious and require long and intensive processes in which many people from many different backgrounds participate. Moreover, the goals of these processes are not fixed but may shift during the course of the process; people may also leave the processes, enter at a later stage or be replaced by others.

Systems Engineering in the Fourth Industrial Revolution: Big Data, Novel Technologies, and Modern Systems Engineering, First Edition. Edited by Ron S. Kenett, Robert S. Swarz, and Avigdor Zonnenshain.

The problems tackled by transdisciplinary processes typically cannot be solved by one person despite the different types of knowledge this person may have. Moreover, such problems can also not be solved by only technical disciplines, because the impact of the solution on society or user communities has to be taken into account. Examples can be found in the development of the autonomously driving car with moral and legal considerations when avoiding collisions with different groups of pedestrians. Another examples is the introduction of 3D printing, which not only disperses the production to almost any place in the world but also has an impact on intellectual property protection of designs and processes.

In transdisciplinary processes, knowledge from different scientific communities as well as from practice is needed to reach a solution that is acceptable to the (often many) stakeholders (Scholz and Steiner 2015). Transdisciplinary processes are most often performed in projects with a particular timeline, which may shift over time. Transdisciplinary processes are performed by teams and subteams composed from different disciplines from science (technical as well as social) and practice (both from companies and user or citizen communities). These teams and subteams perform interdependent tasks. The degree of interdependence depends on whether the handling of a specific activity in a working practice influences or is influenced by the handling of activities in other tasks. Interdependence may also comprise tasks handled in the past, present tasks, and future dimensions of tasks (Mathiasen and Mathiasen 2017). Research activities often form an important part of transdisciplinary projects.

In Figure 19.1, the transdisciplinary research (TDR) process is depicted. As indicated previously, both science and practice are involved in the research process. The outcomes should benefit both research and practice. Moreover, not only technical disciplines need to be involved. Involvement of social science disciplines is deemed essential to achieve solutions that can be used in and valuable and acceptable for the people in the context in which the solution is needed. There are still many challenges to be tackled, though, in TDR (Wognum et al. 2019).

In transdisciplinary processes, the old paradigm of scientific discovery (Mode 1) characterized by the hegemony of disciplinary science, with its strong sense of an internal hierarchy between the disciplines and driven by the autonomy of scientists and their host institutions, the universities, is being superseded – although not replaced – by a new paradigm (Mode 2) that is socially distributed, application-oriented, and subject to multiple accountabilities (Nowotny et al. 2006).

Managing a transdisciplinary project is not an easy task. Gaziulusoy et al. (2016) identified challenges that need to be addressed in managing and participating in transdisciplinary projects. Moreover, as a transdisciplinary project does not necessarily proceed according to preset timelines, budget, and goals, it

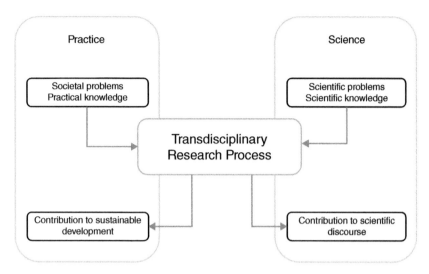

Figure 19.1 Transdisciplinary research process.

is characterized by its emergent behavior and shifting goals. A transdisciplinary project needs to be shaped along its course (Müller and Olleros 2000) and requires leaders that are visionary and flexible.

In engineering contexts, a transdisciplinary approach has not been studied much yet. Many engineering problems, however, can be characterized as large, complex, and ill-defined often with unknown outcomes. New innovations, engineering business development, the adoption of new technology, or the development of a completely new factory, like the smart factory, especially, are examples of such engineering problems. A project approach and multidisciplinary teams with people from science and practice are needed to achieve an acceptable solution to these problems (Peruzzini et al. 2017).

In this chapter, we specifically focus on transdisciplinary processes in the engineering domain. We adopt a systems approach for characterizing and describing transdisciplinary engineering research (TDER) processes. We identify challenges that need to be overcome. We state that a system view will help to understand the complexity of transdisciplinary processes and to anticipate the challenges that need attention. Transdisciplinary systems themselves consist of subsystems, which may each be transdisciplinary also. These subsystems consist of different processes, which need different people and knowledge for their execution. Two examples of such multilayered transdisciplinary systems will be described to illustrate the concept.

The outline of this chapter is as follows. In Section 19.1, we describe challenges that exist in managing and performing transdisciplinary projects.

In Section 19.2, we introduce the concept of a transdisciplinary system. A transdisciplinary system is a complex system, because many different people are involved with their own possibly conflicting goals. In Section 19.3, multiple interacting layers of a transdisciplinary system, which are themselves (possibly transdisciplinary) systems as well, will be described. In Sections 19.4 and 19.5, respectively, two examples of a transdisciplinary system with multiple layers are described. In Section 19.6, some concluding remarks are presented.

19.2 Transdisciplinary Engineering Projects

Transdisciplinary processes are typically performed in projects with a more or less defined deadline and that may last several years. Transdisciplinary projects are especially aimed at solving problems that require a vision beyond the immediate engineering task for their solution. In transdisciplinary projects not only do technical disciplines need to participate but also so, too, do disciplines from social sciences. In addition, knowledge is needed from practice and stakeholder communities, including financers, legislators, sponsors, etc. For example, many construction projects can be considered as projects requiring a transdisciplinary approach. Other examples can be found in the medical and in the aeronautics industries.

TDER projects are performed by teams of people from research and practice. While the literature refers to arrangements for organizing transdisciplinary teams, Beckett and Vachhrajani (2017) suggest it is more appropriate to think of transdisciplinary networks of autonomous agents and mutual interfaces to enable them to interact. This recognizes the fact that individual actors are also connected to external actors who may indirectly contribute to a transdisciplinary project. This way of thinking recognizes that particular actors may be linked in different ways at different times.

Research constitutes a large part of transdisciplinary projects, because standard solutions for the problems addressed do not exist. Collaboration between and coordination of researchers from different disciplines is an important characteristic of transdisciplinary projects. Problems addressed in such projects are practical problems, or at least problems that are relevant for practice. This means that people from practice also need to be involved as well as other stakeholders, like financers and legislators. A transdisciplinary project is similar to other types of collaborative (research) projects but differs in three main characteristics (Gaziulusoy et al. 2016):

1. It is agenda driven.
2. It aims at integration between and alignment of knowledge from different disciplines, as well as theoretical and methodological transformation of each discipline throughout the process of the research.
3. It involves nonacademic participants with significant stakes in the (research) problem and process, as researcher or as informant.

In the literature many challenges that are faced by transdisciplinary teams have been described. Gaziulusoy et al. (2016) grouped challenges in transdisciplinary projects as reported in the literature into three groups:

1. *Inherent challenges.* Challenges that directly rise from the characteristics inherent to a transdisciplinary project.
2. *Institutional challenges.* Challenges that arise from the current structures and procedures of knowledge generation and performance evaluation in academic institutions.
3. *Teamwork challenges.* Challenges that stem from the requirement of collaboration of researchers from different expertise backgrounds and often from different academic institutions with each other and with nonacademic stakeholders in ways to enable transdisciplinary knowledge generation.

Teamwork challenges, more specifically, have been described in a publication on the longitudinal study of a large transdisciplinary project (Frescoln and Arbuckle 2015). Frescoln and Arbuckle have assembled these challenges from literature on complex projects like TDR projects. These challenges are:

- Communication and language barriers.
- Professional cultures and cognitive cultural differences create subgroups among team members, challenging cross-discipline collaboration.
- Differences in methodologies between disciplines.
- Competition for funds.
- Difficulty in reproducing research.
- Different geographical locations of participants.
- Conflicting goals amongst team members.

These challenges are well known for large distributed projects. Managing such projects is an encompassing task. The Project Management Body of Knowledge (PMBOK) (Project Management Institute, Inc. 2017) has provided guidelines for managing different kinds of projects, from small to large, in various application domains. These guidelines, however, cannot be considered as recipes. Transdisciplinary projects are complex (see section "Business Development"), goals shift during the course of the project, destructive forces may be active, sponsors may lose their interest, etc. (Miller and Lessard 2000). Knowledge of and experience with large, complex problems is needed to manage transdisciplinary projects.

The teamwork challenges listed above have also been identified by Gaziulusoy et al. (2016) in a case study of a TDR project. The strict adherence to project deadlines, fixed budgets, and reporting requirements does not attend to the evolving characteristics of a TDR project. In addition, an institutional challenge like career development may be hampered, because the development of new scientific knowledge is not often the only priority in a TDR project. Greater emphasis is put on knowledge for practice and targeted on a wider audience. Leaders of TDR projects have to develop adaptive strategies to manage emergent challenges that may compromise scientific validity and social responsibility of the project (Gaziulusoy et al. 2016). The often large scope of TDR projects also impacts on project management, expertise management, and resource management.

In an engineering context, Miller and Lessard (2000) have emphasized that large engineering projects cannot be fully predicted and designed beforehand. A shaping approach is needed depending on task complexity and the degree of development of institutional arrangements. The task complexity requires exploration and testing, while in the development of institutional arrangement strong coalitions need to be formed. The real-options framework is applicable, recognizing that decisions determining project cash flows in conjunction with exogenous events are not all made at the outset of the project (Miller and Lessard 2000).

Shaping a TDER project requires several management processes (Miller and Lessard 2000):

- Negotiating a project concept or proposition that truly creates value and can be progressively refined in the overarching issue.
- Developing stability for the future of the project.
- Gaining and ensuring legitimacy.
- Achieving shock-absorption capabilities.
- Ensuring capital cost reduction.

These activities clearly transcend engineering activities. They require the involvement of all relevant actors and disciplines. Below, we explore some projects that require a TDER (project) approach.

Open Innovation

In the past, many companies performed their innovation processes in a closed way. In the research laboratory of the company, breakthroughs were sought, products were developed in the company, built in its factory and distributed, financed and serviced from within the four walls (Chesbrough 2006). Open innovation, on the other hand, requires collaboration with other companies, because not all new technology can be developed in-house or new technology from the own research laboratory may not be profitable enough for the own company (Chesbrough 2006). The former case requires the buy-in of new technology or close cooperation with the inventing company, often a small company. The latter may result in spin-off companies that are required to collaborate with other, often larger companies.

New inventions are not merely given away. Often they are protected by IPR (Intellectual Property Rights), giving a company a means to gain revenues by licensing an invention to other companies to develop and manufacture, or by leasing a name, logo, or slogan to other companies (Jolly 2010). Companies may also get an equity stake in companies that further develop and produce their invention (Chesbrough 2006).

Open innovation requires collaboration between different companies, while involvement of legal people and business people is needed to investigate business opportunities and the legal limits and options (Rauch et al. 2017). In addition, knowledge of potential markets is necessary to build a viable business model. A true transdisciplinary approach is needed, because the process evolves over time and needs to be shaped. A visionary leader is also needed to buy in commitment and support. His or her role is to guide, facilitate, manage, and control the innovation process from idea screening to launch (Aas and Vavik 2015).

Business Development

After a new technology has been developed, a new business may need to be set up, involving possibly the company in which the technology has originated, but more often a new start-up company or spin-off company. The new business may be a technology service provider or a manufacturing company that will produce a new product. A whole new sociotechnical system has to be set up in developing the new business.

In setting up a new business or changing an existing one, many different aspects need to be investigated, like economic feasibility, patenting, licensing, location demands, waste disposal, etc. In addition, resource demands and availability are important to consider, in particular financial resources, knowledge and experience of the employees, and management capabilities.

Especially with the demands on sustainability, the 3 Ps need to be taken into account as well: people, planet, profit. The new business needs to provide a good environment for its workforce, it needs to care for the environment with respect to its inputs, outputs and waste during and after the process. A trade-off needs to be made between investments on the short term and revenues on the longer term (Wognum et al. 2011).

It is clear that business development requires a transdisciplinary (research) approach, because the process may take some time, goals may shift as insights grow, stakeholder values are at stake, and investments are large. In addition, the process is multidimensional, requiring people from both science and practice as well as people from both technical and social science disciplines. Knowledge exchange needs to be extensive and lead to new knowledge and insights, academically and practically.

Adoption of New Technology

Disruptive technology like 3D printing leads to many new business opportunities but also triggers new legislation and copyright and IP protection measures. Ownership of design, printfile or final product need to be redefined.

In setting up a new 3D printing service all that has been indicated in the previous section needs to be taken into account. In addition, new technology is needed to protect products against plagiarism (Holland et al. 2018). Although already incorporated in law, e.g. paragraph 54 of the German Copyright Law, counterfeiting and plagiarism are still possible, especially in the B2B area. Holland et al. (2018) defined four categories of counterfeit protection: internal security, external security, product labeling, and legal safeguard. In their paper, they discuss product labeling more extensively, like visible and invisible tagging and the introduction of marker particles.

This example shows that the adoption of new technology is not only an engineering or technical task but involves other disciplines as well. The process of new technology adoption may also take quit a long time, because new insights and unexpected consumer or client behavior may trigger the need for additional protective measures and business redesign.

Toward Industry 4.0

With the development of new technology and cloud computing a totally new concept of production facilities has become possible, the so-called smart factory. Smart factories are an instantiation of the Industry 4.0 concept (Rojko 2017). The concept Industry 4.0 has been introduced by the German government and is aimed at industrial production systems. Industry 4.0 is the name for the current trend of automation and data exchange in manufacturing technologies. It includes cyber-physical systems (CPSs), the Internet of Things (IoT), cloud computing, and cognitive computing (Hermann et al. 2016, 2017). Data are typically stored in the cloud.

In a smart factory, products, processes, and machines have both a real and a virtual presence. They can be called "smart," because at any point in time their status, progress, and activities can be identified, monitored, and planned. The data are continuously updated and used during a production process and during product and machine life. Factories are becoming "smart" and "adaptive," because of the new intelligence that has been embedded in machines and systems. They are able to share data and support enhanced functionalities at a factory level and include collaborative and flexible systems able to autonomously solve problems that arise during the process (Hermann et al. 2016).

Smart products, processes, and machines can be considered an instance of the IoT. CPSs monitor physical processes and create a virtual copy of the physical world to make decentralized decisions (Zhong and Ge 2018). Over the IoT, CPSs communicate and cooperate with each other and with humans in real time and via cloud computing.

Industry 4.0 is radically changing the way people interact with machines, systems, and interfaces. Many different skills will be required in the new context. Lower-skilled repetitive tasks will be replaced by tasks that require competences in software development, and IT technologies. The Boston Consultancy Group recently reported a set of examples to illustrate the possibilities for deployment and the implications for the workforce in Industry 4.0 contexts (Boston Consultancy Group 2018a, b). For example, companies will need algorithms to analyze real-time or historical quality control data, identifying quality issues and their causes, and pinpointing ways to minimize product failures and waste. The application of big data analysis will reduce the number of workers specialized in quality control, while increasing the demand for industrial data scientists.

Other consequences of adopting Industry 4.0 are:

- Robots will replace humans, because they can be easily trained to take on new tasks, in contrast to humans. A new job may be the robot coordinator.
- Automated transportation systems navigate goods intelligently and independently within the factory. They replace logistics personnel. Increased need will grow for skilled controllers and programmers.
- Production line simulation prior to installation will increase the demand for industrial engineers with production management knowledge and simulation experts.

New jobs will be more cognitive and complex. The business model, in addition, will also change, including the markets that can be served, because the range of products and the degree of customization will change. A flexible lay-out is required because production needs change frequently. Resource management needs to adapt to the changing situation, because workers need other skills and knowledge. New IT systems, like CPSs, are needed to manage the physical world and interact with the virtual world represented in the cloud. These systems need to be able to cross organizational borders.

Implementing the Industry 4.0 concept in a company clearly requires a TDER approach. Changing the business is not only a technical task but involves the whole business as well as sponsors, legislators, and financers. The people in the company as well as existing and potential markets play an important role too. The change process may take many years, with a step-by-step approach, in which the goals to be achieved may shift over time.

Ecosystems

Building on the turbulent experiences of the past decade, global companies have adapted their strategy and structure accordingly. Driven by the increasing complexity of products and processes, as well as the ever-increasing dynamism imposed by the market and society, companies are increasingly focusing on their core competencies, resulting in much greater flexibility. The additional demand for goods and services resulting from outsourcing is covered by a supplier pool. Together with customers and other partners, this creates a

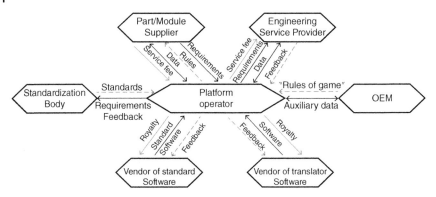

Figure 19.2 Ecosystem of OpenDESC.com.

dynamic, flexible network as a new form of corporate culture, sometimes called an ecosystem. This is a specific characteristic of transdisciplinary systems engineering.

Even innovations that are considered to be the biggest drivers of economic development are increasingly emerging in network structures, not only because this type of cooperation enjoys large political support in most countries (OECD 2015). As a result, the innovation process also includes the capture and use of existing knowledge, machinery, equipment, infrastructure, training, marketing, design, and software development, thus requiring a transdisciplinary approach.

The increasing importance of services in many areas of economic and public life further underscores this development. In ecosystems, synergies are much easier to raise, because the tailoring of labor among the individual members is part of the self-image of one. Thus, an ecosystem provides a first-class means of first establishing innovations and then marketing them on a long term and sustainable basis. For an outsider, a functioning ecosystem is an important sign that the vendor has created a balance between market dynamics and its range of services.

Ecosystems are particularly widespread in the IT industry. They usually combine (i) a disruptive technological development in a field that is very attractive to customers, (ii) the added value created by applying new software or service to existing processes and processes, (iii) a broad range of potential users in different fields, and (iv) a stable customer base through market knowledge and long-term customer relationships. Thus, ecosystems are typical transdisciplinary systems that include technical, economic, and social aspects.

A central component of an ecosystem is often a platform that is built for an economic purpose and depicts complex sociotechnical processes. Figure 19.2 shows the platform OpenDESC.com with its stakeholders in its own ecosystem, which serves as a hub for data communication in the global automotive industry. Arrows show different types of mutual relationships (OpenDESC.com).

It is important to realize, however, that much knowledge from the supply chain and network management domain is needed to properly organize an ecosystem, especially for creating and maintaining the required flexibility, while guaranteeing the quality desired by the customer. Flexibility and strict quality management regimes often require conflicting arrangements. A trade-off between flexibility, network procedures, contracts, and the degree of interdependence between parties is needed (Wever et al. 2012). Proper arrangements are important to make the ecosystem function well.

19.3 Introduction to Transdisciplinary Systems

Systems and systems thinking take a predominant place in current practice and research. The concept of systems-of-systems is used quite frequently in the literature. Systems are encompassing concepts with different structures, aspects, and layers. It is often not clear, though, what actually is meant with systems and whether the concept is used consequently and consistently in academic and industrial circles. Below, we introduce the system concept in more depth. Then the concept of complex system is defined, after which we describe a transdisciplinary system.

The System Concept

The concept of system is widely used in theory and practice. However, in many cases it is not very clear what really is meant by system. In an attempt to give a formalized account of a fundamental theoretical issue in general systems research Marchal (1975) has given a very elementary definition of a system:

S is a system only if $S = \{E, R\}$

where E is an element set and $R = \{R_1, ..., R_n\}$ is a relation set, i.e. $R_1, ..., R_n$ are relations holding among the elements of E.

This definition is a very generic one, but can be given content in any domain and on any level. Even systems-of-systems can be characterized here, when systems on a lower level are seen as the elements of the higher-level system. Relations between elements of a system can be of any kind, e.g. part-of or functional, but also fixed, like in natural systems, or intentional, i.e. created by somebody and existing as long as needed (Caws 2015).

Any object, artificial or natural, can be viewed as a system. Every such system has a function in its context, like a stone, putting weight on the surface it lays upon or storing and disseminating solar heat. A house is a system with many different functions, depending on the context in which the system is considered.

General Systems Theory (GST) (von Bertalanffy 1951) has emerged in the 1950s and describes a level of theoretical model-building that lies between highly generalized constructions of pure mathematics and specific theories of specialized disciplines (Boulding 1956). Mathematics abstracts away from content and context. On the other hand, disciplines, like physics, chemistry, biology, psychology, etc., have their specific theories and correspond to a particular segment of the empirical world.

GST is the result of a quest for a systematic theoretical construct that describes the general relationships of the empirical world. It is not a single, self-contained general theory of practically everything that replaces the special theories of particular disciplines (Boulding 1956). As Boulding claims, such a theory would be without content. GST seeks a place between the specific without general meaning and the general without specific content. The objectives of GST can be defined with varying degrees of ambition and confidence. At a low level of ambition, but with high degree of confidence, GST can point out similarities between theoretical constructs of the different disciplines. At a higher level of ambition, but with possible lower confidence, it aims to develop a spectrum of theories – a system-of-systems. Like the periodic table of elements, it may show gaps in theoretical models, which direct research to filling those gaps. This ambition, however, is still not achieved.

The merit of system theory can be found in specifically framing and defining the focus of attention. This can be disciplinary, like a waste treatment model, but also interdisciplinary, combining two or more different disciplinary systems, like the waste treatment model and the ecosystem (Nelson 1976). Of course, such an integrated model is less acceptable to each of the disciplines, but is a compromise to support communication and the search for transdisciplinary solutions. *Trans*disciplinary systems add a level of analysis, which does not exist on the level of each of the disciplines (Hofkirchner and Schafranek 2011).

Complex System

Much discussion can be found in the literature on the concept of complex systems. When we simply count the number of elements, systems with a large number of elements may appear to be rather simple, like the solar system (Simon 1976), because only a limited number of the pairs of interaction appear to be of significance. In addition, systems that occur in nature are mostly hierarchic and nearly decomposable. Approximations on higher levels are often made possible (Simon 1976). The weather system, on the other hand, is still hard to simulate and predict. Much of our perception of complexity may be due to the fact that we base our models on wrong assumptions, like in forecasting models to guide economic policy (Simon 1976).

For the purpose of this chapter, system complexity as defined by Nelson (1976) is useful. Nelson defines a complex system as a system having at least two conflicting goals. Such a system always contain human beings, otherwise there would not be goals. Systems functioning without persons have functions to reach the goals of human beings, possibly assembled in societies. The central idea here is intentionality (Nelson 1976).

In the context of TDER, complex systems, as defined above, are the organizational systems in which multiple disciplines and multiple organizational roles work together to develop new products or services. Such a system consists of many subsystems, which may not be transdisciplinary but may still be complex. For example, in developing an electromechanic product, the subsystems are the electronic design department and the mechanical design department, each with its own processes, its own goals, people, equipment, and knowledge. In the transdisciplinary system they have a separate, integrated, process, shared people, shared equipment, shared knowledge, and, above all, shared goals. These goals may possibly conflict, requiring negotiation and possibly adaptation of the goals, process, people, equipment, and knowledge. Other subsystems may not be complex, in the sense defined above, like information systems used to manage product and process information.

In the next section we explore the concept of a transdisciplinary system and, in particular, its multilayered nature.

19.4   Transdisciplinary System

A transdisciplinary system is a complex system as defined above, because people are needed to perform the system processes. Moreover, a transdisciplinary system is an organizational system. People perform processes to achieve a certain goal, which is the attainment of a solution to a problem in the environment of the system, often a market or society. The solution is important for many stakeholders in the environment. The execution of processes is made possible by structure and culture of the system, together called the organizational arrangements or governance. The structure comprises hierarchy between people, including management, operational, and support people. Support consists, for example, of the financial administration of the project and creation of project documentation. The culture consists of the norms and rules that together come from the many different organizations and departments the people originate from. Culture is often a hampering factor in collaboration projects (Wognum et al. 2004).

In Figure 19.3, the system of transdisciplinary engineering is depicted (based on Wognum et al. 2016). It shows the transdisciplinary system as the central element of the system. A transdisciplinary system most often consist of multiple subsystems. The development subsystem, shown in Figure 19.1, is aimed at solving the problem that was the trigger of the transdisciplinary project. The development system is a transdisciplinary system in the overall transdisciplinary system, because of the different disciplines, social as well as technical, needed to solve the problem. The solution system, also shown in Figure 19.1, is

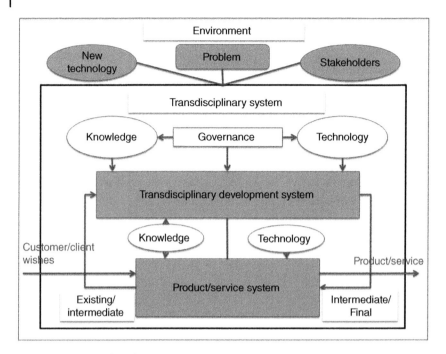

Figure 19.3 A transdisciplinary engineering system.

the outcome of, but also input to, the development system. The solution system can be a transdisciplinary system, but it does not necessarily need to be one.

The process in the development system is performed by and involves many different disciplines, such as engineers from different disciplines, designers, marketing and sales people, people from social sciences, people from practice. Also people from the solution system are involved. Together these people have a lot of different types of knowledge. Stakeholders, like financial institutions, governments, legal bodies, and certification bodies, may have a strong influence on the process but are often not directly involved. The transdisciplinary process uses technology, like information systems, and many different tools, technologies, and methods. The process may also need new technology that has been developed elsewhere.

The outcome of the process is the solution system, a product/service subsystem that performs product and/or service processes with the necessary people, tools, and techniques. It is intended to solve the problem. The solution system may itself be a transdisciplinary system. The solution system can be a preexisting system that needs to be changed to solve the problem. The solutions system consists of the value-adding processes, transforming customer or client wishes into products and/or services.

Input to the solution system consists of the customer or consumer needs. Output of the solution system is the product and/or service desired. Processes in the solution system are performed by people that are often different from those in the development system, because different knowledge is needed. A subset of the people may have participated in or contributed to the development system. Also, the technology used in the solution system is different from the technology used in the development system, although parts may overlap. The solution system is supposed to function independently of the development system after the project has stopped. However, some maintenance and update processes may remain active and can be considered to be part of the development system.

It is clear that the system depicted in Figure 19.3 is a complex system as defined in the section "Business Development". It is a system in the sense that it is an element set and a relation set as defined previously. Many elements, however, are complex systems, while the relationships are many and highly different in nature. Other elements of a transdisciplinary system need be not complex in the sense defined above, like information systems.

Information systems can be large, with many elements and relationships. They are not complex, however, because output can be predicted from input provided. As soon as humans are involved the system in which the information system is used is complex. Humans may not sufficiently understand the system, thus using the system in a way not intended. In addition, the user interface may be difficult to comprehend and use, making users reluctant to use the system in a proper way. Also, users or organizations may have their own goals with the system, like forcing a particular way of working or gaining more power.

There have been attempts to develop frameworks for studying and analyzing complex systems. One example is the soft systems approach by Checkland and Holwell (1998). Another approach is the process model of organizations (Wognum et al. 2004). Both approaches emphasize that systems thinking supports framing the system of focus. Such models are useful to depict a complex system, to support communication between stakeholders of the system, and to identify problems that require further analysis and definition. The system descriptions are not sufficient for problem solving as such but help to understand the complexity, structure, and context of the problem. Often, problems concern only a subset of the system under study but may have an impact on the system as a whole. The systems approach helps to see the relationships between problems and between problems and the behavior of the system. Additional theories and methods are then needed to dive into the problem to come up with ideas for solutions. The context of a system is not depicted in Figure 19.1, but is very important to consider, because system behavior depends on its context as well as impacts upon and influences its context.

Coming back to transdisciplinary systems, it is important to clearly distinguish the boundaries of the system at hand in its context as well as the internal structure of the system. For example, one may want to focus on a particular phase of the process, for example the research phase. In this phase, a subset of stakeholders is involved with a more limited number of functional roles and coming from a more limited number of departments or companies. Still, the system under study is complex. The context of this system consists of the preceding and subsequent phases and the transdisciplinary system as a whole.

Referring to Figure 19.3, a transdisciplinary system consists of at least two subsystems:

1. The development system.
2. The product and/or service system that is the outcome of and input to the development system. It is also the implementation of the solution as aimed for in the development system. The solution system produces the products and/or services desired by customers or consumers. It is the value-adding part of the overall transdisciplinary system.

While the transdisciplinary project stops at a certain point in time, the development system may continue to maintain and update the product/service subsystem. The solution systems is an independent organization in most cases that produces products and/or services for the environment in which the problem has existed and for which the products and services are meant to be a solution.

In every transdisciplinary process such different layers of subsystems can be distinguished that might require separate attention during the course of a transdisciplinary project. It is important to identify the needs of a subsystem in the context of the overall system, because subsystem actions and outcomes do have an impact on the overarching whole and also on other subsystems.

In the following two section examples of transdisciplinary systems are presented to illustrate the concept of a multilayered transdisciplinary system.

19.5   Example 1: Online Hearing Aid Service and Service Development

Mo and Beckett (2018) have studied the development of a service system to provide online hearing aid solutions. Traditionally, the most common way of providing hearing support is to have a patient undertake a simple frequency response test at an audiology clinic. The test measures the patient's response to sound at different frequencies, i.e. the hearing profile, which shows where degradation should be compensated for. Then, a hearing aid, tuned to suit this profile, is sold to the patient in a package deal that includes ongoing support and tuning. The hearing aids are produced by specialist manufacturers and the audiology laboratory clinic acts as their sales agent. The package deal can be quite expensive, hence imposing financial constraints to access of this technology. Also, some patients may have difficulty in attending a clinic for either the initial testing, subsequent fine tuning of the hearing aid or both, due to mobility issues.

Blamey Saunders Hears (BSH) recognized the complexity and inconvenience to patients in the traditional hearing aid packaging process and application pathway. The researcher founders first registered a company in 2007 aiming to develop an alternative system that could offer online hearing enhancement solutions. They allocated minimal effort to product promotion in the first few years as the focus was on Beta testing with initial clients in collaboration with the Bionic Hearing Institute in Melbourne, Australia, where the cochlear implant technology was originally developed. The Bionic Hearing Institute is close to an Eye and Ear Hospital and surrounding specialists' rooms – a kind of technology cluster precinct. From 2010, BSH moved to its own premises in the same area, maintaining knowledge-sharing connections with expertise from multiple disciplines.

In this case, the process of hearing enhancement was taken as the basis of the new service system. BSH examined the processes in hearing solutions development, including experience with online customer engagement practice with an associated firm, America Hears. It was clear that due to the nature of process, social interaction was an indivisible element to determine the characteristics of the final hearing aid system. Through several years of research, they designed the style of front office engagement and back office support infrastructure. The system comprised steps that were identified, as performed by the initiator and associated partners (Beckett et al. 2016):

1. *Recognition of hearing impairment.* This step is heavily patient dependent, involving the patient's social and work environment and some form of screening. The new system makes use of the Internet to assist a personal decision to seek a more detailed assessment.
2. *Assessment of hearing impairment.* This step involves one or more tests where quantitative and qualitative medical data are collected over the Internet.
3. *Enhancement solution identification.* This considers cost–benefit trade-offs and some experimentation with options. With the hearing data, the BSH team can assist the patient to select a suitable hearing aid specific to the patient's case.
4. *Enhancement solution implementation.* This step involves fitting and tuning the hearing aid selected at the BSH facility prior to delivery to the patient.
5. *Solution refinement and patient learning.* The patient needs a period of adjustment to use the new hearing aid in a variety of day-to-day situations. This step is to provide support to the patient on a continual basis to assist the patient in this process over the Internet.

6. *Ongoing review and adjustment.* The hearing aid system tends to become a life-long device for the patient. This step involves advice from BSH in monitoring patient progress, routine re-assessment, and consideration of technology advances, including upgrading to a different kind of hearing aid.

The solution, as produced by the solution subsystem, to hearing-impaired people is some combination of technology (a hearing aid) and support services. Input conditions are shaped by the patient's perceived hearing capability and associated acoustic environments. The desired output is an enhanced patient sensing and discrimination capability. Practical experience also suggests there are two other possible outcomes: rejection of the opportunity to participate at some point (e.g. on cost grounds) or the abandonment of a particular solution after an initial trial period (e.g. the hearing aid is too difficult to use).

The BSH's operation is definitely a transdisciplinary system that intertwines between system development and production plus services. It has all characteristics depicted in Figure 19.3. Within the Environment, identification of requirements of new technology is based on the hearing enhancement process, which works traditionally but with undesirable social difficulties (patient mobility) and supply chain complexity (clinics as the sales agent). It is clear that there are many sources of knowledge required to support the system. Transformation of the traditional clinical-centered process to an online process has many innovative design features of the system's architecture, which makes use of research knowledge of how the Internet-based hearing assessment can be performed accurately and reliably.

In the development phase, BSH has identified several strategic partners in development of the solution system. These include the Bionic Hearing Institute, the Eye and Ear Hospital, the specialists' rooms around the medical precinct, the Internet developers, and the association with American Hears. Involvement of the specialists' rooms ensures a range of patients with different levels of hearing impairment to be accessible. The online testing and feedback can be implemented with specialist knowledge. While the development of Internet access is relatively straightforward, installation of information channels and advertisement on the web is critical to delivery of the products (tuned hearing aid) to the customers. It is worth noting that BSH does not manufacture hearing aids, nor is a sales agent of any of the hearing aid brands. This fact helps BSH to be independent when searching for the best solution for the client.

In summary, BSH is a transdisciplinary system with a development subsystem and an solution subsystem, the product/service subsystem, that are intertwined, as said above. The development subsystem is different from the product/service subsystem with different actors and different knowledge involved. While the development subsystem is a transdisciplinary system, the question is whether the solution subsystem, the product/service subsystem, is a transdisciplinary system also. The answer depends on the degree of involvement of science and practice, and the degree of involvement of the different, not only technical, disciplines.

Several subsystems can be distinguished in the BSH case:

- The development subsystem consists of four interacting subsystems:

    1. *Development of the online hearing profile assessment subsystem.* This subsystem has all features and functional blocks of a transdisciplinary system as shown in Figure 19.3. New Internet-based delivery of the assessment test requires substantial research and development on the capabilities of the equipment under remote control. The sound system needs to be precise, accurate, and consistent.
    2. *Development of the clinic subsystem.* There are some transdisciplinary actions taken in the development of the clinic subsystem, primarily in the social interface during the hearing profile assessment test. The relationship between the clinician and the patients has changed somewhat due to the online test arrangement. However, the delivery and after-post delivery service of the hearing aid involves transdisciplinary actions.
    3. *Development of the tuning and adaptation of the hearing aid.* This part does not require a transdisciplinary approach, because the existing approach can be used.
    4. *Development of the information subsystem.* All disciplines including engineering, medical, IT have worked together collaboratively to build the necessary databases and links.

- The product/service subsystem consists of four subsystems, that are not transdisciplinary:

    1. *Profile assessment subsystem.* With this subsystem the patient hearing profile is assessed.
    2. *Clinical subsystem.* This subsystem provides the interface between clinician and patient.
    3. The tuning and adaptation of hearing aid can be done readily when the hearing profile is known, as before. Not much has changed.
    4. *The information subsystem.* This subsystem supports the other subsystems.

As will be clear from this example, the development system is different from the solution system. Both systems, though, are mutually dependent. The success of the solution system depends on the vision of the people that have undertaken the development and on the efforts they have spent to, for example, select the right people and technology for executing the necessary processes in the solution system. Conversely, the solution system success also depends on the continuous efforts in the development system to maintain and update the solution system. In the end, the solution system is the value-adding part of the whole system.

19.6   Example 2: License Approach for 3D Printing

In this section, we present a novel approach which overcomes the limitation of the use of 3D printing and builds an ecosystem for the wide exploitation of this production technology anywhere in the world.

Challenges and Legal Background

The rise of Additive Manufacturing (AM) not only creates tremendous chances for a disruptive shift in the area of manufacturing, but also opens the door for many threats by plagiarism and product piracy. Like a bubble-jet printer usage yesterday, the 3D printer has become typical equipment of almost each engineering office today. The integration of AM procedures into the production process and the complete product life cycle incorporates significant challenges regarding authorized access to product data, assured supply of the agreed quantity, distinction of original parts from counterfeits as well as prevention of intellectual property, product liability, and warranty (Stjepandić et al. 2015).

Copyright in the consumer area, according to §53 German Copyright Law, also applies to parts additively manufactured by the end user and allows copies for private use without the permission of the author. However, a few conditions have to be taken into account: The number of copies must not exceed a maximum of seven copies, which can be passed on to friends and relatives free of charge. Hence, the printer operator may not receive remuneration for the printed pieces, as the parts otherwise would then be sold for profit, being plagiarism. Furthermore, the copy may not originate from an obviously illegal source.

In the field of B2B, it is important to address the need for IP and counterfeit protection for each product and take corresponding protective measures (Chen et al. 2016). Although there will never be a 100% protection, the barrier has to be set as high as economically justifiable for the copyright holder, such that it is not feasible for a pirate to produce counterfeits. The subject of counterfeit protection is bound to a company-wide concept for product and know-how protection. Measures for counterfeit protection require a typical transdisciplinary approach that can be divided in four categories (internal security, external security, product labeling, and legal safeguards).

Particular attention has to be paid to the usability in court when selecting the right procedure for an individual application (Holland et al. 2018). Usability in court means recognition and admission of a procedure by the court. This might be a crucial factor in case of a defense against a product liability claim or against unjustified usage warranty claims.

Within the additive manufacturing process chain, the preparation of a geometry, determination of process parameters or manufacturing of components is often done by external partners with whom the copyright questions have to be answered. In the case of a service provider preparing the geometry model for printing and subsequently creating the print template with a slicing software, he may eventually have created a work according to copyright law, §3 section 1 No. 1 or No. 7. The author is then granted protection by preparing the file. Thus, to protect the work, it does not have to be registered.

The conditions required to classify it as authentic work are that it has to be created by a human and also that it is an "intellectual creation." In this case, the resulting work must not be copied and distributed without approval of the copyright holder. Public availability needs the approval of the author as well. Furthermore, the original product manufacturer could be restricted by amendments to the prepared geometrical model. Thus, the rules for legal boundary conditions must be defined clearly when entrusting service providers with the creation of a print template, because printing a template means copying it (Holland et al. 2018).

In case the printing file is passed on to a service provider for production, he has no property rights with regard to the protected work. The reason is that in case of the mere process of the printing order the intellectual creation is missing.

For all above mentioned reasons, the success of AM is dependent on a secure procedure to prevent misuse of the original data (Liese et al. 2010). One way is the introduction of a license procedure for a print controlled by a strict procedure called "Chain of Trust."

Technical Solution

The goal of the envisioned solution is to reduce risks to a minimum by using cryptographic approaches to secure the authenticity of printing data and prevent unauthorized use of it. Encoding and licensing of data by means of Blockchain Technology provides an opportunity. With this technology, the relevant data are encoded and identification of the print template and licensing of the printing process is enabled. Blockchain Technology, however, may be used as well for the application of transactions in terms of franchising. Contrary to Bitcoins, the license allows printing of a certain number of a component (Holland et al. 2018).

A so-called Smart Contract files the license information in the Blockchain and secures that only the recipient has the permission to update the license, e.g. register a printer part to it. The recipient's printer verifies the license before starting to print. Additionally, the serial numbers of the separately printed components can be written into the Blockchain to prove type and quantity have been printed in accordance with the license terms. To ultimately close the Chain of Trust, the machine and automation suppliers have to be taken into account. Similar concepts as those of manufacturing copiers can be realized. In this way, a complete Chain of Trust can be built-up from copyright holder to service provider (Holland et al. 2018). Other ways to improve trademark protection are certified partners and the use of trusted printers ("Blockchain Ready").

A consistent Chain of Trust for Additive Manufacturing Procedures for a commercial purpose is realized, from development of digital 3D printing data via the exchange with a service provider of 3D printers trusted by specific secure elements, up to labeling of printed components by means of RFID-Chips. In addition to the available encoding mechanisms, a digital license management based on Blockchain Technology is integrated into the data exchange solution OpenDXM GlobalX of PROSTEP AG. The interface for the exchange of certification and license data between copyright holder and receiver is Industry 4.0 Standard OPC-UA. Figure 19.4 illustrates the System Architecture of SAMPL (Secure Additive Manufacturing Platform).

Business Implementation

After several years of implementation, eight categories of Blockchain projects have been formed (Nussbaum 2018). The application presented here falls into the category "Shared Data," which comprises, amongst others, the use of Blockchain in supply chains. Initial Blockchain efforts could have a quick impact by transforming even a small portion of the supply chain, such as the information needed for the individual, decentralized, manufacturing of spare parts, which used to gather dust in warehouses waiting to be used. There are many similar possibilities, such as the "open data platform," which has been a popular startup idea for a few years now, with several companies achieving great success with this model. Because business rules and smart contracts can be built into the platform, a Blockchain ecosystem can evolve as it matures to support end-to-end business processes and a wide range of complementary activities.

Transdisciplinary System Model for 3D Printing

In summary, referring to Figure 19.3, the creation of a 3D printing service and facility incorporated many different fields and actors.

The development subsystem proceeds in parallel with the development of the 3D printing service, which is the solution system. When problems are identified in the print service activity, like illegal copying and counterfeit, they require action on the development level. At this level, legal bodies, manufacturers, 3D print experts, and certification bodies at least need to be included in the processes to create a safe and secure 3D printing service. Business developers and business owners also play a role, while proper external that are willing to and capable of providing the necessary technology, material, and tools parties need to be searched for and selected. Moreover, these external actors need to agree

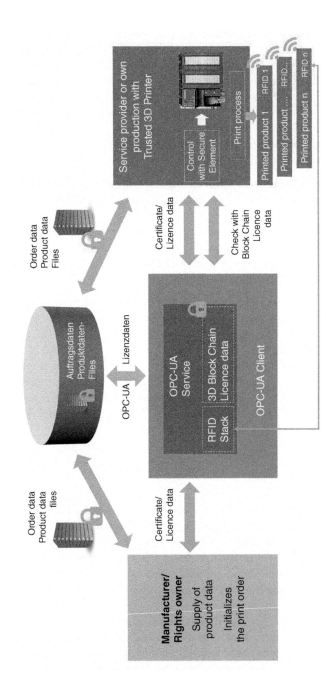

Figure 19.4 Secure additive manufacturing platform system architecture.

on copyright protection measures. In addition, the search for a suitable technology is needed to support the whole supply chain in acting according to the copyright protection agreement, in this case blockchain technology.

It is evident that the development system is a transdisciplinary system, because many different actors from different disciplines need to bring and exchange their knowledge mutually. Moreover, some activities in the system require the input from different sciences, like business and legal disciplines and technical disciplines.

The operational product/service subsystem, which is the solution system, consists of the following actions, actors, agreements, and technology:

1. Preparation of geometry and process parameters is often performed by an external actor. This actor needs to settle the copyright protection matter with the partner(s).
2. Creation of the print template is also often performed by an external partner. This partner automatically is granted protection by the copyright law.
3. Production of the print may be performed by an external actor. This actor does not have property rights.

It is recommendable to incorporate only certified partners to secure the supply chain from copyright theft and counterfeiting. Suitable technology is used to support and secure the supply chain, like the Chain of Trust for Additive Manufacturing based on blockchain technology.

It is not clear yet whether the solution system, the 3D printing service, is a transdisciplinary system. It is definitely a complex system but does not necessarily require the input from science for its operation. Though science may not be involved in the solution system itself, it is still involved in the development system to improve or change the solution system.

19.7 Summary

In this chapter, a systems view on TDER has been presented. Transdisciplinary processes are most often performed in large, complex, projects, aimed at creating a solution for a problem that cannot be solved by one person, nor by one discipline alone. Collaboration is needed between science and practice and involves multiple disciplines, not only technical ones. The solution for this problem is expected to have a large impact on the social environment in which the problem exists.

In the system view on TDER processes, two major intertwined (sub)systems can be distinguished: the development system and the solution system, which is the outcome of and input to the development system. The solution system is often a product/service subsystem offering the desired products and services for the (social) environment that can use or need the products and services.

Each subsystem operates with its own set of people, knowledge and technology, and governance. Elements of these sets may overlap.

Both (sub)systems are complex systems, because the people acting in the systems may have their own, often mutually conflicting, goals. Moreover, people are not only acting in these systems but also in other systems in their own environment, thus bringing different cultures, experiences, knowledge, and technology into the subsystem.

Viewing a transdisciplinary process from a systems perspective may help to identify a subsystem or aspect system that requires specific attention, while taking into account the context in which this system exists. In this way the impact of the system context on the system under consideration as well as the impact of changes made to the system on its context can be better identified and taken into account.

The cases presented here provide an excellent opportunity to study in-depth transdisciplinary processes and all their aspects, technical as well as social. Methods and tools from different science communities can be used in coherence to create rich knowledge of transdisciplinary processes and their complexities. This knowledge is needed to successfully manage transdisciplinary projects.

References

Aas, M. and Vavik, M. (2015). Group coaching: a new way of constructing leadership identity? *School Leadership and Management.* 35 (3). Published online 30 April 2015.

Beckett, R.C. and Vachhrajani, H. (2017). Transdisciplinary innovation: connecting ideas from professional and user networks. *Journal of Industrial Integration and Management* 2 (4): 1750016. https://doi.org/10.1142/S2424862217500166.

Beckett, R.C., Saunders, E., and Blamey, P. (2016). Optimizing hearing aid utilisation using telemedicine tools. In: *Encyclopedia of E-Health and Telemedicine* (eds. M.M. Cruz-Cunha and I. Miranda), 72–85. Hershey, PA: IGI Global. ISBN: 13: 9781466699786.

von Bertalanffy, L. (1951). General systems theory: a new approach to unity of science. *Human Biology* 23: 303–361.

Boston Consulting Group, Industry 4.0 – The Future of Productivity and Growth in Manufacturing Industry, 2018a, https://www.bcg.com/publications/2015/engineered_products_project_business_industry_4_future_productivity_growth_manufacturing_industries.aspx (Last accessed 3 July 2019).

Boston Consulting Group, Man and Machines in Industry 4.0 – How Will Technology Transform the Industrial Workforce Through 2025? Accessed 29

March 2018b, https://www.bcg.com/de-de/industries/engineered-products-infrastructure/man-machine-industry-4.0.aspx (Last accessed 8 July 2019).

Boulding, K.E. (1956). General systems theory – the skeleton of science. *Management Science* 2 (3): 197–208.

Caws, P. (2015). General systems theory: its past and potential. *Systems Research and Behavioral Science* 32: 514–521.

Checkland, P. and Holwell, S. (1998). *Information, Systems, and Information Systems: Making Sense of the Field*. New York, NY: Wiley.

Chen, Y., Dong, F., and Chen, H. (2016). Business process and information security: a cross-listing perspective. *Journal of Industrial Integration and Management* 1 (2): 1650009. https://doi.org/10.1142/S2424862216500093.

Chesbrough, H.W. (2006). *Open Innovation: The New Imperative for Creating and Profiting from Technology*, vol. 2006. Boston, MA: Harvard Business Press.

Frescoln, L.M. and Arbuckle, J.G. Jr., (2015). Changes in perceptions of transdisciplinary science over time. *Futures* 73: 136–150.

Gaziulusoy, A.I., Ryan, C., McGrail, S. et al. (2016). Identifying and addressing challenges faced by transdisciplinary research teams in climate change research. *Journal of Cleaner Production* 123: 55–64.

Hermann, M., Pentek, T., and Otto, B. (2016). Design principles for industry 4.0 scenarios. In: *Proceedings of the 2016 49th Hawaii International Conference on System Sciences (HICSS)*, 3928–3937. Washington, DC: IEEE Computer Society.

Hermann, M. Pentek, T. and Otto, B. (2017) Design Principles for Industry 4.0 Scenarios: a literature review. http://www.snom.mb.tu-dortmund.de/cms/de/forschung/Arbeitsberichte/Design-Principles-for-Industrie-4_0-Scenarios.pdf (Last accessed 3 July 2019).

Hofkirchner, W. and Schafranek, M. (2011). General systems theory. In: *Philosophy of Complex Systems* (ed. C. Hooker), 177–194. Elsevier (North Holland).

Holland, M., Stjepandić, J., and Nigischer, C. (2018). Intellectual property protection of 3D print supply chain with Blokchain technology. In: *Proceedings of the IEEE International Conference on Engineering Technology and Innovation (ICE/ITMC)*. 8436315, 971–978. IEEE.

Jolly, A. (2010, 2010). *The Innovation Handbook. How to Profit from your Ideas, Intellectual Property, and Market Knowledge*, 2e. London: Kogan Page Publishers.

Liese, H., Rulhoff, S., and Stjepandić, J. (2010). Securing product know-how by embedding IP-protection into the organization. In: *2010 IEEE International Technology Management Conference, ICE*, 7477025, 1–8. IEEE.

Marchal, J.H. (1975). On the concept of a system. *Philosophy of Science* 42 (4): 448–468.

Mathiasen, J.B. and Mathiasen, R.M. (2017). Practicing transdisciplinary engineering in a global development context: the transferring, translating and transforming approaches. *Journal of Industrial Integration and Management* 2 (4): 1750017. https://doi.org/10.1142/S2424862217500178.

Miller, R. and Lessard, D. (2000). *The Strategic Management of Large Engineering Projects: Shaping Institutions, Risks, and Governance.* IMEC Research Group. Cambridge, MA: MIT Press.

Mo, J.P.T. and Beckett, R.C. (2018). Architectural knowledge integration in a social innovation context. In: *Transdisciplinary Engineering Methods for Social Innovation of Industry 4.0. Proceedings of the 25th ISPE International Conference on Transdisciplinary Engineering*, 763–772. IOS Press, Amsterdam, The Netherlands (3–6 July).

Müller, R. and Olleros, X. (2000). Project shaping as a competitive advantage. In: *The Strategic Management of Large Engineering Projects. Shaping Institutions, Risks, and Governance* (eds. R. Miller and D. Lessard), 93–112. Cambridge, MA: MIT Press.

Nelson, R.J. (1976). Structure of complex systems. In: *PSA: Proceedings of the Biennal Meeting of the Philosophy of Science Association*, vol. 1976. Volume Two: Symposia and Invited Papers, 523–542. University of Chicago Press.

Nowotny, H., Scott, P., and Gibbons, M. (2006). Re-thinking science: Mode 2 in societal context. In: *Knowledge Creation, Diffusion, and Use in Innovation Networks and Knowledge Clusters. A Comparative Systems Approach Across the United States, Europe and Asia* (eds. E.G. Carayannis and D.F.J. Campbell), 39–51. Westport: Praeger.

Nussbaum, J. (2018). *Blockchain Project Ecosystem: Market Map and Musings on the State of the Ecosystem.* https://medium.com/@josh_nussbaum/blockchain-project-ecosystem-8940ababaf27 (Last accessed 3 July 2019).

OECD (2015). *Frascati Manual 2015: Guidelines for Collecting and Reporting Data on Research and Experimental Development*, The Measurement of Scientific, Technological and Innovation Activities. Paris: OECD Publishing http://dx.doi.org/10.1787/9789264239012-en.

Peruzzini, M., Gregori, F., Luzi, A. et al. (2017). A social life cycle assessment methodology for smart manufacturing: the case of study of a kitchen sink. *Journal of Industrial Information Integration* 7: 24–32.

Project Management Institute, Inc. (2017). *Guide to the Project Management Body of Knowledge (PMBOK Guide)*, 6e. Newtown Square, PA: Project Management Institute Inc. (PMI).

Rauch, E., Dallasega, P., and Matt, D.T. (2017). Distributed manufacturing network models of smart and agile mini-factories. *International Journal of Agile Systems and Management* 10 (3/4): 185–205.

Rojko, A. (2017). Industry 4.0 concept: background and overview. *International Journal of Interactive Mobile Technologies (iJIM)* 11: 77–90.

Scholz, R.W. and Steiner, G. (2015). Transdisciplinarity at the crossroads. *Sustainability Science* 10: 521–526.

Simon, H.A. (1976). How complex are complex systems? In: *PSA: Proceedings of the Biennal Meeting of the Philosophy of Science Association*, vol. 1976. Volume Two: Symposia and Invited Papers, 507–522. University of Chicago Press.

Stjepandić, J., Liese, H., and Trappey, A.J.C. (2015). Intellectual property protection. In: *Concurrent Engineering in the 21st Century: Foundations, Developments and Challenges* (eds. J. Stjepandić, N. Wognum and W.J.C. Verhagen), 521–552. Springer International Publishing.

Wever, M., Wognum, P.M., Trienekens, J.H., and Omta, S.W.F. (2012). Supply chain-wide consequences of transaction risks and their contractual solutions: toward an extended transaction cost economics framework. *Journal of SCM* 48 (1): 73–91.

Wognum, P.M., Krabbendam, J.J., Buhl, H. et al. (2004). Improving enterprise system support: a case-based approach. *Advanced Engineering Informatics* 18: 241–253.

Wognum, P.M., Bremmers, H., Trienekens, J.H. et al. (2011). Systems for sustainability and transparency of food supply chains – current status and challenges. *Advanced Engineering Informatics* 25: 65–76.

Wognum, N., Wever, M., and Stjepandić, J. (2016). Managing risks in knowledge exchange: trade-offs and interdependencies. In: *Transdisciplinary Engineering: Crossing Boundaries. Proceedings of the 23rd ISPE International Conference on Transdisciplinary Engineering*. Curitiba, Brazil (3–7 October), 15–24. IOS Press.

Wognum, N., Bil, C., Elgh, F. et al. (2019). Transdisciplinary systems engineering, implications, challenges and research agenda. *International Journal of Agile Systems and Management* 12 (1): 58–89.

Zhong, R.Y. and Ge, W. (2018). Internet of things enabled manufacturing: a review. *International Journal of Agile Systems and Management* 11 (2): 126–154.

20

Entrepreneurship as a Multidisciplinary Project
Arnon Katz

Synopsis

Entrepreneurship is a creative process, called on to design and manage innovative models, products, services, or businesses. An entrepreneurship initiative is a creative complex project, incorporating all the activities required to form and manage it. The execution plans and related outcomes are derived from this entity formed by the specific ecosystem, market, opportunities, team, and other essential characteristics. This chapter elaborates on a holistic perspective to entrepreneurship, with details on its various components and interrelations. It presents a systemic model for the planning and operation of an entrepreneurship initiative in context of the Fourth Industrial Revolution. It begins with an introduction to entrepreneurship, emphasizing its uniqueness and background. Then, a definition of such multiple disciplinary projects is discussed, highlighting the role of risk and uncertainty. The presented concepts are supported by data and practitioners' experience. They provide the background for guidelines and a checklist for a systemic approach. This approach is illustrated by a systemigram, a diagram presenting the systems' mutual influence and reliance within the overall system. It is used to emphasize the need for a careful tailored solution aiming at optimal management, according to the unique conditions of entrepreneurship.

20.1 Introduction to Entrepreneurship

Entrepreneurship is a dynamic, ever-changing entity in constant interaction with its ecosystem. It begins with an idea sparked by the identification of an opportunity within the abilities of the team. The main challenge is identifying an opportunity and approaching the matching market by materializing this idea into a business success, where the outcome may be different from

Systems Engineering in the Fourth Industrial Revolution: Big Data, Novel Technologies, and Modern Systems Engineering,
First Edition. Edited by Ron S. Kenett, Robert S. Swarz, and Avigdor Zonnenshain.

the initial idea. The essence is finding that opportunity, exciting the market and raising demand, planning the scenario and delivering a winning solution, while navigating within a risk presented by uncertainty. This entity exists in a challenging ecosystem consisting of highly competitive market conditions and requirements, such as the need for faster development and delivery of new and differentiated products, services, value, and ever-growing customers' expectations.

An initiative is characterized by a combination of factors, each of which is a complex process by itself, subject to uncertainty and high risks. The leadership and management team of the entrepreneurship initiative is running a "system-of-systems" project, requiring an embedded systematic approach. Its main goal is to provide a feasible and worthwhile solution to a pain, problem, or situation, where the current existing responses are insufficient. One can view an entrepreneurship initiative as a design, development, and materialization with a need to be focused on the process's success and not on its separate components.

Most experts identify the team composition as the most important success ingredient. The market and opportunities may change, as may the solutions also, yet a qualified, synergistic, and able team is a true asset that can adapt to a dynamic environment. Founders and entrepreneurs are evaluated by their qualities, strong confidence in themselves, their ideas, and abilities to execute their dream. Yet the unique start-up environment also raises tough questions: Are they stubborn enough to challenge the risky roller coaster called "entrepreneurship"? Are they able, experienced, and flexible ready to lead a team under challenging conditions, and decisive enough to make fast, determined resolutions under uncertainty? Can they manage in an unfriendly and unsupportive environment?

The environment, its organizational cultural support and empathy toward innovation are key factors in the emergence of entrepreneurships. If it forms a sustainable local ecosystem, it is able to accept, absorb, and nourish exceptional and nonconservative concepts, approaches, and operations. This tendency, combined with the practice of appreciating calculated risk-taking and tolerance to failures, forms an encouraging incubator for various initiatives. These are only a handful of the issues to be considered.

This holistic perspective to entrepreneurship initiatives requires an integrated systematic approach, focusing not only (or mainly not) on the various components and factors. Challenging problems should assimilate an overall approach, avoiding concentrating on details per se. A state of mind of dealing with a complex situation under dynamic ecosystem is required, while often timing is of the essence. One should not disregard all relevant aspects of all factors involved, including human factors. Reasoning the main systemic components and carefully planning the way under a systemic concept may vastly increase the chances for the success of an entrepreneurship.

20.2 Entrepreneurship as a Project

Many definitions exist for a project. Integrating and summarizing such definitions to a comprehensive, all-inclusive and simple format can be formulated as follows: "A managed endeavor (including design, planning, and control) with a beginning and an end, formed by a set of interrelated tasks integrated to a defined mission, not carried out previously."

As such, entrepreneurship initiatives are considered to be of relatively high uncertainty and/or economic risk. They consist of an activity to be executed over a fixed period of time and within certain costs and other limitations, intended to create a unique product, service, or result, elaborated progressively.

Viewing entrepreneurship in light of this definition emphasizes the following characteristics of an entrepreneurship:

- *An endeavor that results in an unique activity.* Under specific conditions – the combination of suitable ecosystem, conditions, market, and opportunity.
- *Consisting of a set of interrelated tasks.* As a whole, provides the setting and ability to form the specific activities and provide the multidisciplinary environment to produce the planned results.
- *Not carried out previously.* As it realizes and implements an innovative idea, as a solution to an existing problem, pain point, or unsatisfactory situation, or forming a new reality.
- *To be executed over a fixed period, with a defined beginning and end.* An entrepreneurship is most often limited in time by an opportunity window. These are imposed by the current ecosystem and market, the relevant and updated competition, resources and many other factors.
- *Intended to result in a unique product, service, or result.* Represents the breakthrough and innovation of the entrepreneurship.

Entrepreneurship as a Multiple Disciplinaries Entity

An entrepreneurship will always involve several disciplines as it refers to the real world in which the problems and solutions need to be approached considering various perspectives. It is a combination of knowledge domains, gathering, and teaming experts to compete with the challenge.

The following points define the expressions often used for describing relations between disciplines to enable comprehension of similar meanings, but practically of completely diverse significance:

- *Intradisciplinarity* refers to approaching an issue with a single discipline. This is the simplest approach and easy to analyze, yet it is theoretical as does not comply at all with the real world.

- *Cross-disciplinarity* presents the understanding that there are always a few disciplines to approach an issue, and it refers to viewing one discipline from the perspective of another. However, this does not present a concept for combining all parts to one complete entity.
- *Multidisciplinarity* draws on knowledge from different disciplines. It describes different disciplines working together, each on its disciplinary knowledge, yet stays within the boundaries of the specific knowledge domain. It can be referred to as an additive concept for viewing different disciplines.
- *Interdisciplinarity* analysis combines and unifies links between disciplines into a coordinated and coherent whole. Practically it integrates knowledge and methods from different disciplines, using a real synthesis of approaches. It can be referred to as an interactive concept for viewing different disciplines.
- *Transdisciplinarity* integrates the disciplines in a harmonized context, creating a unity of frameworks beyond the disciplinary perspectives and transcends the traditional boundaries. Practically it means crossing disciplinary boundaries to create a holistic approach. It can refer to concepts or methods that were originally developed by one discipline, but are now used by other disciplines.

The objectives of multiple disciplinary approaches are to resolve real world or complex problems, to provide different perspectives on problems, and to create a comprehensive solution – facilitating a systemic way of addressing a challenge. Multiple disciplinary teamwork is suitable for complex problems, and such is the case requiring the involvement of multiple dimensions disciplines – such as issues of technology, management, human relations, marketing, etc.

Entrepreneurship as a Systemic Project

A simple, yet accurate definition of "system" can be found in Wikipedia – "A system is a regularly interacting or interdependent group of items forming a unified whole. Every system is delineated by its spatial and temporal boundaries, surrounded and influenced by its environment, described by its structure and purpose and expressed in its functioning."

Thus, "Systemic" means that changes or updates (e.g. in the inputs, the environment, resources, etc.) affect the whole and not only a component or part of the system. Clear criteria to that are the influence of the various components on the sustainability of the entity, since sustainability is a systemic issue.

An entrepreneurship is a transdisciplinary entity (as described before), originated to provide a holistic solution to a mission, that practically performs as a system within the environment. Yet the result of this endeavor is also a system, as it is a product providing a multiple disciplinary whole.

A project has five main stages – (i) initiating, (ii) planning and design, (iii) executing, (iv) monitoring and controlling, and (v) completion and closing. Thus, the practice of the "project" entity is based on a systemic concept, and the origin of the name is the project management foundation. The word *project* originated from Latin (*prōiectus* – "before taking an action") and indicates that there are activities to be done prior to performing or carrying out an activity. The beginning is in learning, setting objectives and a state of mind, checking the possibilities and planning.

An elaborated description of "project" (Mesley 2017) combines these ingredients into a complete "whole" by synergizing all to a sequential and continuous process, leading to the conclusion that practically every project has to be managed systematically:

A project consists of a concrete and organized effort motivated by a perceived opportunity when facing a problem, a need, a desire, or a source of discomfort. It seeks the realization of a unique and innovative deliverable, such as a product, a service, a process, or in some cases, scientific research. Each project has a beginning and an end, and as such is considered a closed dynamic system. It is developed along the four Ps of project management: Plan, Processes, People, and Power (e.g. lines of authority). It is bound by the triple constraints of calendar, costs, and norms of quality, each of which can be determined and measured objectively along the project life cycle. Each project produces some level of formal documentation, deliverables, and some impacts, which can be positive and/or negative.

Within this description, the essence is highlighted: a project is required to follow a systemic approach and is managed according to a systemic concept (Miklosik 2014). An entrepreneurship fully matches the description of a project, according to the following guidelines:

- It is an organized effort motivated by an opportunity.
- It is challenged by a problem, pain, need, or desire, to which it considerably has an idea and/or know-how to seek an innovative, deliverable product, service, or process.
- It has a beginning and will end with a certain solution. It may be expanded, elaborated, or improved later, but it is limited in time due to the opportunity window, competition, and resources.
- Due to the vulnerability and dynamics of the resources, environment, market, and competition is it required to be flexible and agile within itself (PMI 2013).
- It is developed along the four P's of project management: Plan, Processes, People, and Power.
- It is bound by the triple constraints that are calendar, costs, and norms of quality.

- It is obliged to produce formal documentation for proof of concept, and most possible registered IP.
- At the end of the day, the entrepreneurship is judged by its deliverables (if it survives to that point in the process), which can have a positive and/or negative impact.

In today's world, there is a growing change in the understanding of projects at whole, and especially on issues of project management. The swift changes in technology and know-how, the growing requirements and demands for innovation, unique solutions and, customization, and the resulting complexity of projects, impose a growing focus on issues of uncertainty and risk. The environment, market, and technology do not develop any more in predictable ways, and time constants are shrinking dramatically. These characteristics, along with cost sensitivity and rising competition, increase the need for risk taking (and vulnerability) in this ecosystem which – typical for entrepreneurship – highlights the need for a systemic approach to risk and uncertainty.

20.3 Approaching Change, Risk, and Uncertainty Systematically

Change is an integral part of a project, while risk is a corresponding vital part. Though quality planning and management are able to predict potential changes, additional unexpected changes will almost surely occur and a management system for eventual control is essential. Changes can be of many types – such as technology, environment, markets, teams and partners, economic situation shift, *force majeure*, and others, and all can be considered risks and uncertainties, which may have a crucial impact.

Risks and uncertainties should be clearly distinguished, as they are of a different nature and should be dealt with separately. A risk is a probability and can be mitigated as a statistical issue, and defined "insurable." It is to be differentiated from an uncertainty, the probability of which is not statistical (Knight 1921), and therefore can be referred as "uninsurable." Another way to describe uncertainty is as a perceived mismatch between the known data and the desired situation (Livio 2017). A common and popular way to refer to uncertainty is by "luck" – meaning the possible results are unexpected due to the fact that the data are not available. Uncertainty is a state of doubt about the future or about what is the right thing to do.

This assumption about uncertainty is debatable: the conclusion is that if we know more, probably the uncertainty (or luck) will be less dominant. So, the more planning and preparations, the more relevant data can be collected and thus reduce the risks of "luck." Related and suitable mechanisms to deal with these are essential and exist (Ben-Haim 2015), and identification and immediate treatment should be monitored and controlled upon discovery.

Black Swans in the Entrepreneurship Ecosystem

A Black Swan is considered a highly improbable event with three characteristics (Taleb 2007):

- It is unpredictable.
- It has a unique, fundamental, and substantial impact.
- After appearing, various explanations and excuses will be made to claim that it could have been more predictable than it practically was.

In the entrepreneurial world, "Black Swan" events often ignite the innovation. These unexpected events provide some of the unique opportunities to enable innovation – may it be a natural event, a surprising invention or novel technology, a war or a natural disaster – all of them create new situations, pains, requirements, demands, and the related opportunities for new solutions. Entrepreneurs who have the ability to foresee or estimate the impact of the occurrence, and react swiftly (utilizing their abilities) to the opportunity such an event creates, have preferable prospects to gain value and become winners.

Decision making under uncertainty is a critical part in the functioning of an entrepreneur (Knight 1921) as represented in "Black Swan" events. However, this is not the sole challenge or the utmost demand from an entrepreneur.

Challenging "Black Swan" Events by an Entrepreneurial Systemic Approach

An entrepreneurship can be defined as "an activity that involves the discovery, evaluation, and exploitation of opportunities to introduce new goods and services, ways of organizing, markets, processes, and raw materials through organizing efforts that previously have not existed." (Shane 2005).

One of the unique advantages of an initiative is its flexibility and pivoting qualities. Locating opportunities and reacting swiftly to utilize the occasion can be done only by an open minded, flexible entity, ready to put almost all in risk for the prospects of success in a daring adventure. This activity cannot be carried out by a big or established company: their procedures are too heavy; previous obligations are there to be accomplished first; state of mind and concepts are established and take time to be changed, if at all; and there is much to lose if it cannibalizes an existing solution or fails in the endeavor. So, generally speaking, if the required preparations are taken care of and the leading concept is systemic, the challenge is worthwhile taking. If the opportunity exists but the systemic approach is not taken, the risk of loss (of the opportunity, of time, of money, of credit) is many times over.

Katz (2018) provides an allegory: An entrepreneurship is like surfing. You have the will, you know the effort is worthwhile, and you just want to go there and do it, catching the wave for the longest ride (an exit …); however, you need the opportunity for the "best wave ever." So you have to inquire where to look for it (location) and the best timing (season of the year, time of the day). These are uncertain but collecting the data in advance and being prepared accordingly, can dramatically improve your chances, and you wait patiently for the right moment to arrive. Yet being in the right place at the right time, and if the opportunity occurs, is not sufficient. Long before you should have prepared for the moment: practicing the correct training, acquiring suitable equipment, and setting all that is required to be there. This is a symbolic demonstration of an entrepreneurial systemic approach in a nutshell.

Real life is much more challenging as the situation becomes more complex. With increasing complexity, there are many more variables, risks, and uncertainties to be considered, as referred to in Section 20.5

20.4 The Need for a Systemic Transdisciplinary Concept – Conclusions of Case Studies and Experience

An entrepreneurship is a complex and complicated entity due to the many ingredients involved and to the reciprocity between them, especially in a dynamic environment characterized by uncertainty, threats, and risks. However, the common approach and much of the professional literature has been different, and the widespread practice avoided involvement in multiple disciplinary complex environment (Shane 2005) – most likely due to the difficulties in competing with a complex structure consisting of multiple disciplines (Cuervo et al. 2007).

Observing an entrepreneurship from a holistic perspective and embedding it in a systemic approach has been discussed generally for years but dealing with a methodology to materialize such an approach has not been challenged. A partial list of authors dealing with this includes Chang and Wang (2011), Lepnurn and Bergh (1995), Ries (2011) and Senor and Singer (2009).

A few resources present the broad perspective but lack details (Drabenstott et al. 2003). Other references focus on the assimilation and implementation of systemic thinking as an approach that can be directly connected to entrepreneurship success (Chang and Wang 2011, Kariv 2011, Katz 2018).

Mandatory Attributing Systemic Comprehensions

Many disciplines and components pave the way to success by referring to mandatory systemic attitudes:

- Entrepreneurship begins with a vision, followed by a process. Setting an assignment and a materialization plan of the mission, with the obligation for the ultimate obligation are a must. The strategy follows, and the "Value Proposition" is practically the distribution of the assignment to principles of progression and preferabilities (Byers et al. 2011; Reynolds and Curtain 2007). These should be followed by definitions – of the objectives, guidelines of the business plan, and principles of execution.

- The entrepreneurs' function commences with the ability to identify and isolate proper opportunities, to advance innovation afterwards by initiating and establishing a business entity, whereas the drive and motivation of the entrepreneur are critical factors (Maital 2016). The drive is practically the life force of the initiative, derived from persistence, insistence to continue, devotion and responsibility, backed by belief in the success. All these can be defined as the intangible system of the entrepreneur.

- Financing is one of the first issues to be considered and of essential importance always. Though the final destination is providing a significant end value to the market, the best innovation available will not lead the entrepreneurship to success if not holding an intrinsic economic value (noticeable return on investment [ROI] or worth) to the market (Christian 2006). Many of the entrepreneurs are not aware to that nor do they place the financial issue on top of their consideration list, yet swept away by other priorities (such as technology). Finance and its planning are crucial, as an entrepreneurship is in constant need of funding. Taking care of obtaining financing and financial resources is an ongoing activity. The source of the money has its importance, so financial planning is not only obtaining the money, but also considering from whom, at which stage, and how (Drabenstott et al. 2003; Hofer et al. 2010).

- Planning and Design are necessary in order to help deal better with complex situations. Entrepreneurs should precede current planning procedures with an iterative, conversational design process based on *systems thinking*. This process is important to build a systemic understanding of the situation, so a course of action emerges intuitively, informed by an explicit design that provides a governing logic for the operation. Then a subsequent planning can proceed effectively. The underlying concept is that understanding a problem leads to an almost self-evident solution to it. It is important to distinguish carefully between designing and planning, defining the former essentially as problem setting and the latter essentially as problem solving.

- Assimilating the dynamic capabilities perspective refers to allocation of resources in an organization in order to maximize its competitive value (Lepnurm and Bergh 1995, Kariv 2011, Boccardelli and Magnussen, 2006). This is a systemic concept referring to balancing, matching, and adjusting of the limited available resources dynamically. It optimizes compatibility to the changing requirements, ecosystem and market, which is one of the main challenges confronted by entrepreneurships.

- Flexibility and abilities for fast dynamic response are the main sustainable advantage of entrepreneurships (Byers et al. 2011). The dominant factor is the search for a suitable market, when the changeability and dynamics of resource allocation are main parameters. This matches some of the principles and milestones of the "Lean Startup" approach (considered to be one of the main guidelines for entrepreneurship behavior).
- NABC (Need, Approach, Benefit, Competition) approach presents clearly the main concept of the systemic characteristics of an entrepreneurship. The concept is based on finding suitable answers to basic questions, which practically evolve to a systemic strategy. A definition of the needs and requirements challenges the response and enables to build an approach for facing the competition and building the business plan, based on analysis and diagnosis of advantages, the market and alternatives for this project. The main question is not "Are things being done right?" but "Are the right things being done?"
- Understanding criteria for investing in entrepreneurships is a systemic analysis from the investors' point of view. Money is the basic need for an entrepreneurship's existence and this need will be a lifesaving factor all along the life cycle, till it turns to be profitable. This results in the need to be attractive for investors and funding parties to survive. Understanding this principle and that investors are aware to the fact that entrepreneurships success prospects are low, the entity should present all along the way an impressive market value that should result, to the best evaluation, in a noticeable financial profit at the end of the day. This requires a systemic schedule to convince that there is a constant progress in advancement toward the goal.
- Internalizing and integrating the challenges is a necessary yet not obvious insight (Drabenstott et al. 2003). Entrepreneurships confront many challenges, but the need is to deal with them as a whole and find the best reflected combination, not one by one. Examples for challenges to be combined in the strategic approach:
  - Quality entrepreneurships utilize innovations to create opportunities, and do not settle in idea per-se.
  - An entrepreneur entity is established to materialize and bring its unique innovation to the market (and not settle solely with the knowledge or IP). The resources should serve to transform the innovative idea to a product, process or service, and to implement it in the relevant market(s).
  - Globalization opens a wide range of opportunities, as ecosystems and markets are not identical. Avoiding the inferred meaning and derivative planning minimizes drastically success prospects.
- The team is probably the most complex system (Katz 2018), including tangible and intangible assets, a variety of individuals required to integrate

and synergize in order to compete with all the above mentioned. The team may need to change according to the phase and dynamics of the entrepreneurship state and life cycle. There is a variety of potential partners to the entrepreneur, but the initial, burdensome and fateful responsibility is his. Who is suitable, who to choose and with whom to partner in order to initiate the activity. His first dilemma is to select between friends and reliable acquaintances (probably from similar background) to a heterogenic team, or any combination. It is essential to consider and plan the required qualities and teaming in advance, as they will have to work together in difficult situations and to rely one on the other. The building of the team is actually choosing the ingredients to a melting pot of experience, professionalism, background, culture, synergetic abilities and will with an emphasis on ego-less consent, and a mutual understanding that the outer world should be exposed to one firm and homogenous voice.

Team management and leadership is the engine, and the recommended configuration is a team, but minimal. Up to three members with a defined division of power and responsibilities, but with an agreed mechanism for decision making. Mentioning leadership means not only to create enthusiasm in the internal team, but to excite and lead all stakeholders – customers, investors, and consultants, analysts, etc. As the team is small and the dynamics and requirements almost unlimited, it is a must that all, especially the leaders, will be multidisciplinary and able to "change hats," functioning in a variety of positions and tasks. Last, but not least, is the very sensitive issue of the suitability and dignity of the entrepreneur. Though he was the initiator, he may not be the best figure to manage the activity, the same as a CEO may be the right person for a certain phase but not for another. It should be understood by the specific person, and if not intuitively than conveyed to him or her gently and honorably. Many times, this situation is clear to all, except to the person himself. Many entrepreneurships were terminated due to lack of understanding on behalf of the entrepreneurs that the time has come to let go, if the dream is to be materialized.

Causing Failure Due to Abstention from Systemic Attitude

It is of interest to view the negative effect of not adopting and embedding a systemic perspective. If this concept was not of essence, it would be expected to find a prime or crucial reason for failure of entrepreneurships; however, meta-analysis of failures pointed explicitly that there is no sole noticeable reason for failure of entrepreneurships but a combination of accumulated and integrated components (Drabenstott et al. 2003, Moran 2016, Carle et al. 2012). The main reasons to be identified as the leading combination of factors for failures was: impractical business model, shortage of cash, insufficient financial planning, absence of relevant experience, and qualifications of the entrepreneurs.

In case of technological entrepreneurship, the dominant failure factors added were focusing on technology and not on the market and customers' requirements, lack of harmony, synergy and coordination within the entrepreneurship group, wrong timing and negligence of cost and pricing aspects.

The main conclusion regarding the criticality and exclusiveness of a systemic concept for the survival and success of entrepreneurships was verified and confirmed by a wide scope of experts in the field, in multiple interviews (Katz 2018). The long list of interviewees included CEOs of the Israel innovation authority, serial entrepreneurs, CEO's of large Hi-Tech companies which raised dozens of spin-offs, University professors – innovation and entrepreneurship experts, Venture Capital funds partners, and directors of innovation centers and hi-tech incubators. All emphasized the importance of the holistic approach vs. the common narrow-band access that characterizes many of the entrepreneurs.

Elaborating the main systemic issues for failure leads to the following:

- Many entrepreneurs start and continue solely with a limited idea and a technology. Basic questions of what is the general direction, what is the mission, what is the related strategy to reach, and how to materialize are not considered. Add a common approach (mainly of technological entrepreneurs) of sticking to their idea or technology and avoiding openness or considering other existing solutions or ideas, i.e. the NIH approach (Not Invented Here – a common behavior of contempt or reduction of other solutions) and this can be defined as a systemic fixation.
- Narrow minded and disqualified management – as the CEO (usually the entrepreneur) may focus on one activity only. May it be technology, development, marketing, finance – this reflects a misunderstanding of the fact that an entrepreneurship is a small size complete company with a need to foresee, approach and refer to all issues in a holistic preference. This interrelates with a malfunctioning of not setting priorities, and delay of the long range and strategic planning and activities due to immediate tasks. Another recognized symptom is the lack of willingness and/or ability to make severe and fateful decisions in the short time required. As an entrepreneurship faces an ongoing survival struggle, management must be aware to the necessity, and the price of wrong dynamic resource allocation decision making procrastination.
- Negligence of the criticality of finance and money. "Money makes the world around," and at the end of the day people, endeavors, products, and marketing need to be financed. It cannot be assumed that on the right time an investor will pop up, or that when money is needed it will be obtained. Despite strong strategy and execution, all activities and operations can be liquidated due to lack of financial planning, road mapping, and funding. Part of the financial management is searching and obtaining the suitable investors.

The funding origin is of importance and most entrepreneurships are not checking all abilities and running a tender for the money, but set along with the first funder available, usually consenting with intolerable conditions and restricting terms.

It is important to notice that though the systemic concept is fundamental, there is no single combination of factors that can be recommended. Each entrepreneurship has its unique composition of conditions, ecosystem and market, and the components to be considered as most important or having the highest impact may vary.

20.5 Assimilating System Concepts in Entrepreneurship Management

The purpose of the entrepreneurship is to offer a noticeable added value (new solutions to existing problems and needs or for future utilizations) based on innovation. Accordingly, the process commences by locating difficulties and problems in existing solutions (if they exist) and questioning them. Approaching with "out of the box" concepts contravenes conservativeness and fixation, and therefore requires addressing each issue with a suitable and specifically tailored preparation phase for this unique endeavor. A professional approach should include the identification of prospective opportunities, analysis of the worthiness and profitability in execution and building a suitable activity plan. Risk and uncertainty reduction can be managed by utilizing task assigned tools and readiness for agility and pivoting. Beyond identifying and characterizing crucial components required, there is a need for operational insights (as contingency plans and pre-defined alternatives) and conceptual insights (such as agility, flexibility, and decisiveness).

The management should be based on the SWOT model (Strengths, Weaknesses, Opportunities, Threats). When this initializes the process, it provides a restrained, objective consideration of the situation and prospects before getting involved - financially, emotionally, exploiting resources and addressing investors. The issues to be analyzed will include the SWOTs of the inclusive market, the idea (and/or technology), the team, the initiative entity, and the preferred choice of a market segmentation. Utilizing the systemic approach induces that the focus should be on the synthesis of the various ingredients due to the connectivity, influence, and reciprocity. The real value, competitiveness compatibility and sustainability rely on the holistic summary and cannot be determined by any isolated components.

Dealing with and managing such a complex project requires special attention to the control and current management, not only for follow up and verification of advancement in light of the vision, but to react in real time to confront disturbances, changes, and shifts (Shane 2005), (Knight 1921).

Systems and Disciplines to Be Considered

Emerging from the point of view that the complete picture is built from many components, it is obvious that some of them are more influential than others. Practical reference should highlight the important; however, it is important to declare that referring to a collection of elements does not set a fixed template for the relative importance of each. The adjusted weight depends on the unique characteristics of each project, and may (and probably will) change frequently throughout the life cycle of the project, according to the state, timing, and conditions.

Before detailing and elaborating it is important to set the guidelines for consideration:

- Every component in the system influences the others and the system as a whole.
- The influence of any component on the system in whole depends on and is conditioned upon all other components, with all related reciprocally.
- Dependency and reciprocity are such that disassembly to sub-systems does not enable disregarding the interdependence.
- All said results in the conclusion that functional examination and operation are possible only in a synchronized activity of the system as a whole.

An overview of the systemic approach that refers to nine systems is presented in Figure 20.1.

Though it is supposedly easy to differentiate between the "external world" (top three issues, hereinafter "ext") and the "internal world" (of the entrepreneurship, hereinafter "int"), both are connected and interrelated, and

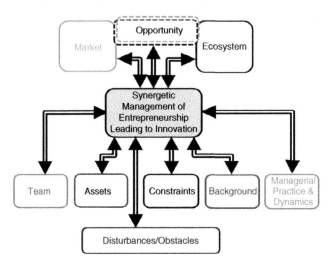

Figure 20.1 Overview of systemic approach to entrepreneurship.

these two "worlds" cannot be practically separated. Before elaborating on the systems and disciplines, the complexity will be described by two examples:

- A critical change (ext) e.g. an innovative, disruptive technology, or major *force majeure* as a critical environmental disaster, can immediately be reflected as new constraints and/or disturbances (int.), open a new opportunity (ext) and if taken timely and correctly by the entrepreneurship (int.) open a new market and imply a change on the ecosystem (ext).
- The management decided to let go of most of the team (int) due to financial constraints (int) and hire specialists. These came with a unique idea – a new asset, IP (int). Utilizing managerial marketing experience and assets – connections (int) the IP rose much interest in the market (ext) and opened an opportunity (ext) which enabled extreme funding, thus removing financial constraints (int) and providing financial asset (int) to hire a new team (int), develop, and produce a new asset – a product (int), presented to market (ext) and accepted enthusiastically it practically changed the ecosystem (ext).

Systems to be considered and unique disciplines within:

- The broad Ecosystem in which the entrepreneurship functions. Its factors set the basic and general operation conditions system, such as the surrounding economic situation (local, regional, national, global), political environment, nature, and importance of related industry. An additional factor will be ongoing technological and IP developments and innovations. It is important to mention that there may be a big variance in part or all of above mentioned due to different conditions in different locations.
  - Economics and financial situation, in the environment of activity of the entrepreneurship and in its Domain of activity. This includes the investment and support market, typical trends and orientations, status of investors and attractiveness in said time frame of the entrepreneurship disciplines.
  - Environmental culture in all circles – the "organizational culture," to include approach toward entrepreneurship in general in in the specific field of activity, the popular outcome toward risk taking and competing with failures, and openness to novel ideas and products.
- Market, in which the entrepreneurship is active – considering the potential market, the specific market and the segmented market. This is a combined and synergic system which includes customers, partners, competitors, various stakeholders and unique characteristics of the relevant market(s). It should be noted that the practical relevant market(s) is not necessarily the initial market to which the entrepreneurs aimed. Many times the pivoting and deviation from the original market (and possibly a change in the product and its utilization) to the relevant one (after checking and comprehensive sensing and experience) enabled existence, progress, and success of

the initiative, which otherwise could have failed if persisted with the original direction.

- Customers and users need to be involved in defining the needs and requirements. The best progress is when the entrepreneurship has a partner from the market that acts as a β-site and represents the real market, enabling updates and trial-run in real life conditions.
- The nature of the market imposes the type of investment and funding required, time frame and risks.
- Regulations and standards are major considerations in many markets, especially where standardization is compulsory, and due to an expensive and weary approval phase. In new and innovational segments there is much value for the entrepreneurship to be involved in the regulation and standardization process, to influence it and to avoid unpleasant surprises later on.
- Partnering is a position of power and a multiplier. Careful investigation may minimize competition, as many times partnerships can be formed of market segmentation between potential competitors.
- Go to Market, the marketing strategy, focused operational plan to approach the market and customer the best and obtain a competitive advantage.

• Opportunity – an outcome of combining the ecosystem (including technology, inventions, and innovations, unexpected events and nature disasters) and the market, the potential platform for realization of the uniqueness of entrepreneurship. An entrepreneurship segregates itself in abilities to identify opportunities that the dynamics and changes in the ecosystem and market requirement (the known and the concealed) create in its discipline of expertise, and to utilize them. It should excel in the ability to pivot and adapt to the changing conditions, and in swift response to suit the time frame of the opportunities. Occasionally unique assets of the entrepreneurship (such as a revolutionary idea, innovative technology or materializing a pioneering concept) create the opportunity itself, relying on a supportive market and suitable ecosystem, open to it.

- There is a strong dependency on an integrative, synergetic team capable to fast reaction and operability.

• Team, and especially the leadership team – the human composition system, the collection of expertise and skills, with high priority to the "soft" skills of the team function, synergy, and compliance between the individuals. This complicate system includes all ingredients: leadership, group dynamics, inertia, persistence, dedication, devotion, can-do attitude, etc. Many see this system as the most complicated, sensitive, and important to reach the goal.

• Assets – properties of the entrepreneurship (as an entity) that provide its unique value proposal. Some of them may be ideas, patents, and other IP; Some of them team qualities such as experience, expertise, professionalism, personal qualities; Some tangible like money, equity, and other material assets, and some intangible yet very valued – like mentors, consultants, connections and relationships, team synergy and culture.

– The importance of networking cannot be exaggerated. It is significant, though mostly not referred at its' real value. Good networking can expose the entrepreneurship to the market, investors, the ecosystem, experts, potential team members and other potential stakeholders, high valued attributes that would not approachable otherwise.

– Mentors and experienced consultants concentrate and refine the important relevant experience, and will probably impose a unique value of shortening times, saving efforts and resources and preventing typical mistakes.

• Constraints define the system of limitations within which the entrepreneurship functions. This system includes shortage of finance, reduced team, unavailable equipment, swift, and unexpected changes in the ecosystem and market, etc. This system can be seen as a collection of typical dynamic challenges.

– Restrictions of Time to Market with the products and supplies required, under competition, market demand, time frame of the opportunity or lack in required abilities.

– Time constraints in dealing with the challenges can be viewed in two meanings:

  o Timing is inappropriate: whether the market is not yet mature to the innovation, or the timing is not suitable, due to economic conditions, war, natural catastrophe, etc.

  o Insufficient availability (according to the request or demand) of resources, team, required experts. In spite the opportunity exists, it is impossible to accept the challenge due to a lack in required abilities.

Apparently above mentioned cases can be considered as improper management, and with a systemic approach could probably be avoided.

• Background and experience of all team members, including their education, culture, experience (professional, organizational, initiative, team work and leadership, improvisation and more), all which provide the human infrastructure for the entrepreneurship activity. Each team member with his qualities, and as a team – value of the integrated background and experience.

– One of the important contributions is from a supportive cultural background – encouraging decision making, entrepreneurial processes, taking calculated risks and accepting failures as a way of life.

– Previous experience in entrepreneurship is of value. Confronting uncertainty, risk management and operations under pressure in problematic conditions – all are unique background to identify opportunities, obstacles, and dynamic competing in a changing environment that characterizes entrepreneurships.

- Managerial qualifications and dynamics – the pooling of all the managerial resources and abilities required to initiate, check, verify, plan, schedule, manage, operate, control, and advance the entrepreneurship as a trans-disciplinary complex project. The combination of all qualifications to optimize processes, activities, and resource utilization while operating in an ever-changing environment with the flexibility to provide exceptional solutions and ad-hoc improvisations, under limiting constraints.
  - Vision, high level imagination but materialization abilities of the dream. Long range envision of the mission and the obstacles along the way, and planning progress according to constraints and abilities.
  - Project and complex process management and insights, proven managerial abilities in leading a team embedding working habits of neat, organized, and accessible documentation.
  - Solution oriented, holistic concept, managerial, and bricolage improvisation.
  - Risk management includes the ability to bear and survive failures.
  - Marketing orientation, ability to learn, feel and understand the market and trends, with proven abilities to maintain connections and partnerships.
  - Agility and flexibility, fast adjustment to changes. Professional planning, including alternatives and fallback optimization, adapting dynamic abilities and a pivoting nature.
- Disturbances and Obstacles, referring to the internal processes and conceptions which are common within entrepreneurial entities. The issue may begin as typical characteristics but if not judged carefully and with firm and reliable risk management process may doom the entity. At certain points these factors need to be neutralized or at least diminished to enable the existence of the entrepreneurship in the real life market. Samples can be emphasizing technology instead of living with and adapting to the market, disregarding technological current innovations, and break-through, negligence or lack of financial conservative planning, etc.

In this section, nine systems were defined, but dozens of disciplines influence the complex project of the entrepreneurship, which can be segmented under the suitable system. Though the significant general disciplines were mentioned above it is hereby mentioned that this is not a general template or format, but a checklist to which many additional disciplines can be added, considering the case and conditions.

Demonstrating a Decision Support System

The previous paragraph refers to systems in the project and various influential factors. This paragraph deals with the spatial vision of the situation. There are only a few sources referring to decision making and resource allocation in the process of innovation development (Chang and Wang 2011), yet none found (till now) that refer to the systemic view of entrepreneurship or start-ups. Verifying that an entrepreneurship is practically a complex project, it is natural to view these issues in light of management decisions in complex projects (Zonnenshain and Stauber 2015).

Practically this entity is composed of a few systems, so should be referred to as a "system-of-systems." Each of these systems actuates a few functions (each related to another discipline), all synergetic with the development and activity of this entrepreneurship project. It is mandatory to perceive that the features and adjusted weight of these components are unique for each project, but the foundation is similar.

A systemigram seems to be the optimal description tool for the process (Figure 20.2).

A systemigram is a tool to describe a system without sacrificing the details required to accomplish clarity. The value of this description is the possibility to provide insight of the complex project process by defining the relations, reliance, and reciprocity in a simplified graphic conceptual presentation. The system is displayed clearly using nodes and links, focusing on the nodes as concepts and the links as defining the relationships (including the nonlinear) between those concepts. The process starts at the upper left and ends in the lower right. This mainstay defines the purpose of the system.

The project (entrepreneurship) is an outcome to a managerial act, utilizing an opportunity that arises due to certain conditions formed by the relations between the ecosystem and the market (the external world domain). This does not cancel or reject the possibility that an entrepreneurship can begin with

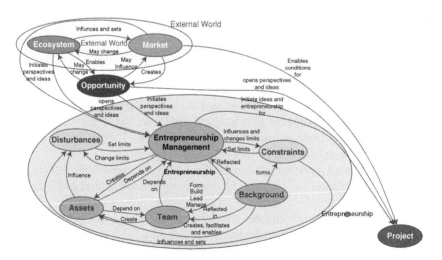

Figure 20.2 Systemigram of the entrepreneurship project process.

an idea, invention, discovery, or new technology, but rather states that it is required to be practical and support real needs and customers.

It is clear that though the environment imposes the conditions that form the market and produce the opportunities, each of these three systems can impact the other, e.g. the market can raise demand and influence the ecosystem, and opportunities can set options and activities that impact both the market and the ecosystem.

Similar reciprocity exists in the entrepreneurship domain, e.g. present assets set limits for the management, but quality management can improve or create assets, and the same for the reciprocity of the management and the team.

All possible interrelations set the need for the systemic thinking and approach. An environment in which various components can compensate or damage others, the total systemic overview, and holistic value should be considered within the transdisciplinary system. Given the external world conditions, if the entrepreneurship can identify and utilize the opportunity to initiate an activity that is welcome, anticipated, and accepted by the market, it can be defined as a multiple disciplinary systemic project.

Entrepreneurship strategy is not an entrepreneurship nor a strategy but a combination of them. It is part of the decision support system, functioning as the strategic processes in use to identify and utilize opportunities. It can be referred to as the business strategy, considering the market and competition, in search of potential opportunities.

An effective decision support system (to differ from an efficient one) should support the decision regarding the added value of the operation. The principle question in establishing an entrepreneurship is not if it is possible and how to do it, but if it is truly required – meaning does it bring with it a true estimated ROI, and according to which measures.

An important issue usually laid aside is the definition of criteria – criteria for success and criteria for failure. Management is obliged to refer to this issue before commencing the activities due to a few reasons:

- It is difficult to admit failure or lack of success (not the same). Human nature, being willing and optimistic, may long for additional chances and hope for the best. Setting firm definitions for failure and success limits the gray zone, and in case of failure avoids further efforts in vain.
- Considering in advance initiates a need for evaluation combined with initial planning and resource allocation, and enables setting milestones and benchmarks.
- Considering in advance enables true objective criteria with a long-term observation. Once the activity begins it is a common habit that rigid criteria and requirements are eroded due to the ongoing activities and shortsighted perspective.

20.6   Overview of Entrepreneurship Elements

Insights for Execution – Managing Entrepreneurship Project-Wise

The real world cannot be referred to as a collection or combination of stand-alone effects and events. System thinking enables the right perspective, and system engineering provides the tools and approach to plan and manage successfully in conditions of complexity and uncertainty. It is the ultimate concept to deal with entrepreneurship. The multidisciplinarity of this type of system (the entrepreneurship) exceeds beyond the common understanding of technological complicated systems. Additional factors to be considered include leadership, team integrity, team members "soft" skills and qualities, risk management, and the environment of state-of-art technologies and concepts. All challenged by the competition on markets, ideas, resources, financing, and people.

Highlights, to be converted to abilities and considered when directing to success of the entrepreneurship project are:

- *Vision, strategy, and planning.* As in any activity, all operations should be proactive. The entrepreneurship needs to recognize its strengths and weaknesses to define its potential field of activities. Any scheduled activity is triggered by the suitable opportunity (as follows). The prospects for success of any project are deeply reliant upon the primary stage of strategy and planning, while normally at least 30% of the activity is done prior to the first execution activities. An entrepreneurship project is not excluded. On the contrary, as risks and uncertainty are higher than in common projects, the attention dedicated to the initial analysis and planning phase should be greater.
- *Identification of the right opportunity.* Prior to initiating a project, efforts should be invested in setting the requirements, defining the destination, and deciding how to approach the execution. The opportunity is derived from the ecosystem and the market, and the main issues to be examined are the capability of the entity to withstand with the requirements, the business worthiness, and the profitability. The threats should be considered while executing a thorough and swift survey of the conditions and the potential market.
- *Involvement and dependability in the market and ecosystem.* The project needs to serve the customers; its survivability and future depend on providing a fitting and appropriate product to an anticipating and rewarding market. Many entrepreneurships are extinguished due to disengagement from the true customers' requirements or mismatch with the real needs. Inappropriate market segmentation, lack of involvement of potential consumers, bad timing, lack of preparation for a winning *Go To Market* – all these and more can be stated as a faulty examination of the situation and lack of recognition of the market.

- *Economic considerations and financial management.* An entrepreneurship is expected to be a profitable project; investors see this as their prime reason for investment (excluding a few exceptional prefinanced projects). The financial planning and considerations are essential, and to differ from a "regular" project the financing is an ongoing task, being checked, verified, and limiting all along the way. As the risks and uncertainty are significant and operations can be terminated due to change of revenue success prospects, the financial management and ROI considerations are constantly at stake.

- *Team and personnel characteristics.* Leadership qualities and integrated team abilities are considered to be among the most fundamental factors for success. In this activity one would look for the openness of mind, ability to withstand hardness and difficult challenges, virtuosity, devotion, and result driven orientation, as individuals and as a team. Creativity and synergy are a must, all bond by the functioning and leadership of the team leader (not necessarily the original entrepreneur).

- *Created and accrued assets.* Referring to unique added values of the entrepreneurship as an entity, to be created, generated, or acquired throughout the activity, to be materialized as IP or products. Additional perspectives of added values are management and financial abilities, and a set of relations and connections. All of these are reflected in the Value Proposition of the entrepreneurship.

- *Flexibility, agility, and responsiveness.* These are required to enable fast and appropriate reaction to changing markets, conditions, technologies, requirements, and all other challenges that characterize the changing reality of the entrepreneurship ecosystem. The ability for fast concept alteration and shifting of resources (defined as "pivoting") are critical as a compliance

- *Awareness and consciousness to constraints and obstacles.* Any and every activity goes forward alongside obstacles and unique challenges. Constraints, risks, and uncertainties should be considered to avoid their transformation into obstacles, which may happen if they are neglected, overlooked, or disregarded. It may happen that some of an entrepreneurship advantages may turn into obstacles if not treated properly and in the required proportions. Therefore, a holistic systemic perspective that integrates both constraints and obstacles (practical and potential) is required, which also includes firm objective success and failure criteria. These need to be discussed, analyzed, and decided upon before the process is initiated, and strictly referred to on an ongoing basis and in a timely manner. Typical obstacles and issues of awareness can be pointed (as in any project), such as concentrating in technology, miscalculating the importance of the market, not understanding the critical significance of timing, etc.

- *Entrepreneurship and luck.* There are entrepreneurs who tend to connect their difficulties or failures to sheer luck. "Luck" is practically the collection of uncertainties that challenge the entrepreneurship, to differ from "risks," which can be calculated and confronted statistically. Though uncertainty cannot be completely eliminated, it can be minimized. A systemic approach that commences with comprehensive preparations, planning, learning the ecosystem and the conditions, and synergizing the ingredients minimizes "luck" to a large extent. Competitive methods to cope with uncertainties exist (Ben-Haim 2015). The context of "luck" is referred to wisely by Lucius Annaeus Seneca, a Roman statesman and philosopher: "Luck is a matter of preparation meeting opportunity."

Integrating Entrepreneurship Systemic Concept in the Fourth Industrial Revolution

The Fourth Industrial Revolution (4IR) is characterized by a range of new technologies that are fusing the physical, digital, and biological worlds (collectively referred to as cyber-physical systems) and impacting all disciplines (Schwab 2017). It is critical to collaborate across geographies, sectors, and disciplines to grasp the opportunities this revolution presents.

As it seems at present this revolution has four main effects:

1. Product enhancement based on innovation.
2. Disruption of processes, routines, and standards.
3. Raising customer expectations, leading to intense involvement of customers and consumers and collaborative innovation.
4. Reflection of all above said in organizational form changes.

Emphasis will be directed on improving products and customer services, including customization of products, all enhanced by digital, analytical, and communication capabilities to increase value.

The 4IR will, at the end of the day, disrupt most or almost every industry in every country (Schwab 2018). These changes agitate the transformation of entire systems of production, management, and governance. Practically speaking, one can refer to this process as entrepreneurship on an enormous scale. Accordingly, main features and insights of regarding entrepreneurship as a multidisciplinary project can be naturally integrated into the 4IR, as follows:

- Embedding and executing a holistic concept, developing the understanding and abilities to think transdisciplinarily, and evaluating qualities and results on an overall systemic result.
- The ability and know-how to detect, refine, and evaluate the relevant opportunities, and analyze and select the feasible and profitable operations.
- Widening the exposure and reciprocal relations with the markets. Understanding and assimilating the growing importance of involvement of all stakeholders, especially customers and consumer engagement in the development, production, and go to market process, to provide better immunity to the shortening time-to-market delay.

- A growing demand for skills, abilities, qualifications, and talents, combined with know-how, is forming to compete with the complexity and challenges. Due to the professionalism and specialism required in a multidisciplinary environment, there is a growing need for transdisciplinary abilities, which require an elaborated teamwork and agile team management and leadership.
- The 4IR emphasizes the rate of change and acceleration of knowledge and technology innovation, which toughens competition. These processes elaborate the criticality of IP, thus leading to a need for severe and high-quality IP management and security.
- Comprehension that innovation and technology are rapidly changing, advancing, and growing exponentially. Thus, company and management agility, pivoting, and swift actions turn out to be significant and necessary in improving the quality, speed, and price that form the value proposition in the competitive environment, setting fast adaption to novel concepts and technologies.
- Since the dynamics of innovation and the velocity of disruption are difficult to predict, this situation forms an ongoing state of lack of clarity, surprise, and need to pivot. As mentioned, uncertainties should not be disregarded but approached systematically, by means of proper and careful preparations and various suitable methodologies.
- The new and fast-changing environment necessitates a constant need for forming (or re-inventing) suitable standards. These are likely to impose flexible regulations that must continuously change to adapt the changing world, and will require close collaboration of all stakeholders.

20.7 Summary

An entrepreneurship initiative, resulting in an innovative project, operates in a difficult and sometimes hostile environment. It is encouraged by the vision and dream to support a market searching for a solution to an existing problem or challenge. As a project, it is unique and focuses on an outcome not achieved before, thus confronting risks and uncertainty:

- Limited by available resources and in constant struggle to optimize utilization.
- In a constant rivalry with improved or added means.
- Restricted in a time frame and challenged to produce results in a timely manner (which is defined by the conditions of the ecosystem and market).

Many entrepreneurs are not aware of all the factors involved in and influencing their initiative, nor of the characteristic of their transdisciplinarity project (see Chapter 19). Many of these components may offset or compensate each other, or accumulate to a crucial situation (positive or negative), thus effecting the success of the initiative. Practically certain components can be referred to as tiebreakers due to interactions with others. As a rule, the whole is greater than the sum of its parts, and a project is not accomplished until all tasks have been finalized successfully and in the right order.

People usually abstain from multidimensional competitiveness, inasmuch as they are accustomed to confronting challenges one-by-one. Approaching this systematically is complex, requires multidisciplinary involvement, and, especially, understanding the interfaces and reciprocal influences; yet this is the nature of any project.

References

Ben-Haim, Y. (2015). Dealing with uncertainty in strategic decision-making. *Parameters, US Army War College Quarterly* 45 (3): 63–73.

Boccardelli, P. and Magnussen, M.G. (2006). Dynamic capabilities in early-phase entrepreneurship, knowledge and process management. *Knowledge and Process Management* 13 (7): 162–174.

Byers, T.H., Dorf, R.C., and Nelson, D. (2011). *Technology Ventures – From Idea to Enterprise*. New York, NY: McGraw-Hill.

Carle, P., Kervarc, R., Cuisinier, R. et al. (2012). Simulation of systems of systems. *AerospaceLab* 4 (May): 1–10.

Chang, L. and Wang, T.J. (2011). The development of the enterprise innovation value diagnosis system with the use of systems engineering. In: *Proceedings of the International Conference on System Science and Engineering (ICSSE)*, Macao, China (8–10 June). IEEE.

Christian, N. (2006), The NABC Method from Stanford Research Institute. SRI International Best Practice, EDUC303X.

Cuervo, A., Ribeiro, D., and Roig, S. (2007). *Entrepreneurship: Concepts, Theory and Perspective*. Springer. ISBN: 978-3-540-48542-1.

Drabenstott, M., Novack, N., and Abraham, B. (2003). Main streets of tomorrow: growing and financing rural entrepreneurs – a conference summary. *Economic Review* 88 (3): 73–84.

Hofer, A., Potter, J., Fayolle, A., et al. (2010), From Strategy to Practice in University Entrepreneurship Support: Strengthening Entrepreneurship and Local Economic Development in Eastern Germany, OECD Local Economic and Employment Development (LEED) Working Papers.

Kariv, D. (2011). *Entrepreneurship – An International Introduction*. New York, NY: Routledge.

Katz, A., (2018), A Systemic Approach as a Perception for Managing to Success of Entrepreneurships in the High-Tech Industry, BMGC (Bernard M. Gordon Center for System Engineering), Technion (Israel Institute of Technology).

Knight, F.H. (1921). *Risk, Uncertainty and Profit*. Reprinted 2002. Washington DC: Beard Books.

Lepnurm, R. and Bergh, C.D. (1995). Strategic management and entrepreneurial orientation in sick, marginal and healthy small businesses. *Journal of Small Business and Entrepreneurship* 12 (8): 8–19.

Livio, M. (2017). *WHY? What Makes US Curious?* NY: Simon and Schuster Inc.

Maital, S. (2016). *Innovate Your Innovation Process*. Singapore: World Scientific.

Mesly, O. (2017). *Project Feasibility – Tools for Uncovering Points of Vulnerability*. New York, NY: Taylor and Francis, CRC Press.

Miklosik, A. (2014). Selected aspects of systemic approach to project management. *Actual Problems of Economics* 155 (5): 195–202.

Moran, D. (2016). *100 Doors – An Introduction to Entrepreneurship*. Tel-Aviv: Yedioth.

PMI (Project Management Journal) (2013). *Agile Project Management: Essentials from the Project Management Journal*. Hoboken, NJ: Wiley.

Reynolds, P.D. and Curtain, R.T. (2007). Panel Study of Entrepreneurial Dynamics II, Program Rationale and Description. *SSRN Electronic Journal* https://doi.org/10.2139/ssrn.1028256.

Ries, E. (2011). *The Lean Startup*. London: Penguin Books.

Schwab, K. (2017). *The Fourth Industrial Revolution*. London, UK: Penguin Books Ltd.

Schwab, K. (2018). *Shaping the Future of the Fourth Industrial Revolution*. London: Penguin Books Ltd.

Senor, D. and Singer, S. (2009). *Start-Up Nation*. NY: Grand Central Publishing.

Shane, S. (2005). *A General Theory of Entrepreneurship, the Individual-Opportunity Nexus*. Cheltenham, UK: Edward Elgar Publishing.

Taleb, N. (2007). *The Black Swan: The Impact of the Highly Improbable*. NY: Random House.

Zonnenshain, A. and Stauber, S. (2015). *Managing and Engineering Complex Technological Systems*. Hoboken, NJ: Wiley.

21

Developing and Validating an Industry Competence and Maturity for Advanced Manufacturing Scale

Eitan Adres, Ron S. Kenett, and Avigdor Zonnenshain

Synopsis

The Fourth Industrial Revolution presents exceptional challenges and opportunities for the developed industrial countries and the leading industrial companies. Through advanced and "smart" technologies like massive digitization, big data analytics, and 3D printing, it is now possible to design innovative products and systems and to produce and maintain them in very productive and flexible ways. In many countries, governments have launched national programs for advanced manufacturing, like *Industrie 4.0* in Germany, Catapult in the UK, Factories of the Future (FoF) in Europe, and the National Network for Manufacturing Innovation (NNMI) in the US. In Israel, the Ministry of Economy and Industry published a strategic plan for advanced manufacturing and initiated, with the Israel Innovation Authority, the Institute for Advanced Manufacturing in the North of Israel. To support these trends and initiatives we present here a model that is used to assess the status and the maturity of advanced manufacturing implementation in companies, in industrial sectors, and in regional industrial parks. In this work we describe the process of building and validating a model for assessing the maturity level of manufacturing plants in advanced manufacturing implementation. The assessment tool is based on four dimensions (content areas) comprising 14 subdimensions that are relevant to advanced manufacturing. The questionnaire includes statements designed for assessing the status and the maturity of a manufacturing unit relative to Industry 4.0 ideals. The outcomes of this work provide a helpful tool and roadmap for promoting advanced manufacturing at the company level, the sectorial and regional level, and the national level.

Systems Engineering in the Fourth Industrial Revolution: Big Data, Novel Technologies, and Modern Systems Engineering, First Edition. Edited by Ron S. Kenett, Robert S. Swarz, and Avigdor Zonnenshain.

21.1 Introduction to Industry Competence and Maturity for Advanced Manufacturing

During the last decade, industries in advanced economies have experienced significant changes in their engineering and manufacturing practices, processes, and technologies. These changes have the potential to create a resurgence in their engineering and manufacturing activities. This phenomenon is often referred to as the Fourth Industrial Revolution or Industry 4.0. It is based on advanced manufacturing and engineering technologies, such as massive digitization, big data analytics, advanced robotics and adaptive automation, additive and precision manufacturing (e.g. 3D printing), modeling and simulation, artificial intelligence (AI), and the nano-engineering of materials. This revolution presents challenges and opportunities to the systems, manufacturing, and process engineering disciplines. Several authors have discussed approaches to assess organizational readiness to advanced manufacturing challenges (President's Council of Advisors on Science and Technology 2014, PwC 2016, McKinsey and Co. 2016, Schuh et al. 2017). Under Industry 4.0, systems have access to large types and numbers of external devices, and to enormous quantities of data, which have to be analyzed through advanced data analytics (Kenett and Shmueli 2016; Kenett et al. 2018).

To help companies advance on the roadmap toward Industry 4.0, we developed an assessment tool that assesses the current maturity level of a specific or group of companies and sketches a set of focused areas for the companies to pursue in their effort to deploy Industry 4.0 methods. We call this model Industry competence and Maturity for Advanced Manufacturing (IMAM). The next section is an introduction to IMAM.

21.2 Maturity Levels Toward the Fourth Industrial Revolution

Companies who want to improve the way they are doing their business using opportunities implied by the fourth industrial revolution, need to assess (formally or informally) their maturity levels in different business and operations areas. Such an assessment can also serve the companies, as a tool for benchmarking. The IMAM framework consists of an assessment tool based on the Software Engineering Institute's (SEI) Capability Maturity Model Integration (CMMI) approach. It is specifically designed for assessing the maturity level of a company in the area of advanced manufacturing and engineering. For our CMMI maturity level assessment in systems and software development see Kenett and Baker (2010). Appendix 21.A reviews such maturity models in general. The model deals with different application areas relevant for advanced manufacturing and engineering, including:

1. Strategy and long-term planning for advanced manufacturing
2. Human resources for advanced manufacturing
3. Communication with customers and the market
4. Processes in manufacturing
5. Processes in engineering
6. Business processes
7. Processes in maintenance
8. Logistics processes
9. Processes in the supply chain
10. Processes in product life cycle
11. Information and knowledge management
12. Processes in cyber assurance
13. Investment in infrastructure and equipment
14. Actual improvement outcomes and results.

In each of these areas, several possible actions and activities can be considered by companies aiming toward the advanced maturity level. For example, in the area Processes in Engineering, the following activities can be considered:

- Our company has an engineering planning system based on information technologies.
- Our company's tools of engineering design are computerized.
- Our company's processes of engineering design include modeling and simulations in the framework of model-based engineering (MBE).
- Our company uses simulations for statistical design of experiments (DoE) as part of the design and engineering processes.
- Our company smartly uses 3D printing for fast prototyping and for designing molds and dies.
- During the development and engineering of new products and systems, our company considers the use of new, advanced, and innovative materials which are improving the products, the systems and the manufacturing processes.

Similar maturity level assessment tools can be developed for systems engineering areas in companies aiming to upgrade their systems engineering processes so that they meet the challenges of the fourth industrial revolution. Our model supports companies in assessing their strengths and weaknesses and helps prepare an improvement plan. It also provides companies with a tool for assessing their actual improvements and achievements. In addition, it is an effective benchmarking tool.

To the best of our knowledge, the current literature offers no measure to capture the latent construct of a single IMAM. Our first step was to develop the content of the model and validate its construct.

The scale is derived from the model and can be used for a self-assessment by management. The scale can also serve as an assessment tool for professional consultancy activities involved in the process.

21.3 The Dimensions of Industry Maturity for Advanced Manufacturing

At the specific industrial company level, we consider the competence and maturity for advanced manufacturing as a multidimensional latent construct that can be characterized along four dimensions: (i) value chain, (ii) infrastructure, (iii) monitoring and control processes, and (iv) engineering processes. These dimensions, together, make up an individual-level characteristic that can be understood as a single industrial organizational construct, reflecting the competence and maturity for advanced manufacturing implementation.

Each of these four dimensions consist of subdimensions that were identified through an extensive process based on literature, consultation with experts, and reports by international entities working on promoting the implementation of advanced manufacturing.

Fourteen subdimensions evolved from the research, hence characterizing the content of the questionnaire. For each subdimension, we developed a self-report questionnaire, based on statements (items) measured on a five-level Likert-type scale. The 11th subdimension is about information and knowledge management and builds on the information quality framework presented in Kenett and Shmueli (2016) and Reis and Kenett (2018). We added a concluding item stating: "It may be said that, in general, the advanced manufacturing status of our company is in level..." (1–5 in Likert scale).

The subdimensions are:

1. Strategy and long-term planning for advanced manufacturing (sample item: "Our company defined quantitative strategic goals for the development of advanced manufacturing.").
2. Human Resources for advanced manufacturing (sample item: "Our company identified needs/gaps for developing the human resources for advanced manufacturing.").
3. Communication with customers and the market (sample item: "Our information on market needs is managed by an updated, active, and controlled digital repository.").
4. Manufacturing processes (sample item: "Our company has a production planning system based on information technologies.").
5. Engineering processes (sample item: "Our company smartly uses 3D printing for fast prototyping, and for designing molds and dies.").
6. Business processes (sample item: "Our company uses data and findings of the Business Intelligence (BI) system for business decisions.").
7. Maintenance processes (sample item: "Our company uses data on the machines status for Predictive Maintenance (PM) or Conditioned Based Maintenance (CBM).").

8. Logistic processes (sample item: "Our company smartly uses robots and automation equipment for logistics, storage and transportation.").

9. Supply-chain processes (sample item: "Our company's information communication with suppliers is computerized and Internet based.").

10. Product life cycle processes (sample item: "Our company has an information technologies-based system for analyzing the behavior of products families.").

11. Information and knowledge management (sample item: "Our company effectively uses information and data for making managerial and business decisions.").

12. Cybersecurity processes (sample item: "Our company has a plan for coping with cyber threats and risks.").

13. Investments (sample item: "Our company has an investment program for infrastructure and equipment for advanced manufacturing which is covering most of the needs.").

14. Actual improvement results (sample item: "Our company measures actual improvement of its productivity.").

The full IMAM questionnaire is presented in Appendix 21.B.

21.4 Validating the Construct of the Scale

External Content Validation

To validate the questionnaire, we conducted an experts' survey with 18 international experts from the USA, Italy, Israel, Holland, Portugal, Germany, and France taking part. We first posed the following open question:

"How would you characterize an Industry 4.0 mature industrial organization? Please kindly indicate all the various dimensions of the maturity status that should be included when examining such an organizational status (a short, bulleted answer will be appreciated)."

The responses we received were mostly about infrastructure (43%) and value chain (32%). Less frequent responses were about monitoring and control (13%) and engineering (12%). We conducted text analytics assessments to these free text responses and found, overall, good coverage of the questions on the questionnaire question. Using association rule analysis (Kenett and Salini 2008; Kenett and Raanan 2011), we found the *confidence* (proportion of comments where the second term followed the first term in the item set pair) in the item set "collect data" to be 100% with a *lift* of 5.5. The item sets "product equipment" and "technology use" had confidence and lifts of (45%, 10.1) and (38%, 3.9), respectively, indicating the "stickiness" in these terms (a lift higher than 1 means that they go together much more often than by chance alone). Overall, we found good coverage of the questionnaire's 81 items by these responses.

Additional directions indicated by outlying comments in a singular value decomposition (SVD) of the document term matrix (DTM) were considerations of smart tools with forecasting abilities and maturity of practice aspects.

Secondly, we asked the experts (after their answer to the first question) "Would you agree that the following dimensions characterize Industry 4.0 mature organizations?" followed by the list of the 14 dimensions of the questionnaire. The answers were on a four-level Likert scale (totally disagree; disagree, agree; totally agree). The average agreement levels, for all 14 dimensions, ranged between 3.26 (lowest) and 3.79 (highest), i.e. between "agree" to "totally agree." The average agreement levels of Value Chain was 3.43, Infrastructure 3.45, Monitoring and Control 3.51, and Engineering 3.53.

The external content validity evaluation by the experts revealed high agreement to our questionnaire dimensions and items and did not indicate any missing dimension. This process established the external content validity of the IMAM questionnaire.

Internal Validation

Following development of the IMAM questionnaire, we contacted members of the executive committee of the Manufacturing Association of Israel in Northern Israel and asked then to fill an Excel file with the questionnaire. Each respondent represented a separate industrial company. The companies significantly vary in number of employees, facilities, revenues, and field of activity. Fifteen companies participated in the process. We followed DeVellis (2012) for the validation process, while observing the conditions in which exploratory factor analysis (EFA) can yield good quality results for N below 50, as stated by De Winter et al. (2009).

Validating the Scale Construct

In this validation exercise, we followed Adres et al. (2016), and De Winter et al. (2009). We fist checked the reliability of the items of each subdimension. We found that Alpha Cronbach levels in all subdimensions were high (more than 0.858 for thirteen and 0.703 for one). The reliability for each of the 14 subdimensions was, therefore, determined to be very good.

Secondly, we calculated the mean of the scores for each subdimension to be the score for the subdimension.

Thirdly, we conducted an EFA that converged to the four expected dimensions, (i) value chain, (ii) infrastructure, (iii) monitoring and control processes, and (iv) engineering processes. The loadings are much higher than the accepted level of 0.4. The results of the EFA are shown in Table 21.1.

Fourthly, we conducted a reliability test for the items of each of the four dimensions. The results are shown in Table 21.2 and indicate a very good reliability.

Table 21.1 Exploratory factor analysis.

Rotated component matrix	Component			
	1	2	3	4
Market communication	0.782			
Business processes	0.789			
Logistics	0.776			
Supply chain	0.966			
Product life	0.642			
Information and know-how	0.742			
Strategy		0.963		
HR		0.822		
Manufacturing		0.643		
Investment		0.757		
Maintenance			0.942	
Cybersecurity			0.745	
Improvement			0.742	
Engineering				0.979

Extraction method: Principal component analysis. Rotation Method: Varimax with Kaiser normalization.

Table 21.2 Dimension reliability.

Dimension	Cronbach's alpha
Value chain	0.919
Infrastructure	0.936
Control	0.786
Engineering	One item

Fifthly, to see if all four dimensions are loaded onto one single second-order factor, we conducted the analysis shown in Table 21.3. All four dimensions converge to one multidimensional latent construct to be the IMAM scale. We also checked the reliability between the four dimensions that showed a high Alpha Cronbach index of 0.751.

Conclusions from Validation

We validated the construct of the IMAM, which comprises four dimensions: (i) value chain; (ii) infrastructure; (iii) monitoring and control processes;

Table 21.3 Second-order factor analysis.

Component matrix	Component 1
Value chain	0.886
Infrastructure	0.893
Control	0.763
Engineering	0.488

Extraction method: Principal component
analysis.

and (iv) engineering processes. The subdimensions of value chain are: market and customer communication, business processes, logistic processes, supply-chain, product life cycle, information and knowledge management. The subdimensions of infrastructure are: strategy, human resources, manufacturing processes, investments. The subdimensions of monitoring and control processes are: maintenance processes, cybersecurity processes, actual improvement results.

The total score of is the mean of the scores of these four dimensions. The score for each dimension is the mean of the scores of the relevant subdimensions. This construct is very useful as it gives the "big picture" along with specific subdimensions scores that need to be dealt with.

Following De Winter et al. (2009) we infer that the variance among the 15 companies that participated in the study, together with the high loadings in the EFA and the limited number of dimensions, indicate good quality results.

21.5 Analysis of Assessments from Companies in Northern Israel

The IMAM Scale – A General Assessment

We found significant correlation between the self-general assessment of a company and the general IMAM score, as shown in Figure 21.1. The Pearson Correlation coefficient is 0.745 ($p = 0.002$). Interestingly, for 12 companies the general score is higher than the self-general assessment.

Comparing the Subdimensions Scores of each Dimension for the Fifteen Companies

Figure 21.2 shows the subdimensions of the value chain dimension. Figure 21.3 shows the subdimensions of the infrastructure dimension. Figure 21.4 shows the subdimensions of monitoring and control processes. Figure 21.5 shows

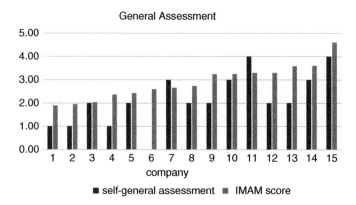

Figure 21.1 General assessment score.

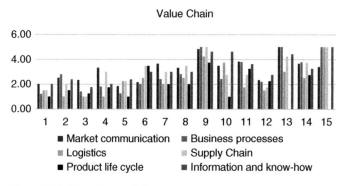

Figure 21.2 Value chain subdimensions scores.

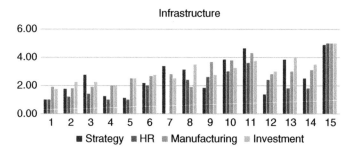

Figure 21.3 Infrastructure subdimensions scores.

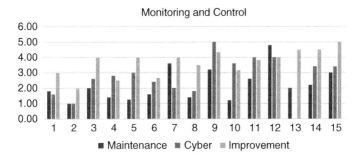

Figure 21.4 Monitoring and control subdimensions scores.

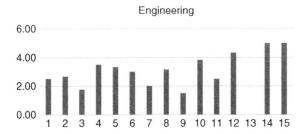

Figure 21.5 Engineering scores.

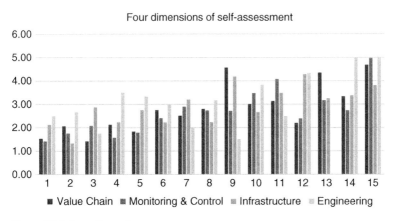

Figure 21.6 Four dimensions scores.

engineering processes. Figure 21.6 shows comparison of the four dimensions for the 15 companies.

Descriptive Statistics

The analysis of means, medians, and standard deviation for self-general assessment, IMAM score, and the four dimensions is shown in Table 21.4. We must

Table 21.4 Descriptive statistics.

	Self-general assessment	IMAM score	Value Chain	Infrastructure	Monitoring and control	Engineering
N Valid	14	15	15	15	15	14
Mean	2.29	2.8998	2.8174	2.6728	2.9311	3.1488
Median	2.00	2.7328	2.7500	2.7125	2.8667	3.0833
Std. Deviation	0.994	0.74718	1.04668	0.96802	0.82522	1.10852
Range	3	2.72	3.27	3.56	2.93	3.50
Minimum	1	1.89	1.41	1.41	1.33	1.50
Maximum	4	4.61	4.68	4.97	4.27	5.00

bear in mind that self-assessment may be affected by overcriticism on one hand and by social desirability on the other hand.

However, it should be noted that all scores of means and medians are less than the mid-point of 3 (except for engineering). This pinpoints the need for general improvement in the industry, hence this finding calls for planning a national policy.

21.6   Identifying Strengths and Weaknesses

In analyzing the data, we use a methodology and tools developed to emphasize areas for improvement and areas of excellence (Kenett and Salini 2011). A basic element in this analysis is the computation of the proportion of "1" + "2'"rating, labeled Bot1 + 2 and the proportion of "5," labeled TOP5. The analysis shown below builds on two standard statistical methods, control charts for proportions (p-charts) and analysis of variance (ANOVA) with Student's t-tests for paired comparisons, controlled for multiple comparisons. The analysis was performed with JMP version 14.2. The control charts show 2 sigma and 3 sigma zones, the ANOVA plots use circles to indicate the outcome of the paired comparisons. Data groups considered equivalent are marked using nonoverlapping circles. By clicking the top circle, one can identify the groups with are outstandingly high. For more on this analysis see Kenett and Zacks (2014).

BOT1 + 2 represents item with low scores, very high BOT1 + 2 indicates a weakness. TOP5 measures areas with high implementation level, very high TOP5 indicates a strength.

By analyzing BOT1 + 2 and TOP5 proportions across dimensions, subdimensions and items, one can identify areas of strength and weaknesses. Figure 21.7 presents the TOP5 for the 81 items, across all respondents.

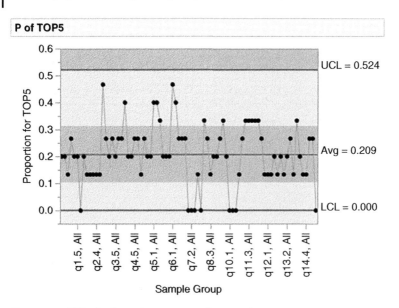

Figure 21.7 TOP5 for the 81 items, across all respondents.

Figure 21.8 presents TOP5 aggregated by subdimension with significantly high values marked in red. These correspond to relative strengths.

Figure 21.9 presents the BOT1 + 2 for the 81 items, across all respondents.

Figure 21.10 presents BOT1 + 2 aggregated by subdimensions with significantly high values marked in red. These correspond to relative weaknesses.

Based on this analysis of 15 respondents representing industry in Northern Israel, we identify as strengths:

S1: Communication with customers and the market.
S2: Processes in engineering.
S3: Processes in the supply chain.
S4: Information and knowledge management.

The weaknesses identified are:

W1: Processes in maintenance.
W2: Processes in product life cycle.

21.7 Summary

In this chapter, we present the process of building and validating a model for assessing the maturity level of plants in advanced manufacturing: IMAM. The assessment tool is based on four dimensions (content areas) comprising 14

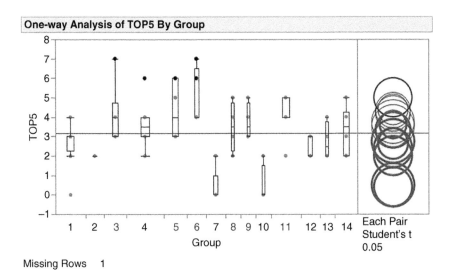

Missing Rows 1

Figure 21.8 TOP5 aggregated by sub dimension with significantly high values.

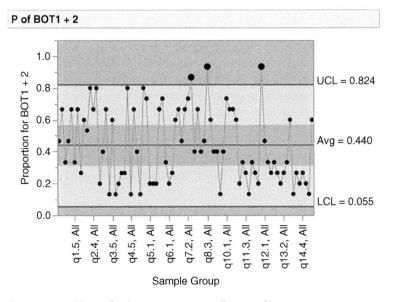

Figure 21.9 BOT1 + 2 for the 81 items, across all respondents.

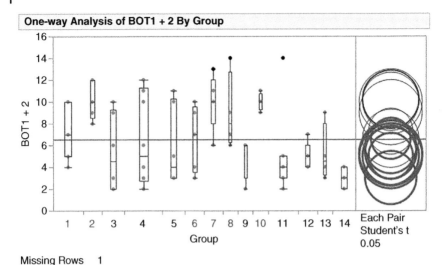

Figure 21.10 BOT1 + 2 aggregated by subdimensions with significantly high values.

subdimensions that are relevant to advanced manufacturing. The questionnaire includes statements designed to assess the status of a manufacturing unit relative to Industry 4.0 ideals. To validate this model, we performed a pilot with 15 companies; this assessed their advanced manufacturing status with the survey questionnaire listed in Appendix 21.B.

Based on findings from this data, we validated the model by conventional methods. We therefore offer a validated model that can be used to assess the maturity level of companies for advanced manufacturing performance. It is a tool that can be used for implementing advanced manufacturing at the plant level and, also, at the sectorial, regional, and national levels. The findings of this pilot study provide us with an assessment of the overall strengths and weaknesses of the companies which participated in the survey. These are listed in Table 21.5.

As stated above, with the questionnaire and the model we developed, it is possible to achieve two major goals in efforts to implement advanced manufacturing.

1. Goal 1: Assessing the organizational maturity level of a specific company and positioning its level on a 1–5 maturity ladder. IMAM also helps companies design an advanced manufacturing program based on its strengths and weaknesses; it helps, too, the company assess its progress along the maturity ladder.
2. Goal 2: Identification of regional strengths and weaknesses in the dimensions of advanced manufacturing. This can be done at the regional level as well as at the national level and for different industrial sectors.

Table 21.5 Strengths and weaknesses of industrial group with respect to Industry 4.0 implementation.

Strengths	Weaknesses
Communication with the customers and the market	Strategy and long-term planning for advanced manufacturing
Engineering processes	Human resources for advanced manufacturing
Processes in the supply chain	Processes in maintenance
Information and knowledge management	Processes in product life cycle

The validated questionnaire can be completed individually, by management team members, or by consensus in a group discussion. Besides self-declared questionnaires, companies can use experts to help facilitate the self-assessment or to provide an independent assessment.

The analysis shown above can be implemented with various statistical analysis software platforms, our use of JMP is only one such example.

The same approach of maturity assessment can be used to assess the maturity level of advanced systems engineering in companies. The same methodology with different dimensions can be recommended and applied.

Acknowledgments

This research was carried out at the Samuel Neaman Institute, Technion, Israel and did not receive any specific grant from funding agencies in the public, commercial, or not-for-profit sectors.

References

Adres, E., Vashdi, D., and Zalmanovitch, Y. (2016). Developing and validating the new individual's level of globalism (ILG) scale. *Public Administration Review* 76 (1): 142–152.

De Winter, J.C.F., Dodou, D., and Wieringa, P.A. (2009). Exploratory factor analysis with small sample size. *Multivariate Behavioral Research* 44: 147–181.

DeVellis, R.F. (2012). *Scale Development: Theory and Applications.* Newbury Park, CA: Sage Publications.

Kenett, R.S. and Salini (2008). Relative linkage disequilibrium applications to aircraft accidents and operational risks. *Transactions on Machine Learning and Data Mining* 1 (2): 83–96.

Kenett, R.S. and Baker, E. (2010). *Process Improvement and CMMI for Systems and Software*. Taylor and Francis, CRC Press.

Kenett, R.S. and Raanan, Y. (2011, 2011). *Operational Risk Management: A Practical Approach to Intelligent Data Analysis*. Wiley.

Kenett, R.S. and Salini, S. (2011). *Modern Analysis of Customer Surveys: With Applications Using R*. Wiley.

Kenett, R.S. and Shmueli, G. (2016). *Information Quality: The Potential of Data and Analytics to Generate Knowledge*. Wiley.

Kenett, R.S. and Zacks, S. (2014). *Modern Industrial Statistics: With Applications in R, MINITAB and JMP*. Wiley.

Kenett, R.S., Zonnenshain, A., and Fortuna, G. (2018). A road map for applied data sciences supporting sustainability in advanced manufacturing: the information quality dimensions. *Procedia Manufacturing* 21: 141–148.

McKinsey and Company (2016). Industry 4.0 at McKinsey's model factories: Get ready for the disruptive wave. https://www.mckinsey.com/business-functions/operations/how-we-help-clients/capability-center-network/overview (Last accessed 8 July 2019).

OSD (2011). Manufacturing Readiness Level (MRL) Deskbook. https://www.dodmrl.com/MRL_Deskbook_V2.pdf (Last accesssed 8 July 2019).

President's Council of Advisors on Science and Technology (2014). Accelerating U.S. Advanced Manufacturing [Report to the President]. https://www.manufacturingusa.com/sites/prod/files/amp20_report_final.pdf (Last accessed 5 July 2019).

PwC (2016). *Industry 4.0: Building the digital enterprise*. https://www.pwc.com/gx/en/industries/industries-4.0/landing-page/industry-4.0-building-your-digital-enterprise-april-2016.pdf (Last accessed 5 July 2019).

Reis, M. and Kenett, R.S. (2018) Assessing the value of information of data-centric activities in the chemical processing industry 4.0, *AIChE Journal*, 64 (11). https://onlinelibrary.wiley.com/doi/abs/10.1002/aic.16203 (Last accessed 5 July 2019).

Schuh, G., Anderl, R., Gausemeier J., ten Hompel, M., Wahlster, W. (Eds.) (2017). Industrie 4.0 Maturity Index: Managing the Digital Transformation of Companies [acatech study]. https://www.ptc.com/-/media/Files/PDFs/IoT/acatech_STUDIE_Maturity_Index_eng_WEB.ashx?la=en&hash=C58F5B5372F90785F5C152FFD2F1B7409F584840 (Last accessed 5 July 2019).

Westerman, G., Bonnet, D., and McAfee, A. (2014). *Leading Digital. Turning Technology into Business Transformation*. Boston, MA: Harvard Business Review Press.

21.A A Literature Review on Models for Maturity Assessment of Companies and Manufacturing Plants

21.A.1 General

Maturity assessment of companies and manufacturing plants, in different areas, is a common practice in management. Such and assessment can be an introductory step for general or specific improvement initiative. It can help in locating areas for improvement, and it assess periodically progress of the improvement process. There are several families of maturity models, such as:

- Models that are based on CMMI which was originally developed by the Software Engineering Institute (SEI) for assessing the maturity levels of software engineering in companies.
- Models that are based on assessing readiness level which were originally developed by the DOD (US Department of Defense).
- Models which are based on programs of national quality awards like the American Malcolm Baldrige National Quality Award (MBNQA) or the European Foundation for Quality Management (EFQM).
- Models for assessing the maturity of organizations in specific areas like production, engineering, and computers

This literature review presents short descriptions of the various models, with a focus on the area of advanced manufacturing.

21.A.2 CMMI – Capability Maturity Mode Integration

The CMMI model was originally developed as the capability maturity model (CMM) for assessing the maturity levels of software engineering in companies on a five-level scale. It was developed by the SEI at Carnegie Mellon University (CMU). CMM was used to develop the managerial and professional awareness that software engineering should include structural processes, so that the developed software will meet the functional requirements and will meet budget and schedule constraints. This model is used today for different professional areas, like: systems engineering, project management, human resources (people CMMI), etc. More details on the CMMI model can be found elsewhere (Kenett and Baker 2010).

In this model there are two main parts:

- Definition of focus and process areas, which are included in the maturity assessment, like: basic project management as focus area, and requirements management, project planning and project monitoring and control as process areas.

- Definition of maturity levels in each area by appropriate statements for the corresponding six maturity levels: incomplete, performed, managed, defined, quantitatively managed, optimizing.

The maturity assessment can be performed as self-assessment by internal experts' team, and by external assessors. It always includes collection of supporting data and evidences.

Over time, hundreds of organizations over the world went through a CMMI assessment as a maturity evaluation. The IMAM approach builds on this methodology for assessing the maturity level of organizations in advanced manufacturing implementation.

21.A.3   Models for Assessing Readiness Levels

These models were developed by the acquisition system of the American DOD for defining in acquisition tenders the needed readiness level from potential suppliers in manufacturing and development projects.

The first model that was developed in this framework is the Technical Readiness level - TRL which is aimed at projects of development and system engineering. The model defines nine levels of basic principles observed and reported for readiness of the project – TRL1 through TRL9, such as:

TRL1 – Basic principles observed and reported.

TRL2 – Technology concept or application formulated.

TRL3 – Experimental and analytical critical function and characteristic proof of concept.

Using the same framework, a model for assessing the readiness for manufacturing was developed. The Manufacturing Readiness Level (MRL) defines 10 readiness levels for manufacturing:

MRL1 – Basic manufacturing implications identified.

MRL2 – manufacturing concepts identified.

MRL3 – Manufacturing proof of concept developed.

MRL4 – Capability to produce the technology in laboratory environment.

MRL5 – Capability to produce prototype components in production relevant environment.

MRL6 – Capability to produce a prototype system or subsystem in a production-relevant environment.

MRL7 – Capability to produce systems, subsystems, or components in production environments.

MRL8 – Pilot line capability demonstrated; Ready to begin Low-Rate Initial Production.

MRL9 – Low-rate production demonstrated: Capability in place to begin Full Rate Production

MRL10 – Full Rate Production demonstrated and lean production practices in place

The MRL framework describes and demonstrates how to apply it through the whole system life cycle by the project managers and the acquisition authorities. The control of the model is included in system reviews.

Details on this framework are included in: MRL Deskbook, Version 2.0 May 2011, OSD Manufacturing Technology Program (OSD 2011).

21.A.4 Models for Assessing the Digital Maturity of Organizations

The fourth industrial revolution includes essential efforts for digital transformation of companies and plants. The digital transformation is defined as the process of business and organizational substantial and continuous change using digital technologies. Its aim is to improve the business performances, to improve the customers experience, and to implement new and innovative business models. Based on applied researches the success of digital transformation is based on the digital maturity of the organization. The digital maturity is an organizational index that indicates the amount, the depth, and the effectiveness that the organization is applying digital technologies in doing business and creating its competitiveness edge. This index demonstrates the stage of the organization in implementation of digital technologies, the quality of the skills and capabilities of the managers and workers in response of the challenges of the digital era. The dimensions of the digital maturity index are:

- The digital vision and strategy.
- The organizational culture.
- The customers experience.
- The business processes.
- The technological skills.
- The technology architecture.

Several consulting companies, such as McKinsey, PwC, IDC, Gartner, Altimeter, and Capgemini (with MIT), have developed models for digital maturity. Some of them reported lessons learned from applying their models with companies. It is clear that the maturity models are helpful for the companies in preparing the roadmap for digital transformation and in applying successfully this transformation. More details can be found in Westerman et al. (2014).

21.A.5 National Models and Standards for Assessing the Readiness of Industry

As stated previously, in many countries the transformation of industry to advanced manufacturing is a strategic trend. In this context, many countries prepared national programs for advanced manufacturing. For example, Singapore joined this effort and prepared a national index, the Singapore smart industry readiness index. This index supports the Government in Singapore in an effort to coordinate the national program for advanced manufacturing, The Government of Singapore hopes that the index will be a catalyst for companies, industries, workers, and Government to come together to prepare for and create Singapore's future in this new era of Advanced Manufacturing. The Singapore's Index includes:

- 3 Building Blocks: Process, Technology, Organization.
- 8 Pillars: Operation, Supply Chain, Product Life Cycle, Automation, Connectivity, Intelligence, Talent Readiness, Structure and Management.
- 16 Lateral Dimensions: Vertical Integration, Horizontal Integration, Integrated product Lifecycle, Shop Floor, Enterprise, Facility, Workforce Learning and Development, Inter and Intra Company Collaboration, Leadership Competency, Strategy and Governance.

The companies are assessed based on these dimensions by self-assessment or by external expert's assessment. Using the index is done in the LEAD framework:

- LEARN the key concepts of Industry 4.0.
- EVALUATE the current state of the companies.
- ARCHITECT a comprehensive transformation roadmap.
- DELIVER and sustain transformation initiative.

The national quality programs like the MBNQA in the US and the EFQM in Europe include criteria for management for excellence. These criteria are also used as maturity models for organizations assessing their level of management quality. The seven categories for performance excellence in the MBNQA framework are:

1. Leadership
2. Strategy
3. Customers
4. Measurements, Analysis, and Knowledge Management
5. Workforce
6. Operations
7. Results

Each category is defined by several dimensions that describe the organizational excellence in the relevant category. These dimensions can be applied for self-assessment and external assessment of the maturity of the organization in this category. The document of the criteria and framework for assessment of organizational excellence is included in the Baldrige excellence framework – A Systems Approach to Improving Your Organization's Performance (https://www.nist.gov/baldrige).

In similar way, the EFQM program includes nine categories for organizational excellence:

1. Leadership
2. People
3. Strategy
4. Partnerships and Resources
5. Processes, Products, and Services
6. Customer Results
7. Society Results
8. Business Results

The description and the characteristics of excellence in each category are presented by statements that can be applied for maturity assessment of organizations for management for quality and excellence. Details on the EFQM Framework are available at https://en.wikipedia.org/wiki/EFQM.

In the ISO management systems standards, such as ISO9001, ISO 14001, and ISO9004, there is a special section for self-assessment of the maturity of the organization in each of the standards requirements. It includes five maturity levels described by statements like those in the CMMI framework.

21.B The IMAM Questionnaire

1. Strategy and Long-term Planning for Advanced Manufacturing
    1.1 Our company has strategic plan for advanced manufacturing.
    1.2 Our company defined quantitative strategic goals for the development of advanced manufacturing.
    1.3 Our company defined qualitative strategic goals for the development of advanced manufacturing.
    1.4 The implementation of the strategic plan for advanced manufacturing is led by the senior management.
    1.5 Our company has a long-term plan that fosters advanced manufacturing.
    1.6 Our company has an investment program in infrastructure and equipment for advanced manufacturing.

1.7 Our company has an investment program in human resources for advanced manufacturing.

1.8 Advanced manufacturing is an important element in the innovation and entrepreneurship of our company.

2. Human Resources for Advanced Manufacturing

2.1 Our company's managers and workers have the knowledge and experience in the areas of advanced manufacturing.

2.2 Our company identified needs/gaps for developing the human resources for advanced manufacturing.

2.3 Our company has a management development program for advanced manufacturing.

2.4 Our company has a workers training program in the areas of advanced manufacturing.

2.5 Our company has a recruitment program for workers with relevant capabilities and experience in the areas of advanced manufacturing.

3. Communication with Customers and the Market

3.1 Our information on customers is managed by an updated, active, and controlled digital repository.

3.2 Our information on competitors is managed by an updated, active, and controlled digital repository.

3.3 Our information on market needs is managed by an updated, active, and controlled digital repository.

3.4 Our connections and interactions with customers (orders, deliveries, TT, feedback, etc.) are executed in digital channels.

3.5 Our information and data on the customers and the market are analyzed by quantitative and statistical advanced tools.

3.6 Insights from the information and data analysis on the customers and the market are used for business decisions.

4. Manufacturing Processes

4.1 Our company has a production planning system based on information technologies.

4.2 Our company has a real time production control system (for scheduling, configuration, and quality).

4.3 Our company's manufacturing equipment is computerized and automatic as needed.

4.4 Our company rationally uses robots and automation in the production lines.

4.5 Our company's manufacturing and operation specifications are computerized.

4.6 Our company uses a digital visualization (through cellular or wireless) to the manufacturing and operation specifications.

4.7 Our company has a computerized production quality control system.

4.8 Our information and data on the performances of the production lines (outputs, quality, and schedule) are analyzed by advanced quantitative and statistical tools which leads to continuous improvement initiatives.

4.9 Our company has a sensor system networked through the Internet of Things (IoT) for managing and controlling the production.

4.10 Our company analyses the return on investment (ROI) for the use of 3D printers in manufacturing of complicated parts.

5. Engineering Processes

5.1 Our company has an engineering planning system based on information technologies.

5.2 Our company's tools of engineering design are computerized.

5.3 Our company's processes of engineering design include modeling and simulations in the framework of Model-Based Design (MBD).

5.4 Our company uses simulations for statistical design of experiments (DoE) as part of the design and engineering processes.

5.5 Our company smartly uses 3D printing for fast prototyping, and for designing molds and dies.

5.6 During the development and engineering of new products and systems, our company considers the use of new, advanced, and innovative materials which are improving the products, the systems and the manufacturing processes.

6. Business Processes

6.1 Our company has a system for planning and controlling the business system based on information technologies – Business Intelligence (BI).

6.2 Our company uses data and findings of the Business Intelligence system for business decisions.

6.3 Our company has an information and data system on business partners (current and potential).

6.4 Our company has a system for tracking the business trends in the market.

6.5 Our company analyses business information by advanced quantitative and statistical tools, which lead to business decisions

7. Maintenance Processes

7.1 Our company has an information technologies-based system for maintenance planning of the machines and the infrastructure.

7.2 Our company smartly uses sensors for monitoring the machines and infrastructure status for predicting deterioration.

7.3 Our company uses data on the machines status for predictive maintenance (PM) or condition-based maintenance (CBM).

7.4 Our company has a monitoring system for energy use for improving the efficiency use of energy resources.

7.5 The information and data on the maintenance performances are analyzed by advanced quantitative and statistical tools, and it leads for continuous improvement initiatives.

8. Logistic Processes

8.1 Our company has an information technologies-based system for planning the logistics and transportation (internal and external).

8.2 Our company has a real-time control system of the logistics and transportation.

8.3 Our company smartly uses robots and automation equipment for logistics, storage, and transportation.

8.4 The information and data on logistics (outputs, quality, and schedule) are analyzed by advanced quantitative and statistical tools and leads to improvement initiatives.

9. Supply Chain Processes

9.1 Our company has an information technologies planning system for supply chain processes.

9.2 Our company has an information-based data and information system on suppliers and subcontractors.

9.3 Our company's information communication with suppliers is computerized and Internet based.

9.4 The information and data on suppliers and subcontractors (outputs, quality, and schedule) are analyzed by advanced quantitative and statistical tools and leads to improvement initiatives.

10. Product Life Cycle Processes

10.1 Our company has information technologies-based system for planning the lifecycle of products and systems.

10.2 Our company has an information technologies-based system for tracking and controlling the products through their life cycle (e.g. products under warranty, customers service, etc.).

10.3 Our company has an information technologies-based system for analyzing the behavior of products families.

10.4 The information and data on products life cycle (reliability, quality, and service) are analyzed by advanced quantitative and statistical tools and lead to improvement initiatives.

11. Information and Knowledge Management

11.1 Our company has information technologies based on advanced information and data systems, implemented in key processes.

11.2 Our company effectively uses information and data, through implementation of advanced analytics methods (statistical and quantitative).

11.3 Our company effectively uses information and data for making managerial and business decisions.

11.4 The information in our company is under configuration control, kept and available for real time use.

11.5 Our company has knowledge management and use of collected knowledge processes, based on information technologies and artificial intelligence (AI).

11.6 Our company has a process of organizational learning and lessons learning from successes and failures.

11.7 Our company has an effective process for improving the information base and the information systems based on smart adaptation of advanced information technologies.

11.8 Our company uses the Information Quality (InfoQ) model for assessing the status of information and data in the company and defined a road map for improvement.

12. Cybersecurity Processes

12.1 There is identification and mapping of cyber threats and risks on the processes in the company

12.2 Our company has a plan for coping with cyber threats and risks.

12.3 Methods and technologies for security as response to cyber threats are implemented according to assurance program.

12.4 There is continuous control on system withstanding cyber threats.

12.5 Cyberattacks events are analyzed, lessons are learned, and changes are implemented in the assurance systems.

13. Investments

13.1 Our company has an investment program for infrastructure and equipment for advanced manufacturing which is covering most of the needs.

13.2 Our company implements its investment program.

13.3 Our company currently analyzes the effectiveness of investment (like ROI).

13.4 Our company surveys innovative solutions for advanced manufacturing (in the market, in academia, in start-ups, etc.) and assesses relevancy to its needs.

14. Actual Improvement Results

14.1 Our company measures actual improvement of its productivity.

14.2 Our company measures actual improvement of its response time.

14.3 Our company measures its market share and its growth.

14.4 Our company measures customers' satisfaction and its actual improvement.

14.5 Our company measures the quality of its products and services and its actual improvement.

14.6 Our company measures its business performances and its improvement.

15. General

15.1 It may be said that in general the advanced manufacturing status of our company is at level…

22

Modeling the Evolution of Technologies
Yair Shai

Synopsis

Technologies never seem to cease changing and developing at an increasingly growing rate. System engineers have been challenged by this trend which affects advanced manufacturing and Industry 4.0 challenges. This chapter presents a holistic approach as the basis for scientific and engineering modeling. The holistic approach recognizes that new properties and behaviors emerge when broadening the perspective and obtaining a macroscopic observation. Holism (from Greek ὅλος, holos, a word meaning all, whole, entire, total) is the idea that "high-level" properties of a system, whether physical, technical, biological, chemical, social, economic, etc., cannot be satisfactorily determined, explained, and described in an inductive fashion, i.e. from an inductive reasoning that makes generalizations based on the behavior of the "individuals" (component parts). In contrast, a deductive "holistic" reasoning determines how the particular system of interest behaves and performs. Again, this also applies to advanced manufacturing systems and their related supply chains. It is the system as a whole that determines the way its parts behave or perform, not the other way around. The term *holism* was coined by Jan Smuts in 1927 as a concept that regards objects, both animate and inanimate, as "wholes" and not merely as assemblages of elements or parts and makes the existence of "wholes" a fundamental feature of the world. We apply here general principles of holistic reasoning to the reliability engineering and condition-based maintenance field and treat particular areas of engineering, from the reliability standpoint, on a broad holistic basis of the performance of engineering objects. This chapter stretches the boundaries of reliability engineering and system failure modeling as a new line of research capable of explaining high-level performance, life cycle characteristics, and evolutionary properties of current technologies.

Systems Engineering in the Fourth Industrial Revolution: Big Data, Novel Technologies, and Modern Systems Engineering, First Edition. Edited by Ron S. Kenett, Robert S. Swarz, and Avigdor Zonnenshain.

22.1 Introduction to Reliability of Technologies

Over the last few centuries, since the first industrial revolution, human society has harnessed science and engineering to an extensive technology race. Having synergetic relations with science, economic growth and welfare, technology and its artifacts have expressed a measure of human progress. The objective of this chapter is the quantification of some of the characteristics of technologies, while subjugating some known mathematical and engineering fields and considering their impact on Industry 4.0. The term "technology" is used vastly in modern discourse having numerous contexts and expressions. In many cases it is used to qualitatively describe a relatively advanced (newer) state compared to an inferior (older) one: technological achievement; technological gap; technological advantage; high-tech; technological breakthrough; state of the art technology; etc. These imply the common perception that newer is better; the use of a current technology dwindles and diminishes while another becomes accepted as a more desirable solution.

Even more so, new technologies and new generations appear at an accelerating pace, so much so that products, nowadays, "outlive" the technologies they represent. Common knowledge of history suffices to notice the exponential decrease in the time duration between new technological appearances. History seems to identify and record the major leaps of technological progress, i.e. the revolutions. However, the dynamics of technological change is so complex that no leading theory, until today, has been accepted as the unifying model to encompass all the characteristics of these processes in a continuous time explanatory manner.

The seemingly sudden appearance of new technologies raises fascinating questions about their origins. One may seek to identify the stimulators of innovations; wonder whether they are consequential or spontaneous, emergent or evolutionary; try to scale and measure the gap, or leap size, to the new technology or identify a transition period announcing the upcoming change. Above all, it is of primary interest to conclude whether the process is predictable, controllable or, at the least, traceable. In the same manner, once a new technology is born, our interest turns to the identification of patterns or models of the technology distribution, of the conditions for its absorption by the users and of the resistive forces participating in the process. In addition, interactions with other technologies, signs for the diminishing vitality of the technology, and the process of its extinction occupy innovators and researchers.

Many ideas, models, and theories about the evolution of technologies exist in a variety of scientific disciplines: economy, sociology, physics, biology, mathematics, and philosophy. One may notice the tendency to an interdisciplinary collaborative effort by, for example, applying classic evolution theories to the development of technologies, combining economic or social motivations with the innovative effort or considering epidemiologic models as a description of the spread of new technologies. Many of the references relate to the growing complexity of technologies and of the involving processes of innovation. Thus, ideas such as emergence, self-organization, or even chaos become plausible. There are also contradicting theories like reductionism vs. holism (Jan Smuts, 1927) regarding the behavior of new innovations – those which promote the understanding of nature through the interaction of all fundamental parts of a system vs. those who are in favor of the competition of (inanimate) species as entities of complex systems.

The sources reviewed in this chapter show the absence of a unifying scientific approach to the philosophy which identifies Technology as a unique entity, as the "product" to which all measures, interactions, processes, and behaviors apply. We notice that a technology is born, strives to spread and circulate, reaches maturity, and dies. Alternatively, a technology emerges or evolves, is adopted, self organizes to a temporal equilibrium, declines, or loses its domination, and becomes extinct. This life cycle phenomenon applies to technologies as it does to any other known entity in nature. However, defining the term technology is challenging; some scholars transfer the discussion from technology to innovation, others only relate to the artifacts, as they are the tangible representatives of technologies. This chapter suggests that technology itself is a system – a complex system, with the level of complexity becoming so high that it may even be referred to as a System-of-Systems (SoS). This definition stands hand in hand with the growing adaptability of technological applications (e.g. software-based technologies, autonomous vehicles, etc.) covered by the field of complex adaptive systems.

This new approach gains two major advantages as the starting point of the discussion. First, some well-known ideas and theories are bundled simultaneously to explain the evolutionary characteristics of technology and, second, a known engineering discipline is adopted, one with life cycle as the backbone of its principles. Adopting reliability theory is trivially recommended as a fundamental question of nature is the survivability of species, which inherently relates to life cycle characteristics. Thus, we propose a new and innovative line of research: "*Reliability of Technologies*."

The research aims to provide an adaptation of the reliability theory – from the very basic expression of life span characteristics of identical population to a holistic utilization regarding the integrated technology as a complex whole. It involves identifying stages of technologies' life cycle with conventional ones, and creating adequate terminology, such as, for example, failure modes of technologies, life duration of technologies, and other reliability inspired terms. Failure modes, for instance, may be the result of either inner forces within the SoS subsystems (e.g. economic forces, social forces, etc.) or outer pressure (e.g. interaction with other SoS, catastrophes on global scale, etc.).

The idea of nonphysical failure modes is of a major importance. We conceive the idea of technology to be a "Man–Machine" system, or rather

"Human–Artifact" system. The human factor plays a major role in the evolution of technologies: humans are the innovators, they are the entrepreneurs, the manufacturers, the dealers, the buyers, the users, they assemble the market, they have passions and fashions, and they interact through a complex social net. Thus, the cultural and sociological features, which are difficult to quantify, predict, and accelerate, compel, or at the least justify, the holistic approach. This chapter uncovers the dynamic processes of technology change in different aspects of the holistic perspective. Three quantitative models are described in Section 22.7, all showing practical and useable utilities for multidisciplinary research of technologies. Finally, our goal is the creation of a new quantitative perception of the term "Technology" and its properties.

22.2 Definitions of Technology

Many different interpretations of the term "Technology" present abstract descriptions. Several recent lexical definitions are:

- Encyclopedia Britannica
  - *The application of scientific knowledge to the practical aims of human life or to the change and manipulation of the human environment. Technology includes the use of materials, tools, techniques, and sources of power to make life easier or more pleasant and work more productive. Whereas science is concerned with how and why things happen, technology focuses on making things happen.*
- Merriam-Webster's Medical Dictionary (2002)
  - *1: The science of the application of knowledge to practical purposes: applied science.*
  - *2: A scientific method of achieving a practical purpose.*
- The American Heritage® Dictionary of the English Language, 4th Edition (2009)
  - *1a: The application of science, especially to industrial or commercial objectives.*
  - *1b: The scientific method and material used to achieve a commercial or industrial objective.*
  - *2: Electronic or digital products and systems considered as a group.*
  - *3: (Anthropology) The body of knowledge available to a society that is of use in fashioning implements, practicing manual arts and skills, and extracting or collecting materials.*

Combining those definitions, technology can be characterized by the existence of a goal *(purpose, objective)* and the use of materials and tools. It identifies the influence of technology on human society *(environment, welfare, fashion)* and notices that families of related products create *subtechnologies* or *species*.

In principle, these definitions emphasize the applicable and practical essence of technologies but generally exclude innovative activities from the integral ingredients and fail to recognize the complex interdependence of human users and products of technology.

Arthur (2007) offers a more precise definition, though comprehensive in concept: "*A technology is a means to fulfill a purpose, and it does this by exploiting some effect.*" He elaborates that the purpose is widely interpreted and may be explicit or hazy. This conceals the human relationship with technology – the purpose is of humans and the effect exploitation is for humans and made by humans. As he explains, the technology is the creation of humans to answer a human need; however, he defines the domain of technology from the point where an effect to be exploited exists, i.e. the invention and/or innovation are not included in the definition.

Another idea is presented by Ray Kurzweil (www.KurzweilAI.net). Here, a technology is "*an evolving process of tool creation to shape and control the environment.*" An evolutionary process, as may be concluded from this definition, is rooted in its own history, it maintains some sort of periodic behavior and claims that the gap between future and past technologies is sensible, or even measurable. This reference, it seems, is aware of the "human–artifact" dynamics of the process: "*The 'genetic code' of the evolutionary process of technology is the knowledge base maintained by the tool-making species.*" The genetic code defines the growing complexity, which is all about human knowledge and perception. From an economist perspective, Mokyr (2005) also states that "*technology is knowledge.*"

Sherwood and Maynard (1982) emphasize the close relations between technology, science, and engineering. They state that technology deals with carrying out engineering applications of scientific understandings. This diagnosis is important since it sketches realistic boundaries to which technologies relate. However, as far as it concerns, technology has an indirect definition: "*technology deals with*" instead of "*technology is.*"

The variety of definitions indicates a strict definition of technology is elusive and incomplete. On these grounds, this work suggests the following distinction: products or systems are merely the tangible representations of technologies, whereas the latter are, in fact, definable complex entities within the science–engineering–user domain that encompass all possible physical and human factors, e.g. social, cultural, and economic motivations.

22.3   The Birth of New Technologies

Arthur (2007), divides technological change into three phases: invention (the creation of new technologies); innovation (the commercial introduction of new technologies); and diffusion (the spreading of new technologies). He interprets invention as a process *"whereby novel technologies came into being."* In other words, radical novelties are born and this very moment is the end of some kind of pre-inventive process, which probably entailed thinking, researching, trials & errors, etc.

Evolutionary Biology

Some sources identify evolutionary behavior in the creation of technologies. The appearance of the first member of a new species somewhat resembles the birth of a technology, hence such ideas are relevant. Evolutionary biology is one of the most debatable disciplines. One such interesting discussion, the punctuated equilibrium vs. phyletic gradualism (Figure 22.1), was introduced in the 1970s by Eldredge and Gould (1972). The dominating paleontological theory at the time determined that new species arise from the slow and steady transformations of the entire population. However, fossil records did not agree with the expectation for unbroken series linking two forms by insensible gradation. Instead, it seemed that new species arise very rapidly in small isolated local populations. This suggested an alternate picture of *"punctuated equilibrium,"* i.e. *"the history of evolution is a story of homeostatic equilibria, disturbed only **rarely** by rapid and episodic events of speciation."*

Of course, they developed a scholarly paleontological argumentation; however, the importance of the theory is in the change of perception toward the records, which no longer need to be thought of as imperfect. For our matter, this theory gives rise to a more realistic description of the appearance

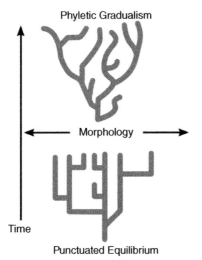

Figure 22.1 Punctuated equilibrium vs. phyletic gradualism.

of technologies. For instance, revolutionary developments of technologies (in difference with evolutionary slow developments) tend to have sudden appearance in some historical time scales.

Bak and Sneppen (1993) came up with a model, the Bak–Sneppen model, which exhibits the punctuated behavior of evolution by what they call Self-Organized Criticality (SOC). It describes the tendency of large dynamic systems to *"organize themselves into a poised state far out of equilibrium with propagating avalanches of activity of all sizes."* The model, which is claimed to be simple and robust, describes the biological evolution of ecologies by the interaction of species. They engage an entire species with a single fitness located in a fitness landscape over which the species performs adaptation. The fitness and even the landscape of a species are affected by the adaptive moves of other co-evolutionary partners in the same ecosystem. Each species quickly evolves to a local fitness maximum, a stable position, and it takes a large enough mutation, or new "better" species, to cross the fitness barrier to the next local fitness maximum. The barrier is a measure of stability and, since higher barriers correspond to higher fitness, it is also a measure of fitness. However, once a barrier is (suddenly) crossed, the entire ecosystem responds with an avalanche of adaptations of other species and more barriers crossing, until the ecosystem rests in a new self-organized critical steady state. In systems with many connections (enhanced complexity), the co-evolutionary dynamics is speeded up, thus the survival time of complex species is smaller according to this model. The authors summarize: *"The mechanism of evolution in the critical state can be thought of as an exploratory search for local better fitness, which is rarely successful, but sometimes has an enormous effect on the ecosystem."*

Hall et al. (2002) developed a new mathematical framework for the evolutionary ecology that they called the "Tangled nature" model (TaNa). It was also later published by Jensen (2004), with more relevance to the Bak–Sneppen model. They attempt to model the evolution of the highly interconnected and interdependent net comprising ecosystems. Whereas the Bak–Sneppen model is defined at the level of species, the TaNa model is an individual-based model. It imitates the laws emerging from microdynamics to the level of macrodynamics; from the reproduction, death, and mutation of individuals to the creation and extinction of species. In each step of the process, the mathematical model enables every individual to reproduce (split), mutate, or be killed. The expression for the probability of reproduction is not simple and considers many mutual influences among individuals. On the other hand, the probability of "kill" is chosen to be constant for all individuals at any position and time. From a reliability point of view, this assumption must be questioned.

At the macrodynamic level, the model shows a punctuated equilibrium behavior: *"periods of stable configurations are separated by fast transitions."*

These periods are named qESS (quasi-evolutionary stable strategies). One interesting emphasis is clear from the model: the system is in a nonstationary state. The number of transitions per time unit between qESS decreases, and accordingly the extinction rate decreases gradually in time. This phenomenon is said to match the observed fossil records and if this finding is correct, it may express a trend of growing adaptability over generations, as may be expected should species become more complex; however, no such indication is found in these papers. On the other hand, as mentioned earlier, the Bak–Sneppen model finds that the survival time of more complex systems is smaller.

Karev et al. (2005) derive an evolutionary regime that they interpret as punctuated-equilibrium-like phenomenon. They present a special solution for the birth, death, and innovation models of genome evolution. N is the number of families and $f(x,t)$ is the number of families of size x at time t. $\lambda(x)$ and $\delta(x)$ are the birth and death rates, respectively, in a family of size x. The dynamic equation for the birth and death process is:

$$\frac{\partial f(x,t)}{\partial t} = f(t, x - 1) \cdot \lambda(x - 1) - (\lambda(x) + \delta(x)) \cdot f(t, x)$$
$$+ \delta(t, x + 1) \cdot f(t, x + 1) \quad x = 0, 1..N$$

The authors argue a numerical solution is known but the high dimensionality is a technical problem. Instead, they offer an approximated continuous space diffusion function under certain assumptions, by applying Taylor's expansion truncated at its second derivative. They recognize this as a Fokker–Planck diffusion equation; however, this solution compels unnecessary restrictions on the model.

Random Walk and Lévy Flight

The power law distribution of the avalanche phenomena is related to random processes known as random walks, mainly the Lévy flight (Bak and Pacsuski 1997). Random walk is a mathematical description of successive random steps that form a trajectory. Many natural behaviors are explained by random walk, for example the route of a molecule traveling in liquid or gas or a path searched by a foraging animal. It is also a fundamental model of time-dependent random processes such as diffusion. The Lévy flight is a special case of random walk. The term "flight" is due to the instantaneous jump of the random process. The step sizes have a heavy-tail distribution and after a large number of steps, the distance from the origin spot tends to a stable distribution. Lévy distributions decrease according to the power law of the form $y = x^{-\alpha}$ when $1 < \alpha < 3$. They are rather widely used in, for example, physics, biology, economics, noise modeling, etc. and related to modeling of clustered data and even to fractals. It seems this kind of process somewhat resembles the appearance of technological revolutions.

Complex Systems

Many reviewed sources perceive the systems they investigate in terms of complex behavioral systems with a large number of participants, competitors, or individuals. Even more so, those populations have complicated co-influential relationships with an ever changing dynamics. This attitude holds for the already mentioned evolutionary biology models, for social psychology theories, for the economy driven theories about technologies, and even for the epidemiologic models of diffusion. They all sense the notion of a compound behavior differing from what might have been deduced in the level of individual ingredients – a sense of wholeness. For example, Rogers (1995) relates to the dynamic diffusion process of innovation within a social net and Hall et al. (2002) state that evolutionary ecology is realized in a web of interacting individuals and seek for the emergent properties of the entire ecology system. Naturally, technologies have salient relevance to this topic: they have huge numbers of products as their representatives, large communities of human users and dynamic processes of implementation and diffusion as well as many competing and interacting technologies. Some relevant expressions are discussed hereinafter based on, for example, Heilighen (2008), Holland (1992, 1995, 1998), Erdi (2008) and Bar-Yam (1997, 2004).

Complexity

The term is used to describe something with a large number of parts having an intricate structure and relationship. A complex system is one that consists of many interconnecting elements that as a whole develop specialized properties which cannot be predicted from the behavior of the individuals. The elements are distinct and connected; autonomous and at the same time dependent. The extent of dependency scales the measure of complexity in between a completely ordered structure (e.g. in a crystal) and a total disorder (chaos). The components of complex systems (e.g. people, firms, molecules, etc.) act as agents and are assumed to be goal-oriented striving to maximize their fitness or utility. They may have the ability to change and adapt due to evolutionary variations or by learning and implementing. Complex systems exhibit nonlinear behavior and tend to self-organize, so that new properties emerge and characterize interactions with the environment at the higher level. Some relevant examples for the evaluation of complexity include *Kolmogorov complexity* of a string x. In information theory, it is the length of the smallest program that outputs x. In other words it can measure the inherent information in x. Another example is the *Computational complexity*, which relates to the classification of computational problems according to their inherent difficulty. The measure of difficulty is either the time required by the best algorithm to solve the problem, or the required size of memory. *Information theory* uses Shannon's entropy as the measure of uncertainty of a random variable. Greater values of entropy correspond to greater uncertainties. This can be used as the measure of complexity, assuming the most certain is in the most orderly structured state. Adami et al. (2000) present an intriguing example for using entropy of information as the measure for evolution of biological complexity. Following a daring assumption that "*an organism's complexity is a reflection of the physical complexity of its genome,*" they calculate the cumulative entropy of DNA sites, each having four possible elements: *C, G, A, T*. In this case the entropy of site *i* becomes:

$$H(i) = - \sum_{j}^{C,G,A,T} p_j(i) \cdot \log(p_j(i))$$

and the complexity *C* of the organism, is given by: $C = l - \sum_{i} H(i)$ for DNA of *l* base pairs.

By using a computer simulation that imitates gradual evolutionary change they present results that show periods of transition in which the entropy drops sharply, the fitness jumps, and the complexity overshoots and then stabilizes.

Self-organization

Self-organization may be defined as the spontaneous emergence of global structures out of local interactions. The process is neither internally nor externally controlled or triggered. This is a collective process and the addition or elimination of components does not disrupt the result. In the realm of technologies, a simple example is the arrangement that developed around vehicular technologies; besides users, it includes manufacturing, distribution, transportation, garages, gas stations, etc.; this whole arrangement has (self) organized to support and distribute the existence of the technology. Bak (1997) extends the principle of SOC to describe many natural complex phenomena, such as earthquakes, mass extinctions, stock market fluctuations, and traffic jams. He uses the example of a sand-pile, which self organizes, while pouring sand grains, to a critical state after which one more grain causes a landslide, an avalanche, which, again, self organizes in a new stable form. The process of change toward organized state propagates through the system to a preferred state in the sense of "natural selection." These processes, however, have unique statistical properties and they also show how scale invariant phenomena such as fractals and power laws emerge at the critical point between phases.

Global Dynamics – Attractors

Not all individual components are able to realize their preferences; the interests of individuals are not necessarily similar. However, the interaction, using a selective "trial and error" process, drives synergetic compromise in which the individuals are mutually adapted. The more individuals are alike, the faster the system organizes to a more perfectly ordered pattern. However, the complex system, as a "higher-level" entity, also performs variations by exploring different trajectories. Thus, a system's self-organization means settling on an attractor, which is a region that the global dynamics chooses or prefers. Each of the many attractors in nonlinear systems corresponds to a particular self-organized configuration. Generally, an attractor consists of a subset of states between which the system is actively (and unpredictably) moving. For example, Bak and Sneppen (1993) claim that the critical self-organized state "*is a global attractor for the dynamics.*"

Emergence

The synergetic coordinated organized structure imposes constraints on the agents. They have to obey new rules restricted by the attractor. The organization has become a whole that is "*more than the sum of its parts.*" This super-agent has emergent properties that cannot be reduced to those of its parts. The human characteristics cannot be derived from the summation of the properties of its body organs and the properties of a metal alloy differ from the properties of its atoms. These complex system emergent properties also interact among themselves and include global, or holistic, aspects such as synergy, coherence, or robustness as regard to environmental fluctuations.

Chaos

The behavior of dynamic systems when highly sensitive to initial conditions. Though seemingly random, the behavior of such systems is deterministic. Simple sets of equations may exhibit chaotic behavior expressed in a strange appearance of trajectories within a "strange attractor."

We are interested in the relevance of chaotic behavior to the birth of technologies. The "sudden" appearance of a revolutionary technology may be the result of a chaotic phase transition that drives the emergence of a new higher level of self-organization.

Wholeness and the Implicate Order

The special, and seemingly unpredicted, event of technological invention is bound to take place in a strictly definable pattern. This is inspired by the famous physicist David Bohm (1980) and his somewhat transcendental ideas of hierarchies of implicate orders. These form the universe as a whole with total order where higher implicate orders organize the lower ones, which in turn influence the higher ones. Evolutionary developments emerge as relatively integrated wholes from implicate levels of reality, i.e. though not randomly, evolution (and, hence, also the birth of new technologies) emerges bound to universal laws.

22.4 Adoption and Dispersion of Technologies

Once a technology is born (e.g. emerges or locks in), its survivability depends on its own abilities to flourish and dominate. It means conquering the potential niches by suppressing its competitors and overcoming constraints by following certain models. Some relevant models are described hereinafter.

Social Psychology and Behavioral Theories

One of the most well-known theories is the Technology Acceptance Model (TAM), presented by Davis (1986) in order to explain, and predict, the computer usage and acceptance of Information Technology (IT). TAM considers the user's acceptance as the decisive parameter of the success or failure of an Information System (IS). Two external parameters consisting of all the system design features were later added (Davis 1989): Perceived Usefulness (PU), i.e. *"The degree to which an individual believes that using a particular system would enhance his/her job performance,"* and Perceived Ease-Of-Use (PEOU), i.e. *"The degree to which a person believes that using a particular system would be free from effort."* These parameters were developed from previous social cognitive theories such as the self-efficacy theory or the cost–benefit paradigm of the behavioral-decision theory. TAM was largely based on Fishbein and Ajzen's (1975) Theory of Reasoned Action (TRA), originated in social psychology to study the main factors of human consciously intended behavior. TRA had shown success in predicting behaviors; it was very general and *"designed to explain virtually any human behavior"* (Ajzen and Fishbein 1980). Finally, TAM adopted TRA with a refinement of adding Behavioral Intention as the integrating parameter feeding the actual use, however omitting the difficult-to-explain parameter of Subjective Norm (Davis et al. 1989). TAM was criticized later on for its incompleteness. Malhotra and Galletta (1999), amongst others, pointed out the model's lack to account for the *"social influence processes on user's behavioral intentions and attitudes toward using the technology."* They suggested Psychological Attachment as an additional external parameter affecting a person's commitment to the use of IS (Figure 22.2). The leading scholars of the theory also assimilated some improvements and performed TAM2 to include social influence processes (e.g. Subjective Norm as per TRA) and cognitive instrumental processes (e.g. job relevance, output quality and result demonstrability). Obviously, the model became very complex.

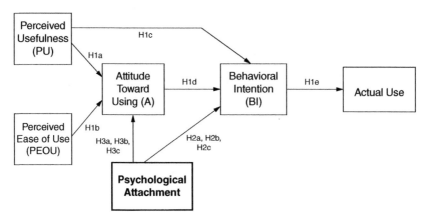

Figure 22.2 TAM with psychological attachment.

A few of the eminent figures of IT acceptance theories published a collaborative research (Venkatesh et al. 2003) which suggested the Unified Theory of Acceptance and Use of Technology (UTAUT). It was the outcome of synthesizing eight competing theories (including TRA, TAM, TAM2, Innovation Diffusion Theory (IDT), and others) to their most influential parameters. UTAUT included four constructs with four moderators that affect both the behavioral intention to use IS and the use behavior. It claimed to gain an improved ability to explain as much as 70% of the variance in the intention. However, although UTAUT turned out to be superior to its predecessors, its thinkers assumed that "*it is possible that we may be approaching the practical limits of our ability to explain individual acceptance and usage decisions in organizations.*" A few years later, Bagozzi (2007) presented an attempt of his own to unify a theory for the decision making of the adoption/acceptance/rejection of IS. He surveyed the shortcomings of TAM and claimed that "*The study of technology adoption/acceptance/rejection is reaching a stage of chaos, and knowledge is becoming increasingly fragmented with little coherent integration.*" As regard to the UTAUT, he noticed dozens of independent variables for predicting intentions and at least eight for predicting behavior. The approach considered group, cultural, and social processes together with a new idea of self-regulated processes of decision making.

A different perspective to this issue was developed by Rogers, a sociologist, and presented since 1983 as the IDT. This theory aims to predict the time frame needed for a technology to be accepted. According to Rogers (1995), "*Diffusion is the process by which an innovation is communicated through certain channels over time among the members of a social system.*" This is a dynamic process occurring within a social net while each member of this system faces subjective innovation-decision through a five-step process:

1) *Knowledge*: a person becomes aware of an innovation and has some idea of how it functions.
2) *Persuasion*: a person forms a favorable or unfavorable attitude toward the innovation.
3) *Decision*: a person engages in activities that lead to a choice to adopt or reject the innovation.
4) *Implementation*: a person puts an innovation into use.
5) *Confirmation*: a person evaluates the results of an innovation-decision already made.

IDT classifies five different groups of populations by the characteristic of Innovativeness, which describes the willingness to adopt new technologies in regard to their stage of maturity (Figure 22.3).

The persuasion step is directly influenced by several perceived characteristics of the innovation. These, in turn, influence the rate of adoption, which is a time-dimension parameter, measured by the number of members, within a system, who have already adopted the innovation within a time period. These characteristics are:

1) *Relative advantage*: may be measured in economic terms; social prestige, convenience, and satisfaction arealso important factors.
2) *Compatibility*: the degree to which an innovation is perceived as being consistent with the existing values.
3) *Complexity*: the degree to which an innovation is perceived as difficult to understand and use.
4) *Trial-ability*: the degree to which an innovation may be experimented with on a limited basis.
5) *Observability*: the degree to which the results of an innovation are visible to others.

The Social system is defined as *"a set of interrelated units that are engaged in joint problem solving to accomplish a common goal."* It is its structure, its norms, and its opinion leadership which may utterly influence the diffusion rate.

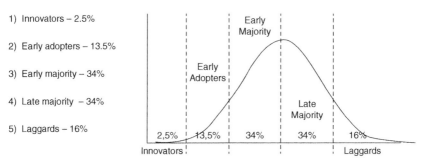

Figure 22.3 Adoption/Innovation curve.

Social Networks

Social networks, and network analysis in general, are relevant to all stages in the life cycle of technologies. The human users assemble an interacting net which, by all means, is the heart and soul of the existence of any technology. One particular interesting utility of the TaNa model for the description of human-centered complex activities is fashion; Shen and Wilde (2005) consider the probability that an individual, holding a strategy, is imitated (by an additional individual in the population) as well as the probabilities that an individual dies or mutates (chooses another strategy). As in the TaNa model, the probability to die is constant for all individuals at any time – obviously a problematic assumption. In comparison, the probability for imitation, based on interactions, is much more complicated.

Business and Economy Driven Theories

Norris and Yin (2008) discuss small and medium enterprises relating their potential global economic power. Because the resources of such enterprises are small, they need to react to market dynamics by implementing new technologies and, especially, IT and Communication Technologies (CT). The authors indicate the failure rate of these technologies which affect the survivability of those firms (i.e. failures in the adoption process). They call for effective technology adoption management and do so by ascribing 22 core constructs (variables), found in previous theories, to four key action-learning areas: commitment, end users, experience, and training. The main feature of this work is the controllability gained by these action-learned areas in order to maintain and support the adoption process of new technologies. Economists are accustomed to associate long-term economic growth with technological progress. For example, Mokyr (2005) discusses the precise connection between the first industrial revolution and the origins of the economic growth. He presents the history of technology progression and its role in the economic growth and concludes that the industrial revolution can be regarded not as the commencement of growth but as the time at which technology began to assume an ever-increasing weight in the generation of growth and when economic growth accelerated dramatically.

Utterback (1994) identifies technological change as a key factor in the creative, as well as in the destructive, forces in the growth of enterprises. He offers strategies to master innovative processes beginning with differentiating product innovation from process innovation. Companies which first innovate in their products must then innovate and improve their processes in order to grow. Although essential to large scale marketing, quality improvement and cost reduction, these process innovations hook the company to procedures, standards, and bureaucracy and cause the diminishing of next stage product innovation. This work and alike led the way to the classic book *"The innovator's dilemma,"* written by Christensen (1997), an economist, who introduced the term "disruptive technologies" as opposed to sustaining technologies. It describes a phenomenon of powerful and successful corporations driven out of business by startups with new ideas. Sustaining technologies are the consequence of incremental improvements in an already established technology. Disruptive technologies, on the other hand, evolve outside the sustaining envelope and their burst is difficult to foresee. They evolve "through" the sustaining technology until their performance is accepted by traditional markets. At this point, the established corporations suffer fatalistic consequence. Christensen distinguishes low-end disruption from new-market disruption. The first relates to customers who do not need the full performance offered in the current market, and the other to those whose needs are not served by the market. Disruptive innovation may also occur in well-established firms, however; the firms' leaders cannot identify the opportunity or do not take the risk, since they are focused in being close to their customers and are tied up by the mechanism of sustaining development.

It makes sense to attribute the theory about the "hype cycle" of technology to disruptive technologies and the way they diffuse. Fenn and Raskino (2008) identify an overenthusiasm phenomena when companies encounter new technologies, then a fall to disappointment, and slow long run adoption. This cycle is a graphic representation (Figure 22.4) of the maturity, adoption, and business application of specific technologies. Fenn and Raskino describe what drives this pattern and what could be done to avoid its danger and plan the timing of investments.

The hype cycle was criticized for not being a cycle at all and for questionable empirical support. However, such a pattern, if it applies, may cause confusion in measuring global parameters of technologies. Moore (1991, 1995) discusses why practical business plans, based on Roger's IDT, encounter difficulties in smoothly propagating through the stages of diffusion. A special gap, in high-tech technologies in particular, known as the "Chasm", causes difficulties

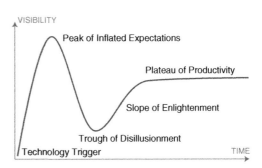

1) Technology trigger

2) Pick of inflated expectations

3) Trough of disillusionment

4) Slope of enlightenment

5) Plateau of productivity

Figure 22.4 Five stages of the hype cycle.

to cross over from the stage of early adopters to early majority. This happens due to differences in the attitude of the two populations. Late adopters are more demanding and less enthusiastic, and a mature understanding is needed in order to focus in one group of customers at a time. It should be mentioned that this chasm phenomenon applies mostly, if not only, to disruptive technologies, otherwise technologies are expected to follow the diffusion stages without abruptions. Norman (1998) uses the chasm phenomenon to investigate the possible routes that the computer industry should follow in order to cross over the gap and appeal the mass of consumers. In the view of reliability theory, an item may encounter intensive stresses in the period known as the infant mortality. These stresses may fail the product after a relatively short period of use. Accordingly, the hype cycle and the chasm of diffusion may be appreciated as the infant mortality failure modes of technologies. Other recently published examples examine the strong connection between economy and technology diffusion, and point out international spillovers of technology as the measure for technology diffusion. Eaton and Kortum (1999) suggest international patenting as the channel for diffusion of technologies. Patenting is correlated with R&D expenditure and its data are readily available.

Xu and Chiang (2005) combine patenting and world trade, embodied in capital goods, together with the technology gap between leading and following countries to determine the diffusion rate. The gap is measured by the TFP (Total Factor Productivity) ratio between two countries. Human capital and the level of education influence the catch-up of new technologies. With a very simple mathematical model, they demonstrate correlation of technology diffusion with the source of productivity growth within three income groups of countries.

Epidemic Models

The epidemic modeling approach for the diffusion of technology has been suggested in the literature. Pulkki (2008), for instance, relates to international diffusion of technologies and examines historical examples of steam and motor ship diffusion and adds an extension to two known epidemic models in order to consider the effect of international usage. He suggests that, according to the epidemic approach, the use of new technology increases as information about its existence and features becomes more widespread. He discusses two earlier such models: (i) the Bass model and (ii) the Mansfield model. The Bass model opines the spreading of information is what drives the diffusion. This spreading is achieved by the information flow among the users themselves (i.e. a clue for social networking). The Mansfield model claims that more information reduces the uncertainty about the profitability of the new technology, thus encouraging the diffusion.

An interesting biologically-based model for early epoch simulation for disease outbreak and bioterror attacks is suggested by Adams et al. (2007). The model enables the testing of intervention strategies in the first few days when a small portion of the population is infected. While classic epidemic models present global predictions, this one uses the inverse problem approach to characterize the source of a disease outbreak and, eventually, predicts the spread over a longer time and larger spatial scales. The analogy to technology "outbreak" is clear; such models may be useful for "defense" strategies against new competing technologies. Other interesting epidemic models are presented by Janssen et al. (1999). They consider epidemic models as growth models for phenomena arising abruptly in nature. They show "*new long-range determined universality classes*" of infection spreading by imposing stochastic processes, such as the Lévy flight on the infection distribution. In epidemic models, individuals may be contaminated by infected neighbors and, at the same time, go through processes of spontaneous healing and immunization. However, a long-range infection may be applied, following the examples of when flying parasites contaminate trees almost instantaneously in a widespread manner or when infected passengers (e.g. Ebola or SARS) on transatlantic flights unexpectedly transfer the disease out of the regional source. In our case, long-range processes may refer to mass media marketing of new technologies.

22.5 Aging and Obsolescence of Technologies

In general, the adoption of new technologies may show a negative picture of the extinction process of older ones, if superseding technologies drive the abandonment of their predecessors. This is a somewhat simplistic assumption since the arrangement of users communities of next and prior technologies may differ. Since technologies consist of a large number of end users and products, one may expect negative trends in the number of representatives per time unit or a spatial degradation of their existence. Technologies, being complex systems, do not necessarily lose their global vitality due to local failures. These "local" failures at the technology level can be considered analogous to loosening the molecular ties in matter. It is reasonable, therefore, to examine the dynamics of damage accumulation processes in the physical domain as the presenter of the aging process of technologies.

Aging processes put the older technology under growing stresses and eventually lead to its extinction, i.e. to the inexistence of its tangible representatives. Some indications are found in the cases of competing new technologies (e.g. disruptive technologies or revolutionary leaps); it is obvious that due to the continuous trend of technology progression and competition, there must always be a technology that eventually loses. This loss is expressed in rapid or even sudden reduction of relevant products, or rather, all related species of products as mentioned above in some of the evolutionary theories, which inherently include birth and death of individuals pertaining to a certain species in the overall dynamics of existence and disappearance. There is a point, an event or some kind of critical mass, which disrupts the continuation of the technology revival – in short, a failure mode must exist. Obviously, these failure modes need to be attributed to the domains in which stresses occur. They are expected to be found in the physical domain (e.g. materials, engineering, complexity, environment, etc.) as well as in the human factor domain (e.g. cultural, social, and psychological as per TRA, TAM, and UTAUT models).

Another work is worth mentioning. Ray Kurzweil, an inventor and futurist, coined the term "Technological Singularity" to describe an inevitable event, so he claims, which will cause *"a rupture in the fabric of human history."* In his many books (e.g. Kurzweil 2005) he uncovers a power-law-like behavior in several niches and different aspects of technological progress. He characterizes a "law of accelerating returns" to describe the growing capability of technology to foster its own next generation. This acceleration yields exponential growth of returns, while in some cases even the growth rate increases exponentially. This trend will result in a technological change so rapid and so profound that an event of "explosion" will occur. In his view, this event of singularity, acting like a "black hole" in the technology domain, has no alternative but to create a new high-level paradigm shift to a new era: the age of intelligent machines. If such a singularity occurs, it would be the ultimate extinction of technologies as we know them today. The singularity extrapolates historical events to predict the occurrence of a cataclysmic failure.

Kurzweil's futurology is challenging and a few of his actual examples are dazzling. For instance, he relates to Moore's law (Moore 1965), which predicted that squeezing electronic components onto integrated circuits would double in a constant pace. Kurzweil shows that not only Moore's law holds until today but it should also be considered as part of a larger accelerating process of computation power (i.e. the purpose) for which electronic transistors are just one phase, or paradigm, out of several. Each phase is the realization of a superseding revolutionary computer technology (electromechanical, relay, vacuum tube, transistor, integrated circuit). He presents this idea by measuring the number of calculations per second per $1000 along the years, starting long before the birth of transistors and showing the purpose is satisfied, persistently accelerating even in a logarithmic scale. The evolution of computation power is not disturbed by the revolutionary technological change once in a few decades. This kind of presentation is a genuine attempt to quantitatively predict future life duration characteristics based on historical data of the accelerating technological evolution (and revolution) and, therefore, corresponds to the vision of technology life cycle in a broad perspective.

22.6   Reliability of Technologies: A New Field of Research

Acknowledging *Technology* as a definable complex system is merely conceptual annotation unless expressed by sustainable elements. These elements ought to express the holistic essence of the technology on one hand and permit measurable properties on the other. A genuine suggestion for dividing the term *Technology* down to its fundamental ingredients calls for both simplicity and practicability. The most common expression of technology is the enabling of a production process. This process constitutes the representation of the capabilities of technology and its characteristics in the practical or feasible domain. As such, we may assert that the principal traits of technology are as shown in Figure 22.5.

<table>
<tr><td>(I)</td><td>*Raw material to be manipulated*</td></tr>
<tr><td>(II)</td><td>*Tools for production and usage*</td></tr>
<tr><td>(III)</td><td>*Adequate manpower skills*</td></tr>
</table>

Figure 22.5 Principal traits of technology.

Any technology is the combination of these three traits, which provide a unique set of characteristics, i.e. an Eigen vector. Each technology has its own ID based on different measures of these elements defined in a three-dimensional space. The proposed set is also perfectly suitable for the description of technologies in the post-production period. Users need adequate skills to operate the product, they need special tools to perform maintenance actions, and, in some cases, they need materials to energize the operation. One of the sources suggests a linkage between some purpose or need and the effect that can be exploited to satisfy it. Here, the new model implements the idea that products might be the tangible representatives of a technological effect and, as such, they satisfy a purpose or need of humans. These needs rest in the realm of the soft sciences, e.g. sociology, psychology, economics etc., emphasizing the holistic essence of the discussion. A purpose may bear a long-term role of human society while its measurable scale, i.e. users' satisfaction, changes and adapts with the technological advancement. Some difference in one or more ingredients may be identified as a technology gap. For example: transformation of music record albums to magnetic tapes compelled the change of *raw materials, tools,* and probably *skills*. This is, no doubt, a technological change which has some more advanced features as byproducts. For comparison, the transformation of Bakelite to Vinyl records was mostly, if not only, of the *raw material*. Whether this was a technological change is a debate to be resolved. However, it was a noticeable change. Another example is the recent trend in the automotive industry – internal combustion engines are replaced by electric motors; *raw materials*, facility (*tools*), and know-how *(skills)* utterly change. The principal purpose does not change (mass transportation); however, the impact on society may be dramatic and yet to be measured – environment, energy, efficiency, etc. We notice that new professions (*skills*) become necessary when new technologies emerge due to currently nonexisting required specializations. We may also relate any technology to the time period required for acquiring relevant education, or to the required level of education – if a completely new subject must be learned in order to sustain a new line of products this is clearly a new technology. Finally, we argue that:

1. The three ingredients constitute a closed group of basic building blocks of the "DNA" of technologies, sufficient to quantitatively define any technology as an entity.
2. The extent to which a change of ingredients is perceived as a technological gap depends on the resolution of the measurements, which, in turn, are subjected to interpretation of the observers or users, i.e. "calibrated" by the society.
3. This somewhat biological descriptive approach explains the nature of evolutionary technological advancement versus revolutionary technological leap; it is the minor change in the "chain" of building blocks versus radical rebuilding of a completely new DNA (i.e. technological entity) while abruptly shattering the former structures.

The three organs of technology, when materialized into products, may include minor composition differences (or different values) within different products of the same line (species of products). These species, in turn, may be regarded subtechnologies.

The structure of the currently used reliability function aimed at the assessment of the lifetime of a particular population implies, to be statistically meaningful, extensive life duration measurements for a large sample size (Kapur and Lamberson 1977). In many cases there is, indeed, a need to use an "inductive" approach, when there is an objective to understand the roles and contributions of different factors to the general distribution of interest, i.e. a need for the most accurate and the most well substantiated evaluation of the lifetime characteristic functions of different significance and different nature. This is the case, for example, when one wants to understand the role of different underlying factors affecting a design, a phenomenon or a mission, with consideration of the roles of both technical (uncertain environment, insufficient reliability of instrumentation and equipment, uncertainties in the behavior of the object-of-control, etc.) and human (swiftness of reaction, level of qualification and training, human fatigue, etc.) factors (Suhir 2009, 2012). The attributes and challenges in such an endeavor are well known: the efforts and procedures are time and labor consuming, the accuracy of the accumulated statistics is not always clear, the ability to anticipate and to consider the actual use conditions might be next-to-impossible, and the possibility to develop consistent and effective algorithms for adequate simulations of the cumulative impact of the combined action of various "stresses" might be questionable. There are, however, numerous situations, which, by their very nature, are just the opposite of such a situation. For high-level predictions and applications, thorough consideration of numerous impacts of a large number of stresses (factors) of different significance and of different natures is not a must at all. Only the bottom-line, the holistic, effect is important.

The down-to-earth objective is to indicate some practically useful applications of the holistic approach and to suggest a way for obtaining the high-level performance functions (HLPF), i.e. the holistic reliability functions, from the most general available data for a certain population. The rational for using "Performance" as a holistic descriptive term is explained in the Handbook of Performability Engineering (Misra 2008): "*Performability engineering is a holistic interdisciplinary approach to optimally engineer dependable and sustainable products, systems and services.*"

Although such an approach seems, at the first glance, quite tentative and inaccurate, such an "inaccuracy" is, in effect, an important merit of the approach. It makes it suitable for many "high-level" applications and, owing to the global nature of the approach, there is no need to design and carry out specialized experimentations aimed at the most accurate evaluation of the probabilistic characteristics, such as, e.g., life expectancy. One can also establish, using this high-level approach, that the lifespan of a product increases with an increase in the design maturity and/or in a situation when long-term reliability of a product is predicted and assured. At the system-of-systems level the HLPF-based model becomes a unique high-level measurement tool for life cycle management.

An obvious challenge has to do with the recognition and identification of different kinds of stresses that may cause the seeming, but economically meaningful, "death" of a product, as well as with the definition of what should be considered as the end of a product's life. For instance, many still functionally good and undamaged old computer accessories that, in effect, did not fail, either functionally or mechanically, are often abandoned in old drawers. These might be considered failed products due to the appearance of more attractive innovative technical solutions. Economic prosperity may encourage diminution of used cars in favor of new ones, or the threat of high pollution may force consumers to trade in their still good conventional cars for hybrid ones. Various healthcare related considerations and issues might lead to the avoidance of or, on the contrary, to the preference in, use of a particular product. The imaginary or actual consequences of electromagnetic radiation, for example, may adversely affect the cellular phone market. Hence, time-to-failure is replaced by time-to-end-of-service as the holistic measure of performability.

As far as mass-production of today's consumer products is concerned, the human factor (human preference, ergonomics, fashion, etc.) often plays a decisive role in many nonphysical socioeconomic, cultural, and social failure modes and mechanisms and their interdependencies. On the other hand, more complex products require a more general, birds-eye, holistic view assessment of their dependable life span characteristics in terms of market survivability. This statement is true, not only for some particular models or products, but also for a wide collection of related applications and designs – in other words, "species" of products. This is certainly true for consumer products (household devices, laptop computers, cars, etc.) but might be also true for more complex products of public use (spaceships, satellites, aircrafts, ships, buildings, bridges, etc.), whose failure, unlike in many consumer products, is a catastrophe and might be even associated with the loss of human lives. These products are often so costly and are produced in such large quantities that their cost effectiveness is critical (Suhir 2007). This situation might also be valid for some military objects that are produced in large quantities (shells, warheads, torpedoes, etc.). It is also very likely that the holistic approach is particularly suitable for modern consumer markets that do not lend themselves to traditional survivability assessments: software products, insurance contracts, or various service packages.

22.7 Quantitative Holistic Models

On the basis of the holistic approach, Shai (2015) established some quantitative models which fulfill the essence of the philosophy on one hand and lay the foundation for practical experimentation and research on the other. He presented three models, each of which pronounces a different aspect of the holistic perspective. The first is a temporal, or quasi-static, "snapshot" description of the performance behavior of some species of technologies based on global information regarding the population of the entire technological species. Reliability functions are derived from the balance equation of the entire population while the input data require no information on the life span of individual products. The momentary representation is beneficial for, e.g., comparative analysis of different species or of the same species in different countries. However, the performance parameters constantly change with respect to the technology's life cycle stages. Hence, the second model investigates the dynamics, not only of the time to failure distributions but also, and in particular, of the damage accumulation process, known as the fatigue or aging phenomenon. A reliability distribution function is introduced into an equilibrium equation of generalized strength space. The strength deterioration is the outcome of continuous interactions of the intrinsic product characteristics and the extrinsic environmental stresses. This process is simulated based on a modified statistical model and reveals a multidynamic behavior of the well applied three-parameters Weibull distribution. The last model examines technological revolutions in historical scale. The revolutionary events are analogously considered a series of failure times of nonideal repair actions and used for predicting future revolutionary technological events.

High-Level Performance Functions (HLPF)

As systems become more complex and diverse on one hand and more widespread for human use on the other, a greater need arises to distinguish the reliability performances, not only among specific products but also among groups of related products – species of products.

An acceptable query, for example, may wish to determine whether American vehicles are more reliable than European. Thus, American and European vehicles become the species. Moreover, once the performance of such species becomes a measure of quality, the human perception must be taken into account. An owner's decision to eliminate the use of a refurbishable exemplar is no doubt influenced by cultural, social, or economic motivations, as well as by technical, physical, or safety issues amongst others.

A reliability function $R(t)$ is defined in the reliability theory (Kapur and Lamberson 1977) as the probability to fulfill a certain mission during the time period $[0,t]$ under specified, predetermined conditions. This function is obtained from the probability density function $f(t)$ of the time to failure (TTF). Unlike the existing approaches, which require, and are based on, the statistics of the times-to-failure of individual products, the holistic approach is based on the existing and even routinely recorded and, hence, well known and readily available information (data) of the population of the "species" of interest. These data could be based on the periodic registration of the total amount of active ("living") objects and the delivering (birth) rate of the "newly born" ones. Such input data automatically and holistically contain the aggregated information on the cumulative integrated roles of all kinds of governing factors, maintenance policies, usage profiles, age distributions, and a variety of human factors during the time period of interest.

The amount of active items at the current moment $N(T)$ is just the consequence of the births and deaths of individuals that have already occurred with the delivery rate $\dot{n}(t)$ of new items $(t < T)$. The dynamics is realized in the balance equation:

$$N(T) = N_0 \cdot \int_0^\infty f(\tau) \cdot \frac{R(T+\tau)}{R(\tau)} d\tau + \int_0^T \dot{n}(t) \cdot R(T-t) dt$$

The survivability characteristics are described by the function $R(T-t)$, i.e. the HLPF, convolving with the flow of newborns, reflects the evolutionary properties of the population. The survivability of an individual is determined by both the moment of delivery t and the current time T. Here, $T_0 = 0$ is the point in time, at which the data become available, $N_0 = N(T_0)$ is the total population at the initial moment T_0, $\tau = T_0 - t$, and $f(\tau)$ is the probability density function of the population age at the moment T_0. The task now is the reconstruction of the HLPF $R(T)$ from the balance equation. However, obtaining an analytical or even a numerical solution to an inverse problem is not trivial at all. Hence, Shai et al. (2013) suggest a general numerical solution and state it is justifiable since its variations and statistical noise are small.

The unknown age distribution $f(\tau)$ of the population N_0 at T_0 may be separated to, e.g., two major effective age groups – young and old. This rough description provides the model with tuning parameters, while the actual age distribution is not of any interest. This two-group pseudo-age-distribution comprises three parameters, where $g_{(young)}$ and $g_{(old)}$ are the effective ages accordingly, measured in time units, and α is the weighing coefficient, which determines the relative portion of each age group, where the index i is an integer, and $R_i = R(T_i)$. In practice, the number of groups of effective age is preferred to be the smallest, reducing the number of the unknown parameters. Accordingly, a numerical representation of the model becomes:

$$N_i = N_0 \cdot \left(\alpha \cdot \frac{R_{i+g_{(young)}}}{R_{g_{(young)}}} + (1 - \alpha) \cdot \frac{R_{i+g_{(old)}}}{R_{g_{(old)}}} \right) + \sum_{j=1}^{i} n_j \cdot R_{i-j}$$

Shai et al. (2013) discuss practical cases where the reliability function $R(T)$ as well as the age distribution of N_0 are both unknown. These case studies deal with different species of transportation technologies: military aircraft, commercial aircraft and private vehicles. The data are freely available on the Internet. They elaborate some useful regularizations which help setting reasonable guess values and reducing degrees of freedom, applying feasible truncating criterion based on the noisy delivery data, and determining the root mean square error (RMSE) as the measure of convergence.

Case study I: private vehicles in the state of Israel. The species includes all models of "Private Vehicles" in Israel. Source: Central Bureau of Statistics of the State of Israel.

Case study II: United States air force – Fighters. The species includes models A10, F4, F15, F16, F22, F35, and F117 all together in 1996–2011. Source: United States Air Force (USAF) almanac archive.

Case study III: United States air force – Bombers. The species includes models B1, B2, and B52 all together in 1996–2011. Source: USAF magazine almanac archive.

Case study IV: Commercial Jets. The species includes the worldwide active jet fleet in 2003–2011. Source: World jet inventory, Jet information services Inc.

An iterative process for each case yielded the best set of model parameters allowing the reconstruction of the performance functions while assuring

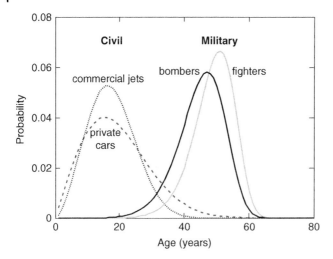

Figure 22.6 PDF of Performance (time to end of life) for different species of vehicular technologies.

adequate population balance. At this point, researchers should be most interested in comparing the species' TTF statistical characteristics. This is achieved by defining the probability density function for each case, i.e. the derivative of each performance function, by applying the relevant Weibull parameters (Figure 22.6). Quite surprisingly, two clusters of characteristics are formed – military aircrafts vs. commercial vehicles. The shape parameter, β, of the two species of military aircraft is relatively high, between 7.5 and 16, and the MTTF ranges from 40 to 50 years, whereas for both jets and private cars it is approximately $\beta \approx 2$ for the shape parameter and around 17–19 years for the MTTF.

One may presume USAF policy is influenced by the tremendous cost of new generation development programs or worldwide challenges and threats. Furthermore, technical malfunctions are overcome and rarely cause the fatal grounding of a military aircraft. On the other hand, commercial vehicles respond to an entirely different set of "stresses"; the variety and multiplicity of users, of use conditions and maintenance policies, and of economic conditions increase the statistical variance of the TTF and, hence, decrease the shape parameter. Still, it seems the physical aging of commercial vehicles has merely a partial effect in the air as well as on ground since technically they could be maintained active for decades longer than their actual MTTF. This clustering is unique for the HLPF approach and is particularly emphasized by the fact that the two species of commercial vehicles represent distinct technologies.

Modified Dynamic Reliability Model for Damage Accumulation

Conventional reliability functions express the engineer's knowledge, experimental or predicted, about specific products and predetermined environmental and use conditions (Zacks 1991). Though these functions describe the change over time of a product's probability not to fail, they perpetuate its current known characteristics. Analogously, the HLPFs, presented above, merely depict the quasi-static performance of technological species. In accordance with the stages of life cycle, reconstructing performance functions in different stages of the same species converge to different values of the model's parameters. We acknowledge this quasi-static result as an incomplete description of the nature of evolving systems. Therefore, our interest tends, at this point, to investigate the origins and nature of the ever changing performance characteristics of species of technologies. By similarity to the latter, one may investigate the dynamic process of strength deterioration in products. The analysis uses a reliability-based, and statistical–mechanics inspired, approach. It is, in fact, a generalized holistic approach, as it combines both intrinsic (strength) and extrinsic (stresses) characteristics along with the nondeterministic nature of the physics of failures.

A damage accumulation process, be it the manifestation of the aging phenomenon, has a fundamental impact on the reliability of materials. A standard reliability function $R(t)$ serves only for explicit evaluation of the element probability not to fail and enables the representation of the corresponding failure rate by the hazard rate function $H(t)$. These, however, could not be used to describe the ongoing element strength deterioration under stress. The simultaneous effect of both processes – the element deterioration and failure – was modeled by a transfer equation for the reliability distribution function (RDF) over the strength space (Ingman and Reznik 1982, 1986, 1991). The model was modified by Shai et al. (2015) to include the matter–environment interaction in terms of strength deterioration under stress. Namely, the higher the stress relative to the strength, the greater is the chance for strength deterioration:

$$\frac{\partial r(x,t)}{\partial t} = -(\lambda(x,y,t) + \omega(x,y,t)) \cdot r(x,t)$$

$$+ \int_{x'>x} \omega(x',y,t)p(x' \to x|y)r(x',t)dx' + \delta(t) \cdot r(x,t)$$

Integrating the RDF $r(x,t)$ over the strength space expresses the probability of finding an element functioning at time t, while being on a nonfailed strength level. The kernel of the model is Smoluchowski's probability of transition from the strength level x to a weaker state x', $p(x \to x')$, while the modification requires a conditioned transition $p(x \to x'|y)$ where y represents the environmental stress level. The modified transition probability is explicitly defined by:

$$
p(x \rightarrow x'|y) = \begin{cases} \dfrac{1}{x} \cdot \left(\dfrac{x}{y}\right)^{b} \cdot \left(\dfrac{x'}{x}\right)^{\left(\left(\frac{x}{y}\right)^{b}-1\right)} & \text{for } th \leq y \leq x \\ \delta(x') & \text{for } y > x \end{cases}
$$

where x is the current strength, y is the instantaneous induced stress, x' is some strength value $0 \leq x' < x$. b and th are model parameters that stand for some intrinsic characteristics. The item fails whenever the stress is greater than the strength, $y > x$, and its strength drops to zero. This probability density over the strength space satisfies the normalization condition (integrating to unit) over $[0, x]$ $\forall x, y, b; b \geq 0; x > y \geq th$. One consequent advantage of the suggested modification is the ability to track accelerating dynamics of strength deterioration along the process – as the strength decreases, the probability of transition to a further lower strength value increases for a certain given stress value.

Solving the transfer integro-differential equation analytically is not trivial since the RDF is a nonstationary nonergodic function. One approximated representation of the equation is the general Fokker–Planck equation with one variable where the term $r(x',t)$ within the integrand is expanded in a Taylor series around x, assuming x falls close to x' for relatively small stresses. However, this solution has restricted drift and diffusion characteristics. Other suggested solutions are far from general. Hence, Shai et al. (2015) implement a simulative approach to represent the full dynamics subject to initial conditions and predetermined model parameters that stand for some intrinsic characterizations of the specimen under test. An initial strength level was set for a large population of material elements in a simulation program followed by iterative steps of the probabilistic response of strength level transitions under given stress distribution. This procedure, imposed on the whole population, one element at a time, entails no need for special assumptions about the rates of transitions; all characteristics of the two processes are included based on the transition probability alone – the time to failure distribution and the strength distribution at any step of the process. The simulation results are depicted in Figures 22.7 and 22.8. The time to failure distribution in Figure 22.7 is shown for two different values of the model parameter b.

A Weibull distribution was fitted to the strengths of unfailed items at each cycle of the simulation; here, a three-parameters distribution. These distributions best represent the deteriorating population after statistically disengaged from the original strength value, namely the daughter distributions. Figure 22.8 shows probability density distributions of strength after different numbers of cycles and the fitted Weibull models accordingly. The bimodal behavior is noticed on the right-hand curve since the original strength value was set to 100 arbitrary strength units.

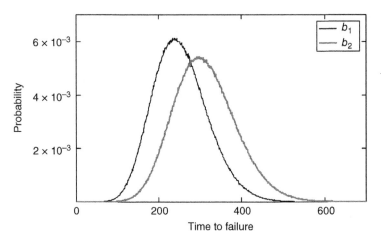

Figure 22.7 Time to failure distribution (simulated).

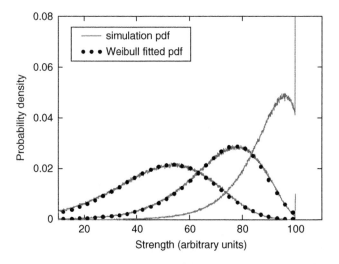

Figure 22.8 Strength distributions after some load cycles (simulated).

Shai et al. (2015) also figuratively describe the statistical dynamics by following the trends in each Weibull parameter along the time axis and as a function of the model parameters b and th. The results clearly show unique paths of change for every different set of stress–strength relationships. They demonstrate the model's applicability based on experiments conducted on an actual population of fiber optics (Ingman et al. 2007) for some behaviors of damage accumulation and reliability.

The fact that pure physical approaches are unable to describe the probabilistic nature of the process is well established. The RDF model combines probabilistic failure occurrence and physical strength deterioration where the statistical properties of which follow a diffusion-like pattern in time. By implementing the modification described above, the dynamics of actual damage accumulation finds expression in a fully dynamic probabilistic behavior, through which all three Weibull parameters of the characterizing strength distribution change in time. This feature is new to previous analyses; the change of the shape parameter represents a continuous change of elasticity characteristics throughout the deterioration process.

Moreover, the modified model is both general and robust in the sense that, on one hand, it considers the interaction of the item strength with the environment at any given time and, on the other hand, it allows the researcher to tune the dynamics according to the material's characteristics and the nature of its unique fatigue/aging mechanism. In other words, the model is both generic and flexible for a variety of environments and materials such as brittle materials, e.g. glass, ceramics, steel, etc., silicon devices, fiber optics, and others. It follows from the discussion that routes of damage accumulation cannot be assumed to be predictable according to other cases; each constellation yields a unique path of deterioration. By analogy, it is suggested this model also applies to the aging process of species of technologies, for which its implementation requires adequate, currently undefined, metrics of strength and stress measures.

From Product Reliability to Reliability of Technologies

A bird's-eye view on the history of technologies identifies them as systems that support needs or purposes. As such, a recorded event of technology swap, instantaneous in historical scaling, may be considered revolutionary. A reliability-based approach suggests such an event is either a critical failure of the older technology (i.e. its "death") or an initiator of an accelerated decaying phase, caused by numerous interrelated stresses of different kinds and origins (physical, social, cultural, technological, etc.). These events are the consequence of numerous aggregated individual user–product decisions regarding the representatives of that declining technology, encapsulating the averaged dissatisfaction with the previous technology. However, the survivability characteristics of the new substituting technology are not independent of history; the appearance of the new technology is analogous to a repair action, not necessarily to the state of "as good as new," i.e. users' expectations do not remain unchanged and they may increase or decrease as a response to the new related technologies. In that sense, consecutive technology swaps, all supporting the same purpose, define a time series of so-called failures. This time series is used to recover the generalized macro-reliability parameters at the technology level. The results of such test cases may be used for further research and philosophical argumentation regarding the way technologies evolve and, in particular, for the prediction of the future timing of technological revolutions. The evolutionary processes, regarding technological change, comply with the holistic approach very well. It contains intrinsic elements of social psychology behavior, in which the expectations and satisfaction of users are adjusted to newer features of the updated technology.

A generic analytical model is shown in the next paragraph. It enables the calculation of the Mean Time Between successive Failures (*MTBsF*) in the case of minimal nonideal repair, i.e. when the system is restored to the state statistically identical to the one prior to the failure event (Rausand 2004), namely "as bad as old." This model, based on the very basic reliability function, is most beneficial for life cycle cost calculations of realistic, generally complex, systems. The model applies to any distribution of time to first failure (TTFF) and shows an elegant Poisson-like formation. This behavior applies to complex and expensive repairable systems (e.g. computers, automobiles, airplanes, ships, etc.), where the contribution of one replaceable new component to the system's reliability is negligible, i.e. the system's age remains the same as it was before it failed.

The *MTBsF* model, together with historical records of technology swap events, allows the reliability characteristics of a series of successive technologies to be restored. Based on the nonhomogeneous Poisson process (NHPP), Epstein and Weissman (2008) and others show expressions for the time-dependent CDF (cumulative distribution function) and PDF (probability density function) of the time to the nth failure. Ulmeanu and Iunescu (2007) take another step and show a closed expression for the expected time between the nth and $(n+1)^{\text{th}}$ failures (*MTBsF*$_n$). A different and genuine approach, free of presumptions and relevant to all TTF distributions, is demonstrated by Shai (2015) constructing the formula for the *MTBsF*$_n$, based on reliability theory, regardless of the NHPP model:

$$MTBsF_n = \int_0^\infty \frac{H(t)^n}{n!} \cdot e^{-H(t)} dt$$

where *MTBsF*$_n$ is the mean time between the nth and the $(n+1)^{\text{th}}$ failures given every repair action restores the system to the state of "*as bad as old*" where $n \geq 0$ is an integer. $H(t)$, $t \geq 0$, is a hazard function of any desired system TTF distribution.

By applying the Weibull TTF distribution $H(t) = (t/\theta)^\beta$ with scale parameter θ and shape parameter β, Shai shows a closed expression, equivalent to the one showed by Ulmeanu and Iunescu:

$$MTBsF_n = \frac{\theta}{\beta \cdot n!} \cdot \Gamma\left(n + \frac{1}{\beta}\right) \qquad n \geq 0; \ \theta, \beta \geq 0$$

This is an elegant expression based on the gamma function and the two parameters θ and β. Note that $MTBsF_n$ decreases with n for $\beta > 1$, whereas for $\beta < 1$ it increases with n. A problem may occur if the time when the process was initiated is unknown. In this case, the index of failure events n becomes a parameter of the model. Hence, for a given time series t_{n+i} and assuming the time to the first failure is distributed according to the Weibull distribution, the model parameters n, β, and θ can be estimated. Shai (2015) describes a simple, tolerant to noisy data, procedure for the best estimate, and though the maximum likelihood approach might yield better results, it is rather complicated due to the gamma function, and in particular since n is unknown. The failure index n is discrete, hence the procedure yields sets of values for θ, β, and n (n = 0, 1, 2, ..., etc.). The most appropriate set of parameters is the one for which the relevant vector of time gaps has the least square distance from the original time gaps.

Shai (2015) presents a few enlightening test cases, based on the model and some historical data while technology swap events are analogously considered nonideal minimal repair actions and the Weibull $MTBsF_n$ model is suggested to describe the time between successive events. These examples demonstrate mathematical practicability and fit well with the historical time series despite the many causes for uncertainty in the data collection. A few test cases showed increasing time between technology revolutions (e.g. old calculating machines and modern computer generations). Others showed decreasing gaps (e.g. lighting devices and computer data storage devices). One such example is detailed here.

This example investigates the technologies of computer data storage devices. A careful identification of the most important milestones marks the dates when new technologies were announced for practical commercial use. On some occasions, a new, but still expensive and not yet widespread, technology is recognized as the precursor of a new period. The list in Table 22.1 is subjectively filtered from all the data.

The very dense time gaps may indicate a major paradigm change in the near future relative to the last data point in 2008. In fact, the analysis of this example was executed during 2012 predicting the next revolution of data storage devices in 2014. One may consider the prediction to fit the announcement of Cloud Datastore as a stand-alone product in 2013 during a Google annual developer conference. Cloud technology is obviously a paradigm change. Identifying the relevant sequences of historical data is not a simple task; it is clearly not deterministic from the historical standpoint, nor is it clear how to distinguish major technological events from those which are merely incremental improvements. Eventually, on the basis of the test cases and under some limitations, one may conclude that the $MTBsF$ model demonstrates reasonable ability to describe some trends of technological change, to evaluate

Table 22.1 Dates when new technologies were announced for practical commercial use.

MTBsF$_n$ (model)	Time gaps (given)	Date (year)	The computer data storage technology
		1890	Punched cards – the first digital data storage device
47.3	47	1937	Punched tapes – longer strings for input and output data
16.96	19	1956	5 MB IBM model 350 with 50 pcs of 24 in. disks
11.5	13	1969	8″ floppy disk – a removable magnetic data storage device
9	7	1976	5.25″ floppy disk – cheap options for private users
7.6	7	1983	SyQuest drive – an early entrance into PC hard drive market
6.6	6	1989	Compact Disc – digital optical disk allows to burn data
5.9	5	1994	Flash memory device – Compact Flash, Smart Media, etc.
5.4	7	2001	DVD-R – affordable 4 GB portable disks
4.9	4	2005	USB stick – a USB format flash memory data storage device
4.5	3	2008	HAMR – Heat Assistance Magnetic Recording allows Terabytes
4.3	**→**	**2014**	**Prediction of next technology revolutionary event**

the characterizing model parameters, to suggest an extrapolation into the past for relevant occasions, and to make predictions of the next technological event, all under good engineering and historical judgment, scientific cautiousness and practical considerations.

22.8 Summary

In his book "*Evolution – The Grand Synthesis*," Ervin Laszlo (1987) examines the patterns of change and transformation in the cosmos, in biological species, and in modern society in order to create a grand evolutionary synthesis. He claims that "*scientific evidence of the patterns traced by evolution,*" in those disciplines, "*is growing rapidly*" and that "*it is coalescing into the image of basic regularities that repeat and recur.*" The author seeks a unifying global dynamic behavior that motivates evolution wherever it is. He also suggests that one kind of evolution prepares the ground for the next; "*out of the conditions created by evolution in the physical realm emerge the conditions that permit biological evolution to take off. And out of the conditions created by biological evolution come the conditions that allow human beings … to evolve certain social forms of organization.*" The Reliability of Technologies, based on its holistic approach and its evolutionary attitude, places the realm of artificial technologies at the next stage of hierarchy, where out of the conditions created by the evolving forms of social organization there sprouts the seed for the creation of the complex adaptive forms of technology. The evolution of technologies does not stand on its own; it is rooted in the evolution of social networks – the complexity of knowledge and ingenuity becomes a prime ingredient of the technological entity. It is also rooted in the evolution of humans, biologically speaking – it is technology that made human life expectancy grow by 100% in just one century; it offers artificial organs, it enables incredible medical intervention, and it supplies advanced medicine. Is this the evolutionary trajectory to an inevitable combination of humans and machines, as some scholars would claim – into a new paradigm of hybrid beings? This chapter does not reject this option however, since the rate of technological leaping events does not always tend to increase and the time between events, in some cases, increases accordingly, there might be more than one possible future as regard to the interaction of humans and technologies.

Laszlo also describes how the two contradicting arrows of time – the downward arrow of classic thermodynamics toward equilibrium and maximum entropy or randomness and the upward arrow of the self-organizing more complex structures of evolution – are settled by the new scientific branch of nonequilibrium thermodynamics. While closed systems near equilibrium obey the second law of thermodynamics and, as their constraints are removed, move toward their only attractor – the state of equilibrium – there are other situations in nature that systems experience. These situations belong to a "third state" – a state far from thermal or chemical equilibrium. Systems in the third state are *"nonlinear and occasionally indeterminate"* and may *"amplify certain fluctuations and evolve toward a new dynamic regime that is radically different from stationary states at or near equilibrium."* It stems from the above that systems in the third state tend to evolve – to increase their level of complexity and organization and become more energetic. It follows that in order to not contradict the second law of thermodynamics, these systems are not isolated, and they are open systems and the entropy within them is not irreversible due to free energy flux from its environment inwards.

The governing sensation in this text is the presence of the evolutionary behavior. Complex adaptive systems, such as modern artificial human habitat technologies, decrease uncertainties within the human society. The same applies to communication and knowledge technologies or to transportation and even to data storage technologies – they all contribute to greater social order due to the global tendency to a simpler use of tools and products on account of the greater complexity of the product itself. The description and quotes above imply that technologies must belong to the third state far from equilibrium. That raises the question where does the energy flow in from? Practical suggestions can basically be found, e.g. in the economic discipline – R&D investments and entrepreneurship programs financed by governments, scientific institutions, and leading companies. However, an interesting exercise can be drawn; what if a society becomes abruptly isolated by, e.g., a massive act of war, a cataclysmic natural disaster or dramatic technological failure such as the fatal loss of national energy resources. In that case, one can imagine, or learn from history, immediate loss of social order occurs and the uncertainty grows until scattered local balances are achieved. High technologies may no longer be valid and simpler means should take their place in order to maintain decreased expectations. The extent of uncertainty gained through such occasions may become a measure of complexity gathered up to the point prior to the event. This measure can be referred to as the *potential uncertainty* of evolutionary progress.

Attributing technologies to those systems with evolutionary behavior may support the "Why" question (why do they evolve?) but not the "How," "Where," and "When" and, in particular, the "How much." The latter refers to the depth of technology transformation – to the extent to which the ingredients of the former technology change. This scaling does not necessarily depend on the rate of change. The test cases in Section 22.7 (Shai 2015) suggest technological change can be described by scalable features of the technology regardless of the time elapsed since the last major technological change. These features are, in effect, the aggregated potential of the new technology to satisfy its purpose. The impression of these features is what gives the common notion of exponential technological growth, but it is not necessarily the relevant impression as regard to the historical rate of technological change. Gradual adaptations, rearrangements of current design or making it more efficient are the realization of the ongoing improvements of the technological features but they are not a leap of change. While patterns and characteristics of technological change events can be studied as part of the science of evolution, with different trajectories of phase transitions, this paper suggests there are some statistically based, modelable, dynamic processes that the Reliability theory, expanded, and adapted, can explain.

This chapter presents the evolutionary dynamic life cycle characteristics of technologies on the basis of reliability theory. The reliability mathematics is successfully implemented on a broader philosophic holistic domain rather than in the engineering domain alone; the classic product becomes the newly defined complex entity of technology or the species of related products. The holistic approach for the description of technological entities and their dynamics is substantiated. It relates to the complex domain of both users and products – of the technical characteristics and all related human behaviors and realized by the use of high-level functions that stem from the life span characteristics at the product's population level. This work also lays the basis for quantitative description of technologies, in general, by defining the very basic group of characteristics required for combining any technology. This description is the product of the holistic approach; however, on this basis, future work may develop new metrics to measure the extent to which technologies differ or to measure their complexity. New models and methods in this work realize the vision of the holistic high-level dynamic perspective while introducing new performance functions, generalizing physical–statistical damage accumulation processes with unique paths or trajectories for strength deterioration in each environment, and finally, a bird's-eye view and prediction of technological revolutions.

References

Adami, C., Ofria, C., and Collier, T.C. (2000). Evolution of biological complexity. *Proceedings of the National Academy of Sciences of the United States of America* 97 (9): 4463–4468.

Adams, B.M., Devine, K.D., Ray, J. et al. (2007). Efficient large-scale network-based simulation of disease outbreaks. *Advances in Disease Surveillance* 2 (191).

Ajzen, I. and Fishbein, M. (1980). *Understanding Attitudes and Predicting Social Behavior*. NJ: Prentice-Hall.

Arthur, W.B. (2007). The structure of invention. *Research Policy* 36 (2): 274–287.

Bagozzi, R.P. (2007). The legacy of the technology acceptance model and a proposal for a paradigm shift. *Journal of the Association for Information Systems* 8 (4): 244–254.

Bak, P. (1997). *How Nature Works: The Science of Self-Organized Criticality*. Oxford University Press.

Bak, P. and Pacsuski, M. (1997). Mass extinctions vs. uniformitarianism in biological evolution. In: *Physics of Biological Systems*, Lecture Notes in Physics (eds. H. Flyvbjerg, J. Hertz, M.H. Jensen, et al.), 341–356. Berlin: Springer.

Bak, P. and Sneppen, K. (1993). Punctuated equilibrium and criticality in a simple model of evolution. *Physical Review Letters* 71 (24): 4083–4086.

Bar-Yam, Y. (1997). *Dynamics of Complex Systems*. Reading, MA: Addison-Wesley.

Bar-Yam, Y. (2004). *Making Things Work: Solving Complex Problem in a Complex World*. NECSI, Knowledge Press.

Bohm, D. (1980). *Wholeness and the Implicate Order*, vol. 1980. Routledge & Kegan Paul.

Christensen, C.M. (1997). *The Innovator's Dilemma: When New Technologies Cause Great Firms to Fail*. Boston, MA: Harvard Business School Press.

Davis, F.D. (1986) Technology Acceptance Model for Empirically Testing New End-User Information Systems: Theory and Results, PhD Thesis, Sloan School of Management, MIT.

Davis, F.D. (1989). Perceived usefulness, perceived ease of use, and user acceptance of information technology. *MIS Quarterly* 13 (3): 319–340.

Davis, F.D., Bagozzi, R.P., and Warshaw, P.R. (1989). User acceptance of computer technology: a comparison of two theoretical models. *Management Science* 35: 982–1003.

Eaton, J. and Kortum, S. (1999). International technology diffusion: theory and measurement. *International Economic Review* 40 (3): 537–570.

Eldredge, N. and Gould, S.J. (1972). Punctuated equilibria: an alternative to phyletic gradualism. In: *Models in Paleobiology* (ed. T.J.M. Schopf), 82–115. San Francisco, CA: Freeman, Cooper and Co.

Epstein, B. and Weissman, I. (2008). *Mathematical Models for Systems Reliability.* Boca Raton, FL: Chapman & Hall, CRC Press.

Erdi, P. (2008). *Complexity Explained.* Berlin: Springer.

Fenn, J. and Raskino, M. (2008). *Mastering the Hype Cycle: How to Choose the Right Innovation at the Right Time.* Boston, MA: Harvard Business Press.

Fishbein, M. and Ajzen, I. (1975). *Belief, Attitude, Intention and Behavior: An Introduction to Theory and Research.* Addison-Wesley.

Hall, M., Christensen, K., di Collobiano, S.A. et al. (2002). Time-dependent extinction rate and species abundance in a tangled-nature model of biological evolution. *Physical Review E* 66 (1): 011904.

Heilighen, F. (2008). Complexity and self-organization. In: *Encyclopedia of Library and Information Science* (eds. M.J. Bates and M.N. Maack). Taylor and Francis.

Holland, J.H. (1992). *Adaptation in Natural and Artificial Systems: An Introductory Analysis with Applications to Biology, Control, and Artificial Intelligence.* Cambridge, MA: MIT Press.

Holland, J.H. (1995). *Hidden Order: How Adaptation Builds Complexity.* Redwood City, CA: Addison-Wesley.

Holland, J.H. (1998). *Emergence: From Chaos to Order.* Redwood City, CA: Addison-Wesley.

Ingman, D. and Reznik, L.A. (1982). A dynamic model for element reliability. *Nuclear Engineering and Design* 70 (2): 209–213.

Ingman, D. and Reznik, L.A. (1986). Dynamic reliability model for damage accumulation processes. *Nuclear Technology* 75 (3): 261–282.

Ingman, D. and Reznik, L.A. (1991). Dynamic character of failure state in damage accumulation processes. *Nuclear Science and Engineering* 107 (3): 284–290.

Ingman, D., Mirer, T., and Suhir, E. (2007). Dynamic physical reliability in application to photonic materials. In: *Micro- and Opto-Electronic Materials and Structures: Physics, Mechanics, Design, Packaging, Reliability* (eds. E. Suhir, C.P. Wong and Y.C. Lee), A571–A594. Boston, MA: Springer.

Janssen, H., Oerding, K., van Wijland, F. et al. (1999). Lévy-flight spreading of epidemic processes leading to percolating clusters. *The European Physical Journal B* 7 (1): 137–145.

Jensen, H.J. (2004). Emergence of species and punctuated equilibrium in the Tangled-Nature model of biological evolution. *Physica A: Statistical Mechanics and its Applications* 340 (4): 697–704.

Kapur, K.C. and Lamberson, L.R. (1977). *Reliability in Engineering Design.* New York, NY: Wiley.

Karev, G.P., Berezovskaya, F.S., and Koonin, E.V. (2005). Modeling genome evolution with a diffusion approximation of a birth-and-death process. *Bioinformatics* 21 (3): 12–19.

Kurzweil, R. (2005). *The Singularity Is Near: When Humans Transcend Biology.* NY: Viking Books.

Laszlo, E. (1987). *Evolution – The Grand Synthesis.* Boston, MA: New Science Library.

Malhotra, Y. and Galletta, D.F. (1999). Extending the technology acceptance model to account for social influence: theoretical bases and empirical validation. In: *Proceedings of the 32nd Hawaii International Conference on System Sciences (HICSS)*, Maui, HI (5–8 January). IEEE Computer Society.

Misra, K.B. (2008). *Handbook of Performability Engineering.* London: Springer, Ltd.

Mokyr, J. (2005). Long-term economic growth and the history of technology. In: *Handbook of Economic Growth*, vol. 1(B) (eds. P. Aghion and S.N. Durlauf), 1113–1180. Elsevier.

Moore, G.E. (1965). Cramming more components onto integrated circuits. *Electronics* 38 (8): 114–117.

Moore, G.A. (1991). *Crossing the Chasm: Marketing and Selling High-Tech Products to Mainstream Customers.* NY: Harper Business.

Moore, G.A. (1995). *Inside the Tornado: Marketing Strategies from Silicon Valley's Cutting Edge.* NY: Harper Business.

Norman, D.A. (1998). *The Invisible Computer: Why Good Products Can Fail, the Personal Computer Is So Complex, and Information Appliances Are the Solution.* Cambridge, MA: MIT Press.

Norris, D. and Yin, R. (2008). From adoption to action: mapping technology adoption constructs for the small to medium enterprise innovation success. *Issues in Information Systems* IX (2): 595–606.

Pulkki, A.M. (2008) Inter-Country Spillovers in the Diffusion of New Technology – An Epidemic Model of Steam and Motor Ship Diffusion. DRUID-DIME Winter Conference, Aalborg, Denmark.

Rausand, M. (2004). *System Reliability Theory: Models, Statistical Methods, and Applications*, 2e. Hoboken, NJ: Wiley.

Rogers, E.M. (1995). *Diffusion of Innovations*, 4e. NY: Free Press.

Shai, Y. (2015) Reliability of Technologies, PhD Thesis, Technion (Israel Institute of Technology).

Shai, Y., Ingman, D., and Suhir, E. (2013). Reliability of objects in aerospace technologies and beyond: holistic risk management approach. In: *2013 IEEE Aerospace Conference*, Big Sky, MT (2–9 March). IEEE.

Shai, Y., Ingman, D., and Suhir, E. (2015). Modified dynamic reliability model for damage accumulation. In: *2015 IEEE Aerospace Conference*, Big Sky, MT (7–14 March), 338–344. IEEE.

Shen, X. and Wilde, P.D. (2005). Analysis and identification of a social interaction model. *Physica A: Statistical Mechanics and its Applications* 350 (2–4): 597–610.

Sherwood, R.S. and Maynard, H.B. (1982). Technology. In: *McGraw-Hill Encyclopedia of Science and Technology*, 6e, vol. 18, 142–146. NY: McGraw-Hill.

Smuts, J.C. (1927). *Holism and Evolution*. London: McMillan and Co. Limited.

Suhir, E. (2007). How to make a device into a product: accelerated life testing (ALT), it's role, attributes, challenges, pitfalls, and interaction with qualification testing. In: *Micro- and Opto-Electronic Materials and Structures: Physics, Mechanics, Design, Packaging, Reliability* (eds. E. Suhir, C.P. Wong and Y.C. Lee), B203–B231. Boston, MA: Springer.

Suhir, E. (2009). Helicopter Landing Ship (HLS) and the role of the human factor. *ASME Journal of Offshore Mechanics and Arctic Engineering* (1): 132.

Suhir, E. (2012). Human in the loop: predicted likelihood of vehicular mission success and safety. *Journal of Aircraft* 49 (1): 29–36.

Ulmeanu, A.P. and Iunescu, D.C. (2007). A closed-formula for mean time between failures of minimal repaired systems under Weibull failure distribution. *International Journal of Performability Engineering* 3 (2): 225–230.

Utterback, J.M. (1994). *Mastering the Dynamic of Innovation*. Boston, MA: Harvard Business School Press.

Venkatesh, V., Morris, M.G., Davis, G.B. et al. (2003). User acceptance of information technology: towards a unified view. *MIS Quarterly* 27 (3): 425–478.

Xu, B. and Chiang, E.P. (2005). Trades, patents and international technology diffusion. *Journal of International Trade and Economic Development* 14 (1): 115–135.

Zacks, S. (1991). *Introduction to Reliability Analysis, Probability Models and Statistical Methods*. NY: Springer.

Acronyms

additive manufacturing (AM)
advanced persistent threats (APT)
agile software development (ASD)
air traffic controller (ATCO)
Akaike information criterion (AIC)
American Iron and Steel Institute (AISI)
American Society of Testing and Materials (ASTM)
analysis of contagious debt (AnaConDebt)
analysis of variance (ANOVA)
application program interface (API)
architectural technical debt (ATD)
architecture significant requirements (ASR)
architecture trade-off and analysis method (ATAM)
artificial intelligence (AI)
artificial neural network (ANN)
augmented reality (AR)
automatic teller machine (ATM)
automation robotics and machines (ARM)
autoregressive integrated moving average (ARIMA)
Bayesian network (BN)
big data (BD)
bovine serum albumin (BSA)
business intelligence (BI)
capability maturity model integration (CMMI)
chemical process industries (CPI)
collision avoidance systems (CAS)
communication technologies (CT)
complex, adaptive systems-of-systems (CASoS)
complex, large-scale, interconnected, open system (CLIOS)
compound annual growth rate (CAGR)

Systems Engineering in the Fourth Industrial Revolution: Big Data, Novel Technologies, and Modern Systems Engineering,
First Edition. Edited by Ron S. Kenett, Robert S. Swarz, and Avigdor Zonnenshain.
© 2020 John Wiley & Sons, Inc. Published 2020 by John Wiley & Sons, Inc.

computed tomography (CT)
computer aided design (CAD)
computer aided manufacturing (CAM)
computer numerical control (CNC)
computer software component (CSC)
computer software configuration item (CSCI)
computer-supported cooperative work (CSCW)
condition-based maintenance (CBM)
controlled drug release (CDR)
corrective maintenance (CM)
cross-industry standard process for data mining (CRISP-DM)
customers, actors, transformation, worldview, owners, and environment
   (CATWOE)
cyber-physical system (CPS)
cyclic dependency (CD)
data analytics (DA)
database management systems (DBMS)
Defense Advanced Research Projects Agency (DARPA)
Department of Homeland Security (DHS)
Department of innovative technologies (DTI)
design evaluation process outline (DEPO)
design for manufacturing and assembly (DFM/A)
design for variability (DFV)
design of experiments (DOE)
design structure matrix (DSM)
digital human model (DHM)
direct energy deposition (DED)
directed acyclic graph (DAG)
document term matrix (DTM)
dynamic data warping (DTW)
dynamic linked library (DLL)
dynamic principal component analysis (DPCA)
empirical mode decomposition (EMD)
European Foundation for Quality Management (EFQM)
European Innovation Scoreboard (EIC)
European Union (EU)
event tree analysis (ETA)
exploratory factor analysis (EFA)
factories of the future (FoF)
failure mode and effect analysis (FMEA)
fault tree analysis (FTA)
Federal Bureau of Investigation (FBI)
Federal Emergency Management Agency (FEMA)

Federal Energy Regulatory Commission (FERC)
finite element analysis (FEA)
first-order reliability method (FORM)
field programmable gate array (FPGA)
flexible contract (FC)
flight management system (FMS)
functionally graded material (FGM)
genetic algorithm (GA)
graphical user interface (GUI)
ground collision avoidance system (GCAS)
Guide to the Expression of Uncertainty in Measurement (GUM)
hazard and operability (HAZOP)
high level performance function (HLPF)
hub-like dependency (HL)
human factors (HF)
human factors and ergonomics (HFE)
human factors engineering (HFE)
human–machine collaboration (HMC)
human-centered design (HCD)
human-computer interaction (HCI)
human-in-the-loop simulation (HITLS)
human–machine interaction (HMI)
human–machine interaction engineering (HMIE)
human–systems integration (HSI)
improving critical infrastructure security (ICIC)
industrial control system (ICS)
industrial process monitoring (IPM)
industrial revolution (IR)
industry competence and maturity for advanced manufacturing (IMAM)
information quality (InfoQ)
institute of systems and technologies for sustainable production (ISTePS)
information technology (IT)
initial graphics exchange specification (IGES)
innovation diffusion theory (IDT)
intellectual property rights (IPR)
intellectual property (IP)
International Atomic Energy Agency (IAEA)
International Council on Systems Engineering (INCOSE)
International Organization for Standardization (ISO)
Internet of Things (IoT)
ionic strength (IS)
istituto dalle molle di studi sull'intelligenza artificiale (IDSIA)

key performance indicator (KPI)
knowledge based system (KBS)
knowledge, information, and data (KID)
Kullback–Leibler (K–L)
Laboratory for Intelligent Systems in Process Engineering (LISPE)
life cycle cost (LCC)
lines of code (LOC)
machine learning (ML)
magnetic resonance imaging (MRI)
maintainability index (MI)
Markovian decision process (MDP)
massive open online course (MOOC)
maximum percentage relative information exploitation (PRIE)
McCabe cyclomatic complexity (MCC)
mean squared error (MSE)
mean time between failures (MTBF)
mean time to failure (MTTF)
minimum percentage relative information gain (PRIG)
mission assurance (MA)
model-based design (MBD)
model-based systems engineering (MBSE)
Monte Carlo simulation (MCS)
multiple input–single output (MISO)
multivariate statistical process control (MSPC)
musculoskeletal diseases (MSD)
national aeronautics and space administration (NASA)
National Cybersecurity Framework (NCSF)
National Institute of Standards and Technology (NIST)
National Network for Manufacturing Innovation (NNMI)
National Security Agency (NSA)
Need, Approach, Benefit, Competition (NABC)
next-generation firewall (NGFW)
nonhomogeneous Poisson process (NHPP)
normal operational conditions (NOC)
Nuclear Regulatory Commission (NRC)
Object Management Group (OMG)
online analytical processing (OLAP)
open services for lifecycle collaboration (OSLC)
Owako Working Posture Analysis System (OWAS)
partial differential equation (PDE)
partial least squares (PLS)
partially observable Markov decision process (POMDP)
particle size (PS)

pattern-oriented software architecture (POSA)
perceived ease-of-use (PEOU)
physical protection system (PPS)
powder bed fusion (PBF)
pressure operated release valve (PORV)
preventive maintenance (PM)
principal component analysis (PCA)
printed circuit board (PCB)
probabilistic risk assessment (PRA)
process analytical technology (PAT)
process systems engineering (PSE)
product use information (PUI)
prognostics and health monitoring (PHM)
project management body of knowledge (PMBOK)
public–private partnership (PPP)
quality by design (QbD)
rapid attack detection, isolation and characterization system (RADICS)
rapid casting (RC)
rapid manufacturing (RM)
rapid prototyping (RP)
rapid tooling (RT)
reliability, availability, maintainability, safety (RAMS)
reliability-centered maintenance (RCM)
remote terminal unit (RTU)
representational state transfer (REST)
resource description framework (RDF)
responsibility, accountability, consulting, informed (RACI)
robust design (RD)
root cause analysis (RCA)
root mean squared error (RMSE)
second-order reliability methods (SORM)
secure emergency network (SEN)
self-organized criticality (SOC)
Shiraryev–Roberts (SR)
single layer (SL)
single photon and positron emission computerized tomography (SPET/PET)
single track (SD)
singular value decomposition (SVD)
software engineering method and theory (SEMAT)
solder paste deposit (SPD)
standard for the exchange of product model data (STEP)
standard triangle language (STL)
statistical process control (SPC)

strengths, weaknesses, opportunities, threats (SWOT)
structurational model of technology (SMOT)
supervisory control and data acquisition (SCADA)
supply chain management (SCM)
Swiss national science foundation (SNF)
system engineering body of knowledge (SEBoK)
system-of-systems (SoS)
systems approach to safety engineering (STAMP)
Systems-Theoretic Accident Model and Processes (STAMP)
systems engineering (SE)
systems-theoretic framework for security (STFS)
standard triangle language (STL)
strenghts, weaknesses, opportunities, threats (SWOT)
systems-theoretic process analysis (STPA)
tangible interactive system (TIS)
targeted Bayesian network learning (TBNL)
technical debt (TD)
technology acceptance model (TAM)
technology, organizations and people (TOP)
technology-centered systems engineering (TCSE)
theory of reasoned action (TRA)
time to failure (TTF)
traffic alert and collision avoidance system (TCAS)
transdisciplinary engineering research (TDER)
transdisciplinary research (TDR)
transmembrane pressures (TMP)
unified theory of acceptance and use of technology (UTAUT)
United Kingdom (UK)
United States (US)
university of applied sciences and arts of southern Switzerland (SUPSI)
unstable dependency (UD)
user interface (UI)
virtual private networks (VPN)
volume of fluid (VOF)
wavelet texture analysis (WTA)
web ontology language (OWL)

Biographical Sketches of Editors

Ron S. Kenett is Chairman of the KPA Group and Senior Research Fellow, Samuel Neaman Institute for National Policy Research, Technion, Israel. He is past professor of operations management at The State University of New York (SUNY) at Binghamton, USA and research professor of mathematics and statistics, University of Turin, Italy. Ron was awarded the *Royal Statistical Society* 2013 Greenfield Medal and the *European Network for Business and Industrial Statistics* 2018 Box Medal for excellence in the development and application of statistical methods. He has co-authored 14 books, including *Modern Industrial Statistics with R, MINITAB and JMP* (Wiley, 2014), *Information Quality: The Potential of Data and Analytics to Generate Knowledge* (Wiley, 2016), *Analytic Methods in Systems and Software Testing* (Wiley, 2018), and *The Real Work of Data Science: Turning data into information, better decisions, and stronger organizations* (Wiley, 2019), and over 250 publications in international journals. His extensive consulting experience includes engagements with organizations such as Amdocs, Nice, HP, Intel, Israel Aircraft Industries, and Tadiran Telecom. For 15 years he taught a graduate course titled *Quantitative Methods in Systems and Software Development* at the Tel Aviv University School of Engineering.

Robert S. Swarz is a Professor of Practice in the Systems Engineering program at Worcester Polytechnic Institute in Worcester, Massachusetts, USA. He holds a PhD in Electrical Engineering from New York University, an MBA from Boston University, an MS from Rensselaer Polytechnic Institute, and a BS, also from New York University . In addition to his academic background, he has engineered systems at The MITRE Corporation, Pratt and Whitney Aircraft, and Digital Equipment Corporation. He is the co-author (through three editions) of *Reliable Computer Systems: Design and Evaluation* (A.K. Peters). His current research interests are in system dependability and security, complex systems of systems, model-based systems engineering, systems thinking, and system architecture.

Systems Engineering in the Fourth Industrial Revolution: Big Data, Novel Technologies, and Modern Systems Engineering, First Edition. Edited by Ron S. Kenett, Robert S. Swarz, and Avigdor Zonnenshain.
© 2020 John Wiley & Sons, Inc. Published 2020 by John Wiley & Sons, Inc.

Avigdor Zonnenshain is Senior Research Fellow at The Gordon Center for Systems Engineering and at the Samuel Neaman Institute for National Policy Research at Technion, Haifa, Israel. He has a PhD in Systems Engineering from the University of Arizona, Tucson, USA. Formerly, he held several major positions in the quality and systems engineering area in RAFAEL, Haifa, and in the Prime Minister's Office. He is an active member of the Israel Society for Quality (ISQ). He was also the Chairman of the Standardization Committee for Management & Quality. He is a Senior Adjunct Lecturer at Technion – Israel Institute of Technology. He is a member of the Board of Directors of the University of Haifa. He is an active member of INCOSE & INCOSE_IL (past president). He is a Fellow of INCOSE. He is the co-author of *Managing and Engineering Complex Technological Systems* (Wiley, 2015).

Index

Systems Engineering in the Fourth Industrial Revolution: Big Data, Novel Technologies, and Modern Systems Engineering,
First Edition. Edited by Ron S. Kenett, Robert S. Swarz, and Avigdor Zonnenshain.
© 2020 John Wiley & Sons, Inc. Published 2020 by John Wiley & Sons, Inc.

Printed and bound by CPI Group (UK) Ltd, Croydon, CR0 4YY

16/04/2025

14658593-0002